ADVANCES IN FISSION-TRACK GEOCHRONOLOGY

Solid Earth Sciences Library

Volume 10

The titles published in this series are listed at the end of this volume.

Advances in Fission-Track Geochronology

A selection of papers presented at the International Workshop on Fission-Track Dating, Ghent, Belgium, 1996

edited by

PETER VAN DEN HAUTE

Geological Institute,
University of Ghent, Belgium

and

FRANS DE CORTE

Institute for Nuclear Sciences,
University of Ghent, Belgium

KLUWER ACADEMIC PUBLISHERS

DORDRECHT / BOSTON / LONDON

A C.I.P. Catalogue record for this book is available from the Library of Congress.

ISBN 0-7923-4904-0

Published by Kluwer Academic Publishers,
P.O. Box 17, 3300 AA Dordrecht, The Netherlands.

Sold and distributed in North, Central and South America
by Kluwer Academic Publishers,
101 Philip Drive, Norwell, MA 02061, U.S.A.

In all other countries, sold and distributed
by Kluwer Academic Publishers,
P.O. Box 322, 3300 AH Dordrecht, The Netherlands.

Printed on acid-free paper

Printed in the Netherlands

Contents

International Workshop on Fission-Track Dating, Gent-1996

Organizing Committee

P. Van den haute	Universiteit Gent (Chairman)
F. De Corte	Universiteit Gent (Chairman)
R. Dams	Unversiteit Gent
G. Stoops	Universiteit Gent
R. Gijbels	Universitaire Instelling Antwerpen
J. Hertogen	Katholieke Universiteit Leuven
P. Pasteels	Vrije Universiteit Brussel
A. De Wispelaere	Universiteit Gent
F. Bellemans	Universiteit Gent
L. Vancraeynest	Universiteit Gent
R. Claeys	Universiteit Gent

Scientific Committee

P.A.M. Andriessen	Vrije Universiteit Amsterdam, The Netherlands
G. Bigazzi	Istituto di Geocronologia e Geochimica Isotopica, Pisa, Italy
A. Chambaudet	Université de Franche Comté Besançon, France
F. De Corte	Universiteit Gent, Belgium
A.J.W. Hurford	University College London, United Kingdom
P. Van den haute	Universiteit Gent, Belgium
G. A. Wagner	Forschungsstelle Archäometrie der Heidelberger Akadämie der Wissenschaften am Max-Planck-Institut für Kernphysik, Heidelberg, Germany

Organized with the Financial Support of

National Fund for Scientific Research
University of Gent
Institute for Reference Materials and Measurements
Canberra Packard Benelux
EG&G Benelux
Olympus-Omnilabo Belgium
L' Air Liquide Belge
Bank Brussel Lambert, Gent
Sidmar, Gent

Preface

The four-yearly, International Workshop on Fission-Track Dating was held in Gent (Belgium) during the week of 26-30 August 1996. This Workshop is regarded as one of the most important meetings for all scientists studying the fission-track method and its applications to the Earth and related sciences. Of the four-yearly Fission-Track Workshops (and omitting the short meetings that took place at ICOG conferences) the Gent-1996 Workshop was the fifth in the row, following Pisa (1980), Troy (1984), Besançon (1989) and Philadelphia (1992). The meeting was organized by the Geological Institute and the Institute for Nuclear Sciences of the University of Gent. It was held in the Aula, a classic university building dating from 1826 and situated in the very lively heart of the city.

About 115 scientists from 24 different countries attended the meeting, during which 59 oral communications and 75 posters were presented. If we compare these numbers to those of the former workshops, it appears that the scientific interest in the fission-track geochronometer is still growing. The amount and the content of the contributions presented at the Gent-1996 Workshop also confirmed the evolution that the fission-track method has gone through during the last fifteen years, from a technique that was debated in most of its facets to an established chronometric tool with qualities that are unique in geo(thermo)chronology. Of the participants on the group photograph of the Gent-1996 Workshop, those who also figure on the photograph with the silhouette of the leaning tower of Pisa pictured in the background, will undoubtedly have experienced this evolution most vividly.

Defining the present status of fission-track methodology and data interpretation and confronting it with knowledge acquired in other scientific disciplines, were the major aims of the Gent-1996 Workshop. This was achieved by introducing the different themes of the conference by keynote speakers who focussed on a critical retrospection of the method and on the significance of fission-track analysis in different parts of the Earth sciences. Keynote addresses were presented on:
the observation and structure of latent tracks by J. Ackermann (Darmstadt),
the calibration of fission-track age determinations by A.J. Hurford (London),
geothermochronometry with the fission track age-length system by A.J.W. Gleadow (Melbourne),
fission-track analysis of continental basements by G. A. Wagner (Heidelberg),
fission-track thermotectonics and geophysical modelling by S. Cloetingh (Amsterdam),
fission-track analysis and the study of the burial history of sedimentary basins by M. Giles (Shell, Rijswijk),
the role of geomorphology in fission-track data interpretation by M. A. Summerfield (Edinburgh).

Compared to the earliest workshops, papers on fission-track methodology were rather limited in number at Gent but some interesting reviews and the introduction of a new calibrated monitor glass certainly merit to be quoted. A number of track stability studies were presented, mainly focussing on zircon and on compositional effects in apatite. The bulk of the communications however dealt with the apatite fission-track thermochronometer applied to uplift and denudation of crystalline basements and orogens and to the thermal evolution of sedimentary basins and their margins. Quite a number of interesting papers were presented on these topics and it would be hazardous to quote some of them above the others. The participants will undoubtedly most lively remember the real time video sequences of thermal histories modelled from fission-track age-length data by Andrew Gleadow and his co-workers. Finally, a number of papers re-emphasized the role of the fission-track method for the age determination of natural glasses.

Following the conference, a post-Workshop field trip was organized on Saturday, August 31. The 32 participants in this trip were taken to the Belgian Ardennes, to visit some of the classic exposures in the native region of European Devonian stratigraphy.

Although the organization of the meeting was in the hands of the geologists and nuclear scientists of the University of Gent, supported by their colleagues of the other Flemish Universities, the members of the Scientific Committee and the panel of keynote speakers ensured that the Workshop was a truly European event. Every member of the committee cooperated in the scientific preparation of the Workshop: from the general planning of the sessions, the screening of the abstracts and the selection of the keynote speakers to the refereeing of the manuscripts submitted for publication in the conference proceedings. Such a cooperation would of course not have been possible without an intensive use of the communication systems that are nowadays available.

With respect to the Proceedings, it was the express wish of the scientific committee that all participants who had presented a paper should be given the opportunity to publish their contribution as a full article. The Journal "Nuclear Tracks and Radiation Measurements" that acted as the middle voice for the publication of the Proceedings of the former Workshops, was not considered anymore by the committee because the emphasis in present fission-track research had shifted away too far from the main subjects treated in this journal, which since 1994 also changed its title to "Radiation Measurements". Therefore, and for the first time in the sixteen year old history of the Fission-Track Workshops, the papers that were accepted for publication have been assembled into a separate volume. This volume, that you have now in your hands has been entitled "Advances in Fission-Track Geochronology". It contains 21 papers out of the 134 that have been presented at the Workshop and includes four papers that were presented as a keynote address. All published papers have been subjected to peer review. In addition, the contributing authors were asked to prepare their revised versions as camera-ready typescripts. In order not to prolong the time required for publication, editorial changes and corrections to the text and lay-out of these final typescripts where only made where really necessary. Hence, the authors remain greatly responsible for the language and lay-out of their final article.

We cannot not hope that this volume will compete with the Pisa volume, which in most labs has largely disintegrated due to heavy usage, but we are convinced that the published papers contain sufficient interesting material for every active fission-track researcher. The papers also cover the various themes treated during the conference quite well and in this way, the present volume, offers a quite reliable view on the present status of fission track geochronology to the interested reader, non-specialist.

We cannot end this preface without expressing our gratitude to the numerous people who have helped in the preparation of this Workshop and its Proceedings. First of all, we would like to thank Prof. J.Willlems, rector of our university for putting the Aula building and all of its facilities at our disposal. We acknowledge Prof. J. Hoste, rector emeritus and former director of the Institute for Nuclear Sciences, who cordially accepted the task of presenting the opening address. We thank the Burgomaster of the city of Gent for the pleasant reception he offered to the participants at the town hall, finished with a touch of local colour. We are also grateful to colleagues, friends and family who helped to make things better than it would have been without their efforts and we are indebted to all institutions and companies that provided the Workshop organization with highly appreciated financial support.

The field trip was guided by J. Verniers (Geological Institute, University of Gent), F. Boulvain (Geological Survey of Belgium) and B. Delcambre (Dpt. Geology, Catholic University of Louvain), whom we gratefully acknowledge for making it to a successful closing event. Unfortunately, we cannot acknowledge the fellow up there, controlling the Belgian weather, who managed to deliver the heaviest rainfall of the century on the day and evening of the conference dinner spoiling our planned aperitif in the gardens of the restaurant bordering the Leie river. But, we do wish to thank the participants in the Workshop for their enthusiasm, which made the task of organizing this event to a pleasure for our entire organizing team. We hope that this volume will evoke good memories to all of them and we are looking forward to seeing them again at the next Fission-Track Workshop in Victoria, Australia.

Peter Van den haute and Frans De Corte

LIST OF PARTICIPANTS

Australia

Belton David,
School of Earth Sciences,
La Trobe University,
Bundoora, Victoria.

Brown Roderick William,
Department of Geology,
La Trobe University,
Bundoora, Victoria.

Foster David A.,
School of Earth Sciences,
La Trobe University,
Bundoora, Victoria.

Gleadow Andrew J.W.,
Department of Geology,
La Trobe University,
Bundoora, Victoria.

Green Paul F.,
Geotrack International,
West Bruswick, Victoria.

Kohn Barry Paul,
School of Earth Sciences,
La Trobe University,
Bundoora, Victoria.

Mitchell Melinda Maree,
Department of Geology,
La Trobe University,
Bundoora, Victoria.

Noble Wayne Peter,
School of Earth Sciences,
La Trobe University,
Bundoora, Victoria.

O'Sullivan Paul Brian,
Australian Geodynamics Coop.
Research Centre,
La Trobe University,
Bundoora, Victoria.

Schwarze Phillip,
Department of Earth Sciences,
Monach University,
Clayton Victoria.

Spikings Richard,
Department of Earth Sciences,
La Trobe University,
Bundoora, Victoria.

Austria

Bojar Ana-Voica,
Institut für Geologie und Paläontologie,
Karl-Franzens Universität, Graz.

Hejl Ewald Roland,
Institut für Geologie und Paläontologie,
Universität Salzburg.

Belgium

Bellemans Frank,
Institute for Nuclear Sciences,
Universiteit Gent.

Boven Ariel,
Geochronology Unit,
Vrije Universiteit Brussel.

Dams Richard,
Institute for Nuclear Sciences,
Universiteit Gent.

De Corte Frans,
Institute for Nuclear Sciences,
Universiteit Gent.

Gijbels Renaat,
Department of Chemistry,
University of Antwerp UIA,
Antwerpen-Wilrijk.

Hertogen Jan,
Department of Fysico-Chemical Geology,
Katholieke Universiteit Leuven.

Ingelbrecht Chris,
EC Institute for Reference Materials and Methods,
Geel.

Pasteels Paul,
Geochronology Unit,
Vrije Universiteit Brussel.

Stoops Georges,
Geological Institute,
Universiteit Gent.

Van den haute Peter,
Geological Institute,
Universiteit Gent.

Canada

Currie Lisel.D.,
Geological Survey of Canada,
Vancouver.

Grist Alexander,
Department of Earth Sciences,
Dalhousie University,
Halifax.

Issler Dale Robert,
Institute of Sedimentary & Petroleum Geology, Geological Survey of Canada,
Calgary, Alberta.

Westgate John,
Department of Geology,
Scarborough College,
University of Toronto, Ontario.

China (PR)

Liu Shunsheng,
Changsha Institute of Geotectonics,
Academia Sinica,
Changsa, Hunan.

Sun Shaohua,
Changsha Institute of Geotectonics,
Academia Sinica,
Changsha, Hunan.

Denmark

Hansen Kirsten,
Geological Institute,
University of Copenhagen, Kobenhavn.

Egypt

Shaaban Mohamed Nasser A.,
Faculty of Sciences,
Alexandria University, Alexandria.

France

Barbarand Jocelyn,
CREGU,
Vandoeuvre les Nancy.

Carpéna Joëlle,
Laboratoire de Géochimie de Cadarache,
Saint Paul lez Durance.

Dorighel Olivier,
Groupe de Géophysique Nucléaire,
Institut Dolomieu,
Université Joseph Fourier, Grenoble.

Igli Hassan
Laboratoire de Calcul Scientifique,
Université de Franche-Comté, Besançon.

Pagel Maurice,
CREGU,
Vandoeuvre les Nancy.

Rebetez Michel,
Laboratoire de Microanalyses Nucléaires,
Université de Franche-Comté, Besançon.

Sobel Edward,
CNRS URA Magma et Volcans,
Clermont-Ferrand.

Sère Virginie,
Laboratoire de Géochimie de Cadarache,
Saint Paul lez Durance.

Toro Gloria,
Institut Dolomieu, CNRS,
Université Joseph Fourier, Grenoble.

Germany

Ackermann Jörg;
Gesellschaft für Schwerionenforschung;
Darmstadt.

Brix Manfred Rudolf,
Institut für Geologie,
Ruhr-Universität-Bochum,
Bochum.

Coyle David,
Forschungsstelle Archäometrie,
Max-Planck-Institut für Kernphysik,
Heidelberg.

Dunkl Istvan,
Institut für Geologie,
Universität Tübingen.

Elias Jurgen ,
Institut für Geologie,
Universität Tübingen.

Glasmacher Ulrich,
Geologisches Institut,
RWTH Aachen.

Jonckheere Raymond,
Forschungsstelle Archäometrie,
Max-Planck-Institut für Kernphysik,
Heidelberg.

Lagerwaard André,
Institute for Transuranium Elements,
European Commission,
Karlsruhe.

Lange Jan-Michael,
Institut für Geophysik und Geologie,
Universität Leipzig.

Lisker Frank,
FB Geowissenschaften,
Universität Bremen.

Meier Stefanie,
FB Geowissenschaften,
Universität Bremen.

Olesch Martin,
FB Geowissenschaften,
Universität Bremen.

Rahn Meinert,
Institut für Mineralogie, Petrologie
und Geochemie, Universität Freiburg.

Schäfer Thorsten,
FB Geowissenschaften,
Universität Bremen.

Thomson Stuart,
Institut für Geologie,
Ruhr-Universität Bochum.

Wagner Günther A.,
Forschungsstelle Archäometrie,
Max-Planck-Institut für Kernphysik,
Heidelberg.

Hungary

Szenes György,
Institute for General Physics,
Eötvös University, Budapest.

India

Sandhu Amanjit S.,
Department of Physics,
Guru Nanak Dev University, Amritsar.

Israel

Feinstein Shimon,
Department of Geology,
Ben Gurion University of the Negev,
Beer Sheva.

Italy

Balestrieri Maria Laura,
Department of Earth Sciences.
Universita di Firenze

Bigazzi Giulio,
Istituto di Geochronologia e Geochimica
Isotopica, CNR, Pisa.

Guglielmetti Alessandra,
Istituto di Fisica Generale Applicata,
University of Milano.

Zattin Massimiliano,
Department of Geological Sciences,
University of Bologna.

Japan

Hasebe Noriko,
Department of Earth Sciences,
Kanazawa University, Kanazawa.

Himeno Osamu,
Department of Mining,
Kyushu University,
Hakozaki, Fukuoka.

Ito Hisatoshi,
Central Research Institute of Electric
Power Industry, Abiko City, Chiba.

Iwano Hideki,
Kyoto Fission-Track Co. Ltd.,
Kyoto.

Kitada Naoko,
Geo-Research Institute,
Osaka.

Koshimizu Satoshi,
Institute for Atomic Energy,
Rikkyo University,
Nagasaka, Yokosuka.

Ohira Hiroto,
Department of Geosciences,
Shimane University, Matsue.

Ono Masako,
Department of Earth and Planetary
Sciences,
Hokkaido University, Sapporo.

Tagami Takahiro,
Division of Earth and Planetary Sciences,
Kyoto University, Kyoto.

Wadatsumi Kiyoshi,
Department of Geosciences,
Osaka City University, Osaka.

Watanabe Koichiro,
Department of Mining,
Kyushu University, Fukuoka.

Watanabe Teruo,
Department of Earth and Planetary
Sciences,
Hokkaido Universty, Sapporo.

Yoshioka Tetsu,
Kyoto Fission-Track Co Ltd,
Kyoto.

New Zealand

Kamp Peter J.
Department of Earth Sciences
University of Waikato, Hamilton.

The Netherlands

Andriessen Paul,
Institute of Earth Sciences,
Vrije Universiteit Amsterdam.

Cloetingh Sierd,
Institute of Earth Sciences,
Vrije Universiteit Amsterdam.

de Bruyne Karen ,
Institute of Earth Sciences,
Vrije Universiteit Amsterdam.

Giles Melvin,
SHELL Research (KESPL),
Rijswijk.

Rohrman Max,
Institute of Earth Sciences,
Vrije Universiteit Amsterdam.

Sanders Carlo,
Institute of Earth Sciences,
Vrije Universiteit Amsterdam.

Stapel Gerco,
Institute of Earth Sciences,
Vrije Universiteit Amsterdam.

Norway

Stiberg Jan Petter,
Institute for Energy Technology,
Kjeller.

Oman

Terken Jos;
Petroleum Development Oman;
Muscat.

Portugal

Pereira Alcides;
Departamento de Ciencias da Terra;
University of Coimbra.

South Africa

De Wit Mike,
Geology Division,
De Beers Cons. Mines Ltd,
Hennopsmeer.

Sweden

Cederbom Charlotte,
Department of Geology,
Gothenburg University, Göteborg.

Tullborg Eva-Lena,
Terralogica AB,
Gråbo.

Switzerland

Seward Diane,
Geological Institute,
ETH Zentrum, Zürich.

Viola Giulio,
Geological institute,
ETH Zentrum, Zürich.

United Kingdom

Carter Andrew,
Department of Geological Sciences,
Birkbeck college,
University of London.

Galbraith Rex F.,
Department of Statistical Science,
University College London.

Gallagher Kerry,
Department of Geolgical Sciences,
Imperial College London.

Harman Rebecca,
Department of Geology,
Imperial College London.

Hurford Anthony J.,
Research School of Geological Sciences,
University College London.

Johnson Christopher,
Department of Geological Sciences,
University College London.

Miller Jack,
Bullard Laboratories,
Department of Earth Sciences,
University of Cambridge.

Miller Marcia,
Bullard Laboratories,
Department Earth Sciences,
University of Cambridge.

Siddall Ruth,
Department of Geological Sciences,
University College London.

Summerfield Michael Anthony,
Department of Geography,
University of Edinburgh.

Upton David,
Department of Geological Sciences,
University College London.

United States of America

Blythe Ann Elizabeth,
Department of Geological Sciences,
University of Southern California,
Los Angleles, California.

Boettcher Stefan,
Deparment of Geological Sciences,
University of Texas at Austin,
Austin, Texas.

Dumitru Trevor A.,
Department of Geological and
Environmental Sciences,
Stanford University, Stanford California.

Fayon Annia,
Department of Geology,
Arizona State University,
Tempe, Arizona.

Fitzgerald Paul G.,
Department of Geosciences,
University of Arizona,
Tucson , Arizona.

Kelley Shari A.
Department of Earth and
Environmental Sciences,
New Mexico Inst. of Min.&Technology,
Socorro.

Ketcham Richard,
Department of Geological Sciences,
Rice University,
Houston, Texas.

Kowallis Bart Joseph,
Department of Geology,
Brigham Young University,
Provo, Utah.

Naeser Charles,
U.S. Geological Survey,
National Center, Reston,VA.

Naeser Nancy,
U.S. Geological Survey,
National Center, Reston, VA.

Roden-Tice Mary K.,
Center for Earth and Environmental
Science, State University of New York-
Plattsburgh, Plattsburgh NY.

Stockli Daniel
Department of Geological and
Environmental Sciences, Stanford
University, Stanford, California.

Tice Steven,
Center for Earth and Environmental
Science, State University of New York -
Plattsburgh, Plattsburgh NY.

Section I

Track Fundamentals, Dating Methodology and Age Calibration

HIGH-RESOLUTION MICROSCOPY OF LATENT TRACKS INDUCED BY HIGH-ENERGY HEAVY IONS

J. VETTER[1], J. ACKERMANN[1], R. NEUMANN[1], L. NISTOR[2], R. SCHOLZ[3]

[1] *Gesellschaft für Schwerionenforschung, Planckstr. 1, D-64291 Darmstadt, Germany*
[2] *Institute for Atomic Physics, P.O. Box MD 6, Bucharest, Romania*
[3] *Max-Planck-Institut für Mikrostrukturphysik, Am Weinberg 2, D-06120 Halle, Germany*

Abstract

On its way through a solid, an energetic heavy ion deposits an enormous amount of kinetic energy per unit path length during an extremely short time interval. The energy transfer induces very fast primary processes which can cause, depending on the specific solid properties, damages ranging from point defects to a total amorphization in the case of crystalline material within a long, thin cylinder-shaped volume along the ion trajectory. Details of size, shape, and morphology of the ion tracks, though representing the final stage of the damage creation, store indirect information about the not simultaneously observable very rapid primary processes. We present images and additional information on the properties of ion tracks in germanium monosulfide, highly oriented pyrolytic graphite, and muscovite mica achieved with transmission electron-, scanning tunneling-, and scanning force-microscopy.

1. Introduction

In numerous insulating or semi-conducting materials an energetic heavy ion creates a zone characterized by a considerable alteration of the original structure along its trajectory. A uranium ion, to take an extreme example, accelerated to a kinetic energy of 11.6 MeV/nucleon, has an energy typical of ion beams provided by the Universal Linear Accelerator UNILAC of GSI. This ion travels along the first tens of microns of its trajectory in a solid with a velocity of about 15% of the velocity of light. At such a velocity, the ion loses its energy mainly by interaction with the electrons of the solid and deposits energy amounts on the order of 30 keV/nm. Assuming a lattice constant of a third of a nm along the ion path, the ion is successively transferring about 10 keV per lattice plane in time intervals of $7 \cdot 10^{-18}$ seconds. This illustrates that the ion passage creates extreme conditions initiating dramatic, very rapidly developing primary processes which cannot be observed directly. However, the details of size, form, and internal structure of the ion track, though representing the final stage of damage creation, contain indirect information about these processes.

Depending upon the radiation sensitivity of the material, the changes range from the occurrence of point defects via, e.g., separate clusters of an atomic component of the solid to continuous damage trails (usually called latent tracks), in which the crystal is fully amorphized. It should be emphasized that a comprehensive understanding of the processes underlying ion track creation can be achieved only on a long-term basis by collecting data from many different materials and accompanying experimental findings

P. Van den haute and F. De Corte (eds.), Advances in Fission - Track Geochronology, 3-18.

with gradually refined modeling. For this purpose, latent tracks have been tackled by numerous groups with a variety of methods: scattering of neutrons and X-rays, or infrared spectroscopy derive the signal from an interaction with a large number of ion tracks. To provide access also to single, individual tracks, these methods were complemented by transmission electron microscopy (TEM), scanning tunneling microscopy (STM), and scanning force microscopy (SFM). The latter techniques are capable of characterizing the objects under study with subnanometer resolution.

This article gives an overview of our results on latent tracks in three different kinds of layered crystals, namely germanium monosulfide, highly oriented pyrolytic graphite, and muscovite mica, obtained by means of TEM as well as STM and SFM. The materials chosen all possess a layered structure that can be easily cleaved parallel to the layers. They are especially sensitive to high-energy ion irradiation, i.e., they store latent tracks with a clear signature. In particular, HOPG and mica are well suited for imaging with STM and SFM, providing atomic resolution even under ambient conditions. Besides presenting images of single track cross sections, we report on subtle track-related effects, such as lattice strain centers, periodic changes of the electron density, and altered friction and surface hardness on a nanometer scale. All samples under study were irradiated with high-energy heavy ions at the UNILAC, which can accelerate ions of all elements throughout the periodic table up to kinetic energies of about 14 MeV/u.

2. Transmission Electron Microscopy

2.1. GENERAL REMARKS

Transmission electron microscopy is capable of visualizing the atomic columns of a crystalline lattice and nanometer-sized objects embedded in a lattice. Therefore, TEM was already being applied in the late fifties and early sixties to fission fragment tracks in mica (Silk and Barnes, 1959; Bonfiglioli *et al.*, 1961; Price and Walker, 1962; See also Fleischer *et al.*, 1975; section 1.2.2), which are objects on a nanometer scale. However, since latent ion tracks tend to fade under the influence of an energetic electron beam, the method was pushed into the background for a longer period of time. In the late eighties, TEM of latent ion tracks was taken up again and applied successfully to a variety of crystalline materials such as yttrium-iron-garnet and others (Fuchs *et al.*, 1987; Houpert *et al.*, 1988; Houpert *et al.*, 1989; Studer *et al.*, 1989). A comprehensive TEM study on tracks in zirkon in 1990 (Bursill and Braunshausen, 1990) was followed by first measurements on heavy ion tracks in germanium sulfide (Vetter *et al.*, 1992; Vetter and Scholz, 1993; Scholz *et al.*, 1993). Recent TEM investigations on latent tracks in high-T_C superconducting crystals should also be mentioned (Wiesner *et al.*, 1996; Yan *et al.*, 1996). The results on GeS having grown significantly since 1993 are briefly summarized in the following section.

2.2. SOLID-STATE PROPERTIES, ION IRRADIATION, AND SAMPLE PREPARATION OF GERMANIUM MONOSULFIDE

Germanium monosulfide (GeS) is a semiconducting IV-VI compound that crystallizes in an orthorhombic structure with the lattice parameters a=0.43 nm, b=1.047 nm and c=0.364 nm. It consists of layers in the (010) plane and can be cleaved easily parallel to these layers by pulling off a thin sheet with the help of adhesive tape. Alternatively, the definition b=0.364 nm and c=1.047 nm is also used in the works cited (Grandke and Ley, 1977; Láznicka, 1975), the plane of the layers being denoted with (001). Concerning the details of the solid-state properties and the crystal growth procedures of GeS, the reader is referred to the following works and further references therein (Grandke and Ley, 1977; Láznicka, 1975; Stanchev and Vodenicharov, 1975; Schönherr and Stetter, 1975).

The samples were irradiated with different heavy-ion species with the kinetic energies ranging from about 5.6 to 13.5 MeV/u. The fluences were typically on the order of 1 to $5 \cdot 10^{10}$ ions/cm^2, large enough to guarantee a facile and fast detection, but still sufficiently small to avoid overlapping of the tracks. The ion beam direction was perpendicular to the (010) basal plane. In some cases, after ion irradiation, GeS flakes were further thinned by means of low-energy ion bombardment (typical parameters, sufficiently low to avoid undesired sample modifications, are: Ar^+, 5 keV, 1 mA) to become suitable for TEM imaging.

2.3. RESULTS AND DISCUSSION

Images were produced with high-resolution transmission electron microscopes (JEOL and Philips). The micrograph of Figure 1 shows the latent tracks of Au-ions with a primary energy of 11.4 MeV/u.

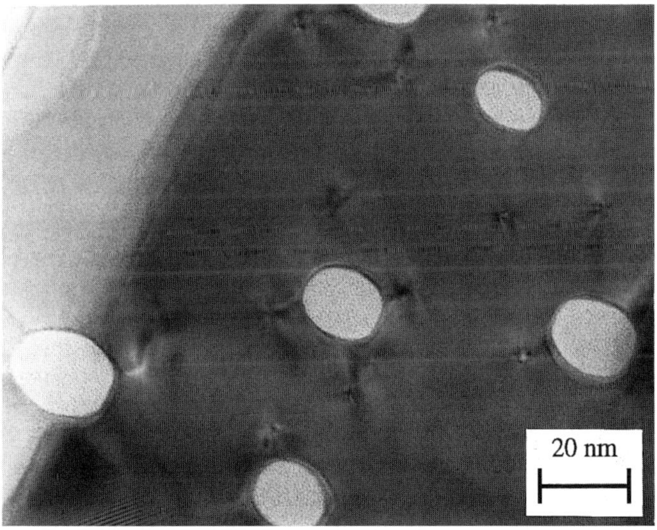

Figure 1. Cross sections of gold ion tracks in GeS imaged by TEM.

The cross sections of the tracks appear as elliptical areas embedded in an undamaged environment. Since the number of tracks per unit area corresponds to the applied ion fluence, they originate without any doubt from the passage of heavy ions. The ellipses are oriented with their long axes parallel to the direction of the lattice constant c. Figure 2 shows a high-resolution image of the cross section of a track caused by a single uranium ion.

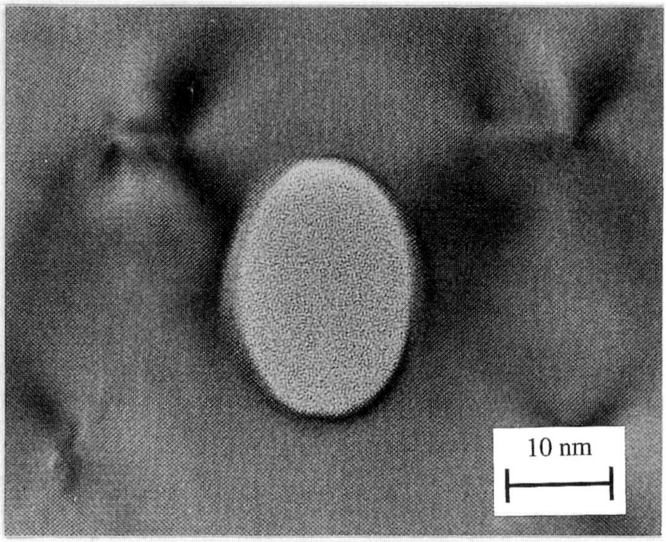

Figure 2. TEM micrograph of a latent track in GeS caused by a single 5.6 MeV/u uranium ion (Vetter *et al.*, 1992).

Each of the dots of the surroundings represents an atomic column perpendicular to the (010) image plane, whereas the material within the track looks completely disordered. This loss of atomic order clearly indicates that the solid has changed into an amorphous phase along the ion trajectory. The amorphized material fills a long and thin volume with the shape of an elliptical cylinder. The amorphous and crystalline phases are separated by a remarkably sharp interface.

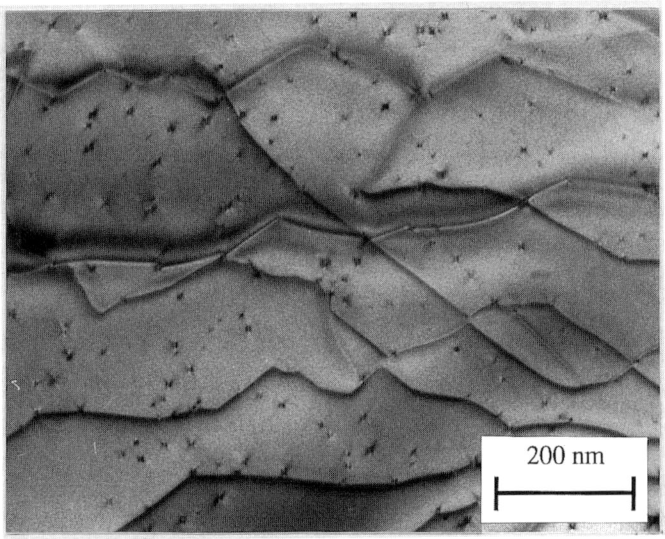

Figure 3. TEM image of zinc ion tracks serving as pinning centers for migrating dislocations in GeS.

The track is surrounded by four symmetrically arranged strain centers that have a preferred orientation. Furthermore, the latent tracks serve as pinning centers for migrating dislocations (for details see Scholz *et al.*, 1993). The phenomenon of dislocation pinning is impressively illustrated in Figure 3, showing a sample irradiated with 11.4 MeV/n zinc ions. Aiming at a quantitative characterization of the ion tracks in GeS, the sizes of the long and short axes of the elliptical cross sections were evaluated for eight different ion species, each species corresponding also to another value of energy loss. Penetration depth and energy loss were calculated with the TRIM (Transport of Ions in Matter) computer code (Ziegler *et al.*, 1985). Mean values of the axes as a function of ion energy loss are plotted in Figure 4.

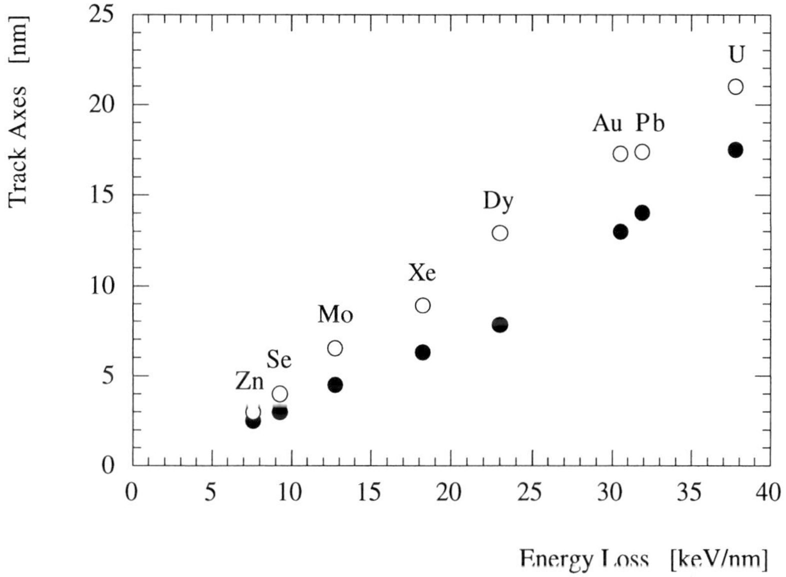

Figure 4. Long and short axes of elliptical latent tracks in GeS as a function of energy loss (standard errors of the data points are ± 0.5 nm).

During the past two decades, the functional dependence of track properties on essential parameters of projectile, solid, and projectile-solid interaction has been the subject of an enormous amount of theoretical work (See, e.g., Katz, 1978; Meftah *et al.*, 1994; and ref. therein). Thus, a sound understanding of the track diameter-versus-energy loss curve will require a detailed comparison of these data with the predictions of available theoretical concepts, such as the so-called thermal-spike model (Meftah *et al.*, 1994; Toulemonde *et al.*, 1992).

3. Scanning Probe Microscopy

3.1. SCANNING TUNNELING MICROSCOPY OF ION TRACKS IN HIGHLY ORIENTED PYROLYTIC GRAPHITE

Highly oriented pyrolytic graphite (HOPG) is a layered crystal consisting of parallel lattice planes with hexagonal ordering of the carbon atoms. The material exhibits a high electric and thermal conductivity within the lattice planes. Since only van der Waals forces are acting between adjacent lattice planes, a much smaller conductivity exists along the normal of the planes, i.e., parallel to the c-axis of the crystal. As a consequence, the HOPG can be cleaved easily parallel to the planes by pulling off a thin surface layer by means of an adhesive tape. In this way, a fresh non-contaminated surface can be created, which is atomically flat over hundreds of nanometers.
During the past decade, damage features in HOPG produced by ion irradiation were studied with STM by a number of working groups (Porte *et al.*, 1989; Coratger *et al.*, 1990; Porte *et al.*, 1991; Kemmer *et al.*, 1992; Bouffard *et al.*, 1993; Yan *et al.*, 1994). We performed studies on HOPG samples irradiated with 15 MeV/u Au^{24+} ions (Kemmer *et al.*, 1992). For most samples, the direction of irradiation was parallel to the c-axis, and in a few cases almost at grazing incidence. In some cases, prior to STM imaging, a surface layer having a thickness of typically several microns was pulled off. For the given initial kinetic energy of 15 MeV/u the ions have a penetration depth of more than 100 μm. Due to the large ion energy, the damage trails in the interior of the crystal, whose cross sections become accessible by the cleavage, are caused exclusively by electronic energy loss.
An STM image containing the cross section of an ion track is shown in Figure 5 (from Probst *et al.*, 1993).

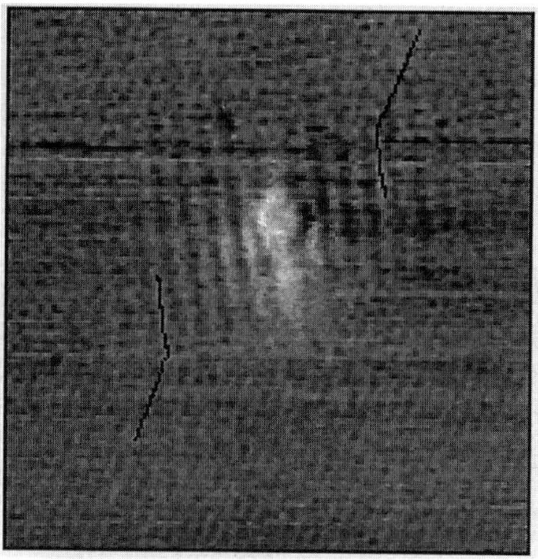

Figure 5. STM image (9·9 nm^2) of a single heavy-ion track in HOPG (Probst *et al.*, 1993).

It is obvious that the crystalline order has been destroyed within the track, whereas the intact surroundings appear with atomic resolution. Interestingly, small hillocks occur at the site of the track cross sections on a freshly cleaved surface. This finding is in agreement with results reported by Porte *et al.* (1989, 1991) and Coratger *et al.* (1990)

for low-energy ions. We interpret the occurrence of hillocks by the release of stress, built up by dislocations of atoms from the track core to the periphery. Wave-like structures of vertical, parallel and equidistant straight lines, starting from both sides of the disordered region are also visible in Fig. 5. The line pattern has a periodicity of 4 Å and an orientation of a ≈ 30° with respect to the HOPG lattice (as indicated by the black bars in Figure 5). We tentatively interpret these oscillations, which occurred repeatedly in our micrographs, as periodic density changes of the two-dimensional electron gas in the hexagonal graphite lattice plane, caused by the scattering of Bloch waves at the ion-induced localized lattice defect (Mizes and Foster, 1989; Probst *et al.*, 1993). Similar features in ion-irradiated HOPG were reported by Porte *et al.* (1989), Bouffard *et al.* (1993) and Yan *et al.* (1994).

3.2. SCANNING FORCE MICROSCOPY OF ION-IRRADIATED MUSCOVITE MICA

3.2.1. *Introductory Notes*

Muscovite mica is an insulating, layered crystalline silicate. Similar to HOPG, it can be cleaved easily parallel to its layers, always splitting between the two neighbouring SiO_4-hexagonal lattice planes. It was early recognized that mica is very sensitive to irradiation with energetic heavy ions. Over several decades, ion tracks in mica have therefore been investigated with a variety of methods. Transmission electron microscopy (Silk and Barnes, 1959; Bonfiglioli *et al.*, 1961; Price and Walker, 1962) and X-ray (Dartyge and Lambert, 1974; Dartyge *et al.*, 1981; Albrecht *et al.*, 1985; Albrecht *et al.*, 1986) and neutron scattering (Albrecht *et al.*, 1985) were followed more recently by SFM (Thibaudau *et al.*, 1991; Bouffard *et al.*, 1993; Hagen *et al.*, 1994; Ackermann *et al.*, 1995; Daya *et al.*, 1995a; Daya *et al.*, 1995b; Ackermann *et al.*, 1996; Daya *et al.*, 1996). The latter technique was also applied to study the unperturbed structure of the SiO_4-planes (Erlandsson *et al.*, 1988; Marti *et al.*, 1990; Sharp *et al.*, 1993). It should be mentioned once more that ion tracks represent objects of very limited size, having diameters of only several nanometers. They are thus not only of interest concerning the analysis of radiation-induced damage features, but are suitable also for the study of friction and elasticity on a nanometer scale.

Samples of mica were irradiated with heavy ions at right angles to the SiO_4-layers. Choosing several ion species with different kinetic energies, a set of data on ion-track diameters versus electronic energy loss could be achieved. The relevant properties of the ion beams applied to mica are listed in Table 1.

Ion Species	Total Energy [GeV]	Energy Loss [keV/nm]	Penetration Depth [μm]
Krypton	0.50	11.8 ± 0.6	47
Xenon	1.47	16.9 ± 1.5	90
Gold	2.64	25.6 ± 1.2	110
Lead	2.37	27.4 ± 1.0	95
Uranium	2.76	31.0 ± 0.6	100

Table 1: Relevant properties of the ion beams applied to mica (Ackermann *et al.*, 1996)

Similar to our work on GeS, the values of energy loss and penetration depth were calculated with TRIM (Ziegler *et al.*, 1985). The measurements were performed with two homebuilt scanning force microscopes (SFM). These microscopes acquire simultaneously the vertical bending (for topographic imaging) and the torsion (for lateral-force imaging) of the force-sensing cantilever. For this purpose, we apply the common technique of reflecting the beam of a laser diode at the cantilever's backside and monitoring its movements on a quadrant photodiode.

3.2.2 *Lateral-Force Microscopy*

The ion energy loss as a function of the penetration depth possesses a broad plateau starting at the target surface. Therefore, a mica sample can be cleaved several times by pulling off a sheet with a thickness of a few microns, before the plateau is exceeded. This cleaving procedure provides a fresh surface free of contaminations and suitable for SFM imaging.

The SFM mode of operation, called lateral-force imaging, records the torsion of the force-sensing cantilever around its long symmetry axis and is particularly sensitive to friction. As reported in a previous study (Hagen *et al.*, 1994), the ion tracks are amorphous zones whose cross sections exhibit a significantly higher friction coefficient than the surrounding intact lattice plane.

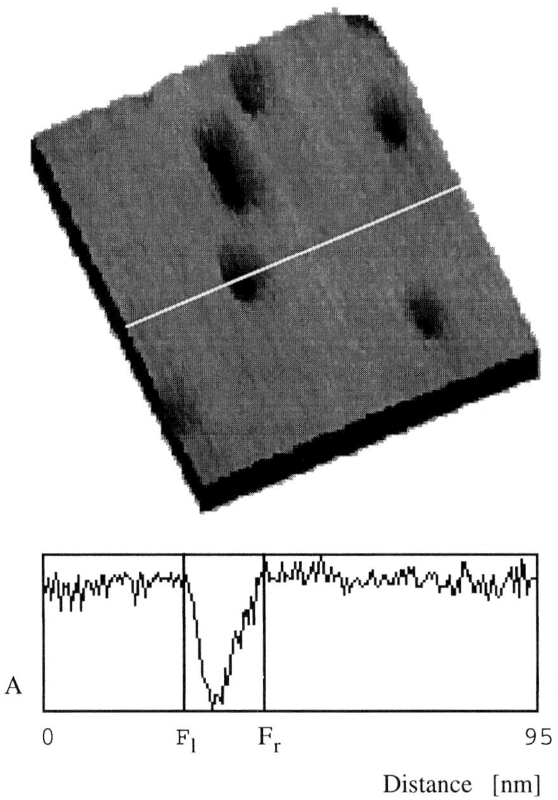

Figure 6a. Lateral-force image (95·95 nm^2) of uranium ion tracks in mica.
b. Line scan through the image from A to B with a left (F_l) and a right (F_r) foot defining the limits of the track diameter.

For the sake of thoroughness, it should be noted that the level of the track cross sections coincided completely with the cleaved surface planes, showing neither hillocks nor craters. Fig. 6a displays the lateral-force image of Au-ion tracks. Each impinging ion created a latent track.
The diameter of an individual track was evaluated by determining a left and a right foot of the line profile as illustrated in Fig. 6b. Mean track diameters for the different ion species were obtained by fitting the diameter distributions with a Gaussian curve. In Figure 7, our result is compared with data from small-angle x-ray and neutron scattering (Albrecht *et al.*, 1985; Albrecht *et al.*, 1986) and from topographic scanning force microscopy (Thibaudau *et al.*, 1991; Bouffard *et al.*, 1993).

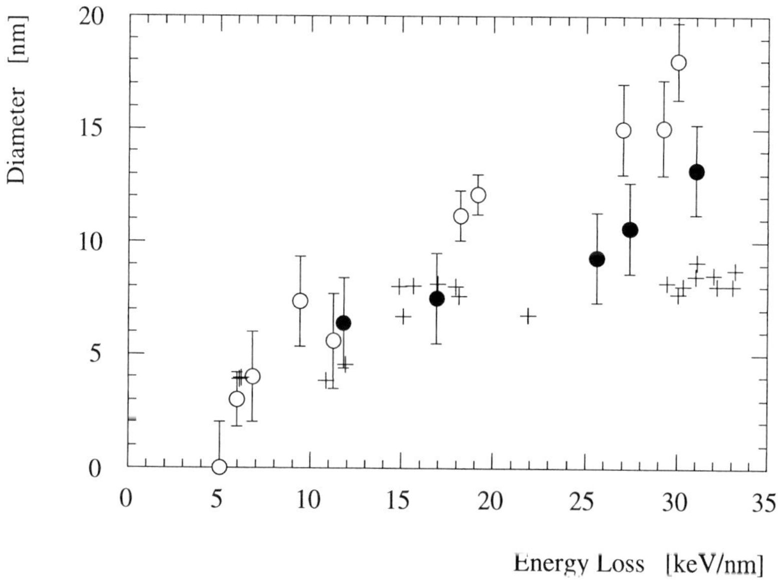

Figure 7. Diameters of latent tracks in mica as a function of energy loss.
(•) Our lateral-force SFM measurements (Ackermann *et al.*, 1996)
(o) SFM (Thibaudau *et al.*, 1991; Bouffard *et al.*, 1993)
(+) X-ray and neutron scattering (Albrecht *et al.*, 1985; Albrecht *et al.*, 1986)

Partial agreement and discrepancies among the three data sets are clearly visible. Further information about the acquisition and evaluation of the data compiled in Fig. 7 can be found in Ackermann *et al.* (1996).

3.2.3. *Force-Modulation Microscopy*

Force-modulation microscopy, which was introduced some years ago (Maivald *et al.*, 1991; Radmacher *et al.*, 1992; Overney *et al.*, 1994; Heuberger *et al.* 1994), allows for the imaging of surface properties, e.g., hardness or elasticity, on a nanometer scale. Recently, our group applied this method for the first time to heavy-ion tracks (Ackermann *et al.*, 1995; Neumann *et al.*, 1996). The microscope was operated in the modulation mode by varying the sample position periodically normal to its surface. The modulation was driven with a piezo crystal at 3 kHz by the reference frequency of a lock-in amplifier, keeping constant an average loading force between the force-sensing tip and the sample via a feedback circuit. Periodic variations of the sample position result in a deformation of the sample surface and a vertical cantilever deflection. This deflection is enlarged (reduced) if the sensor probes a harder (softer) surface area, as illustrated in Fig. 8.

The lock-in filtered signal, derived from the cantilever deflection while scanning the sample, provides a map of the surface hardness. A force-modulation micrograph is shown in Fig. 9.

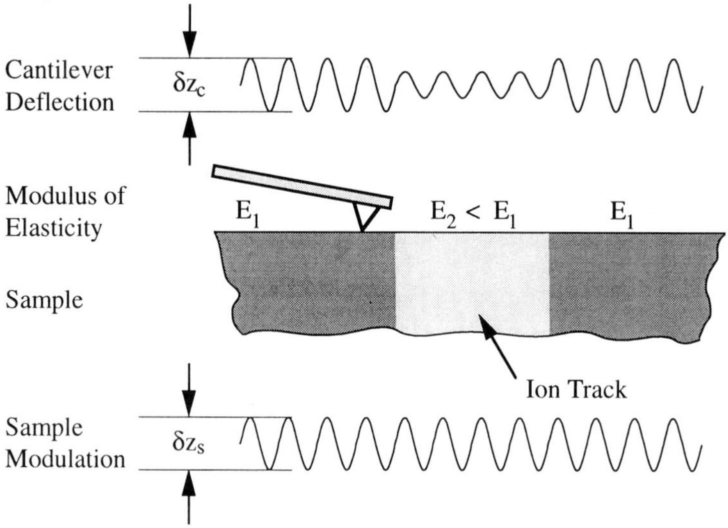

Figure 8. The cantilever deflection δz_c as a function of surface elasticity, driven by a sample modulation of constant amplitude δz_s.

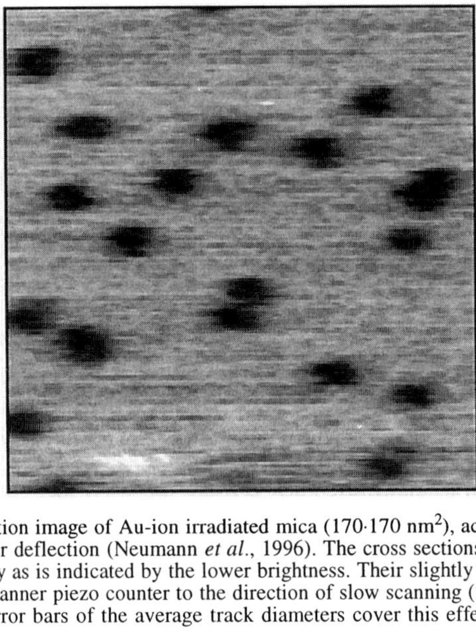

Figure 9. Force-modulation image of Au-ion irradiated mica (170·170 nm^2), acquired by lock-in filtered detection of the cantilever deflection (Neumann *et al.*, 1996). The cross sections of ion tracks are areas of reduced surface elasticity as is indicated by the lower brightness. Their slightly elliptical shape originates from the creep of the scanner piezo counter to the direction of slow scanning (from top to bottom). The error bars of the average track diameters cover this effect.

Mean track diameters were extracted from the images, applying a similar evaluation procedure as for the lateral-force results. Both data sets are plotted in Fig. 10, demonstrating good agreement within the ± 2 nm vertical error bars. From this, we conclude that the changes of friction and surface hardness are both caused by the amorphization.

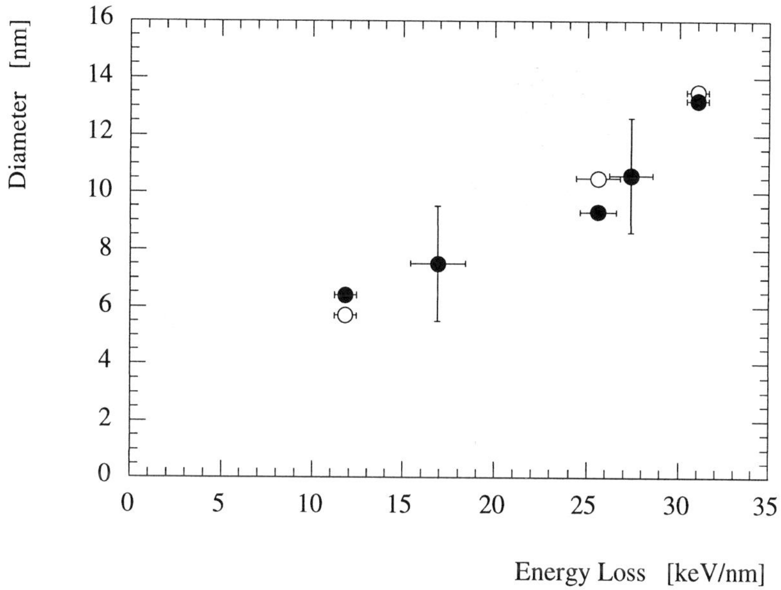

Figure 10. Latent-track diameters in mica versus ion-energy loss (Neumann *et al.*, 1996).
(o) Force modulation measurements of our group
(•) Lateral-force microscopy (Hagen *et al.*, 1994; Ackermann *et al.*, 1996)
The vertical error bars of all data points are ± 2 nm, but have been omitted in all cases where a mutual overlap would occur

3.2.4. *Force Microscopy in an Ultrahigh Vacuum*

The SFM studies described in the previous sections were performed under ambient conditions, where the presence of adsorbates such as a thin film of water and hydrocarbons hamper the reproducibility of quantitative measurements, e.g., of the friction coefficient. Therefore, after ion irradiation, mica samples were transferred into an ultrahigh-vacuum (UHV) chamber, housing a homebuilt scanning force microscope. The optical arrangement, which monitors the deflection of the force sensor, uses movable mirrors driven by inertial sliding mechanisms similar to the setup reported by Howald *et al.* (1993). Our microscope and the UHV system are described in detail in: Ackermann (1994), Sedlacek (1995), Wörtge *et al.* (1994). While baking the UHV chamber for nearly three days, the temperature was restricted to a maximum value of 120 °C to avoid damaging the semiconductor devices (an IR-laser diode and a four-segmented photodiode) of the optical detection system. For this reason, the vacuum obtained was limited to about $1\text{-}2\cdot10^{-9}$ mb, but nevertheless, guaranteed a significant reduction of undesired adsorbates. We used a commercial bar-shaped cantilever with a pyramidal tip at its free end, consisting of amorphous Si_3N_4 and having a bending and a torsion spring constant of k=0.215 N/m and $k=8.4\cdot10^{-10}$ Nm/rad, respectively.

Other than the mica studies described in the previous sections, the following measurements were performed on the original sample surface, as it had been exposed to ion irradiation (Ackermann, 1997). The direction of fast scanning, in the following denoted with y, was perpendicular to the long axis of the sensor bar. Four simultaneously acquired images of the damage caused by the impact of a 11.4 MeV/u uranium ion are shown in Figure 11.

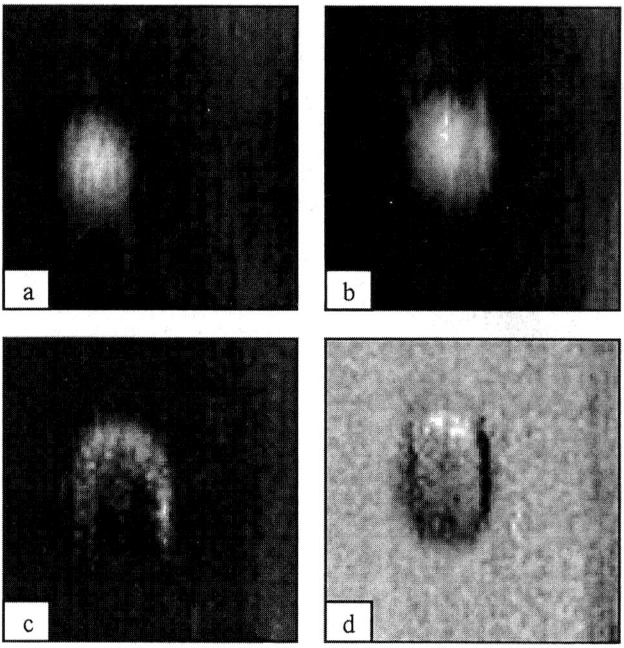

Figure 11. SFM images (45·45 nm^2) of a latent uranium ion track in mica, measured under UHV conditions (Ackermann *et al.*, 1997). (a) and (b) are topographic images, whereas (c) and (d) represent the torsion of the cantilever. The fast scanning motion (along y) occurred in the vertical direction.

The micrographs (a) and (b) were obtained by detection of the vertical bending of the sensor during back and forth scanning and represent the topography, i.e., the spatial shape of the sample surface. Since reversing the scanning direction does not change the vertical bending, images (a) and (b) have a similar appearance. When penetrating the mica surface, the uranium ion created a small hillock roughly 2 nanometers high and with a diameter of about 20 nm. This is in contrast to the absolutely flat cross sections of the ion tracks on freshly cleaved mica surfaces, observed with a SFM in air. The micrographs (c) and (d) were also acquired while scanning back and forth, however, by monitoring the torsion of the cantilever around its long axis. These lateral-force images contain information on topography and friction. They exhibit a different contrast, since the friction force changes its sign when reversing the scanning direction and, therefore, the fraction of contrast caused by friction is reversed.

According to a model presented in: Grafström *et al.* (1993), Grafström *et al.* (1994), the lateral force F_y, directed along the y-axis and effecting the cantilever torsion, can be expressed as:

$$F_y = -(F_L + F_A)\cdot(s_y \pm \mu) \qquad (1)$$

for scanning along $\pm y$. Here, F_L and F_A are the external loading force and the inherent adhesion force; μ denotes the dynamic friction coefficient; and $s_y = \partial s(x,y)/\partial y$ represents the partial derivative of the surface topography $s(x,y)$ with respect to the y-direction. The equation expresses the fact that the topography-induced contrast (s_y) does not depend upon the scanning direction, whereas the friction-induced contrast (μ) reverses its sign. In conclusion, the sum and difference of the torsion micrographs (a) and (b) provide a deconvolution into the pure topography- and friction-induced parts of the total contrast. The sum and difference images are given in Figure 12.

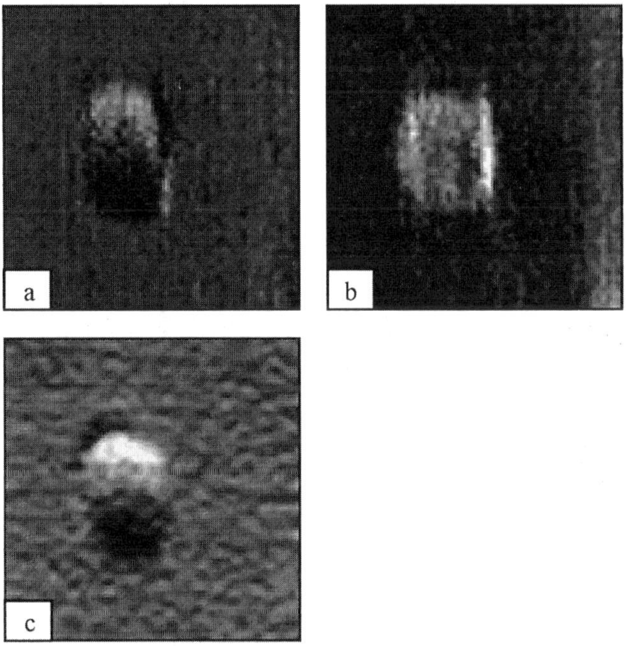

Figure 12. Sum (a) and difference (b) of the images (c) and (d) in Figure 11 (Ackermann *et al.*, 1997). The sum maps the derivative (i.e. the slope) of the topography along the y-direction. The difference represents the pure tribologic contrast. (c) Numerical y-derivative of the image (a) in Figure 11

It should be pointed out, once again, that Figure 12a displays the slope of the topography in the y-direction, rather than the topography itself. The intact mica lattice with zero slope around the damage zone appears as a homogeneously gray background. Areas brighter or darker than this background indicate a positive or negative slope and represent the hillock. This interpretation can be examined by comparing Figure 12a with Figure 12c, which maps the numerical y-derivative of the topographic image in Figure 11a. The agreement of the two images is very satisfying. Figure 12b, containing the pure tribological contrast shows a friction increase (indicated by the higher brightness) on the hillock in comparison with the surrounding mica lattice. This finding is in accordance with previous SFM results on cleaved mica in air, manifesting amorphization of the crystalline lattice along the ion trajectory accompanied by higher friction within the disordered zones.

4. Conclusions

Latent tracks induced by energetic heavy ions in the layered crystals GeS, HOPG, and muscovite mica have been imaged with very high resolution using TEM, STM, and SFM. The micrographs visualize the tracks as disordered regions embedded in an almost intact crystalline environment. For GeS, they further revealed a sharp interface between the amorphous and the crystalline phases; a strict orientation of the elliptical cross sections with respect to the anisotropic lattice; symmetrically distributed and oriented strain centers around the tracks; and a pinning of migrating dislocations at the tracks. The sizes of the track cross sections were evaluated as a function of the ion energy loss. Latent ion tracks on the surface of freshly cleaved HOPG samples were identified by STM as tiny amorphous hillocks. We interpret the occurence of hillocks by the release of stress along the ion trajectory. In addition, wave-like superstructures of parallel and equidistant lines were found, originating from a track and reaching quite far into the unperturbed lattice. They probably represent density oscillations of the conduction electrons due to scattering at the track.
The study of ion tracks in mica by force microscopy provided a data set of track diameters versus energy loss, similar to the results for GeS. It should be pointed out that in both cases our compilations of latent-track data intend to contribute to the experimental basis indispensable for the modeling of track creation processes. Moreover, the track cross sections were used as objects for studies of surface properties on a nanometer scale. By operating the force microscope in different dedicated imaging modes, an increased friction and a reduced elasticity could be demonstrated within the track areas. Finally, first SFM measurements of tracks in mica were performed under UHV conditions. They are considered to be a first step towards in-situ SFM investigations in an UHV chamber flanged to an ion beam line at GSI.

Acknowledgment

L. Nistor thanks GSI for financial support during her stay in Darmstadt in the spring of 1996. The possibility of using a transmission electron microscope at the Department for Materials Science, Technical University of Darmstadt, is also gratefully acknowledged.

References

Ackermann J. (1994): Rasterkraftmikroskopie an Schwerionen-bestrahlten Materialien: Spuruntersuchungen und Entwicklung eines UHV-Kraftmikroskops, *Doctoral thesis*, University of Heidelberg.

Ackermann J., Angert N., Grafström S., Hagen T., Neitzert M., Neumann R. and Trautmann, C. (1995): Force microscopy of heavy-ion irradiated materials. In: *Forces in Scanning Probe Methods* (eds. Güntherodt H.-J., Anselmetti D. and Meyer E.), Kluwer Academic Publishers, Dordrecht, Boston, London, pp. 489-494.

Ackermann J., Angert N., Neumann R., Trautmann C., Dischner M., Hagen T. and Sedlacek M. (1996): Ion track diameters in mica studied with scanning force microscopy. *Nucl. Instrum. and Meth.* **B 107**, 181-184.

Ackermann J., Grafström S., Hagen T., Kowalski J., Neumann R. and Sedlacek M. (1997): Scanning force microscopy of latent heavy-ion tracks in ultrahigh vacuum. In: *Micro/Nanotribology and its applications* (ed. Bhushan B.), Kluwer Academic Publishers, Dordrecht, Boston, London, pp. 261-267.

Albrecht D., Armbruster P., Spohr R., Roth M., Schaupert K. and Stuhrmann H. (1985): Investigation of heavy ion produced defect structures in insulators by small angle scattering. *Appl. Phys.* **A 37**, 37-46.

Albrecht D., Balanzat E., and Schaupert K. (1986): X-ray small angle scattering investigation of high energy Ar-tracks in mica. *Nucl. Tracks Radiat. Meas.* **11**, 93-94.
Bonfiglioli G., Ferro A., and Mojoni A. (1961): Electron microscope investigation on the nature of tracks of fission products in mica. *J. Appl. Phys.* **32**, 2499-2503.
Bouffard S., Cousty J., Pennec Y. and Thibaudau F. (1993): STM and AFM observations of latent tracks. *Radiat. Eff. Def. Sol.* **126**, 225-228.
Bursill L.A. and Braunshausen G. (1990): Heavy-ion tracks in zirkon. *Phil. Mag.* **62**, 395-420.
Coratger R., Claverie A., Ajustron F. and Beauvillain J. (1990): Scanning tunneling microscopy of defects induced by carbon bombardment on graphite surfaces. *Surf. Sci.* **227**, 7-14.
Dartyge E. and Lambert M. (1974): Formation de défauts dans les échantillons dc mica muscovite irradiés par des ions de grande énergie. *Radiat. Eff.* **21**, 71-79.
Dartyge E., Duraud J.P., Langevin Y. and Maurette M. (1981): New model of nuclear particle tracks in dielectric minerals. *Phys. Rev.* **B 23**, 5213-5229.
Daya D.D.N.B., Hallén A., Håkansson P., Sundqvist B.U.R. and Reimann C.T. (1995a): Swift-heavy-atomic-ion-induced surface tracks on mica probed by scanning force microscopy. *Nucl. Instr. and Meth.* **B 103**, 454-465.
Daya D.D.N.B., Hallén A., Eriksson J., Kopniczky J., Papaléo R., Reimann C.T., Brunelle A., Della-Negra S. and Le Beyec Y. (1995b): Radiation damage features on mica and l-valine probed by scanning force microscopy. *Nucl. Instr. and Meth.* **B 106**, 38-42.
Daya D.D.N.B., Reimann C.T., Hallén A., Sundqvist B.U.R. and Håkansson P. (1996): Latent (sub-surface) tracks in mica studied by tapping mode scanning force microscopy. *Nucl. Instr. and Meth.* **B 111**, 87-90.
Erlandsson R., Hadziioannou G., Mate C.M., McClelland G.M. and Chiang S. (1988): Atomic scale friction between the muscovite mica cleavage plane and a tungsten tip. *J. Chem. Phys.* **89**, 5190-5193.
Fleischer R.L., Price P.B. and Walker R.M. (1975): *Nuclear tracks in solids.* University of California Press, Berkeley.
Fuchs G., Studer F., Balanzat E., Groult D., Toulemonde M. and Jousset J.C. (1987): Influence of the electronic stopping power on the damage rate of yttrium-iron garnets irradiated by high-energy heavy ions. *Europhys. Lett.* **3**, 321-326.
Grandke T. and Ley L. (1977): Angular-resolved uv photoemission and the band structure of GeS. *Phys. Rev.* **B 16**, 832-842.
Grafström S., Neitzert M., Hagen T., Ackermann J., Neumann R., Probst O. and Wörtge M. (1993): The role of topography and friction for the image contrast in lateral force microscopy. *Nanotechnology* **4**, 143-151.
Grafström S., Ackermann J., Hagen T., Neumann R. and Probst O. (1994): Analysis of lateral force effects on the topography in scanning force microscopy. *J. Vac. Sci. Technol.* **B 12**, 1559-1564.
Hagen T., Grafström S., Ackermann J., Neumann R., Trautmann C., Vetter J. and Angert N. (1994): Friction force microscopy of heavy-ion irradiated mica. *J. Vac. Sci. Technol.* **B 12**, 1555-1558.
Heuberger M., Dietler G. and Schlapbach L. (1994): Mapping the local Young's modulus by analysis of the elastic deformations occurring in atomic force microscopy. *Nanotechnology* **5**, 12-23.
Houpert C., Hervieu M., Groult D., Studer F. and Toulemonde M. (1988): HREM investigation of GeV heavy ion latent tracks in ferrites. *Nucl. Instr. Meth.* **B 32**, 393-396.
Houpert C., Studer F., Groult D. and Toulemonde M. (1989): Transition from localized defects to continuous latent tracks in magnetic insulators irradiated by high energy heavy ions: A HREM investigation. *Nucl. Instr. Meth.* **B 39**, 720-723.
Howald L., Meyer E., Lüthi R., Haefke H., Overney R., Rudin H. and Güntherodt, H.-J. (1993): Multifunctional probe microscope for facile operation in ultrahigh vacuum. *Appl. Phys. Lett.* **63**, 117-119.
Katz R. (1978):Track structure theory in radiobiology and in radiation detection. *Nuclear Track Detection* **2**, 1-28.
Kemmer H., Grafström S., Neitzert M., Wörtge M., Neumann R., Trautmann C., Vetter J. and Angert N. (1992): Scanning tunneling microscopy of surface modifications induced by UNILAC heavy-ion irradiation. *Ultramicroscopy* **42-44**, 1345-1349.
Láznicka M. (1978): Investigation of GeS (001). *Czech. J. Phys.* **B 28**, 1031-1035.
Maivald P., Butt H.-J., Gould S.A.C., Prater C.B., Drake B., Gurley J.A., Elings V.B. and Hansma P.K. (1991): Using force modulation to image surface elasticities with the atomic force microscope. *Nanotechnology* **2**, 103-106.
Marti O., Colchero J. and Mlynek J. (1990): Combined scanning and friction force microscopy of mica. Nanotechnology **1**, 141-144.
Meftah A., Brisard F., Costantini J.M., Dooryhee E., Hage-Ali M., Hervieu M., Stoquert J.P., Studer F. and Toulemonde M. (1994): Track formation in SiO_2 quartz and the thermal-spike mechanism. *Phys. Rev.* **B 49**, 12457-12463.
Mizes H.A. and Foster J.S. (1989): Long-range electronic perturbations caused by defects using scanneling tunneling microscopy. *Science* **244**, 559-562.
Neumann R., Ackermann J., Angert N., Trautmann C., Dischner M., Hagen, T. and Sedlacek M. (1996): Ion tracks in mica studied with scanning force microscopy using force modulation. *Nucl. Instr. and Meth.* **B 116**, 492-495.
Overney R.M., Takano H. and Fujihira M. (1994): Elastic compliances measured by atomic force microscopy. *Europhys. Lett.* **26**, 443-447.

Porte L., Phaner M., De Villeneuve C.H., Moncoffre N.J. and Tousset J. (1989): Scanning tunneling microscopy study of single-ion impacts on graphite surface. *Nucl. Instr. Meth.* **B 44**, 116-119.
Porte L., De Villeneuve C.H. and Phaner M. (1991): Scanning tunneling microscopy observation of local damages induced on graphite surface by ion implantation. *J. Vac. Sci. Technol.* **B 9**, 1064-1067.
Price P.B. and Walker R.M. (1962): Observation of charged particle tracks in solids. *J. Appl. Phys.* **33**, 3400-3406.
Probst O., Ackermann J., Grafström S., Kowalski J., Neumann R., Neitzert M. and Wörtge M. (1993): Superstructures in the vicinity of heavy-ion induced defects observed by scanning tunneling microscopy. *GSI Scientific Report 1992*, p. 304.
Radmacher M., Tillmann R.W., Fritz M. and Gaub H. E. (1992): From molecules to cells: Imaging samples with the atomic force microscope. *Science* **257**, 1900-1905.
Sedlacek M. (1995): Rasterkraftmikroskopie an Schwerionenspuren unter Umgebungs- und Ultrahochvakuum-Bedingungen. *Diploma thesis*, University of Heidelberg.
Schönherr E. and Stetter W. (1975): Growth of germanium monosulfide single crystals by sublimation. *J. Crystal Growth* **30**, 96-98.
Scholz R., Vetter J. and Hopfe S. (1993): Observation of latent heavy-ion tracks in GeS by transmission electron microscopy. *Radiat. Eff. Def. Sol.* **126**, 275-278.
Sharp T.G., Oden P.I. and Buseck P.R. (1993): Lattice-scale imaging of mica and clay (001) surfaces by atomic force microscopy using net attractive forces. *Surf. Sci. Lett.* **284**, L405-L410.
Silk E.C.H. and Barnes R.S. (1959): Examination of fission fragment tracks with an electron microscope. *Phil. Mag.* **4**, 970-972.
Stanchev A. and Vodenicharov C. (1975): Dielectric properties of thin germanium monosulfide films. *Thin Solid Films* **29**, L13-L16.
Studer F., Houpert C., Groult D. and Toulemonde M. (1989): Latent tracks in magnetic insulators. *Radiat. Eff. Def. Sol.* **110**, 55-59.
Thibaudau F., Cousty J., Balanzat E. and Bouffard S. (1991): Atomic-force-microscopy observations of tracks induced by swift Kr-ions in mica. *Phys. Rev. Lett.* **67**, 1582-1585.
Toulemonde M., Dufour C. and Paumier E. (1992): Transient thermal process after a high-energy heavy-ion irradiation af amorphous metals and semiconductors. *Phys. Rev.* **B 46**, 14362-14369.
Vetter J., Scholz R. and Hopfe S. (1992): Transmission electron microscope observation of heavy-ion tracks in germanium sulfide. *GSI Scientific Report 1991*, p. 248.
Vetter J. and Scholz R. (1993): High resolution image of a single ion track. *GSI Scientific Report 1992*, p. 283.
Vetter J., Angert N. and Scholz R. (1994): Latent track dimensions of various ions. *GSI Scientific Report 1993*, p. 189.
Wiesner J., Træholt C., Wen J.-G., Zandbergen H.-W. Wirth G. and Fuess H. (1996): High resolution electron microscopy of heavy-ion induced defects in superconducting Bi-2212 thin films in relation to their effect on J_c. *Physica* **C 268**, 161-172.
Wörtge M., Dey S., Grafström S., Hagen T., Kowalski J., Neumann R. and Probst O. (1994): An ultrahigh-vacuum system for STM studies. *Rev. Sci. Instr.* **65**, 2523-2526.
Yan J., Li Z., Bai C., Yang W.S., Wang Y., Zhao W., Kang Y., Yu F.C., Zhai P. and Tang X. (1994): Scanning tunneling microscopy investigations of graphite surface damage induced by gold-ion bombardment. *J. Appl. Phys.* **75**, 1390-1395.
Yan Y., Doyle R.A., Campbell A.M., Wirth G. and Stobbs W.M. (1996): Modulated structure and local oxygen reordering induced by high-energy Au-ion irradiation in $YBa_2Cu_3O_{7-d}$. *Phil. Mag. Lett.* **73**, 299-308.
Ziegler J.F., Biersack J.P and Littmark U. (1985): *The Stopping and Ranges of Ions in Solids*. Pergamon Press, New York.

ZETA: THE ULTIMATE SOLUTION TO FISSION-TRACK ANALYSIS CALIBRATION OR JUST AN INTERIM MEASURE?

ANTHONY J. HURFORD
Department of Geological Sciences
University College London
London WC1E 6BT, UK

Abstract

Because of the lack of unique system calibration coupled with a low threshold for track annealing, the fission-track (FT) method was perceived in the 1970s as an inadequate and imprecise dating method. System calibration had a multiplicity of λ_f ^{238}U spontaneous fission-decay constant values and neutron dosimetry schemes with little inter-laboratory standardisation. Differences in methodology also introduced subtle variations into calibrations which often went unrecognised. Resurrection of a comparative calibration approach whereby the multiple dating of carefully selected age standards derived a calibration factor zeta, seemingly circumvented previous problems. Following consensus agreement at the 1988 Besançon FT Workshop, zeta has been adopted by a majority of FT workers and remains the common calibration approach in most published and commercial applications. With advances in neutron fluence dosimetry and evidence of precise experiments to determine λ_f, the case is advanced that zeta should be re-examined and those factors capable of quantification be separated from less readily determinable procedural factors. The danger of such attempts at scientific advance being perceived as a break-up of the calibration consensus is highlighted. Argument is made that applications of the FT method should continue to use existing calibration schemes, with strategies aimed at resolving zeta into its component parts run as parallel projects with no crossover. If and when significant advance in calibration is achieved, the case should be presented to the FT community for a systematic revision following the model of the 1988 Besançon agreement.

1. Introduction

In July 1971, at a conference entitled *Calibration of Hominoid Evolution* held under the auspices of the Wenner-Gren Foundation for Anthropological Research at Burg Wartenstein, Austria, Fleischer and Hart suggested a comparative approach to fission-track (FT) dating calibration whereby an age t could be derived simply from three track densities ρ_s, ρ_i & ρ_d and a proportionality constant, zeta (ζ):

$$t = \zeta \, (\rho_s / \rho_i) \, \rho_d \tag{1}$$

Zeta was defined comprehensively in terms of: c, the isotopic abundances of ^{235}U and ^{238}U; σ, the ^{235}U neutron capture cross section; λ_f, the ^{238}U spontaneous fission-decay constant; B, a neutron dosimeter constant; R, fission fragment range factors; and η, etching efficiencies, by the relationship:

$$\zeta = \frac{\sigma}{B\,\lambda_f} \, \frac{^{235}c}{^{238}c} \, \frac{^{235}R}{^{238}R} \, \frac{^{235}\eta}{^{238}\eta} \tag{2}$$

P. Van den haute and F. De Corte (eds.), Advances in Fission - Track Geochronology, 19-32.

Zeta was assumed to be a constant, requiring in turn the constancy of each expression. For physical constants c, σ and λ_f this was quite reasonable, and the consistent use of the same dosimeter under the same irradiation conditions implied that B could also be accepted as invariant. Measurement of fission-fragment range factors within the same detector material (see later) argued for $^{235}R/^{238}R$ to be unity. Similarity of the etching efficiency factors $^{235}\eta$, $^{238}\eta$ (i.e. the fractions of total tracks crossing a surface revealed by etching) required that neither spontaneous nor induced tracks be lost by annealing. Accepting these assumptions to be valid, zeta could be given by:

$$\zeta = t\ \rho_i\ /(\rho_s/\rho_d) \tag{3}$$

By a FT dating analysis of a sample whose age is known by independent means, ζ could be evaluated and, provided the independent age is equal to the FT age, ζ may be used subsequently to derive ages for unknown samples *via* equation 1. The possibility of spontaneous track loss by annealing argues strongly for multiple determination of ζ using a range of samples whose ages are established. Fleischer and Hart's seminal study suggested zeta primarily as an experimental device for deriving FT ages, but implicitly assumed that the same result could be obtained as easily by solving the "conventional" FT age equation of Price and Walker (1963) and Naeser (1967):

$$t = \frac{1}{\lambda_D} \ln \left(1 + \frac{\lambda_D\, \sigma\, I\, \Phi\, \rho_s}{\lambda_f\, \rho_i} \right) \tag{4}$$

This assumption overlooked three major uncertainties in the FT method:

- there was no consensus on the precise and accurate measurement of thermal neutron fluence used to induce ^{235}U fission in a sample;
- no unequivocal value for the ^{238}U spontaneous fission-decay constant λ_f existed;
- measurement of spontaneous and induced tracks on different surfaces and in different materials resulted in ambiguity in the proposed necessary corrections.

2. Neutron Fluence Measurement

Two techniques are available for measuring neutron fluence, metal activation monitors and uranium glass dosimeters.

In a *metal activation foil* of gold, cobalt or copper, neutron-induced gamma activity can be measured directly by scintillation counter and, from determination of the mass of isotope present in the foil, the reacting neutron fluence determined. The reactions are:

^{197}Au	(n,γ)	^{198}Au	($t^{1}/_{2}$ = 2.696 d)
^{59}Co	(n,γ)	^{60}Co	($t^{1}/_{2}$ = 5.271 a)
^{63}Cu	(n,γ)	^{64}Cu	($t^{1}/_{2}$ = 12.701 h)

Since neutron beams are not mono-energetic and the reactivity of an isotope will vary according to the energy of the bombarding neutron, a fundamental concern was whether the number of neutrons reacting with ^{235}U in a dating sample was the same as the number of neutrons reacting with a metal activation monitor. Figure 1 shows the variation of activation cross section of the isotopes as a function of neutron energy. Within the energy distribution of thermal neutrons (0 - 0.25 eV, with a maximum at 0.0253 eV at 20.4°C), the ^{235}U, ^{197}Au and ^{59}Co responses are virtually parallel with a 0.5 slope denoting $^1/v$ behaviour. In principle, for well-thermalised fluxes the fluences reacting with these isotopes should be the same. However, in the epithermal neutron energy range this relationship breaks down, with large resonance peaks in the Au capture cross section. Clearly use of a poorly-thermalised reactor facility would result in deviation from the $^1/v$ relationship and complicates comparison of the ^{235}U, ^{197}Au and ^{59}Co responses. Corrections may be necessary, in particular for epithermal activation if the thermal fluence is to be determined from the ^{198}Au activity, and also because the ^{235}U (n, f) response is not truly $^1/v$ in the thermal region. Further, whilst thermal neutrons induce fission in ^{235}U only, epithermal neutrons induce fission in both ^{235} and ^{238}U, whilst fast neutrons induce fission in both uranium isotopes and ^{232}Th.

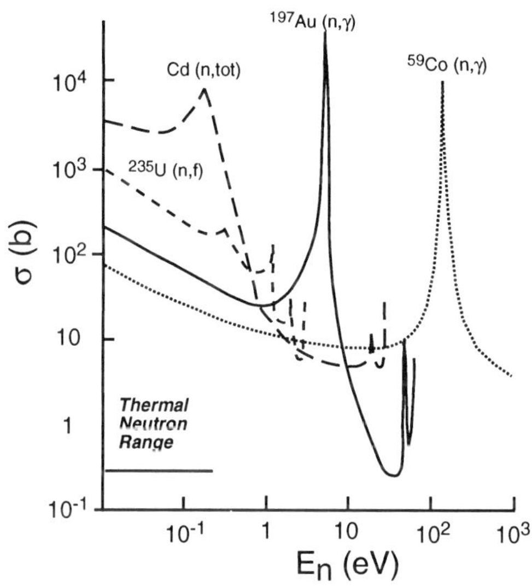

Figure 1. Variation with neutron energy of activation cross-section for ^{197}Au, ^{59}Co and ^{235}U, and total neutron absorption cross-section of Cd. (After Wagner and Van den haute 1992)

Minimum specifications for reactor facility thermalisation were recommended by Green and Hurford (1984) in terms of cadmium ratios of 3 for Au, 24 for Co and 48 for Cu. (Cadmium absorbs thermal neutrons with energies < ~0.5 eV whilst permitting higher energy neutrons to pass through. A cadmium ratio measures the activities of a bare monitor / cadmium-shielded monitor and thus records neutrons of all energies / neutrons with energies >0.5 eV).

Counting neutron-induced tracks in a *glass dosimeter of known uranium* content adjacent to an unknown sample relates directly to the FT method. Absolute calibration of such a procedure is complicated by the need to evaluate absolutely the detection and etching efficiencies of the system, which is difficult empirically. Provided the same dosimeter and experimental conditions are always used, the fluence is related to the measured track density by:

$$\Phi = B \rho_d \tag{5}$$

where:

ρ_d is the induced track density in the dosimeter; and

B is a constant specific to a dosimeter (given by σ N_v c^d R^d η^d (see equation 2) but which factors need not necessarily be evaluated explicitly).

TABLE 1. Specification of various uranium glasses used for thermal neutron dosimetry in fission-track dating analyses. DN = delayed neutron analysis; INAA = instrumental neutron activation analysis; MS = isotope dilution mass spectrometry; FT = fission-track analysis; Pre-WW2 = pre-second world-war uranium salt. SRM 961-964 are the NBS (NIST) pre-irradiated equivalents of SRM 610-616 (see Carpenter and Reimer 1974).

Dosimeter Glass	**Uranium Conc. (ppm by wt.)**	**^{235}U atom%**	**Independent Uranium Determination**	**Reference**
Corning U-1	40.9±0.7	0.3587	MS; FT	Schreurs et al. (1971)
Corning U-2	43±1	0.3587	FT	Schreurs et al. (1971)
Corning U-3	52±1	0.7200 (Pre-WW2 salt)	FT	Schreurs et al. (1971)
SRM 610 & 611 (= SRM 961) (NBS / NIST)	461.5±1.1 442±21	0.2376	MS INAA	NBS Certificate Carpenter & Reimer (1974); Bellemans *et al.* (1995a)
SRM 612 & 613 (= SRM 962) (NBS / NIST)	37.38±0.08 36.5±1.5	0.2392	MS INAA	NBS Certificate Carpenter & Reimer (1974); Bellemans *et al.* (1995a)
SRM 614 & 615 (= SRM 963) (NBS / NIST)	0.823±0.002 0.854±0.049	0.2792	MS; FT INAA	NBS Certificate Carpenter & Reimer (1974); Bellemans *et al.* (1995a)
SRM 616 & 617 (= SRM 964) (NBS / NIST)	0.0721±0.0013 0.070±0.011	0.616	MS; FT INAA	NBS Certificate Carpenter & Reimer (1974); Bellemans *et al.* (1995a)
Corning CN-1	mean: 39.81±0.69	0.7200 (Pre-WW2 salt)	DN; INAA; MS; FT	see Hurford & Green (1983) & Bellemans *et al.* (1995a)
Corning CN-2	mean 36.5±1.4	0.7200 (Pre-WW2 salt)	DN; INAA; MS; FT	see Hurford & Green (1983) & Bellemans *et al.* (1995a)
Moldavite	~2	assumed natural	not determined	M.Wagner (1985)
Corning CN-5	12.17±0.62	0.7200 (Pre-WW2 salt)	INAA	see Hurford (1990) & Bellemans *et al.* (1995a)
Corning CN-6	1.052±0.023	0.7200 (Pre-WW2 salt)	INAA	see Hurford (1990) & Bellemans *et al.* (1995a)

Uranium-doped glasses have provided the most-used dosimeters, usually registering tracks in an adjacent external detector of low-uranium mica or polycarbonate, to preserve the glass for subsequent use and to provide a recorder of higher etching efficiency. Table 1 shows those glasses most frequently used as FT dosimeters. Despite the constancy of the dosimeter factor B, deriving a FT age by use of equation 4 requires an absolute value for the neutron fluence. As an alternative to empirical determination of the track parameters N_v c^d R^d h^d, some workers have calibrated glass dosimeters against metal activation monitors and then used this calibration to derive fluence values in subsequent irradiations. Attempts to evaluate B for glass SRM612 in 74 irradiations in the HERALD reactor, Aldermaston, UK, over a 7-year period (Figure 2) showed ~±20% spread about the mean value, which for such a 2-detector system - SRM612 *vs.* Co wires - should yield the same ratio of response within experimental uncertainty (Hurford and Green 1983). Better precision was found from the relative responses of 2 glasses, especially the Corning glasses CN-1&2, suggesting for routine monitoring a preference for dosimeter glasses over activation monitors, and for Corning glass monitors with natural uranium and freedom from potentially interfering elements. Subsequent comparative analysis of the uranium and thorium contents of NBS and Corning glasses (Bellemans *et al.* 1995a) confirmed a preference for the Corning glasses because their very low Th/U ratio ensures an essentially pure uranium induced track population. The presence of neutron absorbing elements in the SRM glasses can cause neutron shielding with a thermal fluence rate depression of ~1.5% when track densities are compared to those from Corning glasses where such elements are negligible. A further worrying result reported by Bellemans *et al.* (1995a) showed that long-lived isotopes of Eu and Gd, together with the radionuclides such as ^{60}Co, ^{160}Tb, ^{169}Yb and ^{182}Ta create the persistent activity observed in the NBS glasses which, in glasses SRM 610-611, represents a potential health and safety hazard.

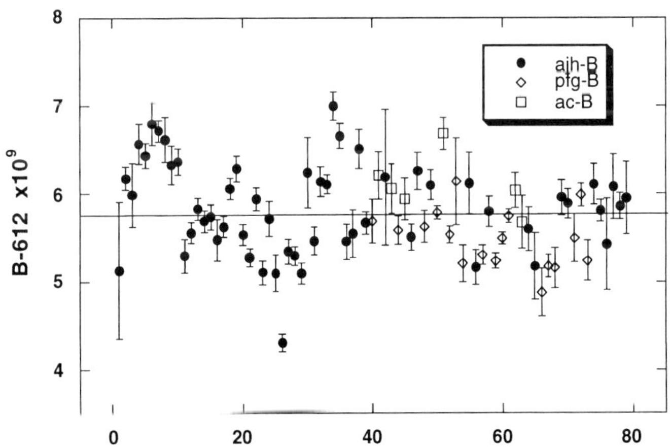

Figure 2. Evaluation of B_{612} in Herald Reactor, Aldermaston, UK over a seven-year period; error bars are 2σ calculated from track densities; horizontal line shows mean B value. (From Hurford and Green 1983).

In 1974, Carpenter and Reimer endeavoured to provide a common fluence calibration system for all FT workers. Batches of SRM glasses (Table 1) were irradiated at NBS to give induced track densities and definitive activation foil baselines. Track densities in these pre-irradiated glasses could be compared with densities obtained in each laboratory from irradiation in the preferred reactor, and a neutron fluence calculated relative to the NBS fluence baseline value. Despite the conceptual excellence of this experiment, the effectiveness was marred by differences of up to 11% in the fluences measured from the Au and Cu monitors of the same irradiation, substantially negating the primary objective of a single baseline value. A consequence of these attempts to assess fluence was that metal activation foil metrology was discredited with many FT workers as being too difficult, too imprecise and not directly relevant to the ^{235}U (n,f) reaction.

3. λ_f: ^{238}U Spontaneous Fission-decay Constant

More than 45 determinations of λ_f have been made since 1940 when Flerov and Petrzhak reported their discovery of the natural fission of uranium. A variety of techniques yielded a diverse range of values with marked peaks around 7, 8.5 and 11.75 x 10^{-17} yr^{-1}. Prior to the mid-1980s, each of the two lower λ_f peak values had been used by different workers for the calculation of FT ages and, although differing by 20%, compensating factors such as neutron fluence measurement reduced (or sometimes enlarged) the resulting discrepancy in calculated age. The disparity of λ_f values has been reviewed comprehensively - see for example Wagner *et al.* (1975); Thiel and Herr (1976); Bigazzi (1981), Hurford and Green (1982); Wagner and Van den haute (1992). Bigazzi (1981) grouped λ_f measurement into four categories:

(i). direct determination of fission events using ion chambers and other particle counting systems;
(ii). radiochemical or mass spectrometric measurement of uranium and fission products;
(iii). detection and counting of fission tracks in solid-state track detectors (SSTDs) or in photographic emulsions; and
(iv). the comparison of fission-track ages with K-Ar and/or Rb-Sr ages, or the FT dating of samples whose ages are supposedly well-known.

(See also Figure 3 of Van den haute *et al.*, this volume).

Results from groups (i) and (ii), independent of any track accumulation or irradiation process, supported the higher λ_f value ~8.5 x 10^{-17} yr^{-1}. In contrast, all FT dating determinations and the original SSTD track accumulation experiments required irradiation to evaluate the uranium content, and each supported the lower λ_f value ~7 x 10^{-17} yr^{-1}. Although the similarity of technique for calibration and unknown analysis is highly desirable intellectually and empirically, the apparent difficulty in defining neutron fluence indicates that λ_f measurements necessitating neutron irradiation cannot be considered apart from the neutron dosimetry. Which λ_f value should then be used for FT analysis?

4. Confusion and Solution?

These uncertainties of neutron fluence measurement and λ_f value meant that the FT method lacked a unique system calibration and that, in the mid to late 1970s, there was minimal inter-laboratory comparison of either calibration or technique. With judicious selection of calibration parameters a wide range of answers could be produced and instances occurred of inappropriate recalculation of data to provide a better fit to an alternative hypothesis. A further problem arose because fission tracks could be annealed easily at upper crustal temperatures frequently yielding mixed ages. For many in geochronology and isotope geology, the FT method was viewed as a Cinderella because of its inadequacy as a numerical dating method. A singular exception was the application of the method to volcanic and volcaniclastic problems where rapid cooling and near surface temperatures often resulted in minimal annealing and acceptable formation ages.

Solution to this apparent impasse came in the early to mid-1980s in two ways. Firstly the importance of track length in interpreting FT ages was re-evaluated leading to a re-orientation of the method as a source primarily of thermochronological information rather than formation ages (see e.g. Gleadow *et al.*, 1986). From this time many workers preferred the title *fission-track analysis* or *fission-track thermochronology* to *fission-track dating* as more precisely reflecting the new major direction of the technique. Secondly the need for a rigorous, consensus system calibration was recognised: re-enter Fleischer and Hart's mystic rune, ζ, to be hotly debated at the 1980 Pisa Workshop together with the need for age standards, defined by Wagner (*in* Hurford and Green, 1981) as samples from accessible, geologically well-documented horizons, of unambiguous age which relate to formation and not to cooling, nor to some inherited component; there should be no corrections for partial track annealing. Zeta now began to be viewed as an alternative calibration procedure which circumvented both the debate about the true λ_f value and the need for absolute quantification of neutron fluence, relying instead on the identification and testing of age standards to derive a ζ value. Interlaboratory distribution of "proven" age standards provided a common calibration baseline if used with the same uranium dosimeter glasses. Zeta values could be derived by each worker from analysis of age standards *via* equation 3 or, more precisely, from:

$$\zeta = \frac{[\exp(\lambda_D t_{std}) - 1]}{\lambda_D (\rho_s/\rho_i)_{std} \rho_d} \tag{6}$$

where t_{std} is the independent age of the standard, and allowance is made for the overall exponential decay of uranium. Ages of unknown samples could be derived subsequently using ζ in equation 1 or, more precisely, *via* :

$$t = \frac{1}{\lambda_D} \ln \left(1 + \lambda_D \zeta \frac{\rho_s}{\rho_i} \rho_d \right) \tag{7}$$

Definition, measurement and testing of four putative age standards (Hurford and Green, 1983) led to the derivation of ~80 zeta values for the more track-retentive mineral zircon and dosimeter glasses SRM612, CN-1 and CN-2 (see Figure 3). Subsequent work by Green (1985) revealed different zeta-values for apatite and sphene, indicating zeta to be mineral specific, and requiring age standards and calibrations likewise to be tailored to the mineral.

A questionnaire on FT calibration and long discussions took place under the aegis of the IUGS Subcommission on Geochronology culminating in 1988 at the

Besançon FT Workshop in the consensus recommendation of two approaches (Hurford
(i). the zeta approach was recommended for all minerals and techniques
(ii). the absolute approach with selection of λ_f and measurement of fluence was recommended for population method analyses of apatite only.

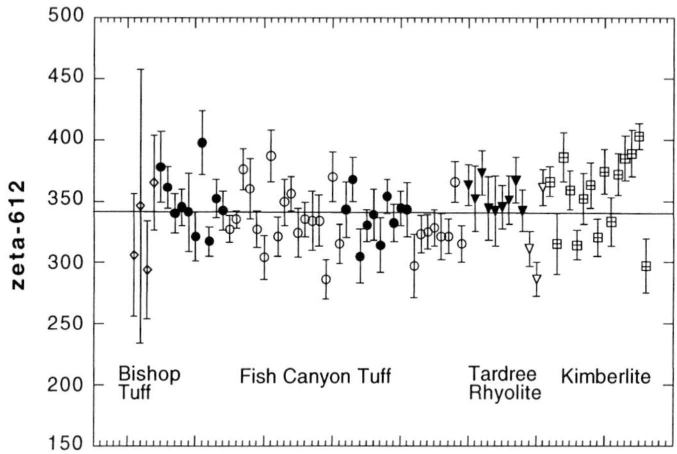

Figure 3 Evaluation of zeta for dosimeter glass SRM 612 from putative zircon age standards. Error bars are 2σ analytical uncertainties. Mean zeta value is shown as solid line. (From Hurford and Green 1983).

A short list of age standards fulfilling the criteria of Wagner was also approved and commended for establishing or testing FT calibration (Table 2).

TABLE 2. Age standards recommended by the 1988 Besançon consensus. (See Hurford 1990).

Apatite		
Fish Canyon Tuff	27.9±0.5Ma	Steven *et al.* (1967); Hurford and Hammerschmidt (1985); Sutter (1986)
Durango	31.4 ±0.5 Ma	McDowell and Keizer (1977)
Zircon		
Fish Canyon Tuff	27.9±0.5 Ma	as above
Buluk Member Tuff (FTBM)	16.4±0.2 Ma	Hurford and Watkins (1987)
Tardree Rhyolite	58.7±1.1 Ma	Hurford and Green (1983); Hurford and Hammerschmidt (unpublished data)
Sphene		
Mt Dromedary Banatite	98.7±0.6 Ma	Williams *et al.* (1982)
Glass		
Moldavite	15.1±0.7 Ma	Gentner *et al.* (1963)

Since 1988 FT workers have almost exclusively followed the IUGS recommendation, deriving thermo-chronological data for a multiplicity of geological (and other) applications; the vast majority of workers have opted for the zeta calibration approach, publishing their zeta determinations or referencing the data in doctoral theses. The 1996 Gent FT Workshop saw 135 presentations dealing with convergent and divergent plate-margin settings, stable cratonic blocks, sedimentary basins, tephrochronology and provenance studies on every continent and in some oceanic settings. The problems that beset the FT method in the 1970s are clearly solved, the method is widely accepted in academe and industry, and the technique contributes unique data to many areas of geology, as evidenced by the contributions to be seen in virtually all geoscience journals.

5. Whither Now?

So what now of the basic calibration of the FT method? The pragmatist might say that the method and its basic calibration, in particular zeta, are working well so why consider anything further? A plausible argument is that discussion about alternative calibration strategies might undermine confidence in the existing calibration and in the method, at worst returning it to the position of the 1970s. Counter to that, scientific argument would maintain that zeta is comprised of many factors: physical constants, variables which could be determined empirically and other variables which are neither easily defined nor quantified. Those factors which could be determined, should be, whilst those more elusive factors - in particular those associated with methodological procedure - should be encompassed within a procedural factor, the Q of Wagner and Van den haute (1992).

5.1 NEUTRON FLUENCE MEASUREMENT

Although zeta primarily served to overcome the real or presumed difficulties of determining λ_f and neutron fluence calibration, each of these factors could be and should be resolved. The imprecise metal activation monitor results from the 1970s and 80s largely curtailed the use of neutron metrology in FT analysis. Theoretical arguments (see Figure 1) demonstrate that for thermal neutron irradiation the response of $^{197}Au = ^{59}Co = ^{235}U$. The earlier-observed variation and inconsistency must have resulted from a combination of poor flux thermalisation and/or methodological imprecision, possibly associated with self-shielding of monitors or detector efficiencies. Specific use of well-thermalised fluxes, with an accurately and precisely known degree of thermalisation, together with diluted activation monitors (e.g. Al-Co alloys) and appropriate accurate and precise monitoring can lead to highly reproducible results, as demonstrated by De Corte *et al.* (1995). 27 irradiations in the Gent THETIS reactor exhibit a high level of consistency between Au and Co monitors, with cobalt 1.8% > than Au, and 1σ of the measurements being $\pm 1.6\%$ (Figure 4).

By including a uranium monitor with Au and Co monitors, the ^{235}U fission activity could be calculated using the Au and Co fluences. γ-spectrometry measurement of the fission products gave an independent measurement of ^{235}U activity. Similarity of the 2 measurements to within a fraction of 1% supported the thesis of similarity of response of ^{235}U to ^{197}Au and ^{59}Co, and provided the initiative to repeat the experiments of Carpenter and Reimer and pre-irradiate a uranium glass for distribution. Bellemans *et al.* (1995b) and De Corte *et al.* (this volume) describe a new dosimeter glass IRMM prepared with 15 wt ppm natural uranium, wafers of which have been irradiated in contact with muscovite detectors using precise Au and Co monitoring. Preliminary results indicate that this experiment represents a very significant advance in FT calibration, possibly providing a new dosimetry baseline for the method.

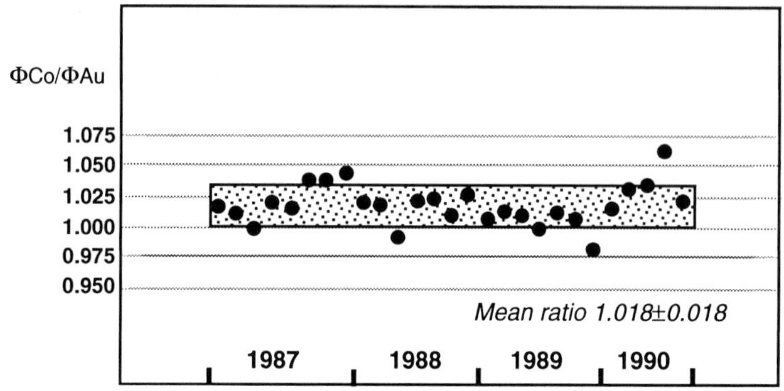

Figure 4. Comparison of fluence values monitored by Au and Co activation monitors in THETIS reactor, University of Gent. (From De Corte *et al.*, 1995).

5.2 λ_f: ^{238}U SPONTANEOUS FISSION-DECAY CONSTANT

Turning to λ_f, the original track accumulation and FT dating determinations necessitated neutron irradiation to determine the sample uranium content and thus the resulting values (most supported $\lambda_f = \sim 7 \times 10^{-17}\ yr^{-1}$) cannot be considered apart from the fluence dosimetry. In contrast the irradiation-free independent measurements have all supported the higher λ_f value $\sim 8.5 \times 10^{-17}\ yr^{-1}$.

Probably the most relevant and convincing λ_f track accumulation experimentation has used a single planar source of uranium atoms. All fissioning fragments should register full length tracks in an adjacent detector since the source thickness is negligible compared to the range of a fission fragment, resulting in virtually 100% detection efficiency. Again because the source is so thin, uranium can be evaluated by alpha-particle counting, area and thickness measurement, and weighing, without recourse to neutron irradiation. Two such experiments gave the disparate λ_f values $(7.2\pm0.2) \times 10^{-17}\ yr^{-1}$ (Emma and Lo Nigro, 1975) and $(11.8\pm0.7) \times 10^{-17}\ yr^{-1}$ (De Carvalho *et al.*, 1982), probably because glass was used to detect the fission fragments: the low etching efficiency of glass would result in track loss at low incident angles. In contrast, tracks from a thin uranium source registered in a mica detector with 100% etching efficiency led Roberts *et al.* (1968) to a λ_f value of $(7.03\pm0.11) \times 10^{-17}\ yr^{-1}$, the most precise of the lower values. Garwin (1985) used an *in vacuo* thin source, separated by an exactly known geometry from the detector, to obtain accurately quantifiable collimation of fission fragments which excluded shallow angle fragments, and avoided particles ranging-out within the source. The interim result of this well-conceived track accumulation experiment was $\lambda_f = 8.7 \times 10^{-17}\ yr^{-1}$. Although sadly the study was never completed, it perhaps points the way to future experimentation.

5.3 OTHER FACTORS

Zeta represents more than just λ_f and neutron fluence calibration, and more than just those factors defined in equation 2. Analysis of the estimates of zeta in any laboratory show a series of variations to be present. For an individual analyst results become more uniform with time reflecting the initial learning curve and subsequently representing an assessment of analytical reproducibility (Figure 5). This representation of 40 ζ_{CN5} measurements on apatite age standards shows the need for multiple determinations of ζ and argues for omission of the initial ζ values from the eventual mean.

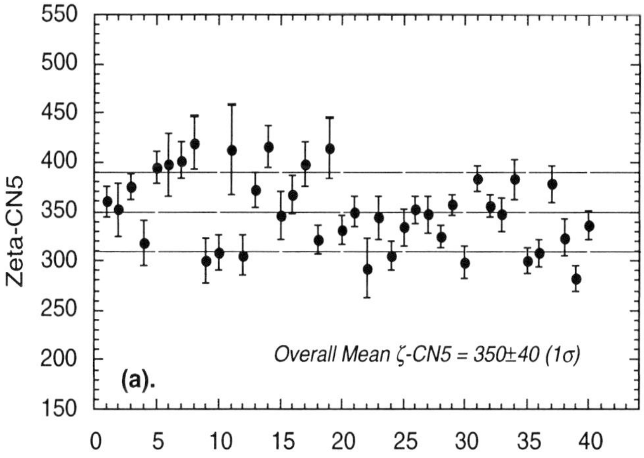

Figure 5. Comparison of zeta determinations by a single analyst over 6 months from apatite age standards.

Results from 13 analysts each show spreads similar to that seen in Figure 5, but also reflects significant differences in mean values despite using the same equipment, same age standard slides and a similar approach to analysis (Figure 6).

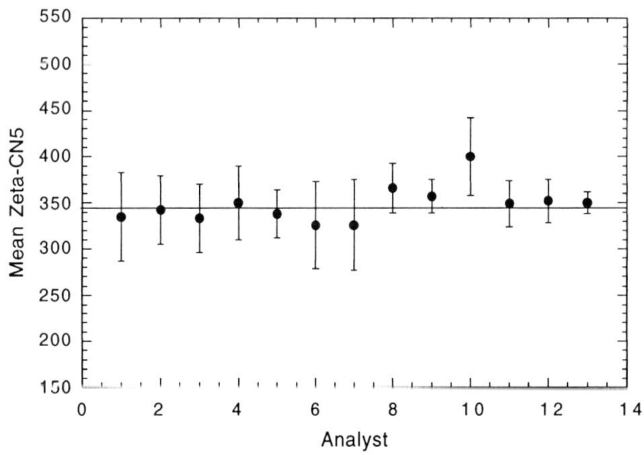

Figure 6. Comparison of mean zeta values measured by 13 analysts on the same Fish Canyon Tuff and Durango apatite age standards. Each mean value represents >15 determinations; error bars $\pm 1\sigma$.

Such variation must arise from factors such as small differences in the size cut-off point for acceptance of a track, the orientation of crystal selection criteria, and the precision of locating an induced image on the detector. Provided the techniques and criteria are maintained for standard and sample, such differences between analysts are of no consequence because the individual ζ values already embody them. Perhaps variation in the personal factor could be removed by automated image processing - but there would still be a need for a machine factor.

Zeta may also be reactor specific, dependent on the level of thermalisation. In the four reactors used by the London FT Group, HERALD, Aldermaston, U.K.; PLUTO, Harwell, UK; HIFAR, Lucas Heights, Australia; and Risø, Denmark, the mean values of multiple zetas from many irradiations determined by a single analyst agree to within ±2%. In less well-thermalised reactors, agreement may not be so close. There may also be slight differences in zeta between age standards: the results of Green (1985) show this to be true between apatite, zircon and sphene, but is it also true for compositionally different samples of the same mineral? Similarly differences of track density can occur as a function of etching (e.g. strong or weak HNO_3 to etch apatite), and revelation conditions (oil *vs.* dry objective). Such track density differences will propagate directly into zeta confirming the factor to be procedurally specific.

6. Conclusions

Zeta represents a calibration procedure which has enabled the fission-track community to produce results which are both directly comparable between laboratories, and which relate to other radiometric ages and to the chronometric scale. Zeta has permitted the production of geologically meaningful, unique thermochronological data serving a broad range of applications and securing the reputation and utility of the method within the geoscience community. But zeta combines a plethora of components, physical constants, some factors empirically determinable and others not. Given the evidence that precision neutron metrology is practicable and that exact determination of λ_f by solid-state track accumulation is experimentally possible, there is a strong argument for revising the concept of zeta. Deconvolution of zeta, with use of an agreed λ_f value and absolute neutron fluence measurement, would leave a composite procedural factor, the factor Q of Wagner and Van den haute (1992). Whether Q could be further resolved into personal, etching, microscope, compositional (and other) factors represents a second-order project. Introduction of the IRMM glasses provides an alternative to the present Corning dosimeters; both glasses should be viewed as preferable to the NBS (NIST) glasses for the reasons given above. Note that the IRMM glasses may be used equally as dosimeters for zeta age determinations as for absolute age calculations.

As hinted at above, the proposition of change to a system calibration which is already highly effective can be a difficult one to introduce: the old adage *"if it ain't broke, don't fix it"* may have a deal of merit. There is also a very definite danger that ill-informed opinion may seize on attempts to derive a second-generation zeta as representing a break-up of the consensus of opinion on calibration, rather than an advance in scientific understanding, with the damaging potential of returning the FT method to the 1970s morass. I would argue, philosophically, that zeta does represent an ***interim*** measure, but a measure which is both highly effective and fully agreed. I would further argue that the FT community ***must maintain*** the agreed IUGS recommendation for all active geoscience projects for the foreseeable future. Parallel with this should be a methodological project aimed at resolving zeta into its component parts. Probably this would be undertaken by only a few laboratories and workers but it should have the active intellectual interest and support of the entire FT community. Attempts to portray such research as weakening the validity of zeta should be fiercely challenged. There should be no cross-over between routine application of the method, and strategies aimed at improving its calibration. When significant advances in calibration as outlined above and in Van den haute *et al.* (this volume) have been achieved the case should be presented to the FT community for a systematic revision of calibration along the lines of the 1988 Besançon agreement.

7. Acknowledgements

Funding of FT research in London has been provided through NERC Grants GR3/7068 and 8261, and through an Elf research grant .

8. References

Bellemans, F., De Corte, F. and Van den haute, P. (1995a): Composition of SRM and CN U-doped glasses: significance for their use as thermal neutron fluence monitors in fission track dating. *Radiat. Meas.* **24**, 153-160.
Bellemans, F., De Corte, F., Van den haute, P. and Ingelbrecht, C. (1995b): Towards a new glass monitor for the determination of the neutron fluence in fission-track dating *Radiat. Meas.* **25**, 527-530.
Bigazzi, G. (1981): The problem of the decay constant λ_f of ^{238}U. *Nucl. Tracks* **5**, 35-44.
Carpenter, B.S. and Reimer, G.M. (1974): Calibrated glass standards for fission track use. *NBS Special Publication* 260-49.
De Carvalho, H.G., Martins, J.B., Medeiros, E.L. and Tavares, O.A.P. (1982): Decay constant for the spontaneous fission process in ^{238}U. *Nucl. Instrum. Meth.* **197**, 417-426.
De Corte, F., Van den haute, P. and Bellemans, F. (1995): The use of uranium doped glasses in fission-track dating. *Radiat. Meas.* **25**, 511-516.
De Corte, F., Bellemans, F., Van den haute, P., Ingelbrecht, C. and Nicholl, C. (this volume): A new U-doped glass certified by the European Commission for the calibration of fission-track dating.
Emma, V. and Lo Nigro, S. (1975): Decay constant for spontaneous fission of ^{238}U and ^{232}Th. *Nucl. Instrum. Meth.* **128**, 355-357.
Fleischer, R.L. and Hart, H.R. (1972): Fission track dating: techniques and problems. In: *Calibration of Hominoid Evolution* (eds W.W. Bishop, J.A. Miller and S. Cole), Scottish Academic Press, Edinburgh, pp. 135-170.
Flerov, G.N. and Petrzhak (1940): Spontaneous fission of uranium. *J. Phys.* **3**, 275-280.
Garwin, L. (1985): Fission track dating and tectonics in the Eastern Pyrenees. Unpublished Doctoral Thesis, University of Cambridge.
Gentner, W., Lippolt, H. and Schäffer, O.A. (1963): Argonbestimmungen an Kaliummineralien-XI: Die Kalium-Argon-Alter der Gläser des Nordlinger Rieses und der bohmisch-mahrischen Tektite. *Geochim. cosmochim. Acta* **27**, 191-200.
Gleadow, A.J.W., Duddy, I.R., Green, P.F. and Lovering, J.F. (1986): Confined fission track lengths in apatite - diagnostic tool for for thermal history analysis. *Contrib. Mineral Petrol.* **94**, 405-415.
Green, P.F. (1985): Comparison of zeta calibration baselines for fission-track dating of apatite, zircon and sphene. *Chem. Geol. (Isot. Geosci. Sect.)* **58**, 1-22.
Hurford, A.J. (1990): Standardization of fission track dating calibration: recommendation by the Fission Track Working Group of the I.U.G.S. Subcommission on Geochronology. *Chem. Geol. (Isot. Geosci. Sect.)* **80**, 171-178.
Hurford, A.J. and Green, P.F. (1981): Standards, dosimetry and the uranium-238 λ_f decay constant: a discussion. *Nucl. Tracks* **5**, 73-75.
Hurford, A.J. and Green, P.F. (1982): A users' guide to fission track dating calibration. *Earth Planet. Sci. Letts.* **59**, 343-354.
Hurford, A.J. and Green, P.F. (1983) The zeta calibration of fission track dating. *Isot. Geosci.* **1**, 285-317.
Hurford, A.J. and Hammerschmidt, K. (1985): ^{40}Ar-^{39}Ar and K-Ar dating of the Bishop and Fish Canyon tuffs: calibration ages for fission track dating standards. *Chem. Geol. (Isotope Geosci. Section)* **58**:23-32.
Hurford, A.J. and Watkins, R.T. (1987): Fission track age of the tuffs of the Buluk Member, Bakate Formation, northern Kenya: a suitable FT age standard. *Chem. Geol. (Isotope Geol. Section)* **66**, 209-216.
McDowell, F.W. and Keizer, R.P. (1977): Timing of mid-Tertiary volcanism in the Sierra Madre Occidental between Durango City and Mazatlan, Mexico. *Geol. Soc. Am. Bull.* **88**, 1479-1487.
Naeser, C.W. (1967): The use of apatite and sphene for fission track age determinations. *Bull. Geol. Soc. Am.* **78**, 1523-1526.

Price, P.B. and Walker, R.M. (1963): Fossil tracks of charged particles in mica and the age of minerals. *J. Geophys. Res.* **68**, 4847-4862.
Roberts, J.H., Gold, R. and Armani, R.J. (1968): Spontaneous fission decay constant of ^{238}U. *Phys. Rev.* **174**, 1482-1484.
Schreurs, J.W.H., Friedman, A.H., Rokop, D.J., Hair, M.W. and Walker, R.M. (1971): Calibrated U-Th glasses for neutron dosimetry and determination of uranium and thorium concentrations by the fission track method. *Radiat. Eff.* **7**, 231-235.
Steven, T.A., Mehnert, H.H. and Oradovich, J.D. (1967): Age of volcanic activity in the San Juan mountains, Colorado. *U.S. Geol. Surv. Prof. Paper* **575-D**, 47-55.
Sutter, J . (1986): Program with abstract. *Geol. Soc. Am. Ann. Meeting*
Thiel, K. and Herr, W. (1976): The ^{238}U spontaneous fission decay constant redetermined by fission tracks. *Earth Planet. Sci. Letts.* **30**, 50-56.
Van den haute, P., De Corte, F., Jonckheere, R. and Bellemans, F. (this volume): The parameters that govern the accuracy of fission-track age determinations: a re-appraisal.
Wagner, G.A., Reimer, G.M., Carpenter, B.S., Faul, H., Van den Linden, R. and Gijbels, R. (1975): The spontaneous fission rate of ^{238}U and fission track dating. *Geochim. Cosmochim. Acta* **39**, 1279-1286.
Wagner, G.A. and Van den haute, P. (1992): *Fission-track dating.* Kluwer Academic Publishers, Dordecht.
Wagner, M. (1985): Spaltspurendatierungen am Bohrkern Urach III - ein Beitrag zur Wärmgeschichte der Geothermischen Anomalie Urach. Unpublished Doctoral Thesis, Ruprecht-Karls-Universität, Heidelberg.
Williams, I.S., Tetley, N.W., Compston, W. and McDougall, I. (1982): A comparison of K/Ar and Rb/Sr ages of rapidly cooled igneous rocks: two points in the Palaeozoic time scale re-evaluated. *J. geol. Soc. London* **139**, 557-568

THE PARAMETERS THAT GOVERN THE ACCURACY OF FISSION-TRACK AGE DETERMINATIONS: A RE-APPRAISAL

P. VAN DEN HAUTE[1], F. DE CORTE[2], R. JONCKHEERE[3] and F. BELLEMANS[2]

(1) Geological Institute, Universiteit Gent, Belgium
(2) Institute for Nuclear Sciences, Universiteit Gent, Belgium
(3) Forschungsstelle Archäometrie der Heidelberger Akademie für Wissenschaften am Max-Planck-Institut für Kernphysik, Heidelberg, Germany.

Abstract Since the generalized use of the ζ-calibration method, research on fission-track age calibration and the parameters that govern the accuracy of fission-track ages has become limited to a few laboratories. The present paper gives a review of the research that has been carried out on these problems in our laboratory during the last decade. The results of these investigations demonstrate that absolute thermal neutron fluence measurements and the determination of the ^{235}U fission rate during sample irradiation no longer present a problem. The only parameter that would seem to remain to be determined is therefore the spontaneous fission decay constant of ^{238}U (λ_f). However, the results of our experiments also re-emphasize the importance of the techniques of track analysis that are used in an FT age determination. They can be responsible for systematic differences up to 15-20% in the final age result if the absolute approach is used. The parameters that play a role are the track registration efficiency and the combined etching-observation efficiency. These parameters have to be properly dealt with if λ_f is determined through the analysis of samples of known age. By applying the same analytical techniques to the age standards and the unknown samples, the ζ-method eliminates the systematic effects of these parameters and intrinsically ensures accurate ages. When a value of λ_f is agreed by consensus and/or as a result of new experimentation, it will still be essential to perform age standard analyses for determining a procedure factor (Q) that yields direct information on the systematic effect of the technique of track analysis. This routine will nevertheless be a significant improvement on the calibration based on the determination of the all-embracing ζ-factor.

1. Introduction

At the International Workshop on Fission-Track Dating held in Pisa, Bigazzi (1981) formulated two alternatives to obtain calibrated fission-track ages:
(a) the choice of a neutron fluence standard coupled to a conventional (working) value for the decay constant of ^{238}U spontaneous fission (λ_f) or,
(b) the choice of one or more age standards.

P. Van den haute and F. De Corte (eds.), Advances in Fission - Track Geochronology, 33-46.

When Bigazzi made this statement, differences of about 20% existed between the results of the available λ_f determinations (more than 40) and several difficulties were encountered in determining both accurate and reproducible absolute neutron fluences (Carpenter and Reimer, 1974; Hurford and Green, 1981). This situation obviously encouraged the development of an approach based on the second alternative i.e. the use of age standards. The ζ-calibration procedure, the principles of which had been outlined by Fleischer and Hart (1972) was introduced in 1982-83, after careful investigation by Hurford and Green and a few years later it proved to be a useful calibration system. At present, through the generalized use of the ζ-calibration a major goal has been achieved: fission-track ages are calibrated and comparable between laboratories (Hurford, this volume).

A consequence of this consensus, as might be expected, was that interest in age calibration and in the parameters that govern the accuracy of the fission-track dating in general, gradually declined and many in the active fission-track dating community, involved in geothermochronometric application might wonder if anything happened in this field since Pisa, 1980. In this paper we will give a positive answer to this question. Moreover, through a re-appraisal of the factors that intervene in fission-track age calibration, we will show that at present, the parameters that are inaccurately known are those related to track registration, revelation and counting and not any more to the experimentally determined thermal neutron fluence ϕ or (more accurately) the uranium fission rate during sample irradiation. This new situation makes it worthwhile to reconsider the use of a more absolute approach in fission-track dating. However, in this approach age standards will still be required to determine a procedure factor Q that deals with the techniques of track registration, revelation and observation that are used for age determination. A major obstacle also remains about the uncertainty of the value of λ_f. This problem needs to be solved before the replacement of the ζ-calibration factor by Q can be considered. Before coming to these conclusions let us first review the present state of knowledge of the parameters that play a role in absolute age calibration.

2. The Thermal Neutron Fluence (ϕ)

In the early days, fission-track dating was based on absolute thermal neutron fluence determinations in several laboratories. The procedure that we used some fifteen years ago relied on Au monitors and had been elaborated by Gijbels and Vanderlinden (Wagner et al., 1975). We were more or less convinced that this procedure led to accurate fluence determinations until a number of experiments including the preparation of the pre-irradiated SRM glasses indicated that accurate and reproducible fluence measurements were very difficult to achieve. This was clearly demonstrated when monitors involving different activation schemes were used (Carpenter and Reimer, 1974; Carpenter, 1984). This situation ended neutron fluence monitoring in fission-track dating in many laboratories and in order to allow any chance for a re-start of absolute age calibration, it was necessary to re-investigate the neutron fluence problem. Such investigations have been carried out in our laboratory (Van den haute et al., 1988; De Corte et al., 1991) and the results showed that if the procedures of modern neutron metrology are respected consistent fluences with the two standard metal monitors Au and Co can be achieved on a routine basis. Later experiments (De Corte et al., 1995; Jonckheere, 1995; Bellemans, 1996) consolidated these results. As an example, Figure 1 shows the Au/Co fluence ratios that have been registered during the pre-irradiation of the new IRMM glass monitor for fission-track dating (De Corte et al., this volume). The glass discs were stacked in three different containers # 1, 2 and 3, for irradiation, each disc being sandwiched between two Al-0.1%Au alloyed foils. At three positions in each irradiation container additional Al-1%Co-foils were added, yielding a total of 9 Au-Co couples. Figure 1 shows that the relative difference between the Au and Co measurements is $\pm$1% or less, which is even better than in our earlier work, due to further improved calibration of the equipment and the use of thin circular foils as monitors instead of wires.

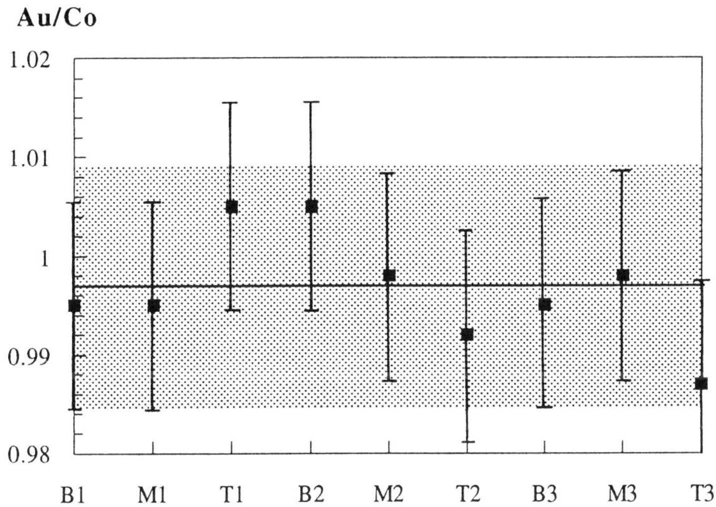

Figure1. Ratios of Au/Co fluences and their error bars registered for the irradiations (Thetis reactor, channel 8) performed in the frame of the preparation of the new IRMM glass (De Corte et al., this volume). The glasses were stacked in three different containers. Au and Co monitor couples were positioned at the bottom (B), middle (M) and top (T) of each container.The average ratio (horizontal line) and the 2s deviation of all measurements (shaded area) are also indicated.

It is well known that it is not just an accurate determination of the thermal neutron fluence that matters to fission-track age calibration but an accurate assessment of the ^{235}U fission rate (R). Indeed, it are the results of this reaction (i.e. the induced tracks) that are counted for the determination of ρ_i in the fission-track age equation. Hence, the question arises whether the neutron fluence value obtained with Au and Co monitors can be directly introduced in this equation. This problem has recently also been re-investigated in our laboratory (De Corte et al., 1995). The experiment involved irradiation of Au, Co (and Mn) monitors together with Al-U (93.2%^{235}U) foils, followed by γ-spectrometry of the monitors and 23 unstable ^{235}U fission products created in the Al-U foil. Details of this experiment are described elsewhere (Bellemans,1996). Figure 2 shows the ratio of the measured value of the fission rate R for each of the fission nuclides to the mean value calculated from the activation monitors. Clearly certain fission products yield slightly discordant results that are not explained by analytical error but the mean ratio is close to 1, demonstrating that fluence values derived from the standard metal monitors do yield an accurate determination of R. The observed discrepancies are most probably related to inaccuracies in some of the val-

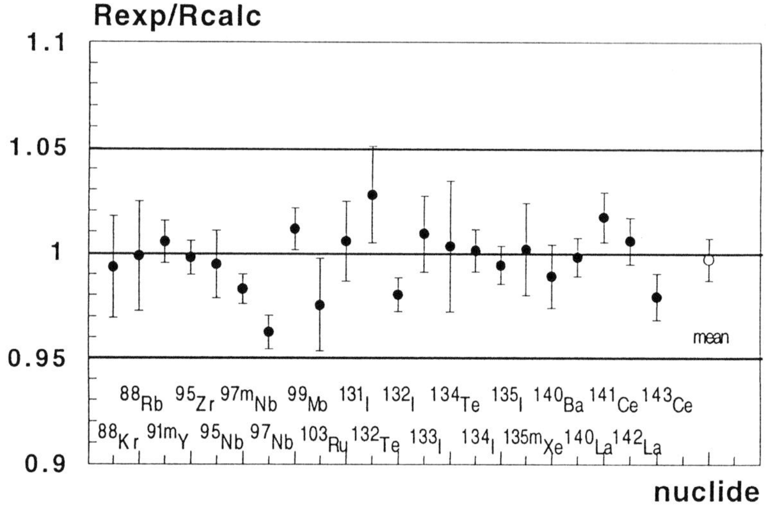

Figure 2. Ratios and 1s error bars of $^{235}U(n,f)$ reaction rates calculated from fluence determinations with Au, Co (and Mn) monitors to the rates measured with γ-spectrometry of 23 fission products (Thetis reactor, ch.8).The weighted mean of all ratios (open symbol) and it 2s error bar is shown at the right (from De Corte et al., 1995).

ues of the relevant nuclear parameters of the fission nuclides. When carried out in a well thermalized reactor facility such as the one used for our experiments (channel 8 of the Thetis reactor; $\phi_{th}/\phi_{epi} = 150$ and $\phi_{th}/\phi_{epi} = 120$), epithermal fission of ^{238}U and fast fission of both ^{238}U and ^{232}Th are quite negligible (Wagner and Van den haute, 1992), which simplified the measurements. Epithermal activation of Co and especially of Au has however to be corrected for even in a facility such as the one used here, something that may have been neglected in the early studies on neutron fluence calibration.

Although a determination of the ^{235}U (n,f) rate can also be performed with the use of a uranium source / track-detector couple we preferred not to use a track method for this study because of possible inaccuracies in the knowledge of the track registration and revelation efficiency of the detector used, even if it is mica.

3. The $^{235}U(n,f)$ Thermal Neutron Fission Cross-Section (σ) and the Decay Constant for Spontaneous Fission (λ_f)

In the early years of fission-track dating a value of 580.2 10^{-24} cm^2 (Hanna et al.,1969) was substituted for σ in the age equation. As stated in Wagner and Van den haute (1992), research work of the last twenty years only slightly altered this value and at present, the generally adopted value is 584.25 10^{-24} cm^2 ($\pm$ 0.19%). What is quite often neglected, is that this value stands for σ_0 i.e. the cross-section for 2200 m/s neutrons and that a correction has to be applied because the variation of the cross-sectionof the ^{235}U (n,f) reaction with neutron velocity is non-ideal (not proportional to 1/v). The correction factor amounts to 0.9832 at 20°C according to a recent report of

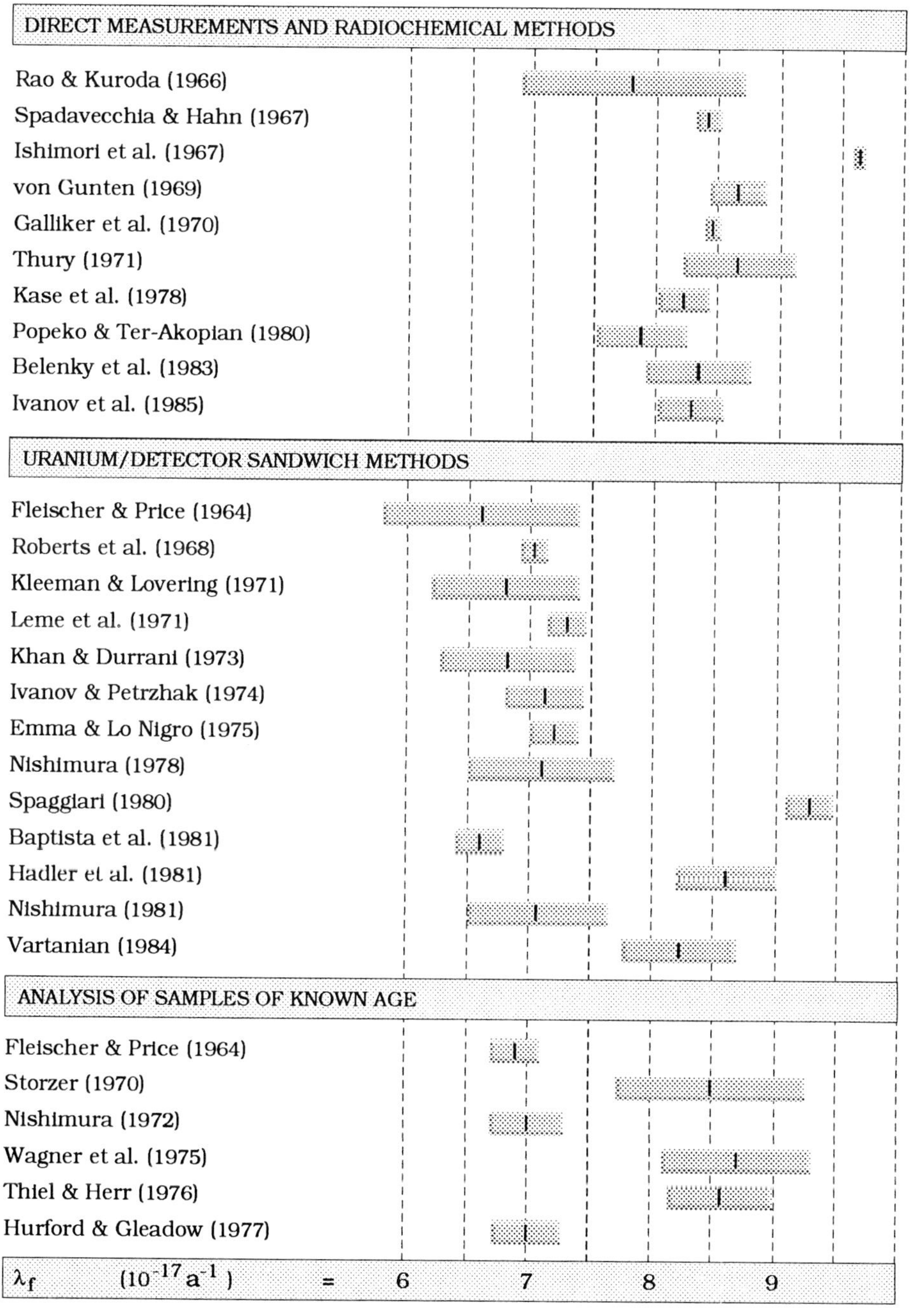

Figure 3. Overview of published λ_f determinations performed since 1960 (after Wagner and Van den haute (1992), which also contains the full references).

Holden (1991) and is slightly dependent on neutron temperature. This brings us to a value of 574.4 10^{-24} cm^2 for σ in the age equation. This value is slightly different from the value of 570.8 10^{-24} cm^2 quoted in Wagner and Van den haute (1992) but rather than focussing on this small difference we prefer to emphasize the high accuracy which the value of σ is known with.

Almost simultaneously with the development of the ζ-calibration system, determinations of λ_f gradually stopped. The latest determination quoted in Wagner and Van den haute (1992) dates from 1985 (Ivanov et al.) and we could not trace any internationally published study of more recent date. This leaves us with the two discrepant (nominal) values of 7.0 10^{-17} a^{-1} and 8.5 10^{-17} a^{-1}.

Classically, emphasis is put on the method of neutron fluence determination that was used in the λ_f determination experiments to explain the observed discrepancy. However, as is suggested by Figure 3, the differences may also be due to other aspects of the methods that were used. One only has to compare the results of direct measurements of fission products with the determinations based on uranium source / detector sandwiches, that use the track method. Fission-track analyses of samples of known age only appear to highlight the discrepancy. Here of course supplementary variables intervene such as fading of natural tracks and variables related to the dating technique that has been used (ED, POP, re-etch etc.). Especially track registration and revelation in the minerals and detectors that are used play an important role as is shown in the following sections.

4. Track Registration

In Fleischer et al. (1975) it appeared quite established that the etchable length of a damage track can be considerably shorter than the true distance travelled by the nuclear particle in the crystal lattice. The difference between the etchable length of the damage track and the full range of a nuclear fragment was called the range deficit. In the case of fission-tracks the difference between the etched length and the combined range of the two fission fragments can simply be called the length deficit. Theoretical calculations of dE/dX curves and combined ranges of the two most probable ^{235}U fission fragments

Table1: Calculated combined ranges (2R) of the two most likely fission fragments and observed etched lenghts of induced tracks. Ranges were calculated according to Lindhard et al. (1963); Northcliffe and Schilling (1970); Biersack and Ziegler (1987).

	combined range (2R) in μm			ratio	
Method of calculation	Mica KAl3Si3O10(OH)2	Apatite Ca5(PO4)3F	Zircon ZrSiO4	Ap/Mica	Zr/Mica
Lindh.,1963	22.6	21.8	15.7	0.96	0.69
North.,1970	23.7	22.9	18.1	0.97	0.76
Biers., 1987	19.6	18.5	15	0.94	0.77
mean	*22.0*	*21.1*	*16.3*	*0.96*	*0.74*
etched length (L)	20-21	16-16.5	10.5-11	0.76-0.83	0.50-0.55

yield values that are different for different mineral species (Table 1). The calculated ranges vary somewhat according to the formulae used but all results are quite consistent when range ratios between minerals are made. We refer to Jonckheere (1995) for more details. In mica the confined track length (L) corresponds approximately to the combined range (2R) of the fission fragments but the lengths of induced tracks in apatite and zircon are shorter than the calculated combined ranges of the ^{235}U fission fragments in these minerals. The length deficit is of specific importance when track densities are determined in different materials as is done in the external detector method. If a mineral-mica external detector pair is irradiated, the ratio of induced tracks in the mica to that in the mineral will in first approximation be given by the relation:

$$\frac{\rho_{Mica}}{\rho_{Min}} = g\,\frac{2R_{Min}}{2R_{Mica}} \cdot \frac{L_{Mica}}{L_{Min}}$$

with g =1/2 for an internal surface of the mineral and g=1 for an external surface. Taking into account the mean combined range ratios and the length ratios of confined induced tracks quoted in Table 1, the above equation leads to a track density ratio of 0.58-0.63 and 0.67-0.74 for a mica-internal apatite and mica-internal zircon surface pair respectively.

Some experimental evidence for the role of the length deficit was reported several years ago by Storzer and Wagner (1985). A number of observations also point at the presence of a length deficit. A first example is the observation of Donelick et al. (1990) who found that the length of induced tracks is longer when etched shortly after irradiation than when etched after a few weeks. Another observation that can be interpreted in the light of the range deficit is that in mica where the calculations yield that $L \approx 2R$, alpha recoil tracks can be visibly etched while this is not so for apatite and zircon.

Iwano et al. (1992,1993) tried to evaluate the importance of the length deficit on observed track densities through careful track to track matching of induced tracks in external zircon and apatite surfaces and their adjacent mica detectors. The values of 0.90 and 0.75 they measured for the Ap/Mi and Zr/Mi track density ratios respectively, quite reasonably match the calculated ratios of 0.78-0.85 and 0.68-0.74 that are obtained when the values in Table 1 are substituted into the above equation (putting g=1). The calculared ratios are somewhat at the low side but it has to be noticed that the density ratios of Iwano et al. are influenced by effects of track revelation and observation. We refer to Iwano and Danhara in this volume for a further discussion.of this matter.

5. Track Revelation and Observation.

Studies of track etching in minerals that have been carried out in different laboratories in the years preceding the Pisa Workshop led to insights that undoubtedly also directed research on age calibration towards the ζ method. The major conclusions of these studies can be summarized as follows.

It was observed that track etching is significantly influenced by the crystallographic orientation of the mineral surface. Tracks are developed as "simple" needle-like holes or as more complex funnel shaped figures, the shape of the track and its surface opening being clearly controlled by crystal symmetry (Wagner, 1968). The crystal surface itself also develops specific etch features and it was described as scratched (S) , pitted (P) or textured (T) (Gleadow, 1978). S-surfaces exhibit easily distinguished needle-like tracks and have an overall smooth appearance, the scratched aspect being due to the sharply etched polishing grooves. P-surfaces exhibit broad polishing scratches and tracks with surface openings. T-surfaces were interpreted as extreme cases of P-surfaces. Bulk etch rates were assumed to increase and track etching efficiencies to decrease correspondingly in the order S-P-T. Typical examples of S-surfaces are the prism faces in apatite and zircon. Because both mineral species are commonly elongated along their crystallographic C-axes, prism faces are the most abundantly exposed in a mineral mount when an aliquot of grains are dropped into the mounting resin.

Information on the track revelation efficiency in apatite, sphene and zircon as compared to external mica or plastic detectors was gathered in an indirect way through geometry ratio determinations (Gleadow and Lovering, 1977; Green and Durrani, 1978). A 5-10% higher efficiency was found for external apatite surfaces compared to internal surfaces, the difference being attributed to the short (<0.5μm) tracks which are not observed. Such tracks occur in internal surfaces but not in external surfaces where the minimum track length amounts to the etchable range of one fission fragment. External detector/internal surface couples yielded geometry ratios close to the ideal value of 0.50 because both have the short track problem. While in apatite the orientation of the crystal surface was not observed to be of any influence, this was not the case for zircon and sphene where only S-surfaces appeared to yield the desired 0.5 geometry ratios.

As a result of these studies the external detector method applied to selected crystal surfaces (the S-surface), became a preferred method, also because it ensured easy and reproducible track counting and a track revelation efficiency close to that of mica. The method offered the additional advantage to date single grains and to deal with populations of grains that are heterogeneous in uranium content.

As the problems of track revelation that are relevant to the accuracy of age determination appeared to be solved, further research on this matter stopped in most laboratories. Some of the conclusions drawn in these works did however not match the lines of reasoning followed in studies on defect etching in crystal physics and stimulated us to continue further investigation (Wagner and Van den haute, 1992; Jonckheere, 1995; Jonckheere and Van den haute, 1996). Etch experiments on different apatite crystal faces led to the observation that the rate perpendicular to the etched surface increases in the order P-S-T,.instead of S-P-T Detailed observations of etched tracks, further revealed that the P, S and T surfaces probably correspond respectively to flat (F), stepped (S) and kinked (K) surfaces of the Kossel crystal model (Figure 4). In this model, etching can be described as a successive removal of unit blocks at the crystal surface and this is most easily done in K (T) surfaces and with most difficulty in F (P) surfaces. The surface openings (funnel mouths) of fission tracks in the overall slow etching F (P) surfaces result from a locally increased bulk etching rate at the track's intersection point and broad polishing scratches instead of sharp ones are symptomatic of a low bulk etching rate.

Variability in track shape influences the efficiency of track identification while bulk etching has a double effect: it deteriorates identification of the tracks situated at low depth under the surface but it also progressively reveals confined tracks from beneath the surface. As a consequence, one cannot expect the revelation and observation efficiencies to be identical for different surfaces or etching conditions.

The revelation and observation factors are not easily quantified precisely but their importance can be estimated with confidence. We will confine ourselves to the prism face of apatite because of its importance in the ED method. As stated above, the tracks

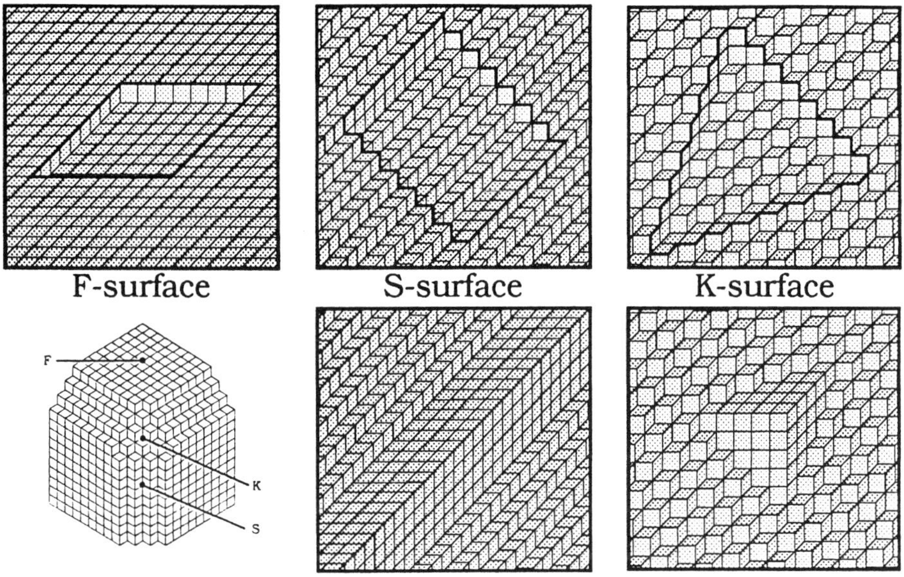

Figure 4. Surface etching according to the Kossel crystal model. According to our observations the pitted, scratched and textured surfaces observed in fission track dating correspond to the flat (F) stepped (S) and kinked (K) faces of this model (after Jonckheere and Van den haute, 1996).

on this surface exhibit the "simple" needle shape, an advantage that has to be regarded as qualitative in the sense that it allows easy track identification. For the etching conditions 2.5%HNO_3-60s-25°C, we found that a layer of about 0.4 μm is removed from this surface. On theoretical grounds, i.e. considering the frequency distribution of vertical depth of etched tracks as being triangular between 0 and 16 μm, this layer contains about 5% of the track population that cuts the polished surface and which may

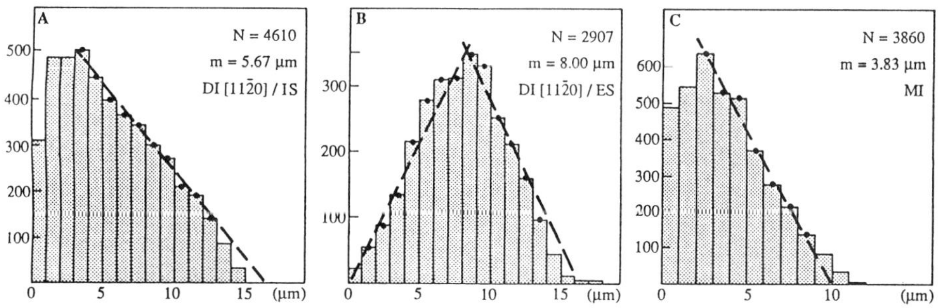

Figure 5. Projected length distributions of induced tracks observed in (a) an internal Durango apatite (1120) prism surface, (b) an external prism surface and (c) a mica external detector irradiated in contact with (a). Regression lines fitted to the central part of these distributions were used to determine the counting efficieny factor in these surfaces (N=number of tracks measured; m=mean projected length).

thus escape observation. That a proportion of the smaller tracks are not observed in practice is indicated by a deficit of short tracks registered in the projected length distribution (Figure 5). The distribution can be used to estimate the relative number of tracks that are missed. If the regression line, fitted to the central part of the length-distribution is taken as the reference line limiting the true distribution, a value of 0.92± 0.01 is obtained. A similar exercise on an external surface, the length distribution of which approximates the shape of an isosceles triangle yields 0.99± 0.01, while a muscovite external detector (40%HF, 20min, 25°C) yields 0.91± 0.01. These values obviously may be only approximations but at least they are informative on the importance of track revelation and observation. Because both factors can hardly be separated, it is reasonable to combine them into a single counting efficiency factor ηq (Jonckheere and Van den haute, 1996).

6. A Dating Exercise

From the above it follows that the cumulative effect of the length deficit and the counting efficiency factor can account for differences of up to about 15-20% in observed track densities. The combined effect of both factors can also be evaluated by applying different techniques of track analysis in a dating experiment. The results of such an experiment that we carried out on the Durango apatite age standard are shown in Figure 6. It was performed on the basal and the prism faces of large single crystals and involved five different analytical techniques: the internal/internal surface (IS/IS) and the internal/external surface (IS/ES) technique both with annealing of the spontaneous tracks before irradiation, the (mica)external detector technique (ED), the re-etch technique (RE) and the re-polish technique (RP). As can be seen in the figure, the age results scatter quite considerably. We do not wish to discuss all results in detail here; in several aspects they confirm the earlier work, such as the consistent results that are generally obtained on prism and basal faces except when external surfaces are used for counting the induced tracks (Green, 1978). Such observations led to the recommendation of the ED and the POP method (for samples with homogeneous uranium composition) for dating apatite minerals (Hurford, 1990). If we confine ourselves to the results obtained with the IS/IS technique (which is similar to the POP technique) and the ED technique we observe that the ED technique yields an age close to the accepted age of 31.4 ± 0.5 Ma with a value of 7.03 10^{-17} a^{-1} for λ_f while for the POP technique the value of 8.46 10^{-17} a^{-1} appears to produce the better result. The difference between the two methods can only be caused by differences in track registration and counting efficiencies. In this respect, the IS/IS method for which the registration and counting efficiencies are the same for induced and spontaneous tracks is expected to yield a more accurate estimate of λ_f than the ED method, for which this is not the case. A possible effect of the reduced spontaneous track length of about 7% in Durango apatite on the age determination is not considered here but it is clear that it can only lower the value of λ_f.

The above observation indicates that the ED technique may not be suitable for λ_f experiments but of course it does not question the use of this technique (in combination with the ζ-method) when it comes to routine dating applications. On the contrary, rocks with different grain populations necessitate the use of a single grain method such as the ED technique.

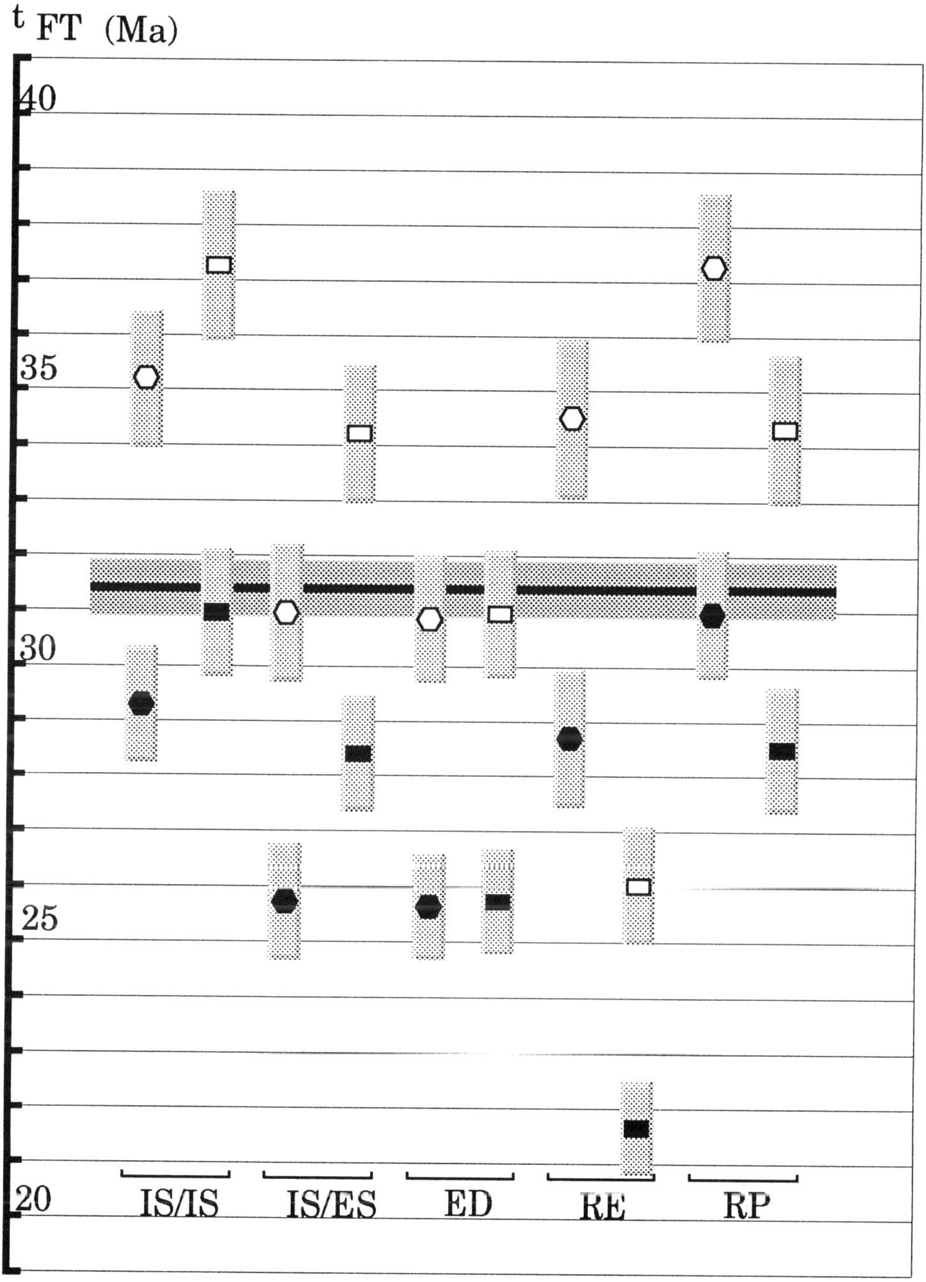

Figure 6. Results of a dating experiment on Durango apatite with 5 different techniques: internal surface/internal surface, internal surface/external surface, external detector, repolish and re-etch. The figures in the centre of each error bar refer to the kind of surface that was investigated, hexagon = basal face, rectangle = prism face. White symbols correspond to ages calulated with $\lambda_f = 7.03\ 10^{-17}\ a^{-1}$, black symbols to ages calculated with $\lambda_f = 8.46\ 10^{-17}\ a^{-1}$. The thick horizontal line corresponds to the reference age of 31.4 Ma (± 0.5 Ma = shaded area).

7. Conclusions and Future Research

The experimental work that we have carried out during the last decade demonstrates that, in contrast to fifteen years ago, an accurate and precise evaluation of ϕ in the fission-track age equation is now possible. A determination of the activity of the unstable fission products in a uranium monitor provides a direct measurement of R (or $\sigma\phi$) but this approach is complex and tedious compared to the much simpler and equally accurate fluence determination through γ-spectrometry of Au and Co monitors. We realize that for a large number of laboratories absolute fluence determinations with metal monitors may pose a considerable problem and this stimulated us to repeat the original initiative of the NBS and to prepare a new pre-irradiated glass for fission-track dating (Bellemans et al.,1995; De Corte et al., this volume). This glass is certified and distributed by the Institute for Reference Materials and Measurements (IRMM) of the EC.

With the ζ-calibration method, track density measurements in an age standard and a glass monitor enables the determination of the combined parameter I σ / λ_f B, (with the factor B relating the track density in the glass to the thermal neutron fluence). With the newly-prepared pre-irradiated IRMM glasses a determination of B becomes possible. As the value of 7.253. 10^{-3} for the isotope abundance ratio I has already been agreed upon twenty years ago (Steiger and Jäger, 1977) and our present knowledge of σ is such that a general consensus on its value (574 $10^{-24}cm^{2}$) should be a formality, only the correct value of λ_f remains to be determined. Our experiments support the higher nominal value of 8.5 10^{-17} a^{-1}, but undoubtedly the fission-track community waits for a new, more precise and reliable determination, if possible with the fission-track method to become convinced. Based on our present knowledge of track registration, revelation and thermal stability this may however not be simple. A zircon age standard of volcanic origin with a homogeneous track density appears to be the right test sample for such a λ_f experiment. Such a zircon will not exhibit a shortening of spontaneous tracks observed with apatite age standards. It also has to be quite young to avoid effects of radiation damage on track revelation and it should still yield isotropically etched tracks. A selective analysis using only prism faces with the population and/or repolish technique could be applied to such a sample and hopefully the results will reconcile the track method with the λ_f determinations based on radio-chemical and ion chamber measurements.

Our work again emphasizes the importance of track revelation and registration as sources of systematic differences between age determinations. The ζ–method originally conceived to tackle the ϕ–λ_f problem, proves to be efficient in handling this problem too. By applying exactly the same procedure of track analysis to the unknown sample and the age standard a systematic deviation in the final age result should normally be eliminated, whatever dating technique is used. Hence, it allows to use techniques other than the ED method, if required.

The power of the ζ–method as a calibration tool is, however, provided by the age standards. Even if all parameters that comprise the ζ-factor will be known, an analysis of age standards will still be required to eliminate the systematic differences related to the technique of track analysis that is used. Nevertheless, once the exact value of λ_f is known, instead of calculating a ζ-value it will be more useful to use age standards for calculating the procedure factor Q as outlined in Wagner and Van den haute (1992). This factor yields more direct information on the combined effects of registration efficiency, revelation and observer bias of the dating technique used. True inaccuracies in the techniques of track analysis will therefore be more readily detected.

A special material in this respect is glass. In glass, the ratio of the bulk etching to the track etching rate depends significantly upon the chemical composition. As this ratio influences track registration and revelation, the use of an age standard does not guarantee an elimination of all systematic errors because track registration and revelation may be different in the unknown and the standard even under the same etching conditions. Analysis of a tektite age standard does not ensure accurately calibrated age determinations of a volcanic glass that can vary from rhyolitic to basaltic in composition. Hence, for dating glasses, the absolute approach combined with a technique of track analysis (e.g. repolishing) that ensures identical registration and ηq factors is the most accurate solution.

Acknowledgements

The authors gratefully acknowledge the Fund for Scientfic Research-Flanders (PVdh, FDC) and the IWONL-IWT (RJ, FB) for their financial support enabling the research reported in this paper.

References

Bellemans F. (1996): Een nucleair analytische bijdrage tot de absolute kalibratie van de fissiesporen-dateringsmethode. PhD thesis, University of Gent, 177pp.

Bellemans F., De Corte F., Van den haute P. and Ingelbrecht C. (1995): Towards a new glass monitor for the determination of the neutron fluence in fission-track dating.**25**, 527-530 (Proceedings of the 17th International Conference on Nuclear Tracks in Solids).

Bigazzi G. (1981) : The problem of the decay constant λ_f of ^{238}U. *Nucl. Tracks* **5**, 35-44.

Biersack J.P. and Ziegler J.F. (1987) : TRIM-91 : The transport of ions in matter. Computer program for range calculations of heavy ions in composed media.

Carpenter S.B. (1984) : Standard reference materials : calibrated glass standards for fission track use. Nat. Bur. Stand. Spec. Publ. 260-92.

Carpenter S.B. and Reimer G.M. (1974) : Standard reference materials : calibrated glass standards for fission track use. Nat. Bur. Stand. Spec. Publ. 260-49.

De Corte F., Van den haute P., Jonckheere R. and De Wispelaere A. (1991): Calibration of the fission track dating method : is Cu useful as an absolute thermal neutron fluence monitor ? *Chem. Geol. (Isot. Geosci. Sect.)* **86**, 187-194.

De Corte F. ,Van den haute P. and Bellemans F. (1995): The use of uranium-doped glasses in fission-track dating. *Radiation Measurements* **25**, 511-516. (Proceedings of the 17th International Conference on Nuclear Tracks in Solids).

De Corte F., Bellemans F., Van den haute P. , Ingelbrecht C. and Nicholl C.(this volume): A new U-doped glass certified by the European commission for the calibration of fission-track dating.

Donelick R.A.,Roden M.K.,Mooers J.D., Carpenter B.S. and Miller D.S.: Etchable length reduction of induced fission tracks in apatite at room temperature (~23°C): crystallographic orientation effects and "intial" mean lengths. *Nucl. Tracks Radiat. Meas.* **17**, 261-265.

Fleischer R.L. and Hart H.R. jr. (1972) : Fission track dating: techniques and problems. In : proc. Burg Wartenstein conf. on calibration of hominoid evolution (W.W. Bishop, D.A. Miller and S. Cole, eds.), Scot. Acad. Press, Edinburgh, p.135-170.

Fleischer, R.L., Price, P.B., and Walker, R.M. (1975): *Nuclear tracks in solids. Principles and applications*, 604 pp. University of California Press, Berkeley.

Gleadow A.J.W. and Lovering J.F. (1977) : Geometry factor for external detectors in fission track dating. *Nucl. Track Detection* **1**, 99-106.

Gleadow A.J.W. (1978) : Anisotropic and variable track etching characteristics in natural sphenes. *Nucl. Track Detection* **2**, 105-117.

Gleadow, A.J.W. (1981): Fission track dating methods : what are the real alternatives ? *Nuclear Tracks* **5**, 3-14.

Green P.F. and Durrani S.A. (1978) : A quatitative assessment of geometry factors for use in fission track studies. *Nucl. Track Detection* **2**, 207-213.

Hanna G.C., Westcott C.H., Lemmel H.D., Leonard B.R., Story J.S. and Attree P.M. (1969) : Revision of values for the 2200 m/s neutron constants for four fissile nuclides. *At. Energ. Rev.* **7/4**, 3-92.

Holden N.E.(1991): Temperature dependence of theWestcott g-factor for neutron capture reactions. In ENDF/B-VI. BNL-46465 Report, 36th IUPAC general assembly, Hamburg, Germany.

Hurford A.J. (1990): Standardization of fission track dating calibration: Recommendation by the Fission Track Working Group of the I.U.G.S. Subcomission on Geochronology. *Chem. Geol. (Isot. Geosci. Sect.)* **80**, 171-178.

Hurford, A.J. (this volume): Zeta: the ultimate solution to fission-track analysis calibration or just an interim measure

Hurford A.J. and Green P.F. (1981a) : A reappraisal of neutron dosimetry and uranium-238 λ_f values in fission-track dating. *Nucl. Tracks* **5**, 53-61.

Hurford A.J. and Green P.F. (1982) : A users' guide to fission track dating calibration. *Earth Planet. Sci. Lett.* **59**, 343-354.

Hurford A.J. and Green P.F. (1983) : The zeta age calibration of fission track dating. *Isot. Geosci.* **1**, 285-317.

Ivanov M. P., Terakopian G.M., Fefilov B.V. and Voronin A.S. (1985) : Study of ^{238}U spontaneous fission using a double ionization chamber. *Nucl. Instr. Meth.* **A234**, 152-157.

Iwano H. and Danhara T. (this volume) : A re-investigation of the geometry factors for fission-track dating of apatite, sphene and zircon.

M., Yamashita T. and Danhara T. (1992) : One-to-one correlation of fission tracks between zirkon and mica detectors. *Nucl. Tracks Radiat. Meas.* **20**, 341-347.

Iwano H., Kasuya M., Yamashita T.,Danhara , T.and Tagami T. (1993) : Track counting efficiency and unetchable track range in apatite.*Nucl. Tracks Radiat. Meas.* **21**, 513-517.

Jonckheere R.(1995): De absolute ouderdomsbepaling van apatiet gebaserd op uranium-fissiesporen:een methodologisch onderzoek. PhD thesis - University of Gent, 504pp.

Jonckheere R. and Van den haute P. (1996): Observations on the geometry of etched fission tracks in apatite: implications for models of track revelation. *American Mineralogist* **81**, 1476-1493.

Lindhard J., Scharff M., Schiøtt H.E. (1963) : Range concepts and heavy ion ranges (Notes on atomic collisions, II). Mat. Fys. Medd. Dan. Vid. Selsk. 34 n°.14; pp. 63.

Northcliffe L.C. and Schilling R.F. (1970) : Range and stopping-power tables for heavy ions. Nucl. Data Tables A7, 233-463.

Steiger R.H. and Jäger E. (1977): Subcommission on Geochronology; convention on the use of decay constants in geo- and cosmochronology. *Earth Planet. Sci. Lett.* **36**, 359-362.

Storzer D. and Wagner G.A. (1985): Bias in the application of the external detector technique. *Geol. Magazine* **122**, 195-196.

Van den haute P., Jonckheere R. and De Corte F. (1988): Thermal neutron fluence determination for fission-track dating with metal activation monitors : a re-investigation. *Chem. Geol. (Isot. Geosci. Sect.)* **73**, 233-244.

Wagner G.A. (1968) : Spuren der spontane Kernspaltung des ^{238}Urans als Mittel zur Datierung von Apatiten und ein Beitrag zur Geochronologie des Odenwaldes. *N. Jb. Min. Abh* **110**, 252-286.

Wagner G.A. and Van den haute P. (1992): *Fission-Track Dating.*
Enke Verlag- Kluwer Academic Publishers , 285pp.

Wagner G.A., Reimer M., Carpenter B.S., Faul H., Van der linden R. and Gijbels R. (1975) : The spontaneous fission rate of U-238 and fission track dating. *Geochim. Cosmochim. Acta* **39**, 1279-1286.

A RE-INVESTIGATION OF THE GEOMETRY FACTORS FOR FISSION-TRACK DATING OF APATITE, SPHENE AND ZIRCON

H. IWANO and T. DANHARA
Kyoto Fission-Track Co., Ltd.
Kyoto, Japan

Abstract - The geometry factors that apply to fission-track dating of zircon, sphene, apatite using mica external detectors were re-investigated. The ideal geometry factor ($2\pi/4\pi$) is described by $(R+d)/2R$, where R is the etchable range of a fission fragment in the mineral and d is an unetchable range related to differences in the registration threshold between the sample and the detector. The geometry factors for zircon, sphene, and apatite, were experimentally determined to amount to 0.68, 0.60, and 0.55 respectively and are consistent with the ideal values for each mineral. Zeta age-calibration factors have been determined for the external detector method using zircon, sphene and apatite age standards and for the population method using apatite age standards. The directly obtained zeta values are quite variable but the geometric corrections yield an identical zeta baseline for the two methods and for the three minerals. Shortening of spontaneous tracks in the apatite standards was also corrected for . These results suggest that fission-track age determinations using the absolute approach may be within reach for both the population method and the external detector method provided that the right correction for the geometry factor is made. It finally follows that the different values that have been found for the decay constant of spontaneous fission of ^{238}U may be related to technical differences between the conventional external detector and population methods.

1. Introduction

The use of fission-track (FT) method as an independent geochronometric method, has been greatly impeded by the uncertainty on the value of the decay constant for nuclear fission of ^{238}U (λ_f), together with problems related to the measurement of the thermal neutron fluence and to absolute track counting efficiencies. At present, an absolute approach with the selection of a λ_f-value and measurement of neutron fluence is accepted only for the population method (Naeser, 1967) analyses of apatite by the I.U.G.S. Subcomission on Geochronology (Hurford, 1990). The method is accepted because it does not require geometric corrections due to the fact that spontaneous and induced track densities are measured on the same material surfaces with identical geometry and identical registration, revelation, and counting efficiencies. In addition, the recent development of thermal neutron fluence determinations with Au and Co monitors has greatly contributed to the performance of the absolute calibration (Van den haute *et al.*, 1988).

P. Van den haute and F. De Corte (eds.), Advances in Fission - Track Geochronology, 47-66.

On the other hand, the absolute calibration approach is not accepted for the external detector method (Gleadow, 1981) for which spontaneous and induced tracks are counted in different materials, requiring a correction for different geometries and different registration, revelation, and counting efficiencies. The relevant correction factor called the "geometry factor" has been an important subject of study for over twenty years (Reimer *et al.*, 1970; Gleadow and Lovering, 1977; Green and Durrani, 1978; Naeser *et al.*, 1980; Storzer and Wagner, 1985). The zeta (ζ) age-calibration (Hurford and Green, 1982; 1983) based on the analysis of age standards intrinsically eliminated the problems related to neutron fluence, decay constant and geometry factor and has been recommended for all FT techniques, including the population method (Hurford, 1990). Adoption of the ζ-approach greatly improved the accuracy of FT age determinations and allowed to compare the FT data from different laboratories, and to relate them to data obtained with other radioisotopic methods (e.g. Miller *et al.* 1990; 1993).

Despite the successful contribution of the ζ-approach to FT age-calibration, it is known that different age-calibration factors (ζ-factors) were found not only by different researchers, but also for different minerals by the same researcher (e.g. Green, 1985) when determined with the external detector method. Because of these inconsistencies, there has been some scepticism regarding the ζ-approach, which was felt to yield personal calibration factors only (Van den haute *et al.*, 1988). An independent dating system should be the eventual aim of all FT practitioners and to be used for all FT techniques if possible. In this context a re-investigation of the geometry factor, both theoretical and experimental, seems essential.

In the present study, we re-investigate the geometry factors for zircon, apatite, sphene, and mica detectors based on recent studies of track registration efficiencies using the one-to-one track correlation technique (Iwano *et al.*, 1992; 1993; Iwano and Danhara, 1994). The zeta age-calibration baselines for the population and external detector methods are compared taking into account the re-determined geometry factors. Finally, the decay constant of ^{238}U fission is briefly reviewed in the light of these results.

2. Framework of Geometry Factors

2.1. REGISTRATION EFFICIENCY

A spatial correlation of induced tracks in a mineral with those in an adjacent detector has revealed that a fission fragment with a certain energy can produce an etchable path in a detector, but not in a mineral, because of differences in the track registration threshold between the two materials (Fig. 1; Iwano *et al.*, 1992; 1993). Results of one-to-one induced-track correlation experiments for zircon, sphene, and apatite using a mica detector are summarized in Table 1.

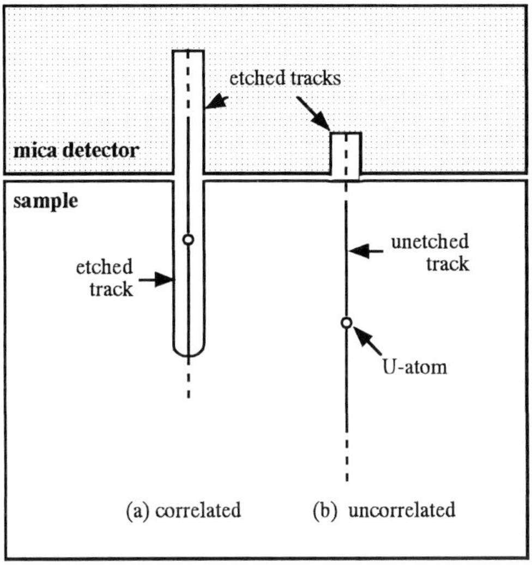

Fig. 1 One-to-one correlations of two pairs of induced tracks between a sample and an external detector with different registration threshold: (a) for a general case that both tracks are etchable and correlated, and (b) for the case that the track is etchable only in the detector and uncorrelated. Solid lines represent the total etchable ranges of a pair of fission fragments in the sample with higher registration threshold, and broken lines represent parts of the total range that are unetchable in the sample but etchable in the detector because of its lower threshold.

For each mineral species, the difference in counting efficiency remains within analytical precision, but the average efficiencies differ significantly between the three minerals. It can be seen that all of the minerals have a lower efficiency than their corresponding mica detector (all the track density ratios are less than unity). This implies that these materials have different registration thresholds in the order: mica < apatite < sphene < zircon.

The difference in the track registration threshold between a mineral and the mica detector correlates with an unetchable range, corresponding to the range deficit (Fleischer *et al.*, 1975), at the end of the latent track in a mineral. The unetchable ranges in apatite, sphene, and zircon were estimated by following the three-dimensional length analysis (Iwano *et al.*, 1992). Relationships between the etchable and unetchable ranges for the three minerals are shown in Fig. 2. It should be noted that these unetchable ranges are minimum values, related to the registration threshold in mica.

Taking into account the mean etchable range (R) and the mean unetchable range (d) of a fission fragment in a mineral, the length ratio of $R/(R+d)$ provides an etchable track registration efficiency ratio of sample/mica. We can use the results of a one-to-one track correlation experiment reported in Table 1 as a measure of this efficiency.

TABLE 1. Summary of data from one-to-one correlation of induced fission tracks between zircon, sphene, apatite, and mica detectors.

Sample	Sample No. of tracks[+]	Mica detector No. of tracks		Ratio (Mineral/Mica)	Reference
Annealed zircon					
Pink Tuff	457	601	(144)*	0.76±0.05	(1)
Buluk Member	654	868	(214)*	0.75±0.04	(1)
Fish Canyon	724	925	(201)*	0.78±0.04	(4)
Unannealed zircon					
Pink Tuff	417	550	(133)*	0.76±0.05	(1)
Buluk Member	396	528	(132)*	0.75±0.05	(1)
			weighted mean:	**0.76±0.02**	
Annealed sphene					
Fish Canyon	757	867	(110)*	0.87±0.04	(3)
Mt. Dromedary	1572	1809	(237)*	0.87±0.03	(3)
			weighted mean:	**0.87±0.02**	
Annealed apatite					
Fish Canyon	489	535	(46)*	0.91±0.06	(2)
Mt. Dromedary	454	506	(52)*	0.90±0.06	(2)
Durango	655	690	(35)*	0.95±0.05	(4)
			weighted mean:	**0.92±0.03**	

note: [+] The number of uncorrelated induced-tracks are less than 1% of the total number and are omitted.
* Numbers in parentheses are the numbers of uncorrelated tracks. (1) Iwano et al. (1992); (2) Iwano et al. (1993); (3) preliminarily reported by Iwano and Danhara (1994); (4) unpublished data

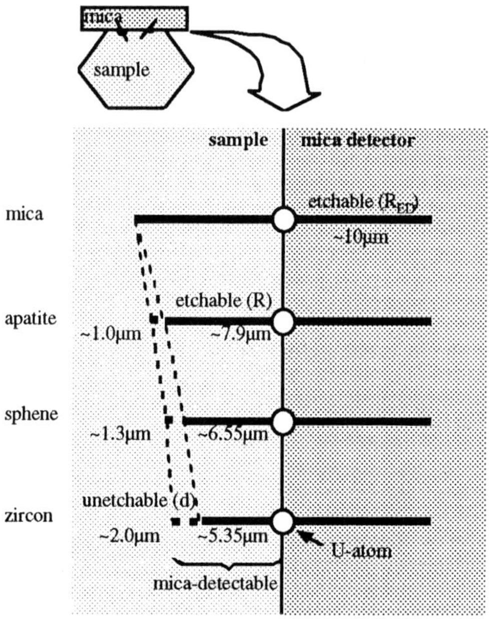

Fig. 2 Schematic illustration of etchable and unetchable ranges in zircon, sphene, apatite, and mica. Unetchable ranges are estimated to be 2.0 μm, 1.3 μm and 1.0 for zircon, sphene, and apatite, respectively, with reference to the registration threshold in mica. The total range (one etchable range and one unetchable range) of a fission fragment in each mineral corresponds to the etchable range in mica.

The conditions for this experiment were kept as follows: (1) *isotropic track revelation* on crystal surfaces with the highest etching efficiencies (e.g. surfaces parallel to *c*-axis of zircon and apatite), (2) *precise track observation* on a high-resolution monitor (e.g. use of alpha-recoil tracks (depth < 0.2 μm) in mica for the minimum criteria of track identification), (3) *perfect contact* between mineral and detector, and (4) *accurate track identification* by a one-to-one track correlation.

2.2. RE-DEFINITION

By combining two of the three geometry surfaces (internal (4π), external (2π) sample-surfaces and an external detector (ED) surface) currently used in FT studies, the geometry factors can be defined as follows:

$$G = \frac{\rho_{2\pi}}{\rho_{4\pi}}\ ;\ g = \frac{\rho_{ED}}{\rho_{4\pi}}\ ;\ g' = \frac{\rho_{ED}}{\rho_{2\pi}}, \tag{1}$$

where $\rho_{2\pi}$, $\rho_{4\pi}$ and ρ_{ED} are induced-fission-track densities on 4π, 2π, and external detector surfaces, respectively (Green and Durrani, 1978). G is theoretically given by

$$G = 0.5 \tag{2}$$

(Wagner and Van den haute, 1992). In the case of g' or g, we must take into account the difference in registration efficiency between a sample and a detector, as discussed above. At first, a total range of (R+d) on the 2π geometry surface can be substituted for the mean etchable range (R_{ED}) on the external detector surface (Fig. 2). Because $\rho_{2\pi}$ or ρ_{ED} is proportional to the radius for each geometry surface, g' is simplified by

$$g' = \frac{R+d}{R}. \tag{3}$$

Combining both of the above relationships (G and g'), we can obtain the geometry factor of g as follows:

$$g = Gg' = \frac{R+d}{2R}. \tag{4}$$

We can now conclude that the geometry factors for FT dating with an external detector have different values for each mineral because of differences in the track registration efficiencies among minerals. The ideal geometry factors of G, g and g' for zircon, sphene, and apatite with reference to the mica detectors are calculated in Table 2.

TABLE 2. Calculation of the ideal geometry factors (see equations (2), (3) and (4)).

Mineral	G	g	g'*
	(ρ_2/ρ_4)	(ρ_{ED}/ρ_4)	(ρ_{ED}/ρ_2)
zircon	0.5	0.66	1.32
sphene	0.5	0.57	1.15
apatite	0.5	0.54	1.09

note: * The mean values of registration efficiency in TABLE 1 were used as $1/g'$.

3. Experiments

3.1. EXPERIMENT ONE

The geometry factors *G*, *g*, and *g'* were measured using pre-annealed samples (all spontaneous tracks erased; see Iwano *et al.* (1992; 1993) for the procedures used) of zircon, sphene and apatite, coupled with low-U muscovite mica detectors. Zircon and sphene samples with an internally homogeneous U-distribution were used and the induced track densities in the samples and external detectors were compared for each grain to eliminate bias introduced by the difference in U-content among the crystals. After measurement of the induced track density on the external (2π) surfaces, zircon or sphene grains were polished to erase all etched tracks and to expose the internal (4π) surface. In the case of apatite, where the U-content was homogeneous between the crystals, each sample was divided into two aliquots, one for the measurements on external surfaces and one for the internal surfaces. The mica detectors were attached to the apatite mount with the external surfaces. Induced tracks in zircon, sphene, apatite and mica were etched in a KOH+NaOH eutectic melt at 225 °C for 55 h, an acid mixture of HF+ $2HNO_3$+ 3HCl at 20°C for ~12 min, 0.7%HNO_3 at 20 °C for ~6 min and 47% HF at 20 °C for 24 min, respectively. Grains with scratched surfaces were selected for the experiment. Tracks were observed using a Nikon® Microphot-FX microscope with a 100X dry objective, a Tokyo Electronic Industry® HVC7110-17 TV camera, and a CHUOMUSEN® THM-14A TV monitor (Iwano and Danhara, 1997). All track counts were made on the screen of the high-resolution TV monitor with a total image magnification of 1900x.

3.2. EXPERIMENT TWO

In this experiment, ζ calibration factors have been determined for zircon, sphene and apatite. The I.U.G.S. age standard and other standard samples listed in Table 3 were used for this experiment. All the minerals were analyzed using the conventional external detector method, in which spontaneous tracks are counted in the internal surface, and the induced tracks are counted in low-U muscovite mica detectors (ED1 method: Gleadow, 1981). External natural surfaces of zircon that have been occasionally used for spontaneous-track counting (ED2 method: Gleadow, 1981) were also analyzed for comparison. The population method was applied to apatite only.

TABLE 3. List of standard samples used for zeta factor determination

standard sample	mineral	reference age (Ma, ±2σ)	reference
Baba Tuff	zircon	11.28±0.08	Odin *et al.* (1995)
Buluk Member Tuff	zircon	16.3±0.2	Hurford and Watkins (1985)
Fish Canyon Tuff	zircon, sphene, apatite	27.8±0.2	Hurford and Hammerschmidt (1985)
Durango Apatite	apatite	31.4±0.5	Green (1985)
Tardree Rhyolite	zircon	58.7±1.1	Hurford and Green (1983)
Mt. Dromedary Complex	zircon, sphene, apatite	98.7±1.1	Green (1985)

Zircon crystals (50-80 grains) were mounted in a PFA sheet (Danhara *et al.*, 1993) with crystal surfaces exposed. Zircons for the ED1 method were tentatively etched in the above-mentioned etchant and were then polished with several diamond pastes until the etched spontaneous-tracks were completely erased and 4π geometry surfaces were revealed. Zircons for the ED2 method were not polished. The zircon mounts were then etched in the above alkali etchant for 29-35 h for the Baba Tuff, ~40 h for the Buluk Member Tuff, 22-28 h for the Fish Canyon Tuff, ~23 h for the Tardree Rhyolite, and ~11 h for the Mount Dromedary Complex. For the Baba Tuff and the Fish Canyon Tuff samples, the shorter etch times correspond to the external surface. experiment. More detailed procedures of zircon FT analysis have been described by Danhara et al. (1991), and Iwano and Danhara (1997).

For sphene, approximately 50 grains of each sample were also mounted in a PFA sheet and polished with diamond pastes. The mounts were then etched in the acid etchant mentioned above for 10 min for the Fish Canyon Tuff and 6 min for the Mount Dromedary Complex.

Apatite samples (50-100 grains) were mounted in epoxy resin (STRUERS® EPOFIX) with prismatic faces exposed. For Durango apatite, three large crystals were sliced along the c-axis of each grain and analyzed for each slice. The mounts were polished with diamond pastes and etched in the above HNO_3 etchant for 4 min for the Fish Canyon Tuff, 5 min for the Mount Dromedary Complex, and 6 min for the Durango apatite samples. For the population method, apatite samples which were annealed at 450 °C for 2 h to completely erase spontaneous tracks were also prepared.

After etching, the mounts were packed in a plastic holder to ensure intimate contact with low-U muscovite detectors. The detectors had been confirmed to produce only a negligible number of background tracks. Each pile of mounts for irradiation was sandwiched between two pieces of uranium dosimeter glass, NIST SRM612, attached with muscovite detectors. Samples were irradiated in the Thermal Column (TC) #5 of the TRIGA reactor of Oregon State University (OSU) where the cadmium ratio for Au is 14 (Facilities Services Manual of OSU). Ten different irradiation runs were made from May 1992 to August 1996.

After irradiation, the muscovite detectors were detached and etched in 46% HF at 20 °C for 24 min (for sample mounts) and 44 min (for dosimeter glass). Track counting was done using the same system mentioned above. Zircon, sphene, and apatite grains with well-etched and isotropically distributed tracks were selected for counting. Photomicrographs were taken for all the grains used for track counts, and the counting area was marked on each photograph. The maximum area of suitable surface was counted in each grain except for Durango apatite, where track counting was made per field of the screen of the monitor (80μmx120μm) within a large crystal.

The ζ-factor was calculated from the standard equation:

$$\zeta = \frac{1}{\lambda_d} \cdot \frac{e^{\lambda_d A_{std}} - 1}{g(\rho_s / \rho_i)_{std} \rho_d}, \tag{5}$$

where λ_d is the total decay constant for ^{238}U ($1.55125 \times 10^{-10} y^{-1}$: Steiger and Jäger, 1977), A_{std} is the reference age of a standard sample, $(\rho_s/\rho_i)_{std}$ is the spontaneous/induced track density ratio for the standard, g is the geometry factor, and ρ_d is the induced track density for a dosimeter glass measured in a mica detector. By convention, all of the ζ-factors are represented by the unit "a x cm^2", which will be omitted hereafter.

For comparison, the ζ-factors were calculated using the geometry factors (1) g=0.5 and g'=1 for convention, or (2) the g and g' measured in experiment 1. Taking into account the track shortening effect (*i.e.* the effect of partial track shortening in apatite age standards: Green, 1980; Gleadow *et al.*, 1986), the length-corrected ζ-factor of apatite was also calculated using Green's method (Green, 1988).

4. Results and Discussion

4.1 GEOMETRY FACTORS

The geometry factors *G*, *g*, and *g'* are summarized in Table 4. The mean values of *g'* for each mineral correspond well with the calculated ideal values in Table 2. However, the mean values of *G* and *g* are slightly (2-6%) larger than the ideal values.

Fig. 3 shows the length distributions of semi-tracks observed on external (2π geometry), internal (4π geometry) and external detector (2π geometry) surfaces using our system. Data are from induced tracks on surfaces parallel to the c-axis of zircon and apatite crystals as well as on [001] the cleavage plane of mica. All tracks in the external surfaces could be counted with relative ease because they are all longer than the half of a complete track (Fig. 3a). All tracks on the mica detectors could also be easily counted because of following advantages: (1) nearly 100% of the etching efficiency of the [001] plane due to extremely low bulk etching (Lakatos and Miller, 1972), (2) nearly 100% optical counting efficiency of the plane (Roberts et al., 1984; Wall, 1986), and (3) extremely low background noise. We can see that there is no counting-loss of tracks less than 1μm (Fig. 3c) when using alpha recoil tracks (depth < 0.2μm) as a minimum criterion for identification of fission tracks in mica.

TABLE 4. Fission track data of geometry factor measurement using zircon, sphene, apatite and mica detectors

Sample code	No. of crystals or fields	External (2π) ρ_i (×105 cm^{-2})	External (2π) (N_i)	Internal (4π) ρ_i (×105 cm^{-2})	Internal (4π) (N_i)	Mica detector(ED) ρ_i (×105 cm^{-2})	Mica detector(ED) (N_i)	Geometry ratio (±1σ) G ($\rho_{2\pi}/\rho_{4\pi}$)	g ($\rho_{ED}/\rho_{4\pi}$)	g' ($\rho_{ED}/\rho_{2\pi}$)
zircon										
Buluk Member	16	7.76	(884)	14.98	(1706)	9.833	(1120)	0.518±0.021	0.657±0.025	1.27±0.06
Kawamoto Gd.	9	18.06	(2947)	35.26	(5754)	24.39	(3980)	0.512±0.012	0.692±0.014	1.35±0.03
						weighted mean:		**0.514±0.010**	**0.684±0.012**	**1.34±0.03**
sphene										
Fish Canyon	9	3.40	(299)	6.34	(558)	3.89	(342)	0.536±0.038	0.613±0.042	1.14±0.09
Mt. Dromedary	9	7.08	(646)	13.41	(1223)	7.99	(729)	0.528±0.026	0.596±0.028	1.13±0.06
						weighted mean:		**0.531±0.021**	**0.601±0.023**	**1.13±0.05**
apatite										
Fish Canyon	100/90/100	6.00±0.17	(3553)	11.5±0.31	(6491)	6.97±0.22	(4062)	0.522±0.020	0.606±0.025	1.16±0.05
Durango	38/40/38	3.38±0.10	(1232)	6.51±0.14	(2500)	3.30±0.09	(1205)	0.519±0.019	0.507±0.018	0.976±0.039
						weighted mean:		**0.520±0.014**	**0.549±0.015**	**1.06±0.03**

note: The number of crystals (or fields) for apatite corresponds to 2π, 4π, ED surfaces, respectively.

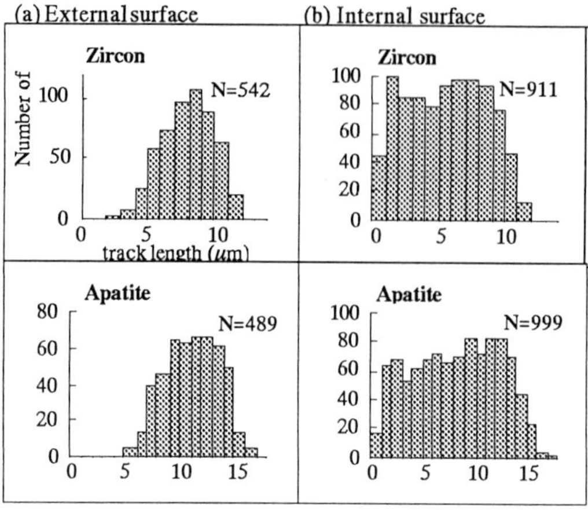

Fig. 3 Length distributions of semi-tracks observed on (a) external surfaces of zircon and apatite, (b) internal surfaces of zircon and apatite, and (c) external detector surfaces of mica. All the tracks are induced ^{235}U tracks measured by using the three-dimensional length measuring system (Iwano et al., 1996). Bulk etching of observed mineral surfaces is not corrected for. A refractive index correction was made but not for dot-shaped tracks. N is the number of tracks measured. Data are after Iwano and Danhara (1997) for zircon and mica, Iwano *et al.* (1993) for external surfaces of apatite, and Iwano (unpublished) for internal surfaces of apatite.

On the other hand, a loss of tracks less than 1μm is observed on the internal surfaces of both zircon and apatite (Fig. 3b). The track identification limits for the internal surfaces of zircon and apatite in our system are estimated to be ~ 0.5μm and ~1μm respectively. Therefore, the 5% larger values of G and g can be reasonably explained by the lowered counting efficiencies of internal surfaces due to the minimum track length cut-off. The minimum length for identification on the internal surfaces of sphene is also expected to be 0.5-1μm.

Consequently, the different geometry factors 0.68 for zircon, 0.60 for sphene and 0.55 for apatite in Table 4 are considered to be adequate and practical values for the conventional external detector technique using internal mineral-surfaces and mica detectors.

4.2. ZETA-BASELINES

The experimentally determined ζ-factors are summarized in Tables 5, 6, 7, and 8, while the weighted mean ζ-factors and the error for each sample/method combination are shown in Fig. 4. Conventional ζ-factors differ significantly between methods and among the three minerals, with apatite ζ-factors obtained with the population method yielding the smallest values (Fig. 4a). Conversely, it seems that ζ-factors corrected for geometry factors (Fig. 4b) or those corrected for both geometry factors and track shortening effects (Fig. 4c) are concordant between all the mineral/detector couples.

TABLE 5. Fission track data of zeta factor measurement using population method (POP) for apatite

Sample code	No. of crystals (n/m)	Spontaneous ρ_s (N_s) (∞10^5 cm^{-2})	Induced ρ_i (N_i) (∞10^5 cm^{-2})	Dosimeter ρ_d (N_d) (∞10^5 cm^{-2})	s' (%)	$1/k$	Zeta (1σ) conventional	Zeta (1σ) length-corrected	Irradiation No.
Fish Canyon Tuff (POP)									
FC3 920327-1	75/50	2.21(1333)	11.0(3333)	5.37 (1373)	7	0.198	259±24	240±22	OSU9205
FC3 931205-1	80/50	2.33(1605)	10.5(4026)	5.07 (1460)	5	0.115	248±18	230±17	OSU9312
FC3 941125-1	100/80	2.34(1249)	11.5(5768)	5.19 (2494)	5	0.188	265±20	246±18	OSU9412
FC3 941125-2	100/90	2.37(1243)	11.6(6491)	5.19 (2494)	4	0.292	262±23	243±21	OSU9412
FC3 950116-1	80/100	2.15(1220)	11.4(5530)	5.22 (2503)	5	0.198	282±22	262±20	OSU9502A
				weighted mean:			**263± 9**	**244± 9**	
Durango Apatite (POP)									
DR1 920327-1	56/56	2.13(1145)	10.9(5840)	5.38 (1376)	1	0.001	298±13	267±12	OSU9205
DR2 931205-1	100/80	1.69(1623)	7.11(5457)	5.09 (1466)	1	0.006	260±11	233±10	OSU9312
DR2 941126-1	80/60	1.66(1274)	6.43(3704)	5.25 (2522)	2	0.001	232± 9	208± 8	OSU9412
DR3 950904-1	100/50	1.68(1611)	8.28(3976)	5.34 (2564)	2	0.007	291±11	260±10	OSU9509A
DR3 950904-2	80/75	1.43(1097)	6.55(4718)	5.37 (2577)	2	0.006	269±11	241±10	OSU9509A
				weighted mean:			**269± 5**	**241± 4**	

notes: (1) *n* and *m* are the number of crystals (or fields) used for spontaneous and induced track counting, respectively. (2) ρ and *N* are the density and the total number of fission tracks counted, respectively. (3) ζ-factors were determined for a dosimeter glass SRM612. (4) Length corrections were made by multiplying conventional ζ by the spontaneous/induced track length ratios of 0.928 for the Fish Canyon Tuff and 0.895 for Durango (Gleadow *et al.*, 1986; Green, 1988). (5) s' is relative standard error of mean induced track count (Naeser *et al.*, 1979). (6) $1/k$ is the coefficient of variation of the distribution of U-content between grains (Galbraith, 1984; Galbraith and Laslett, 1985). (7) The relative error was given by $[1/N_s+1/N_i+1/N_d+1/k(1/n+1/m)+(\sigma_A/A_{std})^2]^{1/2}$ where A_{std} is a reference age of a standard and σ_A is an error of the reference age.

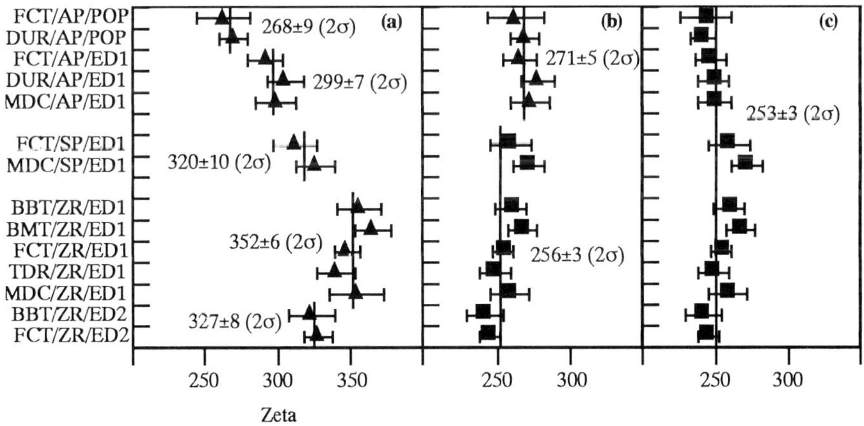

Fig.4 ζ-factors for three different sample/mineral/method combinations: (a) conventional ζ-factors, (b) ζ factors corrected for difference in track registration (geometry factor) among minerals, and (c) ζ-factors corrected additionally for track shortening in apatite. Experimental data are in Table 5, 6, 7, and 8. Triangles represent data biased by either track registration or track shortening effects, or by both. Squares correspond to data corrected for both these biases. The weighted mean values denoted by the solid lines were calculated for each mineral/method couple in Fig.4a, and for each corrected baseline in Fig. 4b and 4c. Error bars represent 2σ. AP, SP, and ZR indicate apatite, sphene and zircon, respectively. POP is the population method. ED1 and ED2 are the external detector methods using internal and external surfaces for spontaneous track counts, respectively. Standard samples are FCT: Fish Canyon Tuff; DUR: Durango; MDC: Mt. Dromedary Complex; BBT: Baba Tuff; BMT: Buluk Member Tuff; TDR: Tardree Rhyolite.

TABLE 6. Determination of zeta factors using the external detector method (ED1) for apatite

Sample code	No. of crystals	Spontaneous ρ_s (N_s) (∞10^5 cm^{-2})	Induced ρ_i (N_i) (∞10^5 cm^{-2})	Dosimeter ρ_d (N_d) (∞10^5 cm^{-2})	P(χ2) (%)	Zeta (±1σ) conven. g=0.5	g-cor.	length-cor.	Irradiation No.
Fish Canyon Tuff (ED1)									
FC3 920327-1	36	2.21 (669)	6.33(1914)	5.41(1385)	41	295±16	268±14	249±13	OSU9205
FC3 930301-1	30	2.11 (539)	6.24(1594)	5.43(1564)	59	304±17	276±16	257±14	OSU9304
FC3 930301-2	30	2.39 (621)	6.51(1693)	5.43(1564)	85	280±15	255±14	236±13	OSU9304
FC3 930301-3	28	2.29 (572)	5.92(1478)	5.07(1460)	18	284±16	259±15	240±13	OSU9312
FC3 931205-1	30	2.40 (656)	6.39(1746)	5.07(1460)	15	293±16	266±14	247±13	OSU9312
FC3 941125-2	30	2.26 (377)	5.65 (941)	5.18(2486)	12	269±17	245±16	227±15	OSU9412
FC3 950116-1	30	2.09 (476)	6.17(1407)	5.17(2481)	40	319±18	290±17	269±15	OSU9502A
				weighted mean:		**292± 6**	**266± 6**	**247± 5**	
Durango Apatite (ED1)									
DR1 920327-1	56	2.13(1145)	5.93(3190)	5.43(1388)	7	323±14	294±13	263±12	OSU9205
DR1 930301-123	31	2.18 (650)	6.07(1807)	5.80(1484)	92	302±16	275±15	246±13	OSU9304
DR2 931205-1	43	1.74 (717)	4.60(1900)	5.09(1466)	73	328±17	299±15	267±14	OSU9312
DR2 931205-2	45	1.57 (678)	4.01(1731)	5.09(1466)	38	316±17	288±15	257±14	OSU9312
DR3 950904-1	50	1.65 (791)	4.35(2086)	6.08(2919)	98	273±13	249±12	223±10	OSU9608A
DR3 950904-2	50	1.46 (701)	4.20(2014)	6.12(2936)	68	296±14	269±13	241±12	OSU9608A
				weighted mean:		**306± 6**	**278± 6**	**249± 5**	
Mt. Dromedary Complex (ED1)									
MDC 930301-1	23	10.3(1013)	8.56 (838)	5.42(1561)	44	304±16	277±15	254±14	OSU9304
MDC 930301-2	30	9.43(1626)	7.49(1291)	5.42(1561)	34	291±13	265±12	243±11	OSU9304
MDC 931205-1	30	9.30(1563)	7.23(1214)	5.05(1454)	56	306±14	279±13	256±12	OSU9312
MDC 941126-1	30	9.57(1561)	7.57(1236)	5.28(2537)	26	298±13	272±12	249±11	OSU9412
				weighted mean:		**299± 7**	**273± 6**	**250± 6**	

TABLE 7. Fission track data of zeta factor measurement using external detector method (ED1) for sphene

Sample code	No. of crystals	Spontaneous ρ_s (N_s) (∞10^5 cm^{-2})	Induced ρ_i (N_i) (∞10^5 cm^{-2})	Dosimeter ρ_d (N_d) (∞10^5 cm^{-2})	P(χ^2) (%)	Zeta (±1σ) conventional (g=0.5)	g-corrected	Irradiation No.
Fish Canyon Tuff (ED1)								
FC3 920319-1	19	9.15 (1101)	27.5 (3314)	5.34 (1367)	10	314±14	261±12	OSU9205
FC3 921127-1	8	9.18 (439)	27.3 (1306)	5.24 (1509)	9	316±19	263±16	OSU9304
FC3 921127-2	9	9.01 (444)	26.7 (1318)	5.24 (1509)	8	316±19	263±16	OSU9304
FC3 921127-3	9	9.86 (359)	29.7 (1080)	5.24 (1509)	28	320±21	266±18	OSU9304
FC3 941125-1	15	8.99 (732)	24.6 (2003)	5.13 (2465)	33	297±14	247±12	OSU9412
				weighted mean:		**311± 8**	**259± 6**	
Mt. Dromedary Complex (ED1)								
MDC 921127-1	7	49.5 (1376)	43.4 (1207)	5.25 (1512)	73	332±16	277±13	OSU9304
MDC 921127-2	6	57.5 (1098)	51.2 (977)	5.25 (1512)	40	337±17	281±14	OSU9304
MDC 921127-3	5	49.7 (1119)	42.4 (955)	5.25 (1512)	10	323±17	269±14	OSU9304
MDC 931206-1	10	51.7 (2924)	43.4 (2456)	4.96 (1429)	4	337±13	280±11	OSU9312
MDC 941126-1	7	57.2 (1995)	45.2 (1577)	5.15 (2472)	43	305±12	254±10	OSU9412
				weighted mean:		**326± 7**	**271± 5**	

(1) ρ and N are the density and the total number of fission tracks counted, respectively. (2) ζ-factors were determined for a dosimeter glass SRM612. (3) P(χ^2) is the upper χ^2-tail probability corresponding to the observed χ^2-statistics (Galbraith, 1981). (4) g-corrected ζs were calculated using the mean values of g or g' in TABLE 4. (5) Length corrections were made by multiplying g-corrected ζ by the spontaneous/induced track length ratios of 0.928 for the Fish Canyon Tuff, 0.895 for Durango and 0.917 for Mt. Dromedary (Gleadow *et al.*, 1986; Green, 1988). (6) The relative error was calculated as $[1/N_s+1/N_i+1/N_d+(\sigma_A/A_{std})^2]^{1/2}$ where A_{std} is a reference age of a standard and σ_A is an error of the reference age.

TABLE 8. zeta factor determinations using the external detector method (ED1 and ED2) for zircon

Sample code	No. of crystals	Spontaneous ρ_s (N_s) (∞10^5 cm^{-2})	Induced ρ_i (N_i) (∞10^5 cm^{-2})	Dosimeter ρ_d (N_d) (∞10^5 cm^{-2})	P(χ^2) (%)	Zeta ($\pm 1\sigma$) conven. (g=0.5, g'=1)	g-cor.	Irradiation No.
Baba Tuff (ED1)								
BBT 910806-1	30	20.4(2875)	31.8(4486)	0.964(1481)	1	365±13	267±10	OSU9608B
BBT 940423-1	30	24.7(3304)	39.0(5209)	0.964(1481)	<1	369±13	270± 9	OSU9608B
BBT 940702-1	25	25.8(2060)	36.0(2880)	0.964(1480)	48	328±13	240± 9	OSU9608B
				weighted mean:		**356± 7**	**260± 5**	
Baba Tuff (ED2)								
BBT 910801-1	30	15.4(1429)	42.8(3961)	0.965(1483)	<1	324±13	242±10	OSU9608B
BBT 910801-2	30	13.7 (886)	38.1(2455)	0.965(1482)	62	324±15	242±11	OSU9608B
BBT 940702-2	30	14.2(1070)	39.2(2943)	0.965(1482)	41	322±14	240±11	OSU9608B
				weighted mean:		**323± 8**	**241± 6**	
Buluk Member Tuff (ED1)								
BMT 910809-1	26	11.5(1482)	68.4(8831)	5.27 (1348)	2	369±15	270±11	OSU9205
BMT 910826-1	30	10.3(1509)	56.0(8210)	5.22 (1503)	71	340±13	249±10	OSU9304
BMT 911024-1	25	10.1(1228)	62.0(7568)	5.23 (1506)	1	385±16	281±11	OSU9304
BMT 910809-2	30	10.5(1781)	58.7(9983)	4.91 (1416)	57	373±14	272±10	OSU9312
BMT 910809-3	30	11.3(2120)	62.1(11612)	4.91 (1416)	34	364±13	266±10	OSU9312
				weighted mean:		**366± 6**	**267± 5**	
Fish Canyon Tuff (ED1)								
FC3 910808-1	24	52.0(3668)	35.0(2469)	1.06 (1637)	33	354±13	259± 9	OSU9502B
FC3 910808-2	12	66.0(2659)	44.2(1781)	1.06 (1641)	10	352±14	257±10	OSU9502B
FC3 910808-3	22	64.7(5348)	40.0(3308)	1.06 (1644)	<1	325±11	238± 8	OSU9502B
FC3 910502-1	15	78.8(2757)	53.1(1856)	1.10 (1689)	78	341±13	249±10	OSU9509B
FC3 911016-1	24	66.4(4450)	44.8(3003)	1.10 (1691)	1	342±12	250± 9	OSU9509B
FC3 920806-1	30	63.6(4961)	40.8(3185)	0.967(1485)	9	370±13	271± 9	OSU9608B
FC3 921127-1	21	62.9(3836)	38.6(2355)	0.966(1484)	10	354±13	259±10	OSU9608B
				weighted mean:		**348± 5**	**254± 4**	
Fish Canyon Tuff (ED2)								
FC3 910808-4	29	33.6(1390)	44.3(1836)	1.07 (1647)	35	344±15	257±11	OSU9502B
FC3 910808-5	29	39.4(1898)	50.9(2452)	1.07 (1651)	72	336±13	251±10	OSU9502B
FC3 910808-6	24	40.2(1455)	49.7(1798)	1.07 (1654)	2	322±14	240±11	OSU9502B
FC3 910826-1	30	35.8(1604)	48.7(2180)	1.10 (1685)	38	344±14	257±11	OSU9509B
FC3 911016-2	24	42.0(1842)	52.8(2316)	1.10 (1687)	16	318±13	238±10	OSU9509B
FC3 920806-2	30	32.8(2414)	37.3(2747)	0.968(1486)	1	328±13	245±10	OSU9608B
FC3 921127-2	24	35.6(1818)	38.5(1966)	0.967(1486)	39	311±13	232±10	OSU9608B
FC3 930309-1	29	35.7(2745)	40.2(3091)	0.967(1485)	30	324±12	242± 9	OSU9608B
				weighted mean:		**328± 5**	**245± 4**	
Tardree Rhyolite (ED1)								
TDR 910807-1	20	61.3(1691)	95.4(2632)	5.25 (1344)	44	350±15	256±11	OSU9205
TDR 910807-2	15	70.4(2120)	19.2 (578)	1.08 (1661)	9	298±16	218±12	OSU9502B
TDR 910807-3	20	75.5(3616)	23.5(1127)	1.08 (1664)	<1	340±15	249±11	OSU9502B
TDR 910826-1	15	76.1(2579)	24.7 (839)	1.10 (1698)	33	349±17	255±12	OSU9509B
TDR 911024-1	10	72.3(1539)	25.0 (533)	1.10 (1700)	4	371±21	271±15	OSU9509B
TDR 920806-1	15	65.6(1836)	17.5 (489)	0.966(1483)	12	325±19	238±14	OSU9608B
				weighted mean:		**340± 7**	**249± 5**	
Mt. Dromedary Complex (ED1)								
MDC 900716-456	7	191(1638)	36.5 (313)	1.10 (1690)	67	346±23	253±17	OSU9502B
MDC 910510-2	9	203(1324)	37.2 (243)	1.10 (1694)	2	332±25	243±18	OSU9502B
MDC 910807-1	9	200(2521)	41.4 (522)	1.10 (1697)	32	374±20	274±15	OSU9502B
MDC 921127-1	13	196(3794)	37.9 (734)	1.11 (1702)	2	347±16	253±12	OSU9509B
MDC 921127-2	8	205(2030)	41.4 (411)	1.11 (1704)	18	363±22	265±16	OSU9509B
				weighted mean:		**354± 9**	**259± 7**	

for explanation of symbols see tables 6 and 7.

Green (1985) has pointed out four possible factors that may introduce differences in age-calibration baselines between zircon, sphene, and apatite when using the external detector method. The factors that may cause differences between methods and researchers, are discussed below.

1. *The influence of the minimum observable track length:* In our system, the minimum track length cut-off for the three minerals has been evaluated to be 0.5-1μm (Fig. 3), which could introduce only a small difference in the track counting efficiencies among the minerals. The small influence of this factor is corrected for when the geometry factors obtained in Experiment 1 are used. On the other hand, track identification criteria strongly depend on each researcher, and, in principle, the longer the minimum length cut-off, the larger the ζ-factor. A review of reported ζ-factors for the same apatite standard samples analyzed with the same technique shows that differences occur in the observed spontaneous track densities up to 10% and more. These differences match those in the reported ζ-factors between researchers taking into account the U-content (which is proportional to ρ_i/ρ_d values) observed for each system (Table 9). When the population method is used (apatite), the influence of the minimum length cut-off is expected to be negligible because the measurement of the spontaneous and induced track densities are carried out under identical conditions on the same material and with the same geometry factor.

TABLE 9. Comparison of fission track data of apatite age standards between Green (1985), Tagami (1987), and this study

Sample	Authors	n	ρ_s ($\infty 10^5$ cm^{-2}, ±2σ)	ρ_i/ρ_d (±2σ)	Zeta (±2σ)	normalized to this study ρ_s	ρ_i/ρ_d	Zeta
Fish Canyon Tuff	Green (1985)	12	1.68±0.08	1.07±0.05	357.7±14.2	0.75	0.91	1.23
	Tagami (1987)	4	1.95±0.18	1.11±0.05	317.3±19.0	0.87	0.94	1.09
	This study	7	2.25±0.09	1.18±0.04	292±12	1	1	1
Mt. Dromedary	Green (1985)	16	6.40±1.00	1.11±0.16	348.6±10.8	0.66	0.76	1.17
	Tagami (1987)	4	9.23±1.69	1.61±0.30	348.5±25.0	0.96	1.10	1.17
	This study	4	9.65±0.45	1.46±0.09	299±14	1	1	1

note: (1) n: number of measurements, (2) ρ_s, ρ_i/ρ_d: mean values for each report, (3) Zeta: sample weighted mean value for SRM612. (4) ρ_i, ρ_d: determined by mica external detector.

2. *The critical angle for fission track revelation:* The critical angles for zircon, sphene, and apatite were estimated from the ratio of the etchable track length to the amount of bulk etching of each mineral-surface (Table 10). The amount of bulk etching, was directly measured with a laser range-finder and/or estimated from the average diameters of the etch-pits in each mineral surface. We emphasize here that it is important to consider the dip angle distribution of semi-tracks to estimate the influence of the critical angle. As can be seen in Fig. 5, both theory and observation show that there are only a small number of near-horizontal and near-vertical semi-tracks in minerals. Based on the theory on the semi-track distributions (Dakowski, 1978), the relative amount of tracks which have shallower dip angles than the critical angle and which are totally erased by bulk

etching is calculated to be less than approximately 2% for the three minerals (Table 10).

TABLE 10. Critical angles and probability of fission tracks for having a dip angle < the critical angle

	Zircon	Sphene	Apatite
amount of bulk etching	~0.5 μm	~1 μm	~1 μm
etchable range	~5.35 μm	~6.55 μm	~7.9 μm
calculated critical angle	5.4_	8.8_	7.2_
probability of tracks less than critical angle	0.9%	2.3%	1.6%

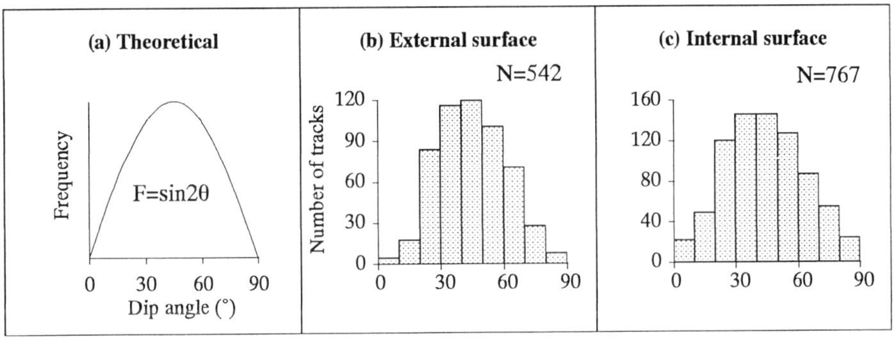

Fig. 5 Examples of dip angle distributions of surface-tracks. Theoretical model (a) is given by Dakowski (1978). Observed data are shown for (b) external and (c) internal surfaces of zircon. The data source is the same as in Fig. 3. Tracks less than 2 μm are omitted in Fig. 5c. *N* is the number of tracks used.

It seems clear that the loss of tracks due to bulk etching is small and cannot account for different ζ-factors among these minerals.

3. *Different registration threshold of the etchable range in a mineral:* we found that this factor causes 10-25% of the biases in geometry factors between zircon, sphene, and apatite when using the mica detectors as described in Section. 2. The geometry factors that we calculated with Eq. (4) were experimentally confirmed. Therefore, the difference in track registration efficiency among the minerals may have a direct impact upon the ζ-factors obtained from the external detector method with the conventional values of g (=0.5) and g'(=1). The biased ζ-factors can be appropriately corrected by introducing the geometry factors evaluated for each mineral/detector couple in Experiment 1. The ζ-factor of the population method is expected to approach the ideal ζ-factor most closely because it is free from the track registration (geometry factor) bias.
4. *The shortening of spontaneous tracks in apatite:* Even rapidly cooled apatite samples, including the age standards, have spontaneous tracks that are 5-10% shorter than induced tracks (Green, 1980; Gleadow et al., 1986). However, whether a track length correction of apatite ages is necessary remains a controversial issue (e.g. Green, 1988). Extrapolation of laboratory annealing studies has shown that the shortening of spontaneous tracks in apatite is produced by annealing under ambient temperatures over millions of years (Green *et al.*,

1986; Laslett *et al.*, 1987). Because track shortening may result in underestimates of the spontaneous track density or the age of the samples, track-length corrections should be done to compare ζ-factors of apatite with that of other minerals without track shortening. The confined track length ratio of spontaneous/induced tracks has been proposed as a measure for this correction (Green, 1988).

In conclusion, the conventional ζ-factors defined by Hurford and Green (1983) give baselines which depend primarily upon factor 1 (researcher or system), factor 3 (mineral and dating technique), and factor 4 (thermal track shortening). As shown in Fig. 4a, our experimental results clearly confirm the different ζ-baselines that were first found by Green (1985).

Considering the baseline corrected for the geometry factor bias (Fig. 4b and 4c), we calculated an overall mean ζ-factor of 252.6 and a standard deviation of 9.5 for the case with track length correction, and an overall mean ζ-factor of 261.1 and a standard deviation of 10.8 for the case without length correction. In our opinion, the value incorporating the track length correction is the most relevant as the age-calibration data become more consistent when this correction is taken into account.

Accordingly, the ideal age-calibration (absolute calibration with physical constants) should be carried out with the baseline corrected for all the biases. The overall weighted mean ζ-factor 253±3 (2σ) of all the mineral/detector couples against the SRM612 dosimeter serves as a reference value for the absolute age-calibration using both the population and external detector methods, in combination with an accurate determination of thermal neutron fluence.

4.3 ON THE DECAY CONSTANT OF ^{238}U FISSION

The value of the decay constant of ^{238}U spontaneous fission (λ_f) is of major importance in the absolute approach of FT age-calibration. As has been thoroughly documented by Bigazzi (1981), many determinations of λ_f did not lead to a unique value, but instead to two groups of values: clustering about 7.0 x $10^{-17}y^{-1}$ and 8.5 x $10^{-17}y^{-1}$ and differing by about 20%. The lower value is supported by most of the Solid-State Track Detector (SSTD) measurements, such as those obtained with the mica-U sandwich method (e.g. $7.03 \times 10^{-17}y^{-1}$: Roberts *et al.*, 1968), while the higher value is strongly supported by most of direct measurements such as those obtained with rotating bubble chambers (e.g. $8.46 \times 10^{-17}y^{-1}$: Galliker *et al.*, 1970).

Values of λ_f obtained by dating of samples of known age agree with either the lower or higher value, and before the ζ method was introduced, FT workers chose between the two for their age determinations. Sometimes different workers using different λ_f-values obtained similar age values for the same sample (e.g. Naeser *et al.*, 1981). Bigazzi (1981) pointed out that the difference in λ_f-values could be related to several factors: (1) neutron-fluence dosimetry, (2) dating techniques and (3) age correction required because of a thermal effect. Among these factors, the interdependence of uncertainties of thermal neutron fluence determination and λ_f-values has been especially emphasized (Hurford and Green, 1981; 1982).

In previous dating studies of well-known-age samples (Table 11), often the higher λ_f-value is supported by studies carried out with the population method, including the subtraction method for apatite or glass, and the lower λ_f-value results from the external detector method for zircon (or apatite). The two values of λ_f may thus depend on the kind of technique that is used for the analysis of the induced tracks.

The present study revealed approximately a difference of about 20% in track registration efficiency between zircon, apatite, sphene, and mica detectors. Such different registration efficiencies may be responsible for the different geometry factors among the minerals when using the external detector method, which in turn may have introduced differences into the age-calibration baselines in the past. We stress here that systematic differences in track registration efficiencies (or geometry factors) approximately counterbalance the difference between the two λ_f-values, with the result that both λ_f-values may yield correct ages depending upon the type of dating technique that is used. Nevertheless, if our line of thinking is continued, it leads to a unique value for λ_f-value that closely approaches $8.5 \times 10^{-17} y^{-1}$. Previous studies, as shown in Table 11, were based on various fluence dosimetries which have not been intercalibrated, as has been mentioned by Bigazzi (1981) but even from those studies one could derive that the true λ_f-value is the higher rather than the lower one, which is biased by three factors (dosimetry, dating technique, and age correction).

TABLE 11. List of dating of known-age samples and supported values of the spontaneous fission decay constant of ^{238}U

Investigator	Year	Material	Fission track technique	Value ($\times 10^{-17} y^{-1}$) 7.0	8.5
Fleischer and Price	1964	tektite etc.	Subtraction	○	
Storzer	1970	U-glass	Subtraction		●
Gentner et al.	1972	tektite etc.	Subtraction		●
Wagner et al.	1975	U-glass	Subtraction		●
Thiel and Herr	1976	U-glass	Subtraction		●
Hurford and Gleadow	1977	zircon	External detector	●	
		apatite	Population	○	
Naeser et al.	1977	zircon	External detector	●	
Naeser et al.	1981	zircon	External detector	●	
		apatite	External detector	○	
		apatite	Population		●
Hurford and Green	1983	zircon	External detector	●	
Van den haute et al.	1988	apatite	Population		○

note: Closed and open circles are data corrected (or needless) and non-corrected for thermal effect, respectively.

5. Summary and Conclusions

1. The geometry factors for fission track dating with external detectors are redefined as $(R+d)/2R$ for $2\pi/4\pi$, where R is the etchable range of a single fission fragment and d is the unetchable range at the end of a latent fission track in a mineral. Using this

equation and the result of track-to track matching studies, the ideal values of the different geometry factors for zircon, sphene, and apatite with reference to the mica detectors were calculated to be 0.66, 0.57, and 0.54, respectively.

2. Experimentally determined geometry factors amount to 0.68, 0.60, and 0.55 for zircon, sphene, and apatite, respectively and are consistent with the ideal values. The experimental values are mainly influenced by the lowered counting efficiencies of internal (4π) surfaces due to the minimum length cut-off for each system (or researcher). Length distributions of semi-tracks in 2π and 4π geometry surfaces were used to evaluate the significance of track identification criteria.

3. The different zeta-baselines among the minerals were re-investigated taking into account four possible factors: (1) the minimum observable length, (2) the critical angle, (3) different registration thresholds (or geometry factors), and (4) the shortening of spontaneous tracks. The correction of all the bias factors leads to the "ideal" ζ-calibration baseline of 253±3 (2σ) for SRM612 glass on which the absolute age-calibration is acceptable for the methods and the minerals.

4. Relating the above results to the problem of the decay constant of ^{238}U spontaneous fission it appears that. the lower value (λ_f=7x$10^{-17}y^{-1}$) which has been supported by dating of zircon with the external detector method may suffer a bias of approximately 20% due to the difference in track registration efficiency between zircon and mica. The higher value (λ_f=8.5x$10^{-17}y^{-1}$) that was mainly obtained by population method analyses is probably more accurate.

Acknowledgments

Special thanks are due to our colleagues, Mr. T. Yamashita and Dr. M. Kasuya, for their technical support. Critical comments from the two reviewers (Drs. P.F. Green and A.J. Hurford) were very valuable. Improvements of the manuscript by Mr. T. Yoshioka and Ms. C. Inoue were helpful. Encouragement by Dr. K. Watanabe was also helpful. Drs. H. Shioda, A.G. Johnson, and B. Dodd helped our thermal neutron irradiation in the reactor at the Oregon State University Radiation Center. Standard samples were kindly provided to us by A.J. Hurford (Buluk Member Tuff and Tardree Rhyolite zircons), P.F. Green (Mount Dromedary Complex zircon, sphene and apatite), A.J.W. Gleadow (Mount Dromedary Complex zircon), M. Kasuya (Durango apatite) and K. Kimura (Kawamoto Granodiorite zircon).

References

Bigazzi, G. (1981) The problem of the decay constant λ_f of ^{238}U, *Nucl. Tracks* **5**, 35-44.

Dakowski, M. (1978) Length distributions of fission tracks in thick crystals, *Nucl. Track Detection* **2**, 181-189.

Danhara, T., Iwano, H., Kasuya, M. and Yamashita, T. (1993) The PFA sheet: An improved mounting material for fission track analysis of zircon, *Nucl. Tracks Radiat. Meas.* **21**, 283-285.

Danhara, T., Kasuya, M., Iwano, H. and Yamashita, T. (1991) Fission-track age calibration using internal and external surfaces of zircon, *Jour. Geol. Soc. Japan* **97**, 977-985.

Fleischer, R. L. and Price, P. B. (1964) Decay constant for spontaneous fission of U^{238}, *Phys. Rev.* **133**, B63-B64.

Fleischer, R. L., Price, P. B. and Walker, R. M. (1975) *Nuclear Tracks in Solids: Principles and Applications*, University of California Press, Berkeley.

Galbraith, R. F. (1981) On statistical models for fission track counts, *Math. Geol.* **13**, 471-478.

Galbraith, R. F. (1984) On statistical estimation in fission track dating, *Math. Geol.* **16**, 653-669.

Galbraith, R. F. and Laslett, G. M. (1985) Some remarks on statistical estimation in fission-track dating, *Nucl. Tracks* **10**, 361-363.

Galliker, D., Hugentobler, E. and Hahn, B. (1970) Spontane Kernspaltung von ^{238}U und ^{241}Am, *Helv. Phys. Acta* **43**, 593-606.

Gentner, W., Storzer, D., Gijbels, R. and Van der Linden, R. (1972) Calibration of the decay constant of ^{238}U spontaneous fission, *Trans. Amer. Nucl. Soc. ANS* **15**, 125-126.

Gleadow, A. J. W. (1981) Fission-track dating methods: what are the real alternatives? *Nucl. Tracks.* **5**, 3-14.

Gleadow, A. J. W., Duddy, I. R., Green, P. F. and Lovering J. F. (1986) Confined fission track lengths in apatite: a diagnostic tool for thermal history analysis, *Contrib. Mineral. Petrol.* **94**, 405-415.

Gleadow, A. J. W. and Lovering, J. F. (1977) Geometry factor for external detectors in fission track dating, *Nucl. Track Detection* **1**, 99-106.

Green, P. F. (1980) On the cause of the shortening of spontaneous fission tracks in certain minerals, *Nucl. Tracks.* **4**, 91-100.

Green, P. F. (1985) Comparison of zeta calibration baselines for fission-track dating of apatite, zircon and sphene, *Chem. Geol. (Isot. Geosci. Sect.)* **58**, 1-22.

Green, P. F. (1988) The relationship between track shortening and fission track age reduction in apatite: combined influences of inherent instability, annealing anisotropy, length bias and system calibration, *Earth Planet. Sci. Lett.* **89**, 335-352.

Green, P. F., Duddy, I. R., Gleadow, A. J. W., Tingate, P. R. and Laslett, G. M. (1986) Thermal annealing of fission tracks in apatite, 1. A qualitative description, *Chem. Geol. (Isot. Geosci. Sect.)* **59**, 237-253.

Green, P. F. and Durrani, S. A. (1978) A quantitative assessment of geometry factors for use in fission track studies, *Nucl. Track Detection* **2**, 207-213.

Hurford, A. J. (1990) Standardization of fission track dating calibration: Recommendation by the Fission Track Working Group of the I.U.G.S. Subcommission on Geochronology, *Chem. Geol.(Isot. Geosci. Sect.)* **80**, 171-178.

Hurford, A. J. and Gleadow, A. J. W. (1977) Calibration of fission track dating parameters, *Nucl. Track Detection* **1**, 41-48.

Hurford, A. J. and Green, P. F. (1981) A reappraisal of neutron dosimetry and uranium 238 λ_t values in fission-track dating, *Nucl. Tracks* **5**, 53-61.

Hurford, A. J. and Green, P. F. (1982) A users' guide to fission track dating calibration, *Earth Planet. Sci. Lett.* **59**, 343-354.

Hurford, A. J. and Green, P. F. (1983) The zeta age calibration of fission-track dating, *Isot. Geosci.* **1**, 285-317.

Hurford, A. J. and Hammerschmidt, K. (1985) $^{40}Ar/^{39}Ar$ and K/Ar dating of the Bishop and Fish Canyon Tuffs: Calibration ages for fission-track dating standards, *Chem. Geol. (Isot. Geosci. Sect.)* **58**, 23-32.

Hurford, A. J. and Watkins, R. T. (1987) Fission-track age of the tuffs of the Buluk Member, Bakate Formation, northern Kenya: A suitable fission-track age standard, *Chem. Geol. (Isot. Geosci. Sect.)* **66**, 209-216.

Iwano, H. and Danhara, T. (1994) One-to-one correlation of fission tracks between sphene and mica detectors, *Abstracts of the Eighth International Conference on Geochronology, Cosmochronology and Isotope Geology.* P.154

Iwano, H. and Danhara, T. (1997) Zeta calibration of fission-track dating system using high-resolution monitor and comparison of reported zeta values, *Fission Track News Letter* no.10, in press.(in Japanese with English abstract).

Iwano, H., Kasuya, M., Danhara, T., Yamashita, T. and Tagami, T. (1993) Track counting efficiency and unetchable track range in apatite, *Nucl. Tracks Radiat. Meas.* **21**, 513-517.

Iwano, H., Kasuya, M., Yamashita, T. and Danhara, T. (1992) One-to-one correlation of fission tracks between zircon and mica detectors, *Nucl. Tracks Radiat. Meas.* **20**, 341-347.

Iwano, H., Yamashita, T. and Danhara, T. (1996) Three-dimensional analysis of fission track length in minerals -a measuring system and its application-, *Fission Track News Letter* no.9, 13-22.

Lakatos, S. and Miller, D. S. (1972) Problems of dating mica by the fission-track method, *Can. J. Earth Sci.* **10**, 403-407.

Laslett, G. M., Green, P. F., Duddy, I. R. and Gleadow, A. J. W. (1987) Thermal annealing of fission tracks in apatite, 2. A quantitative analysis, *Chem. Geol. (Isot. Geosci. Sect.)* **65**, 1-13.

Miller, D. S., Crowley, K. D., Dokka, R. K., Galbraith, R. F., Kowallis, B. J. and Naeser, C. W. (1993) Results of interlaboratory comparison of fission track ages for 1992 fission track workshop, *Nucl. Tracks Radiat. Meas.* **21**, 565-573

Miller, D. S., Eby, N., McCorkell, R., Rosenberg, P. E. and Suzuki, M. (1990) Results of interlaboratory comparison of fission track ages for the 1988 fission track workshop, *Nucl. Tracks Radiat. Meas.* **17**, 237-245.

Naeser, C. W. (1967) The use of apatite and sphene for fission track age determinations, *Geol. Soc. Am. Bull.* **78**, 1523-1526.

Naeser, C. W., Gleadow, A. J. W. and Wagner, G. A. (1979) Standardization of fission-track data reports, *Nucl. Tracks.* **3**, 133-136.

Naeser, C. W., Hurford, A. J. and Gleadow, A. J. W. (1977) Fission-track dating of pumice from the KBS Tuff, East Rudolf, Kenya, *Nature* **267**, 649.

Naeser, C. W., Izett, G. A. and Obradovich, J. D. (1980) Fission-track and K-Ar ages of natural glasses, *U. S. Geol. Surv. Bull.* No. 1489, pp. 1-31.

Naeser, C. W., Zimmermann, R. A. and Cebula, G. T. (1981) Fission-track dating of apatite and zircon: an interlaboratory comparison, *Nucl. Tracks.* **5**, 65-72.

Odin, G. S., Takahashi, M. and Cosca, M. (1995) $^{40}Ar/^{39}Ar$ geochronology of biostratigraphically controlled Miocene tuffs from central Japan: Comparison with Italy and age of the Serravallian-Tortonian boundary, *Chem. Geol.(Isot. Geosci. Sect.)* **125**, 105-121.

Reimer, G. M., Storzer, D. and Wagner, G. A. (1970) Geometry factor in fission track counting, *Earth Planet. Sci. Lett.* **9**, 401-404.

Roberts, J. H., Gold, R. and Armani, R. J. (1968) Spontaneous-fission decay constant of ^{238}U, *Phys. Rev.* **174**, 1482-1484.

Roberts, J. H., Ruddy, F. H. and Gold, R. (1984) Optical efficiency for fission fragment track counting in muscovite solid state track recorders, *Nucl. Tracks Radiat. Meas.* **8**, 365-369.

Steiger, R. H. and Jäger, E. (1977) Subcommission on Geochronology: Convention on the use of decay constants in geo- and cosmochronology, *Earth Planet. Sci. Lett.* **36**, 359-362.

Storzer, D. (1970) Spaltspuren des 238-Urans und ihre Bedeutung für die geologische Geschichte natürlicher Gläser, PhD Thesis, Universität Heidelberg.

Storzer, D. and Wagner, G. A. (1985) Bias in the application of the external detector technique, *Geol. Mag.* **122**, 195-196

Tagami, T. (1987) Determination of zeta calibration constant for fission track dating, *Nucl. Tracks Radiat. Meas.* **13**, 127-130.

Thiel, K. and Herr, W. (1976) The ^{238}U spontaneous fission decay constant re-determined by fission tracks, *Earth Planet. Sci. Lett.* **30**, 50-56.

Van den haute, P., Jonckheere, R. and De Corte, F. (1988) Thermal neutron fluence determination for fission-track dating with metal activation monitors: A re-investigation, *Chem. Geol. (Isot. Geosci. Sect.)* **73**, 233-244.

Wagner, G. A., Reimer, G. M., Carpenter, B. S., Faul, H., Van der Linden, R. and Gijbels, R. (1975) The spontaneous fission rate of U-238 and fission track dating, *Geochim. Cosmochim. Acta* **39**, 1279-1286.

Wagner, G. A and Van den haute, P. (1992) *Fission-Track Dating*, Enke, Stuttgart.

Wall, T. (1986) Use of an alternative neutron dosimetry standard for fission track dating, *Nucl. Tracks* **12**, 887-890.

A NEW U DOPED GLASS CERTIFIED BY THE EUROPEAN COMMISSION FOR THE CALIBRATION OF FISSION-TRACK DATING

F. DE CORTE[1], F. BELLEMANS[1], P. VAN DEN HAUTE[2], C. INGELBRECHT[3] and C. NICHOLL[3]

[1] *Institute for Nuclear Sciences, Gent University, Proeftuinstraat 86, B-9000 Gent, Belgium*
[2] *Geological Institute, Gent University, Krijgslaan 281, B-9000 Gent, Belgium*
[3] *Institute for Reference Materials and Measurements, Retieseweg, B-2440 Geel, Belgium*

Abstract

In this paper we present a new series of unirradiated and pre-irradiated uranium doped glass discs that were manufactured and characterized for use in fission-track analysis. They are issued as Nuclear Reference Material IRMM-540. The certified quantities are the uranium mass fraction [=13.9 mg$\times$kg^{-1}], the $n(^{235}U)/n(^{238}U)$ amount ratio [=0.007277] and the thermal Maxwellian neutron fluence [=1.070×10^{19} m^{-2}]. Additional information is given related to the nominal composition of the glass and its content of thorium and neutron absorbing elements, the uranium homogeneity between and within discs, the neutron fluence monitors, the parameters of the neutron spectrum in the irradiation facility, the thermal stability of the fission tracks in the glass and the properties of the mica that has been used as external detector in contact with the irradiated discs.

1. Introduction

In fission track geochronometry, uranium bearing glasses are commonly used for monitoring the thermal neutron fluence be-it explicitly (in the absolute approach) or implicitly (in the ζ-calibration). Except for some tektite specimens such as moldavite, natural glasses have not been found very suitable for this purpose due to uranium homogeneity and/or the presence of spurious etch pits. Hence, the artificially prepared uranium doped "standard" glasses became the favourite monitors of the fission-track geochronologist (we refer to Table 1 of Hurford in this volume for a list of monitor glasses that are currently in use).

Of these standard glasses, the SRM series 610-617 with nominal uranium contents ranging from 0.02 to 500 ppm, manufactured by NIST (NBS) are the only ones that have been distributed as pre-irradiated discs with a calibrated neutron fluence (serial numbers 961 to 964). Unfortunately, the inconsistency in the neutron fluences quoted for the different metal monitors that were used (mainly Au and Cu) made them less attractive as fluence monitors (Hurford and Green, 1982). In addition, the SRM glasses have also some compositional characteristics which makes tham somewhat less suitable (Crowley,

P. Van den haute and F. De Corte (eds.), Advances in Fission - Track Geochronology, 67-78.

1986; Tagami and Nishimura, 1989, 1992): they are doped with a variety of elements, some of them giving rise to neutron self-shielding upon irradiation and secondly, they are depleted in ^{235}U and contain a significant amount of Th, augmenting the risk of creating fission tracks other than ^{235}U tracks in poorly thermalized reactor facilities. A comparative re-investigation of SRM and the other commonly used CN (Corning) glasses focused on these problems has been reported recently by Bellemans et al. (1995).

Since the International Workshop on Fission Track Dating held in Pisa in 1980, the use of pre-irradiated SRM glasses has been gradually abandoned in fission track dating, together with absolute neutron fluence monitoring in general. Almost simultaneously, uranium doped glasses were given a new role in the ζ–calibration as implicit fluence monitors linking the age standards to the unknowns. When it was demonstrated (Van den haute et al., 1988; De Corte et al., 1991) that thermal neutron fluence determination does lead to accurate and consistent results on condition that the principles of modern neutron metrology are followed (implying the use of the internationally recommended monitors Au and Co), a way was opened to a re-assessment of the absolute approach in the calibration of fission track dating. Taking into account that, at present, the majority of fission track geochronologists use the ζ-calibration system and rely on uranium doped glasses, we realized that such a re-assessment can only be achieved if new, properly designed uranium doped calibration glasses were made available.

In this work we present such a new glass manufactured by the Institute of Reference Materials and Measurements and pre-irradiated with a calibrated neutron fluence in the Thetis reactor of the University of Gent. As a simple uranium doped glass the unirradiated IRMM discs can be used in a straightforward manner in a ζ-calibration procedure. The track densities of 10^5-10^6 cm^{-2} that are induced in the IRMM glass by the neutron fluences that are typically applied in fission track dating, are comparable to those in SRM 612 and CN5 glass. In addition, the pre-irradiated discs allow the introduction of an explicit value for the fluence in the ζ-equation and eventually this may provide a first step in reducing the number of parameters that make up the ζ-factor.

2. The preparation of the IRMM glass and its nominal composition

The glass was prepared at the Institute for Reference Materials and Measurements of the European Commission (IRMM, Geel, Belgium). High purity starting materials in powder form (SiO_2, Na_2CO_3, $CaCO_3$, Al_2O_3, $CaSO_4$ and UO_2) were blended in a mechanical shaker for 5 h.The melting (carried out by the Institut National du Verre, Charleroi, Belgium) was done at 1500 °C in batches of 500 g using platinum crucibles. Each batch was heated for 4 h and homogenized by casting into water to form granules and remelting.This was carried out three times and the glass was finally cast into Inconel moulds preheated to 400 °C and then annealed at 580 °C. The product was 1 kg glass as rods, twelve of which were sliced using a diamond wheel into circular discs of 16 mm diameter and 2 mm thickness.

The nominal composition of the glass was 75% SiO_2, 14% Na_2O, 8% CaO, 3% Al_2O_3 and 15 mass ppm natural uranium. The density of the glass was 2.460 kg.m^{-3}.

3. Characterization of the glass wafers

3.1. THE CERTIFIED URANIUM CONTENT

The uranium content (mass fraction) being a quantity to be certified, three methods were used for its determination: neutron activation analysis (NAA), inductively coupled plasma-mass spectrometry with isotope dilution (ICP-IDMS) and thermal ionization mass spectrometry with isotope dilution (TI-IDMS).

For NAA, 36 glass discs were analysed originating from both ends and from the middle of each of the 12 rods available. The samples, sandwiched between U-standards [0.1 mm thick Al-(0.2±0.004)%U foils (IRMM SP 91005)], were irradiated under Cd cover (to suppress the activities of the matrix elements) in the Thetis reactor of the Gent University. After about one week cooling, activity measurements of the ^{239}U --> ^{239}Np daughter nuclide were performed both on a 38% efficiency high purity Ge detector (for the 228.1 and 277.6 keV gamma-lines) and on an low energy photon detector (for the 106.1 keV gamma-line). In all cases, the counting statistics were of the order of 0.5%. In former work, the accuracy of this determination method was demonstrated by analysing the NIST SRM glasses 611, 613, 615 and 617 (Bellemans et al., 1995).

For the mass spectrometry determinations, eleven glass discs were spiked with a 93.3 (atom)% ^{235}U reference material (NBS SRM U-930) to yield an isotopic mixing ratio, $n(^{235}U)/n(^{238}U)$, of approximately 1. Dissolution was achieved using a combination of HNO_3 and HF. An aliquot of each solution was made up in 0.14M subboiled HNO_3 and analysed by ICP-MS (Fisons VG plasmaquad 2+ with photomultiplier detection) with corrections for mass discrimination made by measuring the reference material, NBS SRM U-500.

Purification of the uranium from silica, sodium, calcium and aluminium salts was then carried out by separation over freshly prepared and conditioned AG1-X4 (Biorad) and U/TEVA (Eichrom, Inc) supported extractant chromatographic columns. TIMS measurements were then performed on 6 solutions using a Finnegan MAT 260m with K-factor correction using the reference material IRMM-187. Calculation of the IDMS results incorporated all known uncertainties in the traceability chain.

The results of the three determination methods were as follows (with expanded uncertainties including a coverage factor k=2):

NAA	(14.10 ± 0.22) mg×kg^{-1}	(n = 36)
ICP-IDMS	(13.76 ± 0.33) mg×kg^{-1}	(n = 11)
TI-IDMS	(13.81 ± 0.18) mg×kg^{-1}	(n = 6)

leading to a certified uranium mass fraction of (13.9 ± 0.5) mg×kg^{-1}, which is the mean of the three mean values. The uncertainty is the expanded uncertainty $U = ku$, with U

determined from a standard uncertainty u = 0.11 mg×kg^{-1} and a coverage factor k = 4.3 based on the distribution n = 2 degrees of freedom, and defines an interval estimated to have a level of confidence of 95 percent.

3.2. THE URANIUM HOMOGENEITY

The U homogeneity between discs was verified by the experiments described above for the determination of the uranium mass fraction via epicadmium NAA. Normalization of the specific count rates obtained for the 106.1, 228.1 and 277.6 keV ^{239}Np gamma-lines was performed versus the specific count rates, measured with the HPGe detector, for the gamma lines of ^{95}Zr (724.2 and 756.7 keV) and ^{97m}Nb (743.3 keV) emitted by 0.127 mm thick Zr foils sandwiching the glass discs during irradiation. As seen in Figs 1 and 2, the relative standard uncertainty of uranium concentration was found to be 0.8% when considering the 36 discs, and 0.6% when considering the 12 rods.

The U homogeneity within discs was verified by fission track counting on four discs (taken from rods nrs. 1, 5, 8 and 11), which were irradiated with a thermal neutron fluence of »1.2×10^{19} m^{-2} in channel 8 of the Thetis reactor. After polishing and etching for 1 min in 16% HF at 25°C, microscopic track counting was performed under transmitted light at 625′ magnification. From the results of Table 1, it can be seen that the track density follows a Poisson distribution, indicative of a homogeneous uranium distribution. Also, the last column giving the track density normalized to the thermal fluence (as measured via a coirradiated Al-0.1%Au alloy - see below), again demonstrates the homogeneity between discs.

In general, it can be concluded from the above that, either between or within discs, any heterogeneity in the uranium concentration is within the certified range.

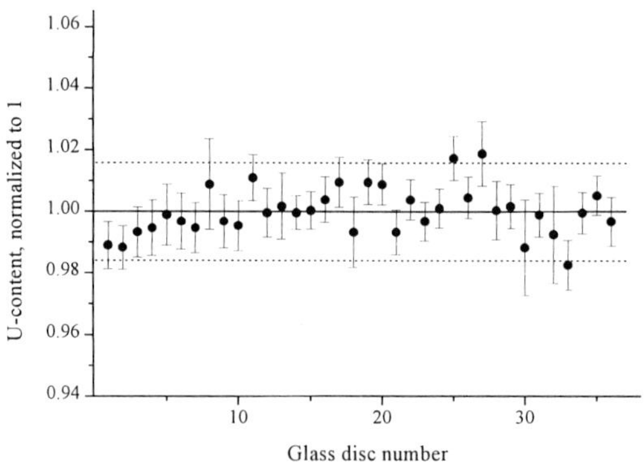

Figure 1. Results of the uranium homogeneity analysis between discs (epicadmium NAA). Thirty six discs i.e. three discs from each of the twelve rods were analyzed. The dotted horizontal lines correspond to the 2s deviation of all measurements.

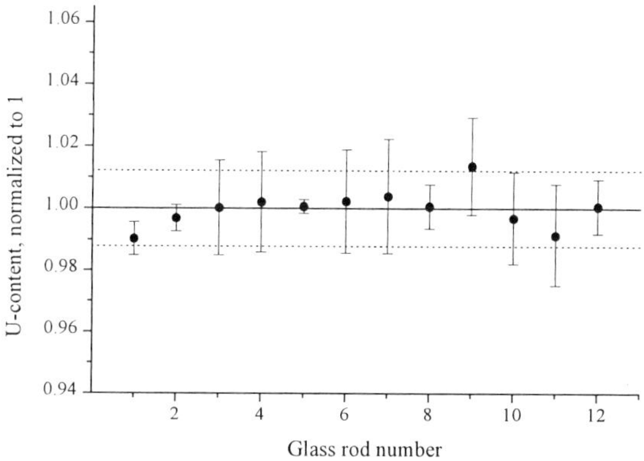

Figure 2. Uranium homogeneity beween the twelve rods (epicadmium NAA). The dotted horizontal lines correspond to the 2s deviation of all measurements.

Table 1: Determination of the U-homogeneity within the discs, by means of fission track counting.

rod	N	ρ	C_E, %	C_p, %	Φ ($x10^{19}$)	ρ/Φ ($x10^{-18}$)	ρ/Φ*
1	6079	76.0	10.6	11.5	1.161	6.58 ± 0.11	0.993 ± 0.017
	6151	76.9	11.2	11.4			
5	6124	76.6	11.2	11.4	1.164	6.61 ± 0.11	0.998 ± 0.017
	6194	77.4	10.5	11.4			
8	6132	76.7	10.8	11.4	1.159	6.61 ± 0.11	0.998 ± 0.017
	6119	76.5	10.5	11.4			
11	2295	76.5	12.5	11.4	1.146	6.70 ± 0.19	1.011 ± 0.028
	4465	77.0	12.1	11.4			

N is the number of counted fission tracks; ρ is the mean number of tracks per unit of counting area (0.16 x 0.16 mm^2); C_E is the experimental variation coefficient of the track density; C_p is the Poisson variation coefficient; Φ is the thermal neutron fluence ($n.m^{-2}$); ± indicates the 2s standard deviation.

* : normalized to 1

3.3. THE CERTIFIED URANIUM ISOTOPIC COMPOSITION

The $n(^{235}U)/n(^{238}U)$ amount ratio was determined by TIMS, calibrated for mass fractionation using the certified reference material IRRM-187 and was certified as 0.007277 ± 0.000007, where the uncertainty is an expanded uncertainty with a coverage factor $k = 2$.

3.4. THE CONTENT OF THORIUM AND NEUTRON-ABSORBING ELEMENTS

Th and the neutron absorbing elements B, Cd, Sm, Eu, Gd, Dy, Ir and Hg (elements with an absorption cross-section for thermal neutrons larger than 300 barn) are unwanted impurities in the glasses, because - if present in too high concentrations - they can lead to fission tracks when irradiated in an insufficiently thermalized neutron spectrum and to thermal neutron self-shielding, respectively. Therefore, their content was determined, using neutron activation analysis and cyclotron induced charged particle activation analysis (for B).

As to NAA, the k_0-standardization method was applied (De Corte et al., 1987), making use of Al-(0.1±0.002)%Au alloy (IRMM-530 certified reference material [Ingelbrecht et al., 1991]) as the single-comparator. Short (10 min) and long irradiations (7 h, involving bare and Cd-covered samples) were performed in several channels of the Thetis reactor, at thermal neutron fluence rates ranging from 1.7×10^{15} to 1.6×10^{16} $m^{-2}s^{-1}$, and were followed by several measurements with Ge(Li) and LEPD detectors after cooling times ranging from a few minutes to a few weeks.

The boron concentration was determined by means of CPAA using the reactions $^{10}B(p,\alpha)^{7}Be$ and $^{10}B(d,\alpha n)^{7}Be$ [T½=53.3 d; E_γ=477.6 keV]. The glass discs were irradiated in the cyclotron of the Institute for Nuclear Sciences, Gent with protons of 10.5 MeV and with deuterons of 12.9 MeV. Using these energies only the reactions $^{7}Li(p,n)^{7}Be$ and $^{7}Li(d,2n)^{7}Be$ had to be considered as interferences, for which a correction was derived from irradiation of pure Li_2CO_3 and H_3BO_3. The latter was also used as the standard material in the actual analysis.

The reliability of the above analysis procedures was formerly demonstrated by analysing the NIST SRM glasses 611, 613, 615 and 617 as reported by Bellemans et al., 1995. For full details on the methodologies applied, the reader is referred to this paper and references therein.

It was found that the concentrations of all neutron absorbing elements were below the detection limit of the applied methods, i.e. (in mg×kg^{-1}): B á 3.4, Cd á 0.55, Sm á 0.020, Eu á 0.011, Gd á 0.79, Dy á 0.21, Ir á 4.1 ´ 10^{-4} and Hg á 0.55. This is in all cases lower than the concentrations leading to 0.1% thermal neutron self-shielding effect. In fact, in the glass discs this effect was estimated to be about 0.35% caused by its major constituents.

Also the concentration of Th is below the detection limit, i.e. á 0.026 mg×kg^{-1}. This ensures a pure U fission population even in poorly thermalized neutron spectra, comparable to the Corning CN5 glasses which, among the formerly available SRM and CN glasses, were found to be of superior quality (Bellemans et al., 1995).

3.5. THE THERMAL STABILITY

Because nowadays, the induced tracks are often analyzed in an external detector, a detailed investigation of the thermal track stability in the IRMM glass has not been carried out. Nevertheless, an annealing experiment has been performed that allows a first evaluation. The experiment involved temperatures between 85 and 230 °C and heating times of 60, 480 and 6000 minutes. In addition, an ageing test was performed over a one-year period at 25 °C. The Arrhenius diagram derived from these experiments is shown in Fig. 3. It is comparable to that of the NIST SRM glasses (Carpenter, 1984; Reimer et al., 1972). It should be remarked that here the long term thermal stability of the fission tracks in the SRM glasses turns out to be better than predicted by the original annealing experiments (Bigazzi, 1993). Nevertheless, taking into account that our results indicate minor track fading for long term storage at room temperature, it is recommended that the new reference samples are stored in a refrigerator at 4 °C.

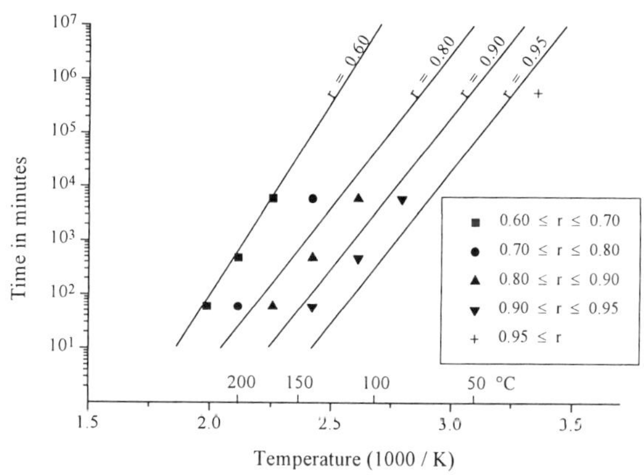

Figure 3. Preliminary Arrhenius diagram for track retention in the IRMM 540 glass (r= retention coefficient).

4. Pre-irradiation with thermal neutrons

4.1. THE CHOICE OF THE THERMAL NEUTRON FLUENCE MONITOR

For the pre-irradiation of the glasses, the thermal neutron fluence was determined with the aid of activation monitors prepared from the certified reference material IRMM-530

Al-(0.1±0.002)%Au as 0.1 mm thick foils which were used as a sandwich for every glass disc. In these foils, thermal and epithermal neutron self-shielding are negligible (Ingelbrecht et al., 1991). The fluence values obtained were verified using the certified reference material IRMM-528 Al-(1.0±0.02)%Co as 0.5 mm diameter wire, for which the epithermal neutron self-shielding correction factor is 0.970 (De Corte et al., 1991) . Corrections were applied to account for the contribution of epithermal neutrons to the activation of the monitors, taking into account the relevant parameters of the neutron spectrum in the chosen irradiation facility (channel 8 of reactor Thetis; see below) and using the nuclear data listed in Table 2. It should be noted that the thermal (Maxwellian) neutron fluence is defined, according to the Westcott formalism, as $n_m v_0 t_i$, where n_m is the Maxwellian neutron density in m^{-3}, $v_0 = 2200\ m \times s^{-1}$ and t_i is the irradiation time in s. For a review of the principles and parameters of this formalism, the reader is referred to De Corte et al., 1994.

Table 2: Nuclear data used for the Au and Co neutron fluence monitors.

	$^{97}Au\ (n,\gamma)\ ^{198}Au$	$^{59}Co\ (n,\gamma)\ ^{60}Co$
Molar mass ($g.mol^{-1}$)	196.96654	58.9332
Isotopic abundance (%)	100	100
Half life	2.695 d	5.271 a
Gamma intensity	411.8 keV: 95.56%	1173.2 keV: 99.88% 1332.5 keV: 99.982%
σ_0(2200 $m.s^{-1}$ cross section, barns)	98.65	37.13
s_0 (Westcott formalism)	17.24	1.765
E_r (effective resonance energy, eV)$^+$	5.7	136
$g(T_n = 21^{\circ}C)^{++}$	1.0066	1

$^+$ for the conversion of s_0 to the working value $s_0(\alpha)$ [$=s_0.E_r^{-\alpha}.1eV^{\alpha}$, with α the epithermal shape factor (see below)].

$^{++}$ Maxwellian neutron temperature in channel 8 of the Thetis reactor (see below).

The method that has been followed corresponds to the one explained in Van den haute et al (1988) and De Corte et al. (1991). Recent investigations carried out in our laboratories (De Corte et al., 1995; Van den haute et al., this volume) show that the monitors used provide accurate measurements of the induced ^{235}U fission rate. The correctness of our methodology is further substantiated by the results of a similar study in a poorly thermalized neutron spectrum (channel 3 of reactor Thetis), where the correction for activation with epithermal neutrons was much more critical (Bellemans, 1996).

4.2. CALIBRATION OF THE GE DETECTOR

The activity measurements of the co-irradiated Al-Au and Al-Co monitors were performed by a 38% efficiency (153 cm^3 active volume) high-purity germanium detector ("AMANDA") with a resolution (FWHM) of 1.77 keV at 1332.5 keV. For dead time and pulse pile-up correction, use was made of a loss-free counting module (Westphal, 1982), which was upon testing found to give results with an accuracy better than 0.5% up to count rates of 50 kHz. Other characteristics such as the stability of the MCA as a function of time and ambient temperature were also controlled and found to be excellent (Bellemans, 1996).

All measurements were performed at 28.9 cm source-detector separation. For this distance the full-energy peak detection efficiency ε_p (needed in calculating the fluence) was measured with absolutely calibrated point sources of ^{22}Na, ^{57}Co, ^{60}Co, ^{65}Zn, ^{109}Cd, ^{133}Ba, ^{134}Cs, ^{152}Eu, ^{226}Ra and ^{241}Am, complemented with relative point sources of ^{110m}Ag and ^{182}Ta. From the thus obtained detection efficiency curve [log ε_p vs log E_γ] the ε_p-values in the gamma-energy range of interest (411.8 keV for ^{198}Au; 1173.2 and 1332.5 keV for ^{60}Co) could be derived with an accuracy of 1-2%. The resulting "point" ε_p-values were then, without significant loss of accuracy, converted to "actual geometry" ε_p-values (thin discs for Al-Au; spirally rolled up wires for Al-Co) by means of the computer code SOLANG based on solid angle calculation (Moens et al., 1981). The accuracy of this efficiency conversion procedure was experimentally established for detector "AMANDA" (De Wispelaere et al., 1992). Finally, in the case of the ^{60}Co measurements, a small correction (of a few tenths of a percent only, in view of the large distance from the detector) was made for true-coincidence effects of the cascading 1173.2 and 1332.5 keV gamma rays.

Spectrum deconvolution and peak area determination was performed with the computer code SAMPO-90 (Aarnio et al., 1988), the choice of which was based on a performance test of several commercially available software packages (Bellemans et al., 1996).

4.3. CALIBRATION OF THE IRRADIATION FACILITY AND THE CERTIFIED THERMAL NEUTRON FLUENCE

The irradiation facility chosen was channel 8 of the Thetis reactor, Gent (250 kW; 5% ^{235}U enrichment; light water/graphite moderated). The parameters of the neutron spectrum (in the Westcott convention) were experimentally determined as outlined in De Corte et al., 1995 and were found to be:

- Maxwellian neutron temperature $T_n = (21.0 \pm 8.2)\ ^{\circ}C$;
- α ($1/E^{1+\alpha}$ epithermal neutron spectrum) $= 0.070 \pm 0.011$;
- thermal/fast neutron fluence rate = 120;
- spectral index r (modified for $\alpha = 0.070$) $= (5.40 \pm 0.18)\times10^{-3}$;

thus guaranteeing a thermalization such that the epithermal neutrons contribute less than 1% to the ^{235}U fission reaction rate and that ^{238}U fission is negligible.

When pre-irradiating the glasses in channel 8, care was taken to minimize the radial fluence gradient, which normally amounts to 6.3%/cm, i.e. some 10% over the diameter of a glass disc. This was done by rotating the packed container during the 2 hours irradiation period. In this way, the residual fluence gradient over the diameter of the glass was, in all cases, less than 2%. On the other hand, no significant axial fluence gradient was observed, so that all (11) glasses (with their adjacent mica detectors, Al-Au monitor sandwiches and Al-Co control monitors) stacked in one container received the same neutron fluence during irradiation. Eventually, the reactor power during the (3) different irradiations could be maintained very constant such that here too no significant differences in thermal neutron fluence could be observed. This made it possible to certify one and the same thermal (Maxwellian) neutron fluence for all pre-irradiated glasses: $(1.070 \pm 0.011)\times10^{19}\ m^{-2}$. The uncertainty quoted is the expanded uncertainty with a coverage factor $k = 2$.

4.4. THE CHOICE AND CHARACTERISTICS OF THE EXTERNAL MICA DETECTOR

The glass discs were pre-irradiated in close contact with a muscovite mica foil as the external detector, for which the Clear Ruby type (0.15 mm thick) of Goodfellow[1] (K603150) was selected. This choice was based on the low contents of U and Th in this material, as found by epicadmium NAA: U á 75 $\mu g\times kg^{-1}$ and Th á 30 $\mu g\times kg^{-1}$. Microscopic investigation showed that in these mica foils, both unirradiated and irradiated (with the same thermal neutron fluence as for the pre-irradiation of the glasses) the fission track density is negligible compared to the one that is induced by the glass. Furthermore, the optical quality of the mica was found to be quite satisfactory.

5. Nuclear reference material IRMM-540 and its certified quantities

The here described uranium-doped fission-track dating calibration glass is issued by the Commission of the European Communities as Nuclear Reference Material (NRM) IRMM-540[2]. Each reference material set consists of three glass discs diameter 16 mm

(1) Goodfellow Cambridge Limited, Cambridge Science Park, Cambridge, CB4 4DJ, England.
(2) Institute for Reference Materials and Measurements, Retieseweg, B-2440 Geel, Belgium.

x 2 mm and a mica foil. One of the glass discs in each set has been irradiated in contact with the mica foil in a thermal neutron field. The certified quantities are:

- the uranium mass fraction: (13.9 ± 0.5) mg$\times$kg^{-1}
- $n(^{235}U)/n(^{238}U)$: 0.007277 ± 0.000007
- thermal (Maxwellian) neutron fluence: $(1.070 \pm 0.011)\times10^{19}$ m^{-2}

In view of its certified U content, the NRM may also be used for uranium determinations in geological or other samples using the same method.

Acknowledgments

The technical assistance of Antoine De Wispelaere (Gent) is gratefully acknowledged. Thanks are due to the National Fund for Scientific Research for financial support (FDC, Pvdh). The authors are also indebted to Karel Strijckmans and Jacques De Waele for the boron determinations.

References

Aarnio P.A., Routti J.T. and Sandberg J.V. (1988): Microsampo-Personal computer based advanced gamma spectrum analysis system. *J. Radioanal. Chem.* **124**, 457-466.

Bellemans F., De Corte F. and Van den haute P. (1995): Composition of SRM and CN U-doped glasses: significance for their use as thermal neutron fluence monitors in fission track dating. *Radiation Measurements* **24/2**, 153-160.

Bellemans F. (1996): Een nucleair-analytische bijdrage tot de absolute kalibratie van de fissiesporen-dateringsmethode. Ph D Thesis, University of Gent, 177 pp.

Bigazzi G. (1993): oral communication.

Carpenter S.B. (1984): Standard reference materials: calibrated glass standards for fission track use. *Natn. Bur. Stand. Spec. Publ.* 260 292.

Crowley K.D. (1986): Neutron dosimetry in fission-track analysis. *Nucl. Tracks Radiat. Meas.* **11**, 237-243.

De Corte F., Simonits A., De Wispelaere A. and Hoste J. (1987): Accuracy and applicability of the k_0-standardization method. *J. Radioanal. Nucl. Chem.* **113**, 145-161.

De Corte F., Van den haute P., De Wispelaere A. and Jonckheere R. (1991): Calibration of the fission-track dating method: is Cu useful as an absolute thermal neutron fluence monitor? *Chem. Geol. (Isot. Geosci. Sect.)*, **86**, 187-194.

De Corte F., Bellemans F., De Neve P. and Simonits A. (1994): The use of a modified Westcott-formalism in the k_0-standardization of NAA: the state of affairs. *J. Radioanal. Nucl. Chem.*, Articles **179**, 93-103.

De Corte F., Van den haute P. and Bellemans F. (1995). The use of uranium doped glasses in fission-track dating. *Radiation Measurements* **25/1-4**, 511-516.

De Wispelaere A., Bellemans F. and De Corte F. (1992): Tuning the geometric parameters of a Ge detector for accurate efficiency calculation via program SOLANG: demonstration with detector AMANDA. *Proceed. Intern. k_0 Users Workshop - Gent, 1992* (F. De Corte, editor), INW/RUG, 45-48.

Hurford A.J. and Green P.F. (1982). A user's guide to fission track dating calibration. *Earth Planet. Sci. Lett.* **59**, 343-354.

ENDF/B-VI (1991): ENDF/B Summary Documentation, fourth edition (ENDF/B-VI). BNL-NCS-17541, ENDF-201, National Neutron Cross-Section Center, Brookhaven National Laboratory, USA.

Holden N.E. (1991): Temperature dependence of the Westcott g-factor for neutron capture reactions. In: ENDF/B-VI. BNL-46465 Report, 36th IUPAC general assembly, Hamburg, Germany.

Ingelbrecht C., Peetermans F., De Corte F., De Wispelaere A., Vandecasteele C., Courtijn E. and D'Hondt P. (1991): Aluminium-gold reference material for the k_0-standardization of neutron activation analysis. *Nucl. Instr. Meth. Phys. Res.* **A303**, 119-122.
Moens L., De Donder J., Lin Xilei, De Corte F., De Wispelaere A., Simonits A. and Hoste J. (1981): Calculation of ε_p for different counting geometries. *Nucl. Instr. Meth.* **187**, 451-472.
Reimer G.M., Wagner G.A. and Carpenter B.S. (1972): The thermal stability of fission tracks in the standard reference material glass standard (National Bureau of Standards). *Radiation Effects*, **15**, 273-274.
Tagami T. and Nishimura S. (1989): Intercalibration of thermal neutron dosimeter glasses NBS-SRM612 and Corning 1 in some irradiation facilities: a comparison. *Nucl. Tracks Radiat. Meas.* **16**, 11-14.
Tagami T. and Nishimura S. (1992): Neutron dosimetry and fission-track age calibration insights from intercalibration of uranium and thorium glass dosimeters. *Chem. Geol. (Isot. Geosci. Sect.)* **102**, 277-296.
Van den haute P., Jonckheere R. and De Corte F. (1988): Thermal neutron fluence determination for fission track dating with metal activation monitors: a re-investigation. *Chem. Geol. (Isot. Geosci. Sect.)* **73**, 233-244.
Wagner G.A. and Van den haute P. (1992): *Fission-Track Dating*. Enke Verlag - Kluwer Academic Publishers, 285 pp.
Westcott C.H. (1962): Effective cross section values for well-moderated thermal reactor spectra. In: Report CRRP-960 of the Atomic Energy of Canada Limited [Nov. 1, 1960 (reprinted 1962)].
Westphal G.P. (1982): Real-time correction of counting losses in nuclear pulse spectroscopy. J. *Radioanal. Chem.* **70**, 387-410.

Section II

Track Annealing

URANIUM-235 FISSION TRACK ANNEALING IN MINERALS OF THE APATITE GROUP: AN EXPERIMENTAL STUDY

J. CARPÉNA

Commissariat à l'Energie Atomique, DCC/DESD
CE Cadarache, St Paul lez Durance, France

Abstract

The name Apatite represents a large family of minerals with general formula $Me_{10}(XO_4)_6Y_2$. Natural apatites are rarely pure and numerous substitutions are encountered on the cation site Me, on the tetrahedra sites (XO_4) and on the anion site Y.
Laboratory isothermal annealing experiments have been performed on thermal neutron induced fission tracks of uranium in apatites of different compositions.
Our data suggest that heavily substituted apatites on the cation site are less resistant to thermal fading of fission tracks than pure phospho-calcic apatites, and confirm that Cl-apatites are more resistant to thermal annealing than F-apatites, as pointed out by Green et al. (1986). Our important finding is that we observe an inverse correlation between the number of substitutions on both cation and anion sites. This may imply that instead of the Cl/F-ratio, the number of substitutions on the cation site is directly responsible for the different thermal stabilities of fission tracks in different apatites.

1. The Apatitic Structure: a Host Structure

Natural minerals provide a means of assessing the long-term performance of actinide host phases in high-level nuclear waste forms. Based on geological observations, the idea for using a nuclear waste disposal medium based on a synthetic analogue of natural mineral is being evaluated at the French Direction du Cycle du Combustible (Commissariat à l'Energie Atomique) (Carpéna, 1992a,1996; Carpéna et Lacout, 1993; Boyer et al.,1997).

Minerals of the apatite structural group form a large family of compounds, natural or synthetic isomorphs that generally crystallize in the hexagonal system (spatial symetry group $P6_3/m$). They have the general chemical formula $Me_{10}(XO_4)_6Y_2$, where Me generally represents a bivalent cation, XO_4 a trivalent anion and Y a monovalent anion. A well known representative term is the phospho-calcic fluorapatite $Ca_{10}(PO_4)_6F_2$.

One of the main characteristics of the apatite structure is that it allows a large number of substitutions at all three sites (Table 1). Some of these substitutions can be realised without any change of cationic or anionic charge. For example, in fluorapatite, the fluoride ions can be substituted with chlorine ions to form chloro-apatite $Ca_{10}(PO_4)_6Cl_2$; the $(PO_4)^{3-}$ ions can also be substituted with $(VO_4)^{3-}$ ions to form

P. Van den haute and F. De Corte (eds.), Advances in Fission - Track Geochronology, 81-92.

vanadinite $Ca_{10}(VO_4)_6Cl_2$; the Ca^{2+} ions can be substituted with Pb^{2+} ions and the F^- ions with Cl^- ions to form pyromorphite $Pb_{10}(PO_4)_6Cl_2$.

Table 1: Main substitutions in the apatitic structure.

10Me:	Na^+,K^+ Ca^{2+}, Sr^{2+}, Ba^{2+}, Cd^{2+},Mn^{2+},Zr^{2+},Pb^{2+},Mg^{2+}, REE^{3+},Al^{3+},U^{4+}, Pu^{4+}
6(XO4):	$(PO_4)^{3-}$,$(VO_4)^{3-}$, $(AsO_4)^{3-}$, $(SiO_4)^{4-}$, $(CO_3)^{2-}$, $(SO_4)^{2-}$, $(CrO_4)^{3-}$, $(HPO_4)^{3-}$
$2Y^-$:	F^-, Cl^-, Br^-, I^-, OH^-, O^{2-}, $(CO3)^{2-}$, S^{2-}, vacancy

It is also possible to substitute the bivalent cation with a monovalent (Na^+, K^+) or trivalent cation (Ln^{3+}, Ga^{3+}, etc.where Ln = Rare Earths = Lanthanides), the trivalent anion $(PO_4)^{3-}$ with a divalent ($CO_3{}^{2-}$), $(HPO_4)^{2-}$ or tetravalent $(SiO_4)^{4-}$, $(SeO_4)^{4-}$ anion, and the monovalent anion (F^-) with a divalent anion ($CO_3{}^{2-}$), S^{2-}, O^{2-}) or a vacancy.

The charge balance is satisfied through coupled substitutions such as $2Ca^{2+}$ == (Ln^{3+}, Na^+), or (Ca^{2+}, $(PO_4)^{3-}$)==(Ln^{3+}, $(SiO_4)^{4-}$) . Another way is to change the anionic charge number in the channel as the following coupled substitution:
(Ca^{2+}, F^-) == (Ln^{3+}, O^{2-}) .

The mineral obtained through (Ca^{2+}, $(PO_4)^{3-}$)==(Ln^{3+}, $(SiO_4)^{4-}$), is designated as Britholite. Silicated apatites are also found in nature. In fact, a solid solution exists between phospho-calcic apatite and britholite (Boyer and al., 1996) but natural silicated apatites rarely contain six $(SiO_4)^{4-}$ ions per molecule. Natural silicated apatites are known to incorporate many lanthanide and actinides into their lattice structure.
Minerals of the apatite group are characterized by:
- a high resistance to chemical corrosion in neutral to alkaline environments,
- the ability to accept substitutions from a large array of elements,
- their potential for restoring self-irradiation damages such as alpha-recoils tracks and fission tracks in the geological environment,
- their ability to be synthesized in processes ranging from high temperature processes (fusion-crystallization; sintering) to low temperature processes (sol-gel procedure or low-temperature crystallization of apatitic cement).

One of the most convincing demonstrations of the stability of the apatite lattice is provided by natural apatites that incorporated significant amounts of actinides, rare-earth elements, metals and halogens millions of years ago. Among the most interesting are the apatites from the fossil nuclear reactor at Oklo (Gabon), as they have incorporated nuclear reaction products such as plutonium, rubidium, strontium and rare-earth elements, as long as two billion years ago (Carpéna and Sère, 1995; Bros and al., 1996). These apatites have been exposed to very high doses of radiation, due to both external irradiation with alpha particles and neutrons and internal irradiation with alpha particles and fission fragments (Sère and Carpéna, 1996). The ability of apatites to anneal their radiation damage (alpha-recoil tracks and fission tracks) with temperature explains how their crystal lattice has been able to remain stable for a so long period of time in spite of high doses of radiation.

A radioactive waste host phase must have good thermal and chemical stability in an irradiating environment. Thus, it is important to select the most suitable apatite composition, , i.e. the one able to sustain the highest dose of radioactivity with a minimum of lasting damage to its crystal lattice. Isothermal annealing experiments have been performed on induced fission tracks of uranium in natural apatites of different compositions to investigate the effect of different substitutions in the apatitic lattice on its potential for self-restoration of radiation damage.

Six apatite samples have been selected for this study. Their chemical compositions were determined by means of electron microprobe analysis (Table 2a). The Ca/P ratio, the percentage of cationic substitutions, the percentage of actinides incorporated into the lattice, and the proportion of anions substituting the fluor site are reported in table 2b,c,d. The unit cell dimensions for each apatite were calculated on the basis of Laue X-ray diagrams. The respective doses of natural radiation to which the apatites have been exposed have been calculated as in Lumpkin and Ewing, (1988).

Table 2a: Locality, provenance rock, crystallization age of the different apatite samples

Sample	Locality	Provenance rock	crystallization age
92-1	Gunlock,Carmel Formation, USA	Volcanic ash	163 +/- 1 Ma (Kowallis and al.,1992)
Du	Cerro de MercadoDurango, Mexico	Volcanic rock,	31,4 +/- 0,5 Ma (Green,1985)
INH 181	In Ouzzal Mole, Algeria	Carbonatites	2100 Ma (Allègre and al.,1972)
INH 171	n Ouzzal Mole, Algeria	Carbonatites	2100 Ma (Allègre and al.,1972)
SF84	Reactor 10Oklo Mine, Gabon	Hydrothermal apatite	1700 Ma (Ruffenach, 1979)
Ju 31	Sao Paulo state, Brazil	Carbonatites	127 Ma (Amaral and al., 1967)

Table 2b: Electron microprobe analysis of the different apatite samples (weight%).

Sample	CaO	P2O5	Cl	F	SO3	SiO2	Fe2O3	Al2O3	SrO	ZrO	Na2O	BaO	MgO	Ln2O3	Total
92-1	53.07	40.76	2.21	1.91	0.28	0.31	0.26	-	0.13	-	0.11	-	0.07	0.57	99.68
Du	53.98	40.52	0.45	3.59	0.41	0.28	0.006	-	-		0.25	-	-	0.96	100.45
INH181	53.77	39.47	0.04	4.52	0.18	0.93	-	-	0.97	-	0.02	-	-	1.52	101.42
INH171	45.21	30.93	0.08	4.63	1.83	4.08	-	-	0.57	-	0.04	-	-	6.32	93.69
SF84	50.85	38.97	0.17	3.20	0.20	0.36	0.10	0.34	0.06	0.95	0.05	-	-	0.14	95.39
Ju31	50.67	35.37	0.01	2.09	0.02	0.01	0.34	0.12	1.09	-	2.36	-	0.06	0.13	92.27

Table 2c: Crystallographic parameters (Angstroms), percentage of cationic substitution (weigh%), Me/X ratio of the different apatite samples

Sample	Crystallographic parameters	Cationic substitutions (weight)	Ca/P
92-1	a=9,525A,c=6,84A	0,52%	2,13
Du	a=9,38A, c=6,88A	1,04%	2,18
INH181	a=9,37A, c=6,88A	1,68%	2,23
INH171	a=9,38A, c=6,96A	6,04%	2,13
SF84	a=9,41A, c=6,88A	1,14%	2,39
Ju31	a=9,38A, c=6,89A	3,70%	2,13

Table 2d: Actinide content, calculated accumulated volumetric fission track density since crystallization, calculated accumulated alpha-recoil tracks since crystallization in the different apatite samples

Sample	Actinide content	Natural fissions	Internal alpha irradiation
92-1	U = 14 ppm	$1,20\times10^{15}$ fissions/m^3	$2,71\times10^{21}$ alpha events/m^3
Du	U = 4 ppm	$6,66\times10^{13}$ fissions/m^3	$1,49\times10^{20}$ alpha events/m^3
INH 181	U =4 ppm, Th = 488 ppm	$4,45 \times 10^{15}$ fissions/m^3	$4,08\times10^{23}$ alpha events/m^3
INH 171	U= 152 ppm, Th= 16 ppm #	$1,37\times10^{17}$ fissions/m^3	$3,17\times10^{23}$ alpha events/m^3
SF84	U = 153 ppm, Th = 1133 ppm	$1,70\times10^{17}$ fissions/m^3	$1,30\times10^{24}$ alpha events/m^3
Ju 31	U = 6 ppm	$4,04\times10^{14}$ fissions/m^3	$9,05\times10^{20}$ alpha events/m^3

2. Experimental Techniques

The apatite samples were annealed for 4 hours at 550°C in a furnace so as to erase all the fossil tracks in the mineral. The apatites were then irradiated with thermal neutrons in the most thermalized column (column P1) of the Orphée reactor (Pierre Süe Laboratory, Commissariat Energie Atomique, Saclay) (Carpéna et Mailhé, 1987) so as to induce fission of the uranium-235 isotope. After irradiation, the apatite samples were split into equal-sized aliquots, and annealed isothermally at temperatures of 260°C, 280°C, 300°C, 310°C for durations ranging from 30 minutes to 672 hours. Each aliquot was mounted in epoxy resin, ground, polished using < 1 μm diamond pastes and etched in 7% HNO_3 (weight), at 21°C. The surface track densities were determined by using a Leitz optical microscope equiped with a 100X dry objective and 10X oculars. For the different apatites the etching times that were used varied from 3 to 25 seconds (Table 3).

Table 3: Neutron fluence, etching conditions and initial induced track density for the apatite samples

Apatite	Thermal neutron fluence (10^{15}n/cm2)	Etching time	Ambiant track density (10^5t/cm2)
92-1	7.99	10s	1.78
Du	3.63	23s	2.38
INH 181	5.27	25s	2.61
INH 171	2.05	10s	49.28
SF 84	0.81	25s	6.35
Ju 31	3.63	3s	4.61

To explain this variation, we have to take in account that several parameters control the etching time such as the chemical composition, the crystallographic orientation, the fission track density (in our case, the induced track density) and the alpha damage. For example, it is well known that carbonate-apatites, hydroxy-apatites and chloro-apatites have higher solubilities than fluoro-apatites (Elliot, 1994). Fission-Track geochronologist have pointed out earlier, that the time needed for optimal track etching is not a constant for a given type of mineral (Naeser, 1978; Gleadow and al., 1976). We are keeping in mind that this also applies to natural apatites of different composition.

The tracks have always been counted on only those surfaces which present sharp polishing scratches and which have a high etching efficiency, moreover a calibration curve has been used for each sample to make the proper choice of the etching time to get the best possible efficiency. To understand the true influence of the chemical composition on the etching time it is better to work on synthetic apatites, i.e. apatites of the same age with only one varying parameter, the chemical composition (work in progress). The thermal neutron fluence received by each apatite, the initial track density, the etching time are summarized in Table 3. The measured track density reduction in each sample is reported in Tables 4 a,b,c,d,e,f. We prefered to measure track densities over track lengths as the optical quality of some samples was bad (Juquia (Ju 31), Oklo (SF 84) and In Ouzzal (INH 171)) due to the presence of many solid or fluid inclusions.

Table 4a: Annealing temperature and duration, measured track densities and calculated track density reduction in Gunlock apatite

Annealing temperature	Annealing time	Tracks counted	Track density (10^5/cm2)	sp (%)	Track density reduction(%)	sp (%)
Ambient	-	805	1.78	3.52	0	7.0
260°C	7 days	451	1.76	4.71	12.0	8.2
260°C	14 days	475	1.59	4.59	11.0	8.1
260°C	21 days	334	1.58	5.47	11.4	9.0
260°C	28 days	290	1.47	5.87	17.8	9.4
280°C	1 hour	280	1.74	5.98	2.6	9.5
280°C	4 hours	487	1.66	4.53	6.9	8.0
280°C	8 hours	161	1.56	7.88	12.4	11.4
280°C	20 hours	740	1.36	3.68	23.6	7.2
280°C	65 hours	157	1.18	7.98	33.9	11.5
280°C	90 hours	259	1.19	6.21	33.4	9.7
280°C	120 hours	412	1.12	4.93	37.4	8.4
300°C	15 minutes	327	1.70	5.53	4.6	9.0
300°C	30 minutes	388	1.68	5.08	5.9	8.6
300°C	1 hour	367	1.47	5.22	17.8	8.7
300°C	8 hours	257	1.40	6.24	21.3	9.8
300°C	24 hours	477	1.32	4.58	25.0	8.1
300°C	72 hours	198	0.87	7.11	51.1	10.6
300°C	7 days	271	0.88	6.07	50.2	9.6
310°C	30 minutes	169	1.55	7.69	13.1	11.2
310°C	1 hour	183	1.02	7.39	43.0	10.9
310°C	2 hours	269	0.96	6.10	45.8	9.6
310°C	5 hours	249	0.93	6.34	47.6	9.9
310°C	8 hours	228	0.92	6.62	48.1	10.1
310°C	24 hours	199	0.78	7.09	56.3	10.6

Table 4b: Annealing temperature and duration, measured track densities and calculated track density reduction in Durango apatite

Annealing temperature	Annealing time	Tracks counted	Track density (10^5/cm2)	sp (%)	Track density reduction(%)	sp (%)
Ambient	-	698	2.83	3.79	0	7.6
260°C	7 days	761	2.37	3.62	16.3	7.4
260°C	14 days	841	2.15	3.45	24.0	7.2
260°C	21 days	878	1.98	3.37	29.9	7.2
260°C	28 days	807	1.75	3.52	38.0	7.3
280°C	1 hour	907	2.76	3.32	2.4	7.1
280°C	4 hours	392	2.25	5.05	20.3	8.8
280°C	8 hours	638	2.13	3.96	24.8	7.7
280°C	20 hours	551	2.06	4.26	27.0	8.0
280°C	65 hours	165	1.95	7.78	30.9	11.6
280°C	90 hours	525	1.67	4.36	41.0	8.1
280°C	120 hours	212	1.50	6.87	46.8	10.6
300°C	15 minutes	718	2.79	3.73	1.1	7.5
300°C	30 minutes	569	2.33	4.19	17.5	8.0
300°C	1 hour	606	2.20	4.06	22.3	7.8
300°C	8 hours	458	2.10	4.67	25.7	8.5
300°C	24 hours	426	1.56	4.85	44.8	8.6
300°C	72 hours	175	1.22	7.56	57.0	11.3
300°C	7 days	262	0.79	6.18	71.7	10.0
310°C	30 minutes	307	2.56	5.71	9.5	9.5
310°C	1 hour	377	1.57	5.15	44.4	8.9
310°C	2 hours	539	1.37	4.31	51.5	8.1
310°C	5 hours	482	1.36	4.55	51.8	8.3
310°C	8 hours	315	1.25	5.63	55.9	9.4
310°C	24 hours	138	0.77	8.51	72.6	12.3

Table 4c: Annealing temperature and duration, measured track densities and calculated track density reduction in In Ouzzal INH 181 apatite

Annealing temperature	Annealing time	Tracks counted	Track density (10^5/cm2)	sp (%)	Track density reduction(%)	sp (%)
Ambient	-	423	2.61	4.86	0	9.7
260°C	7 days	480	2.00	4.56	23.6	9.4
260°C	14 days	381	1.70	5.12	35.0	1.0
260°C	21 days	193	1.56	7.20	40.1	12.0
260°C	28 days	276	1.46	6.02	44.2	10.9
280°C	1 hour	251	2.54	6.31	2.7	11.2
280°C	20 hours	166	2.09	7.76	20.1	12.6
280°C	65 hours	295	1.83	5.82	29.8	10.7
280°C	90 hours	150	1.40	8.16	46.3	13.0
280°C	120 hours	71	1.23	11.87	52.9	16.7
300°C	15 minutes	615	2.86	4.03	0.0	8.9
300°C	30 minutes	445	2.21	4.74	15.2	9.6
300°C	1 hour	516	1.90	4.40	27.0	9.3
300°C	24 hours	497	1.28	4.49	50.8	9.3
300°C	7 days	185	0.67	7.35	74.1	12.2
310°C	30 minutes	207	2.55	6.95	2.4	11.8
310°C	1 hour	78	1.60	11.32	38.7	16.2
310°C	24 hours	70	0.59	11.95	77.1	16.8

Table 4d: Annealing temperature and duration, measured track densities and calculated track density reduction in In Ouzzal INH 171 apatite

Annealing temperature	Annealing time	Tracks counted	Track density (10^5/cm2)	sp (%)	Track density reduction(%)	sp (%)
Ambient	-	8969	49.3	1.06	0.0	2.1
260°C	7 days	6175	34.6	1.27	29.9	2.3
260°C	14 days	9231	32.1	1.04	34.8	2.1
260°C	21 days	3817	30.9	1.62	37.3	2.7
260°C	28 days	6300	28.0	1.26	43.2	2.3
280°C	1 hour	1634	47.1	2.47	4.5	3.5
280°C	4 hours	7938	44.4	1.12	9.9	2.2
280°C	8 hours	5617	41.0	1.33	16.8	2.4
280°C	20 hours	1720	31.9	2.41	37.4	3.5
280°C	65 hours	3262	27.9	1.75	43.4	2.8
280°C	90 hours	550	23.8	4.26	51.7	5.3
280°C	120 hours	1519	20.0	2.57	59.4	3.6
300°C	15 minutes	4436	45.2	1.5	8.3	2.6
300°C	30 minutes	3692	36.9	1.65	25.1	2.7
300°C	1 hour	2651	24.1	1.94	51.1	3.0
300°C	24 hours	2959	16.2	1.94	67.2	3.0
300°C	7 days	1613	10.1	2.49	79.4	3.6
310°C	30 minutes	4228	44.9	1.59	9.0	2.6
310°C	1 hour	3514	25.7	1.69	47.8	2.7
310°C	2 hours	3216	19.0	1.76	61.4	2.8
310°C	5 hours	3040	18.7	1.81	62.0	2.9
310°C	8 hours	1800	15.3	2.36	69.0	3.4
310°C	24 hours	2440	9.21	2.02	81.3	3.1

Table 4e: Annealing temperature and duration, measured track densities and calculated track density reduction in Juquia apatite

Annealing temperature	Annealing time	Tracks counted	Track density (10^5/cm2)	sp (%)	Track density reduction(%)	sp (%)
Ambient	-	770	4.61	3.60	0.0	7.2
280°C	1 hour	339	4.57	5.43	0.8	9.0
280°C	20 hours	142	3.12	8.39	32.3	12.0
280°C	90 hours	98	2.01	10.10	56.4	13.7
280°C	120 hours	109	1.98	9.58	57.0	13.2
300°C	15 minutes	268	4.34	6.11	5.9	9.7
300°C	30 minutes	358	4.38	5.29	5.7	8.9
300°C	1 hour	201	2.48	7.05	46.1	10.7
300°C	24 hours	298	2.10	5.79	54.5	9.4
300°C	7 days	99	8.25	10.05	82.1	13.6
310°C	30 minutes	133	3.58	8.67	22.3	12.3
310°C	1 hour	143	2.00	8.36	56.6	12.0
310°C	8 hours	104	1.26	9.81	72.6	13.4
310°C	24 hours	47	4.56	14.59	90.1	18.2

Table 4f: Annealing temperature and duration, measured track densities and calculated track density reduction in Okloapatite

Annealing temperature	Annealing time	Tracks counted	Track density (10^5/cm2)	sp (%)	Track density reduction(%)	sp (%)
Ambient	-	2401	7.24	2.04	0.0	4.1
260°C	7 days	2076	6.53	2.19	9.8	4.2
260°C	14 days	1034	5.16	3.11	28.6	5.1
260°C	21 days	605	5.34	4.07	26.3	6.1
260°C	28 days	1073	4.36	3.05	39.8	5.1
280°C	1 hour	1162	6.35	2.93	12.3	5.0
280°C	8 hours	1062	6.44	3.07	11.1	5.1
280°C	20 hours	1007	5.22	3.15	27.9	5.2
280°C	65 hours	1199	4.06	2.89	43.9	4.9
280°C	90 hours	599	3.61	4.09	50.1	6.1
280°C	120 hours	494	2.79	4.50	61.4	6.5
280°C	14 days	147	1.58	8.25	78.2	10.3
280°C	27 days	92	1.27	10.43	82.5	12.5
300°C	15 minutes	571	6.21	4.18	14.3	6.2
300°C	30 minutes	1156	6.25	2.94	13.7	5.0
300°C	1 hour	1237	5.83	2.84	19.4	4.9
300°C	8 hours	798	5.70	3.54	21.2	5.9
300°C	24 hours	537	2.45	4.32	66.1	6.7
300°C	72 hours	444	1.51	4.75	79.2	6.8
300°C	7 days	122	0.56	9.05	92.2	11.1
310°C	30 minutes	586	6.96	4.13	3.9	6.2
310°C	1 hour	1692	6.96	2.43	3.9	4.5
310°C	2 hours	744	6.15	3.67	15.1	5.7
310°C	5 hours	509	3.95	4.43	45.6	6.5
310°C	8 hours	364	3.19	5.24	55.9	7.3
310°C	24 hours	236	1.38	6.51	80.9	8.6

Therefore, to measure track lengths becomes much more difficult than to count tracks. For such samples counting has been performed using transmitted and reflected light. The counting data are also reported in Tables 4 a,b,c,d,e and f.

3. Results and Discussion.

An examination of the annealing curves (Figure 1 a,b, c,d) shows that the fission tracks in the different samples fade at different rates. This observation applies to all the annealing temperatures. For example, the Gunlock apatite (92-1) systematically shows the lowest annealing rate at all annealing temperatures. On the other hand, the apatites from In Ouzzal (INH 171) and from Juquia carbonatites (Ju 31) systematically present the highest annealing rates.

In Figures 1e and 1f, the track density reduction has been plotted versus different physical and chemical parameters: the chlorine content, the percentage of substitutions on the cation site, and the unit cell dimensions.

Annealing of fission tracks in apatites is also known to be affected by the amount of chlorine incorporated in the crystal lattice (Green et al., 1986): an increase in chlorine within the crystal structure produces a resistance to annealing. In an attempt to verify this observation, we have plotted the track density reduction for apatite versus its

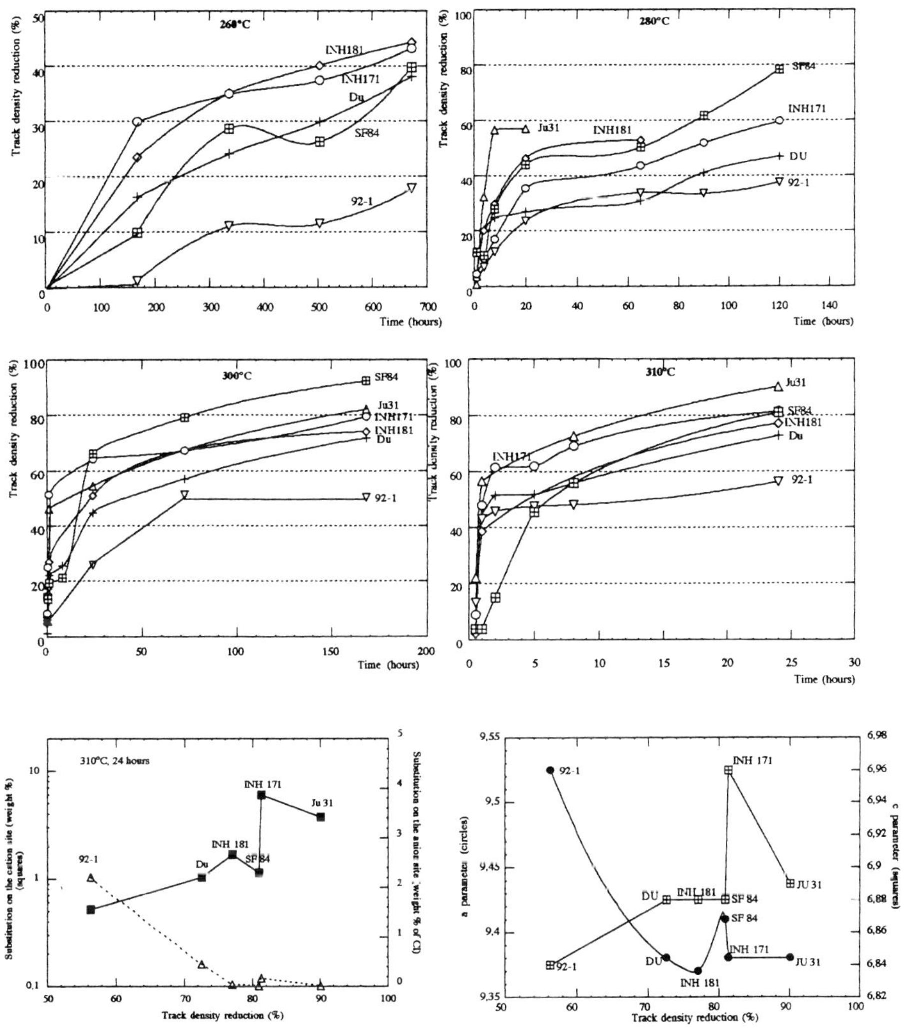

Figure 1: Track density reduction versus annealing time for the 260°C (a), 280°C (b), 300°C (c), 310°C (d) annealing experiment., track density reduction after 24 hours annealing at 310°C versus percentage of substitution on the cation and anion site (weight %) (e) and track density reduction after 24 hours annealing at 310°C versus the unit cell dimensions a and c.

observation, we have plotted the track density reduction for each apatite versus its chlorine content (Figure 1d). This curve shows clearly the correlation between the chlorine content and the annealing rate of tracks and confirms the data obtained by Green.

Apatites are known to tolerate many substitutions on the cation site (metals, rare earths, alcaline elements, Actinides, etc.). For example, apatites from carbonatites show a high rate of sustitution on this site, sometimes as high as 50% (weight) (Ouzegane et al., 1988). Figure 1e shows that a positive correlation seems to exist between the degree of substitution and the track density reduction. The most heavily substituted apatites (INH 171 and Ju 31) also show the highest track density reduction (81.32% and 90.10% respectively, at 310°C for a 24 hours annealing time). Kreidler (1967) has suggested that the unit cell dimensions a and c are influenced by the ionic radii of the substituting ions: c increases when the ionic radius of the ions substituting on the cation site increases. When the substitution occurs on the anion site, a increases with the ionic radius of the substituting ions, whereas c decreases. For instance, when chlorine (1.81 angströms) substitutes the fluor (1.33 angströms), a increases and c decreases.

Our own findings confirm this observation: the Gunlock Cl-apatite (92-1) which has the highest substitution on the anion site and the lowest on the cation site presents the highest a value (9.525 angströms). INH 171 and Ju 31, which have most substitutions on the cation site, show the highest values of c (6.96 and 6.89 angströms respectively). In Figure 1f, the unit cell-dimensions (a, c) for each apatite have been plotted versus the measured track density reduction: the most substituted apatites on the Ca-site are least substituted on the F-site and yield the higher c-values. They present the highest track density reduction, with a maximum for the In Ouzzal (INH 171) and Brasilian (Ju 31) apatites. The F-apatite lattice tends to become distorded when the calcium sites are occupied by small cations, whereas no such distortions occur in Cl-apatites (Kreidler and Hummel, 1970). The most heavily substituted apatites at the cation site contain many different ions and might have a strongly distorded lattice. Such a deformation could be involved to explain why the apatites with the heaviest substitution on the Ca-site show the strongest reduction of fission track density and why Cl-apatites which are less substituted, are more resistant to track annealing.

4. Conclusions

It is not possible at the moment to formulate precise rules for the annealing behaviour of apatites. Our data suggest that the most substituted apatites on the cation site are less resistant to thermal fading of fission tracks than pure phospho-calcic apatites, and confirm that Cl-apatites are more resistant to thermal annealing than F-apatites, as pointed out by Green et al. (1986). Our important finding is that we observe an inverse correlation between the number of substitutions on both cation and anion sites. This may imply that instead of the Cl/F-ratio, the number of substitutions on the cation site is directly responsible for the different thermal stabilities of fission tracks in different apatites. This is certainly an admissible conclusion because substitutions on the cation site directly determine the structural integrity of the apatite lattice, whereas this is less evident for substitutions of the monovalent anions located in channels parallel to the c-axis of the apatite crystal.

Fission-track geochronologists have to be carefull when applying the fission-track method to sedimentary rocks which may contain apatites of different origin , and in areas offering outcrops of plutonic rocks with different chemistry. Apatites from basic rocks such as gabbro, diorite generally have few substitutions, whereas apatites from acid rocks such as granite, pegmatite and alkaline rocks are enriched with actinides, rare earth elements and metals. Discordant fission track ages of apatites and zircons with different chemical compositions had been reported by Carpéna and al. (1988) and by Carpéna (1992b, 1993), suggesting a possible similar dependence of fission track stability on chemical composition in zircon. More heavily substituted zircons from rocks enriched in heavy elements (pegmatite,alkaline granite) would be less resistant to annealing than those from rocks depleted in heavy elements (granulites, basic rocks).

These observations also have to be taken into account in the manufacture of nuclear ceramics if they have to resist intense radiation from high-activity radioactive waste and auto-anneal their radiation damage (Carpéna and Lacout, 1993). The best apatitic ceramics are those which anneal the more easily their tracks, i.e. at the lowest temperature and/or in shorter time. At the moment, more annealing experiments are in progress on synthetically doped apatites to define the most suitable apatite compositions for the different wastes to be conditioned.

References

Boyer L., Carpéna J., Lacout J.L. (1997) Synthesis of phosphate-silicate apatites at atmospheric pressure, *Solid State Ionics, Diffusion and Reactions, 95*, n° 1, 2, 121-129.

Bros R., Carpéna J., Sère V., Beltritti A. (1996) Occurence of Pu and fissiogenic REE in hydrothermal apatites from the fossil nuclear reactor 16 of Oklo (Gabon), *Radiochimica Acta,74*, 277-282.

Carpéna J. (1992a) Utilisation des phosphates dans le stockage et le traitement des déchets nucléaires, *C.E.A. report,NT-SCS, n°92/21.*

Carpéna J. (1992b) Fission Track Dating of zircon: zircons from Mont Blanc Granite (French-Italian Aps), *Journal of Geology, vol. 100*, 411-421.

Carpéna J. (1993) Fission Track Ages in granitic zircons, *Nuclear Tracks and Radiation Measurements, vol.21, n°4*, 598.

Carpéna J. (1996) Conditionnement et confinement des déchets nucléaires. Les matrices apatitiques. Reflexions et propositions de programme, *C.E.A.report,NT-SEP-96/450.*

Carpéna J. and Lacout J.L. (1993) Procédé de conditionnement de déchets radioactifs utilisant des apatites silicatées comme matrices de confinement, *French licence n° 93 08676.*

Carpéna J. and Mailhé D. (1987) Fission Track Dating calibration of the Fish Canyon Tuff standard in French Reactors, *Chem. Geol. (Isotope Geoscience Section), V.66*, 53-59.

Carpéna J. and Sère V. (1995)From the natural to the synthetic analog: interest of Oklo for the synthesis of stable crystalline matrices, *Proceedings fourth joint EC-CEA Final Meeting, Saclay, France*, 225-238.

Carpéna J., Kiènast J.R., Ouzegane K., Jehano C. (1988) Evidence of the contrasted fission-track clock behaviour of the apatites from in Ouzzal Carbonatites (Northwestern Hoggar), the low temperature thermal history of an Archean basement, *Geol. Soc. Amer. Bull., 100*, 1237-1243.

Elliot J.C.(1994) *Structure and chemistry of the apatites and other calcium orthophosphates*, Elsevier.

Gleadow A.J.W., Hurford A.J. and Quaife R.D. (1976) Fission Track dating of zircons: improved etching techniques, *Earth Planet. Sci. Lett.33*,273-276.

Green P.F., Duddy I.R., Gleadow A.J.W., Tingate P.R., Laslett G.M., (1986) Thermal annealing of fission track in apatite: 1- A qualitative description, *Chem. Geol., 59*, 237-253.

Kreidler E.R.(1967) Stoechiometry and crystal chemistry of apatite, *Thesis Pennsylvania University.*

Kreidler E. R. and Hummel F.A.(1970) The crystal chemistry of apatite: structure fields of fluor and chloro-apatite, *Am. Miner., 55*, 170-184.

Lumpkin G.R. and Ewing R.C.(1988) Alpha-decay damage in minerals of the pyrochlore group, *Phys. Chem. Minerals, 16*, 2.

Naeser C.W.(1978) Fission Track Dating, *U.S.Geol.Survey Open-File report 76-190.*
Ouzegane K., Fourcade S., Kienast J.R. and Javoy M.(1988) New carbonatite complexes in the archean In'Ouzzal nucleus (Ahaggar, Algeria): mineralogical and geochemical data, *Contrib. Miner. Petr., 98,* 277-292.

MATHEMATICAL CONVECTION METHODOLOGY USING BERTAGNOLLI AND LASLETT FISSION TRACK ANNEALING LAWS

H. IGLI*, J.-C. MIELLOU*, A. CHAMBAUDET**
and M. REBETEZ**
*Laboratoire de Calcul Scientifique, **Laboratoire de Microanalyses Nucléaires,
U.F.R. des Sciences et Techniques, Université de Franche-Comté,
16 route de Gray, 25030 Besançon Cedex, France

Abstract

In geothermochronology, the interest in fission tracks is mainly to resolve the thermal history from apatite fission track data. To solve what we call the inverse problem, we propose a mathematical convection methodology, that is to say a mathematical formulation formally similar to the standard one used for the thermal convection phenomena in fluid dynamics. This seems to be a basis for an alternative to other statistical ones [Monte Carlo (Lutz and Omar, 1991)] and equivalent time methods (Duddy *et al.*, 1988).
The aim of this paper is to give an account of the use of the convection methodology in thermochronology when applied to standard annealing laws such as Bertagnolli's or Laslett's laws, where the last one is put into the form of a differential equation resembling the former.

1. Introduction

In geothermochronology, let us consider the most often quoted thermal fission track annealing law, namely Laslett's law (Green *et al.*, 1989) as well as an older one: Bertagnolli's law (Bertagnolli *et al.*, 1983). Both of them are based more or less on an empirical expression of the length reduction caused by thermal effect. If we assume that for every one of the produced tracks the considered law represents exactly the annealing phenomenon we are able, by a conservation principle and probability arguments to deal with continuous distribution functions (Miellou *et al.*, 1997). From this conservation principle, we can derive a kind of convection problem, that is to say a mathematical formulation of the equations giving the distribution of tracks with respect to length and time, which is similar to the standard one of the thermal convection phenomena in fluid dynamics. The numerical resolution of this problem allows us to obtain a track length histogram from the temperature history data, and from the knowledge of the track production distribution. This is the direct problem. It is worthwhile to note that the terms "numerical resolution" and "track length histogram" are used in the present paper without involving geological experimental data. Our present point of view is used in Simon *et al.* (1996), in order to study the relationship with geological data.

Conversely, a corresponding inverse problem allows us to go back from such a computed histogram to points in close vicinity to the initially given thermal history curve, provided that the direct convection problem and the inverse integral equation are based on the same annealing law. It is worthwhile to note that in order to apply this kind of methodology to a situation associated to the classical empirical law of Laslett, we must express this law in terms of a differential equation, the form of which is closely related to the usual expression of Bertagnolli's law. Numerical simulations of both the

P. Van den haute and F. De Corte (eds.), Advances in Fission - Track Geochronology, 93-98.

direct and inverse problems are presented within the framework of these two laws. Here is one example of thermal history: slow cooling. Some conclusions for the results obtained by this methodology are presented, which can be summarized by:
i) high precision in the values of the temperature obtained by the resolution of the inverse problem, as far as the considered fading law represents an exact behaviour of track lengths with respect to temperature and time, and the large numbers law allows the description of a typically discrete situation by a continuous model (the last point is described in Miellou *et al.*, 1997);
ii) the number of computed time-temperature points is limited by the histogram's number of classes with the length greater than the one of the non-void class with the smallest lengths.
Point i) above corresponds to the obvious quality of the method. An increase in the number of the histogram steps which should be obtained namely with an increase in the precision of the measurements and a sufficiently large number of tracks in each class of histograms, should allow us to obtain new insights into the difficulty quoted in point ii).

2. Thermal Annealing of a Fission Track in Apatite

NOTATIONS:
The function $\theta(t)$ represents a temperature history where l is the length of the track and L is the total etchable length.

$$\lambda = L - l. \tag{1}$$

$B(\lambda)$, $\alpha(\lambda)$ and $\alpha_0(\lambda)$ are positive functions of λ.
The last equalities just above correspond to the fact that the track produced at time s = t has a length equal to L then:

$$\lambda_s^t = L - l_s^t; \quad \lambda_t^t = L - l_t^t = L - L = 0 \tag{2}$$

BERTAGNOLLI'S LAW:
In the case of Bertagnolli's law, the model which represents the track length reduction is given by this ordinary differential equation (Bertagnolli *et al.*, 1983):

$$\frac{d\lambda}{dt} = c(\lambda, \theta(t)) = (L - \lambda).\alpha_0(\lambda)\exp\left(-\frac{E(\lambda)}{K.\theta(t)}\right) \tag{3}$$

where K is the Boltzmann's constant and $E(\lambda)$ represents a kind of activation energy that depends on λ.
Remark: *We are mainly interested here by the mathematical form of this law which is a rather old model. However, it does give results in accordance with the experimental data of the same order of accuracy as Laslett's one (Simon et al. 1996).*
PUTTING LASLETT'S LAW IN A FORM ANALOGOUS TO BERTAGNOLLI'S LAW:
The relationship model between the fission track length reduction and the annealing time for fanning Laslett's law (Green *et al.*, 1989) is given by:

$$\log(t) = A + \frac{B(r)}{C\theta}\ ;\ B(r) = \left[\left(\frac{1 - r^a}{a}\right)^a - 1\right]^b / b + c_1\ ;\ r = \frac{l}{L} \tag{4}$$

Where A , C, a, b and c_1 are constants. For variable temperatures $\theta(t)$, the model which represents the fission track length reductions is given by way of the equivalent time principle, where the fanning model is taken into consideration (Green *et al.*,1989, Appendix equation (A-2)). By making $\Delta t \to 0$, we obtain the following expression for Laslett's law, which must be compared to (1):

$$\frac{d\lambda}{dt} = c(\lambda, \theta(t)) = (L - \lambda).\alpha(\lambda)\, \theta(t) \exp\left(-\frac{B(\lambda)}{C.\theta(t)}\right) \tag{5}$$

Let us consider now that we must work with a fission track annealing law of the form:

$$\frac{d\lambda}{dt} = c(\lambda, \theta(t); \quad \lambda(s) = 0 \tag{6}$$

where above $\lambda(t)$ represents the reduction of the length at time t of a track produced at time s. For the direct problem, we are interested in the numerical simulation of the track production and annealing phenomena, for an annealing law given by (6).

Let us consider all of the tracks produced between times t_1 and t_2 ($t_1 < t_2$). Because of the annealing effect, these tracks have lengths between l_2 and l_1 ($l_2 < l_1$)., where l_1 and l_2 are the respective lengths of the tracks which should have been produced at the respective time t_1 and t_2 .This gives rise to a conservation property (Miellou *et al.*, 1996) which can be written as:

$$\int_{l_2}^{l_1} w^t(l)\, dl = \int_{t2}^{t1} b(\mu)\, d\mu \,, l_1 = l_{t1}^t \text{ and } l_2 = l_{t2}^t \tag{7}$$

where $b(\mu)$ is the track production speed. The second member of the first equality of (7) represents, by standard argument related to Poisson's process properties, an approximation of the proportion of tracks produced between times t_1 and t_2 with respect to the whole number of tracks produced between times 0 and T. Therefore, $w^t(\lambda)$ represents a density probability of the presence of tracks with a length reduction λ and the first member of the first equality of (7) approximates the proportion of tracks with a length between l_1 and l_2 at time t with respect to the whole quantity of considered tracks. For a justification of the above relation (5) by an argument of probability, see Miellou *et al.* (1997).

With $t_1 = 0$, $t_2 = s$ and $\lambda_s^t = L - l_s^t$, we define the function $u(\lambda_s^t, t)$ by:

$$u(\lambda_s^t, t) = \int_{\lambda_0^t}^{\lambda_s^t} w^t(l) dl = \int_0^s b(\mu) d\mu \tag{8}$$

where $u(\lambda_s^t, t)$ represents the proportion of the number of tracks produced between 0 and s. By derivation with respect to t, we obtain:

$$\frac{du(\lambda_s^t, t)}{dt} = \frac{\partial u}{\partial t} + \frac{d\lambda_s^t}{dt}\frac{\partial u}{\partial \lambda} = 0 \text{ on } 0 < \lambda < L,\ 0 < t < T \tag{9}$$

where by (2) $\lambda_t^t = 0$ and then (8) gives:

$$u(0, t) = \int_0^t b(\mu) d\mu \text{ on } 0 < t < T \tag{10}$$

At time t = 0 there are no tracks. We obtain:

$$u(\lambda, 0) = 0 \text{ on } 0 < \lambda < L \tag{11}$$

Taking (6) and (4) into account, we obtain following mathematical model, which represents the production and annealing of a population of fission tracks over a period [0,T], with a length reduction speed associated to this track annealing law of the form (6):

$$\frac{\partial u}{\partial t} + c(\lambda, t)\frac{\partial u}{\partial \lambda} = 0 \quad 0 < \lambda < L \text{ and } 0 < t < T \tag{12}$$

The above mathematical problem (10), (11), (12) looks like a standard model of thermal convection of fluids. Namely, (12) appears to be a kind of convection equation, (11) corresponds to an initial condition and (10) to one-sided boundary data.

3. The Direct Problem

We suppose that the function $\theta(t)$, the absolute temperature at time t, is known. We considered the track lengths $0 < l_1 < l_2 < L$ and t_1 (resp. t_2) to be the starting times. With respect to the total number of considered tracks, the proportion of tracks, with lengths between l_1 and l_2, is given by:

$$N(l_1,l_2) = u(\lambda_1,T) - u(\lambda_2,T) = \int_{t_1}^{t_2} b(\mu)d\mu; \quad \lambda_1 = L - l_2 \text{ and } \lambda_2 = L - l_1 \tag{13}$$

t_1 (resp. t_2) is the solution for equation (14), where $\lambda(0) = \lambda_1$ (resp. $\lambda(0) = \lambda_2$)

$$\text{Given } t_i \in [0,T] \ (i=1,2), \ \frac{d\lambda}{d\mu} = -c(\lambda, \theta(T-\mu)); \ \lambda(0) = \lambda_i \ ; \ \lambda(T-t_i) = 0 \tag{14}$$

The resolution of this last problem allows us to compute the time interval $[t_1,t_2]$ corresponding to each class of the histogram. Afterward, the value of the ratio considered for each step of the histogram is obtained by (13).

4. The Inverse Problem

For the inverse problem, the temperature function $\theta(s)$ is the main unknown function. In order to simplify the notations, here we can use:

$$u(\lambda) = u(\lambda, T) \tag{15}$$

which is data that we assume to be differentiable with respect to λ.
Now we take λ, which satisfies (1) as an independent variable, to which corresponds a time $s(\lambda)$ of the birth of a track, which at time T is the length l which satisfies (1).
With this independent variable, we can write (8) in the form:

$$u(\lambda) = u(\lambda^T_{s(\lambda)},T) = \int_{\lambda^T_0}^{\lambda^T_{s(\lambda)}} w^T(l)dl = \int_0^{s(\lambda)} b(\mu)d\mu \tag{16}$$

The following age equation (17) gives the function $s(\lambda)$, which represents the starting time of a track with a length l, which satisfies (1) at time T, so

$$u(\lambda) = \int_0^{s(\lambda)} b(\mu)d\mu \tag{17}$$

The temperature function $\theta(s)$ is the solution of the equation (18),

$$c(0,\theta(s)) \exp\left[\int_s^T \frac{\partial c}{\partial \lambda}(\lambda_s^\tau, \theta(\tau))d\tau\right] = \frac{b(s)}{w(\lambda)} \ s = s(\lambda), w(\lambda) = -\frac{du(\lambda)}{d\lambda} \tag{18}$$

For Bertagnolli's model (3), see (19) in Miellou *et al.* (1997). For Laslett's model, where $c(\lambda, \theta(t))$ is given by equation (5), the integral relation (18), where we can replace the independent variable s by the independent variable λ, takes the form:

$$\exp\left(-\frac{B(0)}{C.\theta(s(\lambda))}\right)(L - \lambda)\,\alpha(\lambda)\,\theta(s(\lambda)) \exp\left[-\int_0^\lambda -\frac{B'(\xi)}{C.\theta(s(\xi))}\,d\xi\right] = \frac{b(s(\lambda))}{w(\lambda)} \tag{19}$$

5. Some Simulations

Here in Figure 1, we have presented a numerical simulation in which a computed histogram is associated by the resolution of the direct problem to a continuous curve, which is a given thermal history. Conversely, taking this computed histogram as data, our methodology allows us to calculate temperature values at some points of this thermal history. Namely, let us consider equation (7) in the discrete histogram form:

$$u(\lambda_i) = \int_0^{s_i} b(\mu)d\mu$$

where $s_i = s(\lambda_i)$ corresponds to the histogram's steps where $\lambda_i = L - l_i$. Here, we can solve the Volterra equation (19) (Igli *et al.*, 1995a).
Each point, computed by the resolution of this inverse problem, corresponds to a histogram class. We can only take into account the classes which correspond to the lengths greater than those of the non-void class which includes the smallest ones.
In our numerical simulations, we have considered that the classes of each histogram correspond to a step length of L divided by sixteen (L/16).

The numerical simulations were performed within both the frameworks of Bertagnolli's and Laslett's laws. The corresponding inverse problem was solved by the associated integral equation (18), which in the case of Laslett's law takes the form (19), and the corresponding results are presented in Figure 1. To a given annealing law, assumed to be followed exactly by all the arbitrarily numerous tracks produced inside the samples of apatite and to a given thermal history, we are able to associate a virtual histogram through a computer simulation. This virtual histogram should be very close to the real one obtained nowadays from perfect measurements of the samples. Conversely, the resolution of a corresponding integral equation allows us to go back to the thermal history starting from the data of the histogram, virtual or not. In order to get accurate time-temperature points with respect to the time-temperature curve, this histogram must be issued from two possible situations:
- a "realistic" one in which each of the numerous tracks should have followed exactly the considered annealing law, used to establish the Volterra integral equation;
- a virtual histogram computed on the basis of the same annealing law.

We insist on the fact that in order to perform our calculations a discretisation of the length variable in accordance with the usual need for experimental constraints (L/16) had to be proposed. The consequence is that we can only go back to a finite number less than 16, time-temperature points. It is worthwhile to note here that these points are in strong accordance with the temperature curves. Using smaller step lengths, that is to say with progress in experimental measurements, should allow to get a larger number of time-temperature points.

6. Conclusion

In the exact mathematical modelling of the direct and inverse problems for an interpretation of thermal histories with fission track data:
- the corresponding study for the mono-exponential and multi-exponential annealing laws have been previously studied and published (Chambaudet *et al.*, 1993 ; Igli *et al.*, 1995b).
- this paper shows that it is also possible to apply the convection methodology to describe the more empirical laws of Bertagnolli and Laslett (Bertagnolli *et al.*, 1983; Green *et al.*, 1989).

- the first numerical simulations presented here show that it is possible to get the etchable length histograms from some classical thermal histories, and moreover, return to the (time, temperature) points that were well-positioned on the previous thermal history tested, with quoted limitations. Thus, under:
- a given asserted fading law available for each particular track of a population including an arbitrary large number of measurable tracks,
- length measurements with hypothetically perfect precision, the methodology presented above should be nearly to be almost perfect. Of course physical, geological realities limit our opportunities at the present time. In that respect, it may be necessary to reevaluate this perspective in light of such realities, which is the purpose of Simon *et al.* (1996).

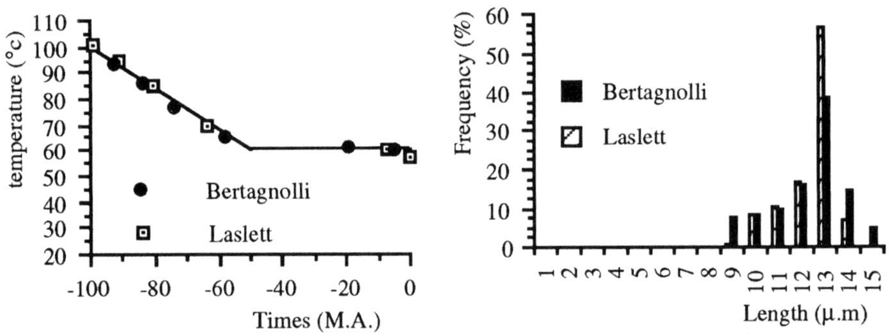

Figure 1. The solid line represents the given thermal history from which the simulated histograms of the volume etchable tracks are computed by the forward convection method. The thermal history is tested with two annealing laws: Bertagnolli's law and Laslett's law. From these histograms, the inverse method allows us to calculate the discrete time-temperature points which are in strong accordance with the thermal histories used as input for the forward calculations.

References

Bertagnolli E., Keil R. and Pahl M. (1983): Thermal history and length distribution of fission tracks in apatite: Part 1. *Nucl. Tracks* **7 (4),** 163-177.

Chambaudet A., Miellou J.C., Igli H., Rebetez M. and Grivet M. (1993): Thermochronology by fission tracks: an exact inverse method associated with the resolution of a single ordinary differential equation (ODE). *Nuclear Tracks Radiat. Meas.* **22 (1-4)**, 763-772.

Duddy I.R., Green P.F and Laslett G.M. (1988): Thermal annealing of fission tracks in apatite 3. variable temperature behaviour. *Chem. Geol. (Isot. Geosci. Sect.)* **73**, 25-38.

Green P.F., Duddy I.R., Laslett G.M., Hegarty K.A., Gleadow A.J.W. and Lovering, J.F. (1989): Thermal annealing of fission tracks in apatite 4. quantitative modelling techniques and extension to geological timescales. *Chem. Geol. (Isot. Geosci. Sect.)* **79**, 155-182.

Igli H., Miellou J.C., Rebetez M., Grivet M. et Chambaudet A. (1995a): Problème inverses pour un modèle de recuit avec energie d'activation répartie. *27e Congrès National d'Analyse Numérique,* Super-Besse, Puy de Dôme, 29 Mai-2 Juin 1995.

Igli H., Miellou J.C., Chambaudet A., Grivet M. and Rebetez M. (1995b): Convection methodology for fission track annealing: direct and inverse numerical simulations in the multi-exponential case. *Appl Radiat.* **46**, 507-508.

Miellou J.C., Igli H., Chambaudet A., Rebetez M. and Grivet M. (1997): Convection methods and functional equations to solve inverse thermochronology problems. *18th International Conference on Nuclear Tracks in Solids.* Cairo, Egypt, 1-5 Sep. 1996 (accepted).

Lutz T.M. and Omar G. (1991): An inverse method of modeling thermal histories from apatite fission track length data. *Earth Planet. Sci. Lett.* **104**, 181-195.

Simon L., Rebetez M., Grivet M. and Chambaudet A. (1996): Mathematical resolution of thermal history: test on some annealing laws. *Abstr. Int. Workshop Fission-Track Dating, Gent-1996,* 106.

REVISED ANNEALING KINETICS OF FISSION TRACKS IN ZIRCON AND GEOLOGICAL IMPLICATIONS

T. TAGAMI[1], R.F. GALBRAITH[2], R. YAMADA[1], G.M. LASLETT[3]

[1] *Department of Geology and Mineralogy, Faculty of Science, Kyoto University, Kyoto 606-01, Japan.*
[2] *Department of Statistical Science, University College London, London WC1E 6BT, U.K.*
[3] *CSIRO MIS, Private Bag 10, Clayton 3169, Australia.*

Abstract

We report data from two new zircon annealing experiments in which spontaneous fission tracks are heated at 398°C and 498°C for 10000 hours. These supplement 29 previous experiments conducted on the same zircon. We fit fanning and parallel Arrhenius models to the 31 experiments using a recently developed modelling procedure, which provides efficient estimates of the kinetic parameters and their precisions. For both fanning and parallel models the fitted mean lengths agree well with the observed ones, although formal statistical tests suggest that the fanning model is a better representation of the data. We discuss the validity of extrapolations of these models to geological time scales, and make comparisons with various geological constraints, including ultra-deep borehole data and fossil partial annealing zones. The model extrapolations are consistent with these constraints, suggesting that the revised kinetics work well, within the limitations of current knowledge. This supports the future use of the zircon fission track system for geological thermal history analysis.

1. Introduction

Zircon fission-track (FT) analysis has been applied in various geological settings to unravel their thermotectonic evolution (e.g. Kamp *et al.*, 1989; Brandon and Vance, 1992; Hasebe *et al.*, 1993; Tagami and Dumitru, 1996). However, while the thermal annealing characteristics of apatite are relatively well-established (e.g. Green *et al.*, 1986, 1989a; Laslett *et al.*, 1987) those of zircon are still a subject of on-going research. A series of laboratory studies of temperature-dependent track retention in zircon has recently been conducted using confined spontaneous track lengths as a measure of annealing, as well as etched track widths for standardisation of track revelation (Tagami *et al.*, 1990; Yamada *et al.*, 1993, 1995a,b; Hasebe *et al.*, 1994). As a result, kinetic parameters that describe annealing at laboratory time scales were established for zircon (Yamada *et al.*, 1995a). In addition,

P. Van den haute and F. De Corte (eds.), Advances in Fission - Track Geochronology, 99-112.

track annealing properties for geological time scales have been better constrained through the zircon FT analysis of a contact metamorphic aureole around a granite pluton (Tagami and Shimada, 1996) and of ultra-deep borehole samples (Tagami *et al.*, 1996).

In the present study we report long-term laboratory zircon FT annealing data that supplement the data of Yamada *et al.* (1995a). We revise the kinetic model of track annealing in zircon, using the new data along with recently developed statistical methods (Laslett and Galbraith, 1996; Galbraith and Laslett, 1997). Finally, the partial annealing zone estimated by extrapolating the fitted model to geological time scales is compared with geological constraints currently available, to infer plausible annealing temperatures in nature.

2. Long-term laboratory annealing data

The sample and experimental methods used in this study are the same as in Yamada *et al.* (1995a). The zircon sample, called NST, has a K-Ar biotite age of ~ 21 Ma and a spontaneous track density of $\sim 4 \times 10^6$ cm^{-2}. These spontaneous tracks have experienced little or no geological annealing since their formation. One aliquot (S_0) of this sample was kept aside, and further aliquots were heated at various temperatures for times ranging from 4.5 min to 1000 h.

We have added two further annealing runs, both for 10000 h, at temperatures of 398° and 498°C, and measured confined track lengths and their angles to the c-axis. The new data are summarised in Table 1 along with some data from the earlier study for comparison, *viz.* the sample S_0 not subjected to laboratory annealing and samples annealed at similar temperatures for shorter times. The new length measurements have shorter means and larger standard deviations than these samples, and show a consistent pattern with the previous data.

3. Fitted models

For reasons given by Yamada *et al.* (1995a), we restrict attention to tracks at an angle of $> 60°$ to the c-axis when fitting models. We fit fanning and parallel Arrhenius models to the data t, T, N, L, sd from all 31 experiments using the method of Laslett and Galbraith (1996). Galbraith and Laslett (1997) applied this approach to the data in Yamada *et al.* (1995a), i.e. excluding the 10000 h experiments, and found that both the parallel and fanning models were tenable. However, when the 10000 h data are included there is a clear statistical preference for a fanning model. Furthermore, for the present data a power transformation of mean length does not improve the fit, so we set the parameter λ of Galbraith and Laslett (1997) to $\lambda = 1$.

TABLE 1. Long-term laboratory annealing data of spontaneous fission tracks in NST zircon: t = heating duration (where S_0 indicates a sample of spontaneous tracks not subjected to laboratory annealing); T = annealing temperature (°C); t_e = etching time (h); N_{all} = number of all measured tracks; N = number of measured tracks > 60° to crystallographic c-axis; L_{all} = mean length of all measured tracks (μm); L = mean length of tracks > 60° (μm); sd = standard deviation of lengths of tracks > 60° (μm); $se = sd/\sqrt{N}$ = standard error of L (μm). Data except for 10000 h annealing are from Yamada *et al.* (1995a), which also has data for 4.5 min, 11 h, 100 h and 1000 h annealing at other temperatures (see Figures 1 and 2).

t	T	t_e	N_{all}	N	L_{all}	L	sd	se
S_0	*	22	107	65	11.14	11.05	0.65	0.08
1 h	395	22	53	32	10.85	10.83	0.89	0.16
11 h	397	22	50	28	10.96	10.89	0.64	0.12
1000 h	398	26	48	33	10.46	10.27	0.55	0.10
10000 h	398	26	60	48	9.98	9.81	0.88	0.13
4.5 min	500	22	51	34	10.85	10.78	0.65	0.11
1 h	500	22	50	35	10.67	10.57	0.73	0.12
11 h	499	26	50	30	10.20	10.13	0.60	0.11
100 h	501	26	50	36	9.37	9.20	0.73	0.12
1000 h	498	26	51	41	8.61	8.59	0.72	0.11
10000 h	498	26	51	39	7.43	7.35	1.34	0.21

3.1. FANNING MODEL

For the *fanning* model, the theoretical mean length μ of tracks (at angle greater than 60°) that have been heated at temperature T Kelvin for time t hours is given by:

$$\mu = \mu_{\max}\left[1 - \exp\left\{c_0 + c_1\,\frac{\log t - \log t_c}{1/T - 1/T_c}\right\}\right]. \tag{1}$$

In this equation, contours of constant mean length are represented by straight lines on an Arrhenius plot ($\log t$ versus $1/T$) which all meet at the point $(1/T_c, \log t_c)$, each line corresponding to a different mean length. The position of the common point $(1/T_c, \log t_c)$ determines the direction and amount of fanning of the lines. The coefficients c_0 and c_1 quantify the relationship between the amount of annealing, temperature and time, and $\mu_{\max}$ represents a theoretical maximum mean length corresponding to zero annealing.

Note that the physical interpretation of $\mu_{\max}$ is somewhat idealised as "zero annealing" refers to conditions that are outside our domain of observation. In particular, the mean length of the sample S_0 does not provide a direct measurement of $\mu_{\max}$ because some natural annealing must have taken place, so that the true mean length for this sample will be less than $\mu_{\max}$ by an unknown amount. We treat the observed mean length for sample S_0 as a censored observation at $T = 20°\text{C}$ and $t = 100$ hours; that is, we assume that if the same tracks had been measured after 100 hours at 20°C they would have a higher mean length.

The five parameters $\mu_{\max}$, c_0, c_1, T_c, and t_c are in principle estimated (by maximum likelihood) from the annealing data. In practice, in order to obtain a stable numerical procedure that produces reliable estimates and precisions, we re-express c_0, c_1, T_c and t_c in terms of b_1, b_2, μ_a and μ_b, given by

$$b_1 = \frac{\log t_a - \log t_c}{1/T_a - 1/T_c}, \qquad \mu_a = \mu_{\max}\left[1 - \exp\left(c_0 + c_1 b_1\right)\right] \tag{2}$$

and

$$b_2 = \frac{\log t_b - \log t_c}{1/T_b - 1/T_c}, \qquad \mu_b = \mu_{\max}\left[1 - \exp\left(c_0 + c_1 b_2\right)\right] . \tag{3}$$

where t_a, t_b, T_a and T_b are two times and temperatures chosen to be within the data region but well separated. We chose $t_a = t_b = 100$, $1/T_a = 0.0012$ and $1/T_b = 0.00145$, which correspond to annealing for 100 hours at temperatures of 560.2°C and 416.5°C, respectively. Here b_1 is the slope of the line on the Arrhenius plot that passes through $(\log t_a, 1/T_a)$ and $(\log t_c, 1/T_c)$ while μ_a is the mean length represented by this line. Similarly b_2 and μ_b are the slope and mean lengths for the contour line through $(\log t_b, 1/T_b)$. Thus we estimate b_1, μ_a, b_2 and μ_b by maximum likelihood and hence estimate c_0, c_1, T_c, and t_c by inverting equations (2) and(3).

For a single annealing experiment, with temperature T and time t, that produces a sample of N tracks, the observed mean length L is assumed to be normally distributed with mean μ and variance given by:

$$\sigma^2 + \frac{1}{N}\exp\left\{-1.04 - 2.4\left[\log(\mu/9) - \sqrt{[\log(\mu/9)]^2 + 0.02}\right]\right\} . \tag{4}$$

The parameter σ represents the "between experiment" standard deviation, which is estimated along with the other parameters. The second term in (4) represents "within experiment" sampling variation and is treated as a known component of variance in the main fitting procedure. The expression in braces has been fitted to the scatter plot of $\log sd$ against $\log L$. For $\mu = 7, 8, 9, 10$ and 11, the within sample standard deviations of individual track lengths calculated from this expression are 1.14, 0.85, 0.70, 0.65, and 0.63 μm, respectively.

In some experiments, annealing is so advanced that no confined tracks are observed. This is a form of censoring that tells us that μ is likely to be small; so in order to use this information in the likelihood function, we need to express the probability that confined tracks are observed as a function of μ. We assume that this probability, $p(\mu)$ say, is given by

$$p(\mu) = \frac{e^{2(\mu-5)}}{1+e^{2(\mu-5)}} \tag{5}$$

which equals 0.5 when $\mu = 5$, 0.01 when $\mu = 2.7$ and 0.99 when $\mu = 7.3$. Thus for an experiment in which no confined tracks are seen, the likelihood function favours values of μ below 7.3 μm, the likelihood increasing as μ decreases, while for an experiment in which tracks are seen the likelihood will be practically zero for μ below 2.7 μm and will increase with μ. In the latter case, the observed value of L will largely determine the likelihood. The range between 2.7 and 7.3 μm where tracks might or might not be seen is chosen to be fairly wide, reflecting our current lack of knowledge. Laslett and Galbraith (1996) and Galbraith and Laslett (1997) discuss the rationale behind this and give technical details for maximum likelihood estimation.

Predicted mean lengths of tracks heated at temperature T for time t may be obtained by substituting t, T and the parameter estimates into equation (1). Prediction intervals may be obtained by choosing appropriate (t_a, T_a) and (t_b, T_b) in equations (2) and (3): we set $t_a = t$, $T_a = T$ and choose (t_b, T_b) within the data region, but on a different contour. The profile likelihood interval for μ_a is then equivalent to that of the predicted mean length μ at (t, T). This takes into account error in the fitted model; when comparing this with a measured mean length, we must in addition allow for the measurement error, which has variance given by (4).

To predict the mean lengths of tracks in a field sample having a known thermal history, one must recognise that tracks accumulate over time so the model predictions should be integrated over the thermal history, as in Green *et al.* (1989b). This will generally lead to a longer predicted mean length than that given by a direct extrapolation of the model equation. In §5 we compute such integrated predictions and intervals for three borehole samples, assuming a thermal history in which T is constant. Even for this simple thermal history the calculation is not straightforward; an easier, but less satisfactory, alternative would be to check whether the prediction obtained by extrapolating the model equation is close to, but less than, the observed mean track length.

3.2. PARALLEL MODEL

For the *parallel* model we have, instead of equation (1),

$$\mu = \mu_{\max}\left[1 - \exp\left\{c_{0p} + c_{1p}\left(\log t - \frac{b_p}{T}\right)\right\}\right]. \tag{6}$$

In this case, contours of constant mean length on an Arrhenius plot are parallel lines that have slope b_p, each line corresponding to a different mean length. For the purpose of fitting equation (6) we re-express c_{0p} and c_{1p} in terms of μ_{ap} and μ_{bp} given by

$$\mu_{ap} = \mu_{\max}\left[1 - \exp\left\{c_{0p} + c_{1p}\left(\log t_a - \frac{b_p}{T_a}\right)\right\}\right] \tag{7}$$

and

$$\mu_{bp} = \mu_{\max}\left[1 - \exp\left\{c_{0p} + c_{1p}\left(\log t_b - \frac{b_p}{T_b}\right)\right\}\right], \tag{8}$$

where we chose the same t_a, t_b, T_a and T_b as for the fanning model. Thus μ_{ap} and μ_{bp} are the theoretical mean lengths corresponding to the lines (with slope b_p) through the 100 h, 560.2°C point and the 100 h, 416.5°C point, respectively. The variance of an observed mean length L is still given by equation (4) and the probability that no tracks are seen is given by (5).

4. Results and interpretation

Estimates and 95% confidence intervals for the model parameters are given in Table 2. The parameter estimates are correlated, so this summary does not quite give the complete picture.

Figure 1 shows the observed mean lengths plotted against temperature along with fitted values for the fanning and parallel models. In both cases the model appears to fit well. The points with down arrows, where no confined tracks are seen, are plotted at the maximum length of observed surface tracks, which was estimated roughly to the nearest micron. In some cases the curve passes slightly below this point, but because the position of the point is approximate the model may still be compatible with the data. Plots of standardised residuals for each annealing time (not shown here) show no systematic patterns, though the residuals are slightly more dispersed for the parallel model.

However, as mentioned earlier, the fanning model is favoured by maximum likelihood theory. This is evident from Table 2 in two ways. Firstly, the maximum log-likelihood ($L_{\max}$) is greater for the fanning model by $32.41 - 23.16 = 9.25$ units, which is a large improvement for fitting just one more parameter. (More formally, the AIC value, defined as $-2L_{\max} + 2p$,

TABLE 2. Maximum likelihood estimates

Fanning model[1] ($L_{max} = 32.41$)				Parallel model[2] ($L_{max} = 23.16$)			
parameter	estimate	95% C.I.		parameter	estimate	95% C.I.	
μ_a	7.82	7.68	7.94	μ_{ap}	7.79	7.62	7.94
μ_b	10.41	10.33	10.49	μ_{bp}	10.54	10.45	10.63
c_0	−6.502			c_{0p}	6.719		
$1000\, c_1$	0.1431			c_{1p}	0.2077		
$b_1/1000$	37.29	35.71	38.84	$b_p/1000$	35.70	33.92	37.79
$b_2/1000$	28.01	25.78	30.81				
μ_{max}	11.35	11.09	11.76	μ_{max}	11.04	10.85	11.26
σ	0.036	0.000	0.119	σ	0.120	0.050	0.202
$\log t_c$	−23.515						
$1000/T_c$	0.4459						

[1]Parameters are given by equations (1) and (4). The units of measurement of μ_a, μ_b, μ_{max} and σ are all microns; T_c is in Kelvin; t_c is in hours; b_1 and b_2 are slopes on the Arrhenius plot; and c_0 and c_1 are found from equations (2) and (3).
[2]Parameters are given by equations (6) and (4). The units of measurement of μ_{ap}, μ_{bp}, μ_{max} and σ are all microns; b_p is the slope on the Arrhenius plot; and c_{0p} and c_{1p} are found from equations (7) and (8).

where p is the number is fitted parameters, is much smaller for the fanning model.) Secondly, the estimates of the two slopes b_1 and b_2 are very different, and their joint 95% confidence region does not intersect (or pass near) the line $b_1 = b_2$ corresponding to a parallel model.

Substituting the estimates from Table 2 into equation (1) gives the following *fanning model prediction formula* for the mean length μ microns after annealing at temperature T Kelvin for time t hours:

$$\mu = 11.35\left[1 - \exp\left\{-6.502 + 0.1431\,\frac{\log t + 23.515}{1000/T - 0.4459}\right\}\right]. \tag{9}$$

The corresponding *parallel model prediction formula* from equation (6) is:

$$\mu = 11.04\left[1 - \exp\left\{6.719 + 0.2077\left(\log t - \frac{35700}{T}\right)\right\}\right]. \tag{10}$$

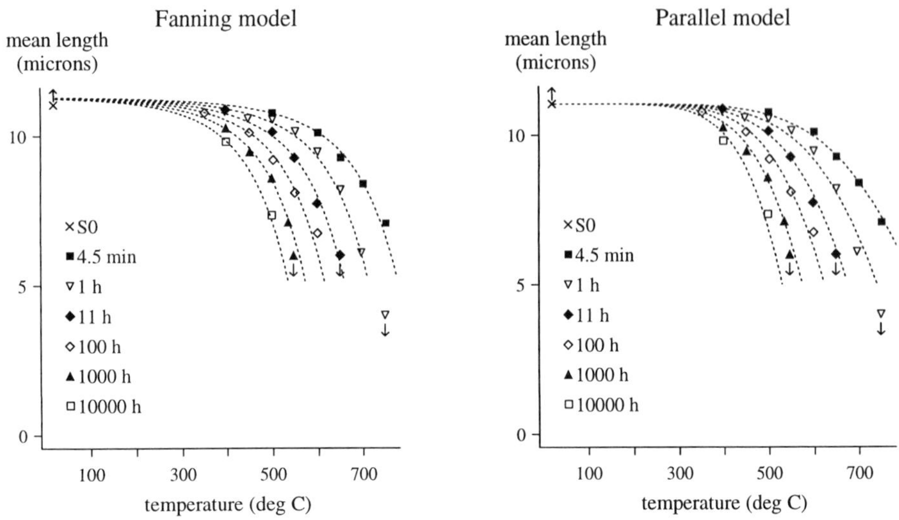

Figure 1. Mean length L versus temperature for each isochronal series. The data are from Yamada *et al.* (1995a, Table 1) plus the 10000 h data from Table 1. The dotted curves are fitted values for the fanning and parallel models (Table 2). For the high temperature experiments where no confined tracks were found, points are plotted at an estimated maximum length (based on lengths of surface tracks) with down arrows to indicate that the fitted curve should pass below the symbol. The mean length for the sample S_0 is plotted at a nominal temperature of 20°C with an up arrow to indicate that the fitted curve should lie above the symbol.

There are several points to bear in mind when interpreting these predictions and their geological implications:

1. Equations (9) and (10) can only be expected to provide sensible predictions for t and T such that the value of μ is close to the observed range of mean lengths, i.e. for 5 μm $< \mu <$ 11 μm, say.
2. Models that give similar fitted values to the data can give very different predicted values when extrapolated. For example, annealing to 6 μm over 100 Ma, requires a substantially higher temperature for the fanning model than for the parallel model (310°C *vs* 288°C), but annealing to 10 μm requires a somewhat lower temperature for the fanning model (200°C *vs* 228°C).
3. There are several sources of uncertainty, including error in the parameter estimates (see Table 2), between and within experiment standard deviations, and uncertainty in the field observations with which predictions are compared. For example, for 100 Ma annealing at 282°C, the 95% interval estimate of μ from the fanning model is from 7.04 to 8.01 μm; while for 100 Ma annealing at 203°C, the corresponding interval is from 9.74 to 10.18 μm, which is understandably tighter, being in a region of less rapid annealing. These intervals are of course much

wider than 95% intervals for μ when t and T are inside the data region (c.f. the 95% confidence intervals for μ_a and μ_b in Table 2).

4. Allowance should be made for several other factors when comparing extrapolated values of (9) and (10) with field data. In particular, in the models derived from laboratory data, all existing tracks were heated at constant temperature T for time t; whereas for field data tracks accumulate over time and T may vary. Also, equations (9) and (10) are derived from tracks at angles greater than 60°, whereas reported field data may include all tracks, and so have slightly longer mean lengths.
5. Etching and measurement conditions may differ between laboratories, producing shorter track lengths for lower temperature of etchant or for etching criteria with narrower track widths (Yamada *et al*, 1995b).

5. Geological implications

Track annealing temperatures in nature can be estimated by extrapolating the fitted kinetic model to geological time scales (Figure 2). Before doing this, however, we need to demonstrate that the track annealing process is the same physical phenomenon under both laboratory and natural conditions. Detailed characteristics of geological track annealing were first described for zircon using a contact metamorphic aureole around a granitic pluton, with estimated heating durations of $\sim 10^5$–10^6 years (Tagami and Shimada, 1996). They found that the observed track length reduction characteristics arc indistinguishable from those in laboratory heating experiments (e.g. Yamada *et al.*, 1995b). It is noted here that the characteristics of track annealing in zircon are very similar to those reported by Green *et al.* (1986) for apatite, and that kinetic models based on laboratory data work well for apatite fission track analysis of geological samples (Green *et al.*, 1989b). These facts suggest that the extrapolation is reasonable for zircon also.

In comparing model predictions with geological constraints, we need to take into account the points enumerated below equation (10). In particular, if track accumulation over time is not taken into account, the model should extrapolate, other things being equal, to a *lower* mean length than the true mean length for a geologically annealed sample, because it effectively assumes that all tracks were formed at time t.

Table 3 summarises several geological constraints which we now discuss. The most direct constraints are obtained from zircon samples collected from ultra-deep boreholes having subnormal geothermal gradients. In such a geological setting, the current geothermal regime can be reasonably assumed to have been stable over a period, allowing us to regard the setting as ex-

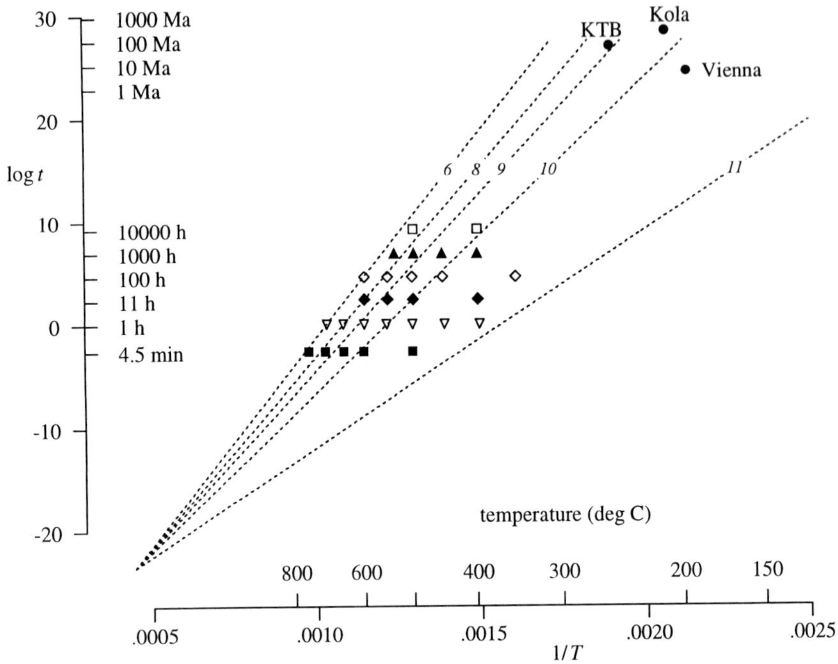

Figure 2. Arrhenius plot showing the design points of the annealing experiments (c.f. Figure 1) and contour lines for the fitted fanning model (Table 2) extrapolated to geological time. Also shown are the approximate locations of the three borehole samples in Table 3.

hibiting natural, long-term isothermal annealing behaviour. Tagami *et al.* (1996) reported detailed age and length data using zircons from sandstone samples in Vienna Basin boreholes. They found that zircons showed no signs of appreciable annealing since sediment deposition throughout the depths of 1.5–7.5 km, the deepest with a present environmental temperature of ~200°C. The present geothermal regime as well as sample burial depths have been near-constant for at least the past 5 Ma according to geology in and around the basin. These samples, particularly the deepest one, therefore give a constraint on the geological track length reduction in the initial stage of annealing.

Track length data from other ultra-deep boreholes have also been reported recently. Green *et al.* (1996) analysed zircons from the Kola Peninsula Superdeep Drillhole SG-3. They found that the mean track lengths vary between 10.8 ± 0.3 μm and 9.8 ± 0.1 μm at depths of 7086 to 12052 metres, the deepest with a present environmental temperature of ~212°C and a heating duration of ~250 Ma. Coyle and Wagner (1996) reported zircon analysis of KTB HB ultra-deep borehole samples. They found a mean track length of 8.9 μm at 8700 metres, approximately 255°C, with a

TABLE 3. Compiled geological constraints on fission track annealing in zircon

Sample	T^1	t^2	L^3 (se)	Age[4] (se)	Reference
Boreholes					
Vienna[5]	197	~5	10.6 (0.14)	186 (24)	Tagami *et al.* (1996)
Kola	212	~250	9.8 (0.10)	271 (18)	Green *et al.* (1996)
KTB HB	255	10–100	8.9 (?)	?	Coyle & Wagner (1996)
Fossil PAZ					
NZ Alps	~300	—	$< 4.4^6$	—	Green *et al.* (1996)
Contact aureole					
Japan	$\sim 50^7$	0.1–1	$8.8 - 4.4^8$	—	Tagami & Shimada (1996)

[1]Estimated annealing temperature (°C)
[2]Estimated heating duration (Ma)
[3]Mean length of measured tracks (μm)
[4]Estimated age of the zircon sample subjected to geological annealing (Ma)
[5]L calculated for tracks > 60° to the c-axis
[6]Apparent length value equivalent to total annealing (Yamada *et al.*, 1995b)
[7]Temperature difference (°C) between top and bottom of observed part of PAZ.
[8]Approximate mean lengths at top and bottom of observed part of PAZ.

plausible heating duration of ~10–100 Ma (personal communication).

The fitted kinetic model shows good agreement with these borehole data. Using equation (9), we have calculated point and 95% interval estimates for the mean length integrated over a thermal history of constant temperature for the relevant geological time (i.e. taking into account the accumulation of tracks over time). These may be compared with the measured mean lengths, bearing in mind that their uncertainties will include a between experiment component σ in addition to the reported standard errors in Table 3. For the Vienna data we used $T = 197°$C and $t = 5$ Ma to obtain a predicted mean length of 10.41 (95% interval 10.23–10.62) μm, which agrees well with the observed L of 10.55 ± 0.28 μm (2se). For the Kola data, using $T = 212°$C and $t = 250$ Ma the predictions are 9.79 (9.59–10.03) μm compared with observed values 9.8 ± 0.22 μm, and for the KTB data, using $T = 255°$C and $t = 55$ Ma the predictions are 9.11 (8.89–9.33) μm compared with the observed values 8.9 ± 0.3 μm, say. Considering that the temperatures and times are themselves uncertain, the predictions are clearly in agreement with the data.

A less direct constraint comes from the study of Green *et al.* (1996), involving comparison of zircon data with vitrinite reflectance (VR) and metamorphic grade. They considered data from the Southern Alps of New

Zealand, where a fossil partial annealing zone (PAZ) has been exhumed during the Pliocene to Recent uplift and erosion, and they found that the onset of significant age reduction correlates with VR values of ~6%, while total annealing is equivalent to ~8% or higher. A metamorphic grade analysis found that total annealing is roughly coincident with the transition between chlorite subzones 2 and 3, which has been assigned a peak metamorphic temperature of 300°C (Table 3). They concluded that "paleotemperatures well above 250°C are required in order to produce significant fission track annealing in zircon".

From equation (9) we may calculate, for a given temperature, the time required to achieve "total" annealing, which we here take to mean that μ is below 4.4 μm. This corresponds to the highest temperature limit of the zircon PAZ, given by $r = 0.4$, where $r = \mu/11.05$ is the track length reduction compared with the sample S_0 (Yamada *et al*, 1995a). For $T = 250$, 280, 300, and 320°C we find that the required t is approximately 4×10^6, 5×10^4, 3480, and 290 Ma, respectively. The corresponding times from the parallel model, equation (10), are somewhat shorter (4×10^4, 940, 100 and 12 Ma). We may regard these calculations as being in agreement with the conclusion of Green *et al.* (1996) in that for T above about 320°C, total annealing would be achieved in a reasonable time, while for T substantially below 300°C the required time would be inordinately long.

A further constraint may be derived from the study by Tagami and Shimada (1996), who used data on a contact metamorphic aureole from Takatsukiyama Gr., S.W. Japan, coupled with thermal modelling of conduction heat transfer, to estimate the temperature difference between two points in the zircon partial annealing zone. They estimated heating durations of 0.1–1 Ma and a difference of ~50°C in maximum temperatures between localities that corresponded to the beginning and completion of measurable partial resetting due to granite intrusion. This resetting zone was recognised by age reduction of outcrop samples and was converted to a track length reduction range of $0.4 < r < 0.8$ (part of the zircon PAZ defined by Yamada *et al*, 1995a, as $0.4 < r < 0.95$). Taking the range of partial resetting to correspond approximately to mean confined track lengths between 4.4 and 8.8 μm (corresponding to $0.4 < r < 0.8$), we may calculate the predicted temperature difference from equations (9) and (10). Fixing $t = 0.1$ Ma, we find this predicted temperature range of fossil PAZ to be 82°C from the fanning model and 55°C from the parallel model. These results are comparable to the interval of ~50°C estimated by Tagami and Shimada (1996) considering the geological assumptions involved.

6. Conclusion

The new 10000 h zircon annealing data reported in Table 1 have been found to be informative. Both fanning and parallel Arrhenius models produce reasonable fitted values and residuals, but the fanning model is now favoured by formal statistical theory. Extrapolations from these models have been found to be consistent with current known geological constraints, insofar as these can be judged quantitatively. However, the extent of available laboratory data is still limited, so it may transpire that a different form of model will emerge in due course, based on study of future laboratory and field observations.

Acknowledgments We thank Paul Green for background information and discussion on the Kola borehole data and S.Alps, New Zealand data. Rex Galbraith gratefully acknowledges support from the University of London Central Research Fund.

References

Brandon, M.T. and Vance, J.A. (1992): Tectonic evolution of the Cenozoic Olympic subduction complex, Washington State, as deduced from fission track ages for detrital zircons. *Am. J. Sci.,* **292,** 565-636.

Coyle, D.A. and Wagner, G.A. (1996): Fission-track dating of zircon and titanite from the 9101m deep KTB: observed fundamentals of track stability and thermal history reconstuction. Abstract of presentation at *International Workshop on Fission Track Dating,* Gent, 1996.

Galbraith, R.F. and Laslett, G.M. (1997): Statistical modelling of thermal annealing of fission tracks in zircon. *Chem. Geol. (Isot. Geosci. Sect.)* **140,** 123-135.

Green, P.F., Duddy, I.R., Gleadow, A.J.W., Tingate, P.R. and Laslett, G.M. (1986): Thermal annealing of fission tracks in apatite, 1. A qualitative description. *Chem. Geol. (Isot. Geosci. Sect.)* **59,** 237-253.

Green, P.F., Duddy, I.R., Gleadow, A.J.W. and Lovering, J.F. (1989a): Apatite fission-track analysis as a palaeotemperature indicator for hydrocarbon exploration. In: *Thermal History of Sedimentary Basins – Methods and Case Histories.* (eds Naeser, N.D. and McCulloch, T.H.) Springer, NY, pp. 181-195.

Green, P.F., Duddy, I.R., Laslett, G.M., Hegarty, K.A., Gleadow, A.J.W. and Lovering, J.F. (1989b): Thermal annealing of fission tracks in apatite, 4. Quantitative modelling techniques and extension to geological timescales. *Chem. Geol. (Isot. Geosci. Sect.)* **79,** 155-182.

Green, P.F., Hegarty, K.A., Duddy, I.R., Foland, S.S. and Gorbachev, V. (1996): Geological constraints on fission track annealing in zircon. Abstract of presentation at *International Workshop on Fission Track Dating,* Gent, 1996.

Hasebe, N., Tagami, T. and Nishimura, S. (1993): Evolution of the Shimanto accretionary complex. A fission-track thermochronologic study. In: *Thermal Evolution of the Tertiary Shimanto Belt, Southwest Japan: An Example of Ridge-Trench Interaction.* (ed Underwood, M.B.) *Geol. Soc. Am. Spec. Pap.,* **273,** 121-136.

Hasebe, N., Tagami, T. and Nishimura, S. (1994): Towards zircon fission-track thermochronology: reference framework for confined track length measurements. *Chem. Geol. (Isot. Geosci. Sect.)* **112,** 169-178.

Kamp, P.J.J., Green, P.F. and White, S.H. (1989): Fission track analysis reveals character of collisional tectonics in New Zealand. *Tectonics* **8,** 169-195.

Laslett, G.M., Green, P.F., Duddy, I.R. and Gleadow, A.J.W. (1987): Thermal annealing of fission tracks in apatite, 2. A quantitative analysis. *Chem. Geol. (Isot. Geosci. Sect.)* **65,** 1-13.

Laslett, G.M. and Galbraith, R.F. (1996): Statistical modelling of thermal annealing of fission tracks in apatite. *Geochimica et Cosmochimica Acta* **60,** 5117-5131.

Tagami, T., Ito, H. and Nishimura, S. (1990): Thermal annealing characteristics of spontaneous fission tracks in zircon. *Chem. Geol. (Isot. Geosci. Sect.)* **80,** 159-169.

Tagami, T. and Dumitru, T.A. (1996): Provenance and thermal history of the Franciscan accretionary complex: constraints from zircon fission track thermochronology. *J. Geophys. Res.* **101,** 8245-8255.

Tagami, T. and Shimada, C. (1996): Natural long-term annealing of the zircon fission track system around a granitic pluton. *J. Geophys. Res. (Solid Earth)* **101,** 8245-8255.

Tagami, T., Carter, A. and Hurford, A.J. (1996): Natural long-term annealing of the zircon fission-track system in Vienna Basin deep borehole samples: constraints upon the partial annealing zone and closure temperature. *Chem. Geol. (Isot. Geosci. Sect.)* **130,** 147-157.

Yamada, R., Tagami, T. and Nishimura, S. (1993): Assessment of overetching factor for confined fission-track length measurement in zircon. *Chem. Geol. (Isot. Geosci. Sect.)* **104,** 251-259.

Yamada, R., Tagami, T., Nishimura, S. and Ito, H. (1995a): Annealing kinetics of fission tracks in zircon: an experimental study. *Chem. Geol. (Isot. Geosci. Sect.)* **122,** 249-258.

Yamada, R., Tagami, T. and Nishimura, S. (1995b): Confined fission track length measurement in zircon.: assessment of factors affecting the palaeotemperature estimate. *Chem. Geol. (Isot. Geosci. Sect.)* **119,** 293-306.

Section III

Thermochronology

DIVINING BURIAL AND THERMAL HISTORIES FROM INDICATOR DATA: APPLICATION & LIMITATIONS

An example from the Irish Sea & Cheshire Basins

M. R. GILES and S. L. INDRELID
Shell International Exploration and Production B.V.

Abstract

Any property of a sediment which depends on temperature or stress can potentially provide a record of the burial and thermal history that the sediment has experienced. Such indicators include the present day temperatures, fission-track age and length data, vitrinite reflectance, spore colour, ^{39}Ar / ^{40}Ar ages from feldspars or diagenetic illites, smectite/illite ratios, biomarker transformations and stress dependent properties such as porosity and sonic velocity.

Almost all known thermal indicators are measurements of irreversible transformations. Thus up until the point at which the transformation is complete these indicators primarily record the maximum temperature. Fission-tracks, however, are a reversible indicator that provides information (over a restricted temperature range), about the heating/cooling rate, maximum temperature reached and timing of thermal events e.g. cooling (uplift). They can potentially give useful information about the uplift history and in favourable circumstances both the burial and uplift history.

A suite of indicators can be used in conjunction with a thermal/stress modelling program to refine the burial and thermal history. Clearly, the more thermal indicators which are included in the analysis, the greater the constraints that can be placed on the thermal history. However, at some time during the burial history even the most resistant indicator may have been reset, and the temperature history prior to this time is thus completely unconstrained. This point can be as recent as the present day for cases where the current temperature is the historical maximum. The situation is even more complex when multiple phases of uplift/burial/thermal events have occurred. Thus reverse thermal models are limited in their ability to see into the past. In order to model the temperature history by forward thermal modelling assumptions have to be made about the unconstrained portions of the burial history i.e. the amount of section missing at unconformities, and the variables controlling the heat flow.

The process of fitting the calibration data is generally carried out on a well-by-well basis adjusting the model parameters to fit the constraining data. However, such a scheme commonly leads to inconsistencies when evaluating the thermal history of a larger area because insufficient attention is paid to the likely range of variation (e.g. the length-scale over which heat flow can vary).

P. Van den haute and F. De Corte (eds.), Advances in Fission - Track Geochronology, 115-150.

A large amount of apatite fission-track data is available for the Northwest of England (Lake District, East Irish Sea Basin and Cheshire Basin). In this area the post-Triassic burial history is very poorly constrained due to erosion and non-deposition. Explanations for the patterns seen in the AFT data include: a Tertiary heat flow pulse, hot fluids circulating through the basin, and several km's of uplift. The use of multiple thermal indicators can be used to show that some interpretations based only on AFT data can be misleading and that there is no unique solution using thermal modelling alone.

In conclusion, interpretations based solely on AFT data cannot supply a unique answer to burial history problems. The use of multiple thermal indicator data improves resolution. However, the inherent limitations on the use of indicator data and the lack of unique solutions from forward modelling ensures that it is difficult or impossible to distinguish between some hypotheses based on indicator data alone. Ultimately the additional data necessary to refine the thermal history can only be found from (regional) geological data, and fundamental geodynamic constraints.

1. Introduction

The calculation of the hydrocarbon volumes and the timing of their expulsion for prospect appraisal purposes requires that the time-temperature history of the source rock interval be known (Tissot and Welte, 1978). A primary input into such calculations is the burial history of the sediment, including the timing and amount of uplift. Any indicator which provides a quantitative measure of the time-temperature history may be used in constraining the thermal model.

The information required in the calculation of the burial history includes the age, depth and compaction history (Sclater and Christie, 1980). However, in some areas significant amounts of time are not represented in the stratigraphic record, making prediction of the timing and location of hydrocarbon generation extremely difficult. An example of this occurs in Oman, where there are several unconformities with stratigraphic gaps of over 50 Myr, making reconstruction of burial history, and hence understanding of the timing of hydrocarbon generation, very difficult.

The amount of missing section and the timing of uplift can be addressed using:

1. Biostratigraphic data
2. Seismic or log correlations
3. Regional geological correlations
4. Sudden jumps in thermal or stress (e.g.vitrinite reflectance, sonic velocity) indicators (see Tables 1 and 2)
5. The presence of surface exposure features such as bauxite or paleokarst.
6. Inclusion of reworked material within a sedimentary unit.
7. The identification of erosional remnants.

The timing of the uplift can be estimated from 1-5 above and in addition apatite fission-track data as discussed below.

2. Constraining thermal and burial history

2.1 HEATFLOW

The starting point of any reconstruction of the thermal history is the present day thermal structure of the sediment column. Fourier's law of heat conduction (Turcotte and Schubert, 1982; Kappelmeyer and Haenel, 1974; Kaviany, 1991; Wangen, 1992) allows the geothermal gradient (dT/dZ) to be related to the heat flow (***Q***) via the thermal conductivity (**k**) i.e.:

$$Q = \mathbf{k} \frac{dT}{dZ} \qquad (1)$$

Down-hole temperature measurements can be used to calculate the geothermal gradient and therefore the heat flow, provided that the thermal conductivity is known or can be estimated. Thermal conductivity varies with lithology, porosity and temperature (Woodside and Messmer, 1961; Sommerton, 1992; Brigaud and Vasseur 1989; Giles, 1997). Although heat flow varies on a relatively long wavelength basis (see below), the geothermal gradient is controlled by local changes in lithology and porosity which determine the thermal conductivity. Similarly, geothermal gradients are expected to vary with depth as a function of thermal conductivity, and hence lithology.

As an example of the relationship between heat flow and geothermal gradient consider an average continental heat flow of 56 mWm^{-2}. The geothermal gradient through a typical shale with a thermal conductivity of 1.4 $W\ m^{-1}\ K^{-1}$ would be 40°C km^{-1}. The thermal conductivity of some shales can be as low as 0.8 $W\ m^{-1}\ K^{-1}$, in which case the geothermal gradient for the same heat flow would be 70°C km^{-1}. In contrast a typical sandstone with a thermal conductivity of 3.4 $W\ m^{-1}\ K^{-1}$ would exhibit a geothermal gradient of only 16.5 °C km^{-1} for the same heat flow.

The temperature reached at any point in the subsurface will depend not only on the geothermal gradient but also the surface temperature. Fluctuations in the surface temperature through time can easily reach 20°C, and will be transmitted in to the subsurface.

2.2 THERMAL INDICATORS

Thermal indicators represent kinetically controlled reactions where the extent of reaction is a measure of the temperature- time path followed by the sediment. Many indicators have been calibrated as given in Table 1. The most commonly used indicator is vitrinite reflectance (VR). The reflectance of the vitrinite group of organic macerals has been shown to vary smoothly and predictably with temperature (Fig. 1). Thus the time-temperature history may be used to predict vitrinite reflectance (Lopatin, 1971; Burnham and Sweeney 1989;

TABLE 1. Thermal History Indicators

Major Indicator	*Example*	*Temperature Range °C*	*Reference*
Vitrinite Reflectance		Continuous	Tissot and Welte, 1978
Thermal Alteration Index (spore colour)		Continuous	Staplin, 1969 Barnard *et al* 1976
Thermal Colour Index		< VRE 2.5	van Gijzel, 1990
Absorption ratio		0-200	Ellacott *et al*, 1994
Solid hydrocarbon reflectance		80->300	Landis & Castãno, 1994
Smectite --> Illite transition in shales		40-200°C	Pytte and Reynolds, 1989
Isomerisation reactions	Hopane Sterane	50-100 60-170	Abbott *et al.*, 1990 McKenzie & McKenzie, 1983
Aromatisation reactions	Sterane	70-170	McKenzie & McKenzie, 1983
$^{40}K/^{39}Ar$ of feldspars		100 - 200	Harrison and Be, 1983
Fluid inclusions	Quartz cement Calcite cement	Continuous but dependent on timing & rate of precipitation.	Roedder 1984 Walderhaug 1994
Fission-tracks	Apatite Zircon	50-125 175-250	Naeser, 1979 Gleadow *et al.*, 1986; Green *et al* 1989 Tagami *et al.*, 1990, 1996; Yamada *et al.*, 1995a, 1995b
Volume of quartz cement		>80	Giles *et al.*, 1997
Temperature measurements		Continuous	

Sweeney and Burnham, 1990). However, given that different macerals can give different results (Bensley and Crelling, 1994) and that vitrinite suppression can result in underestimates of up to 0.25% Ro (Wilkins *et al.*, 1992; George *et al.*, 1994), data quality is a major issue for VR based thermal history studies.

As vitrinite reflectance is a measure of the time-temperature history it follows that a VR-depth plot of data from multiple wells should exhibit a well defined trend where the wells have experienced similar burial and heat flow histories. In an individual well a sudden jump in the VR-depth profile well (Fig. 2) is an indication of the presence of an unconformity with significant missing section. Thus where the VR-depth for multiple wells exhibits scatter, based on reliable VR

Figure 1.
Increase in vitrinite maturity with burial depth for a well which has undergone continuous burial.

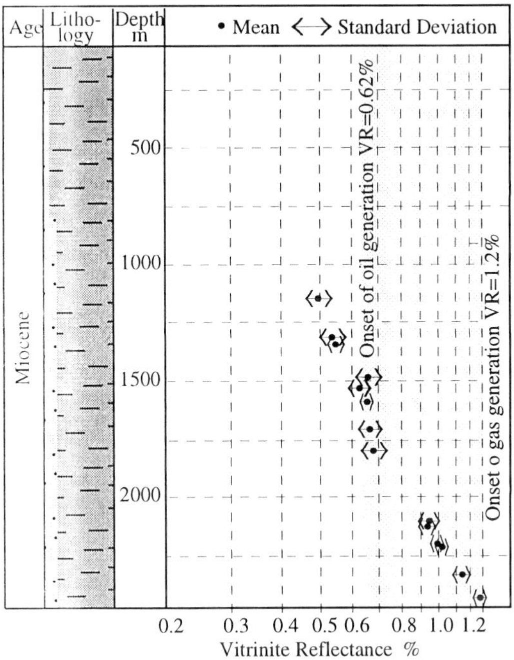

Figure 2.
Vitrinite reflectance profile across a major unconformity, Mazeres borehole, southern France. The unconformity is marked by an abrupt increase in thermal maturity, as indicated by the reflectance. (After Allen & Allen 1990, Robert 1988.)

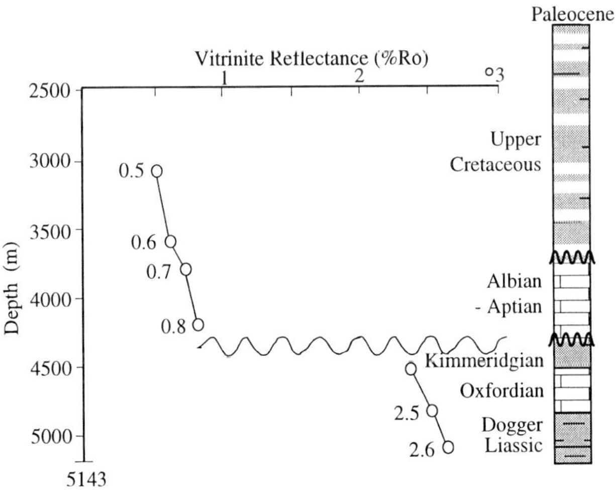

data , then the likeliest explanation is that the burial histories differ significantly between wells.

Biomarkers are organic molecules with a structure which is based on the natural substance from which they were derived. The effect of increasing temperature is to cause transformations in such molecules which can be measured and related to temperature. For some biologically derived substances, only one isomer exists in recently deposited sediments. With increased temperature the isomer is progressively converted to an equilibrium mixture of isomers. Reaction progress is ascertained by measuring the relative proportions of the two isomers (Abbott *et al.*, 1990, McKenzie & McKenzie, 1983). Aromatisation reactions can also be used, for example the loss of a methyl group in the conversion from mono-aromatic to tri-aromatic steroids can also be used as a thermal indicator (McKenzie and McKenzie, 1983). These biomarker transformations are all thought to be first order reactions which are temperature dependent according to the Arrhenius equation. Equilibrium is attained when the temperature is high enough for the forward and reverse reaction rates to balance. The equilibrium (thus maximum) value of the ratio 20S/(20S+20R) for sterane isomerisation is 0.54, while that for hopane isomerisation (22S/(22S+22R)) is 0.61 (Abbott *et al.*, 1990, McKenzie & McKenzie, 1983).

The use of biomarker transformations in addition to VR measurements provides an additional constraint on the thermal history because of the different kinetics (i.e. different time-temperature response). The use of such data in addition to VR can also help identify anomalies in the behaviour of VR (e.g. suppression or poor data).

Vitrinite reflectance and biomarker transformations are irreversible thermal indicators, in that they increase continuously with increasing temperature, up to a maximum value in the case of biomarker. The value of these indicators does not decrease if the temperature falls. As a result such indicators generally "see" the highest temperature reached during the burial history.

Fission-track data are unusual in that they provide a chance to record decreases in temperature as well as increases. Spontaneous fission of ^{238}U within apatite creates damage zones produced by recoil of the daughter nuclei which initially have an etchable length of around 15μm. Fission occurs continuously so that tracks accumulate with time, and the number of tracks is a measure of the crystal's uranium content and its age. However, if the apatite is exposed to high temperatures the tracks heal (anneal) by progressive shortening until the track has healed completely (Naeser and Faul, 1969; Naeser, 1979, 1981; Green *et al.*, 1986; Laslett *et al.*, 1987). At temperatures above ~300 °C the tracks are annealed in a few hours. Total annealing can also occur if temperatures above ~100 °C are maintained for tens of millions of years (Green *et al.*, 1986; Laslett *et al.*, 1987) and shortening of the tracks can occur at temperatures above 50 °C. The age of the tracks in a given sample thus can give an indication of the last time the apatite experienced temperatures within the partial annealing zone (i.e. > 50 °C, < 125 °C for geological time-scales), and the length distribution can indicate how rapidly cooling occurred.

Apatite fission-track data (AFT) unlike other thermal indicators is reversible in the sense that if the temperature exceeds 50 °C then annealing of the tracks begins to reset the sample ages and track lengths. Thus AFT data primarily record information concerning the last cooling event when the temperature fell below those required for annealing. Where the sample temperatures never exceeded 125 °C but have been above 50 °C (the so-called partial annealing zone) then the sample age and track length distribution can reflect multiple cooling and heating events. The accuracy of paleo-temperatures derived by this method can be as good as 10 °C but single grain ages may vary significantly depending on composition, and wildly different ages are sometimes observed in samples from the same outcrop (Green, 1986; McCulloch, 1994).

AFT data are commonly used to reconstruct temperature histories by estimating paleogeothermal gradients based on temperatures calculated from the annealing kinetics (Green *et al.*, in press). By plotting the calculated temperature versus depth it is thought that the palaeogeothermal gradient can be calculated (Fig. 3a). There are a number of reasons why this approach can give erroneous results:-

1. The error in the inferred temperature is at least +/- 10 °C, and with sparse data the range of possible geotherms can be large (see Fig. 3a).

2. Consideration of how geothermal gradients vary with lithology demonstrates that such an analysis is also physically unrealistic (see Fig. 3b).

3. AFT annealing is a kinetic process, and hence a lower maximum temperature maintained over a long period will cause a similar degree of annealing to a high temperature maintained for a short time.

4. This type of analysis also assumes that all the samples from the same area or borehole are uplifted through the annealing zone at the same time. This will not be true for slow uplift where some samples will enter and leave the annealing zone at different times, and over that time the heat flow may have changed considerably.

5. The temperature gradients calculated by this method may depend on the surface temperature chosen. At the time of uplift the surface temperature may have been considerably different from those prevailing today.

Figure 3.
Problems with deriving thermal gradients from AFT data:
a) For a limited number of data, each with associated error bars, a wide range of gradients are possible.
b) Given the changes in thermal conductivity with lithology, porosity & temperature, the thermal gradient is not expected to be linear, but to vary with depth.

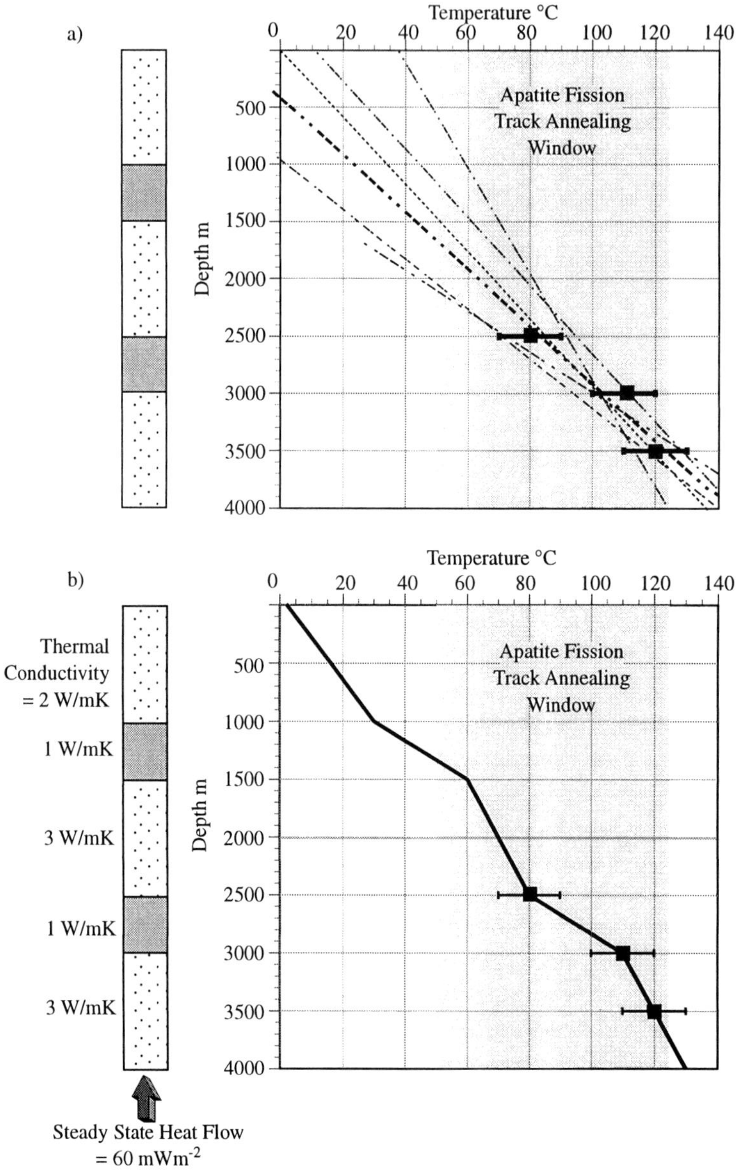

6. The annealing model may be incorrect for apatites with unusual compositions.

Paleo-temperature gradient is better constrained by inversion of several thermal indicators, each with a different time dependence. Paleo-geothermal gradients based on AFT or VR data alone are regarded as unreliable (see below).

2.3 MAXIMUM STRESS INDICATORS

The stress induced by burial can be calculated from porosity and related properties, such as sonic velocity. Porosity can be shown can be shown to be a function of the maximum vertical effective stress (difference between the lithostatic stress and the pore fluid pressure, see Terzaghi, 1923, 1936; Terzaghi and Peck, 1948; Giles, 1997 and references therein). In basins where the pore fluid pressure has remained hydrostatic through time, and the sediments are currently at their maximum burial depth, then the porosity will be a function of burial depth as the latter is related to the vertical effective stress.

Stress indicators (table 2) can be used to estimate the amount of uplift by first establishing a sonic-depth, or a density-depth trend for a given interval in an area where there has been no uplift (Fig. 4a). A vertical shift in this trend can then be used for the same unit (in terms of age and facies) in other, nearby areas to estimate the amount of uplift (Fig. 4b). Where it is impossible to identify an area with no uplift it is still possible to use such a method to obtain the relative amounts of uplift by basing the calibration on the least uplifted area. Note that the trend curves typically have an exponential form, with the result that the uncertainty in the estimated amount of uplift increases as the curvature in the curve decreases. Furthermore, though this technique is useful for estimating the amount of uplift it gives no indication of the timing, although unconformities may be recognized from a sudden increase in the sonic velocity and a corresponding decrease in porosity.

TABLE 2. Stress Indicators

Property	Indicator	Range	Reference
Porosity	Porosity Resistivity Sonic velocity	Continuous, but less accurate at high stresses	Giles, 1997 Alixant, 1989 Lang, 1978, 1994

In overpressured basins (fluid pressure > hydrostatic) it is necessary to reconstruct the fluid pressure history through time in order to assess the maximum vertical effective stress that a given sediment has experienced. Based on this information it is possible to construct maximum vertical effective stress-sonic velocity curves, rather than depth trends.

Figure 4.
Use of sonic interval transit time (or velocity) to estimate uplift.
a) For a continuously buried (and hence compacted) section the interval transit time is expected to decline steadily as porosity is lost. Well data from continuously buried parts of a given area can thus be used to construct a "normal trend".
b) In contrast, uplifted areas will show a vertical shift relative to the "normal trend". The magnitude of this shift can be used to estimate the amount of uplift.
Note that due to the exponential shape of the curves the accuracy of this method decreases downward.

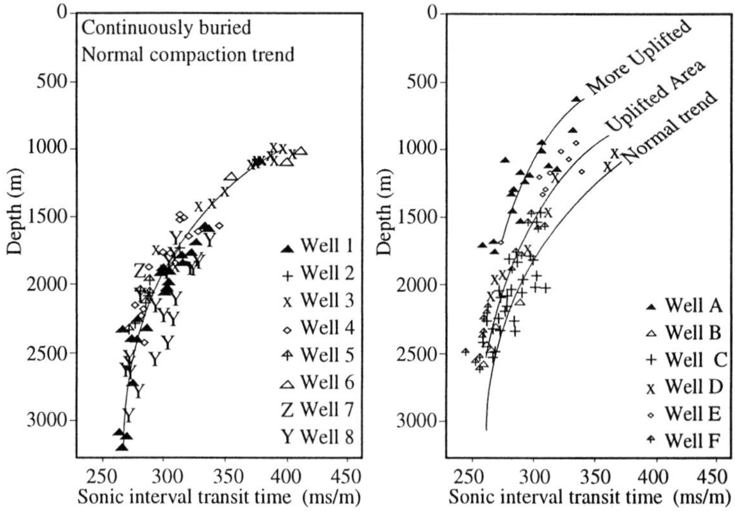

Figure 5.
With seismic or outcrop sections the missing section can often be estimated by simple extrapolation (assuming thickness is either constant or follows a consistent trend between the comparison section and the missing section).
In this case the extrapolation of the white layer to predict missing section is simple. Comparison of the dark coloured unit across the fault to estimate how much of it is missing is more hazardous as thickness may also change across the fault.

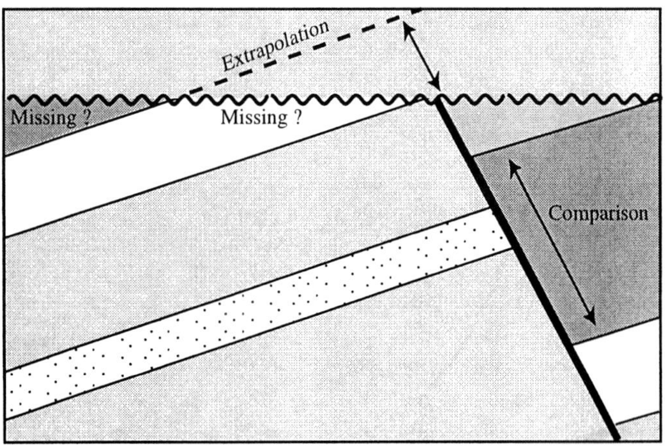

2.4 ADDITIONAL INDICATORS OF MISSING SECTION

Of the other possible indicators of uplift, simple extrapolation of thickness data based on seismic sections (Fig. 5), or well correlations provides a very simple and effective method for estimating missing sections.

Another, approach useful in rift basins, is to use the preserved syn-rift sequence or the extension visible on extensional faults at the syn-rift level to evaluate the stretching factor (so-called beta factor, see McKenzie, 1978). The beta factor in turn allows an estimate to be made of the likely thickness of post-rift sediments that were present prior to uplift.

2.5. PHYSICAL CONSTRAINTS ON BURIAL AND THERMAL HISTORY

Any burial and thermal history model proposed to fit the stratigraphic and indicator data should also be checked to see if it is physically reasonable. For example, the heat flows or thermal conductivities required to explain the temperature should be within known ranges.

As an example consider the impact of advective heat transport (i.e. heat transported by moving water) on the temperature field. Moving fluids have been much called on to explain AFT data in recent years (Green *et al.*, in press; Duddy *et al.*, 1994; Green *et al.*, 1993b). However, little attention has been paid to the physics of the process. Investigations of advective heat flow show that the crucial parameters in this analysis are the cumulative aquifer thickness and the flow velocity of the groundwater (Giles, 1997). For reasonable aquifer thicknesses velocities or 1 m/a or higher are necessary before half of the total heat flow is from advection (Giles, 1997) and thus the temperature field becomes dominated by advective effects. Velocities of this magnitude are on the high side for natural aquifer systems (Giles, 1997). To sustain velocities requires a driving force, either an active hydrothermal system or significant topographic relief. Note also, that as the Darcy velocity is the product of the seepage rate and the porosity, high velocities are relatively less likely in compact (i.e. deeply buried and uplifted) rocks.

3. Thermal models and limitations in the use of thermal indictor data

Each indicator has its own kinetically controlled response to the burial history and can thus be used as an independent constraint on the burial history. *In general therefore the greater the number of thermal indicators employed and their quantity and depth spread, the more reliable will be the thermal reconstruction.* However, each irreversible thermal/stress indicator is only of value until it reaches the limit of its range (see Table 1). In this respect vitrinite reflectance is particularly useful in that it continues to increase with increasing temperature. In contrast, hopane isomerisation is largely complete by a temperature of 100°C. Thus if vitrinite reflectance suggests a temperature in excess of 100°C and then all that can be said of complete hopane isomerisation is that it is not inconsistent with the vitrinite data.

AFT data record primarily information on cooling. The difference in behaviour between fission-track and other indicator data results in each potentially "seeing" different parts of the burial history as outlined in Figure 6. From the view point of hydrocarbon exploration it is feasible for vitrinite data to record a period of deep burial during which hydrocarbons were generated and expelled, whereas AFT data records a recent cooling event (Fig. 6b). As the vitrinite reflectance data records no information on timing it would be possible to erroneously assign the timing of maximum burial to the episode immediately preceding the uplift event recorded by the AFT data. This difference could have important implications if the petroleum play in an area relied on availability of recent charge.

Where the present day temperature is the maximum temperature (generally also the maximum burial depth) all the thermal indicators, with the exception of AFT will reflect *only* the present day temperature. Depending on the temperature AFT data can reflect only present day temperature (wholly annealed) or present day temperature plus any of: maximum burial temperatures, most recent annealing phase, and its predepositional history (e.g. recycling of older apatites into younger sediments, followed by low temperatures will result in preservation of old apatite track ages). The thermal pathway to the present day temperature is completely unconstrained by the irreversible thermal indicator data (Fig. 6d) beyond being a temperature less than the present day value, and other sources of data must be used to define the thermal history. AFT from the shallower units which are not presently annealed may provide useful input.

Figure 6.
Comparison of different burial/uplift and thermal histories.
a) In the case of deep burial and a single uplift event all thermal indicators record the same maximum temperature.
b) In the case of multiple phases of uplift irreversible thermal indicators below the unconformity will see only the event with the highest temperatures (usually the deepest phase of burial). This may result in a jump in thermal maturity across the unconformity (as in Fig. 3). AFT data will see the more recent annealing events. In this case, because the second event takes apatites through the annealing window, AFT data will record only the second event.
c) In the case where the second phase of burial does not result in total annealing of apatites, the AFT data will contain information about both phases of burial, while the irreversible thermal indicators below the unconformity will reflect only the early maximum temperatures.
d) In the case where maximum burial is at the present day all thermal indicators will reflect the present day situation, although shallow AFT data which have not been completely annealed will also provide information about earlier events. With no shallow AFT data earlier phases of high temperatures cannot be inferred except from the geological model. This can have important implications for the timing of oil generation.

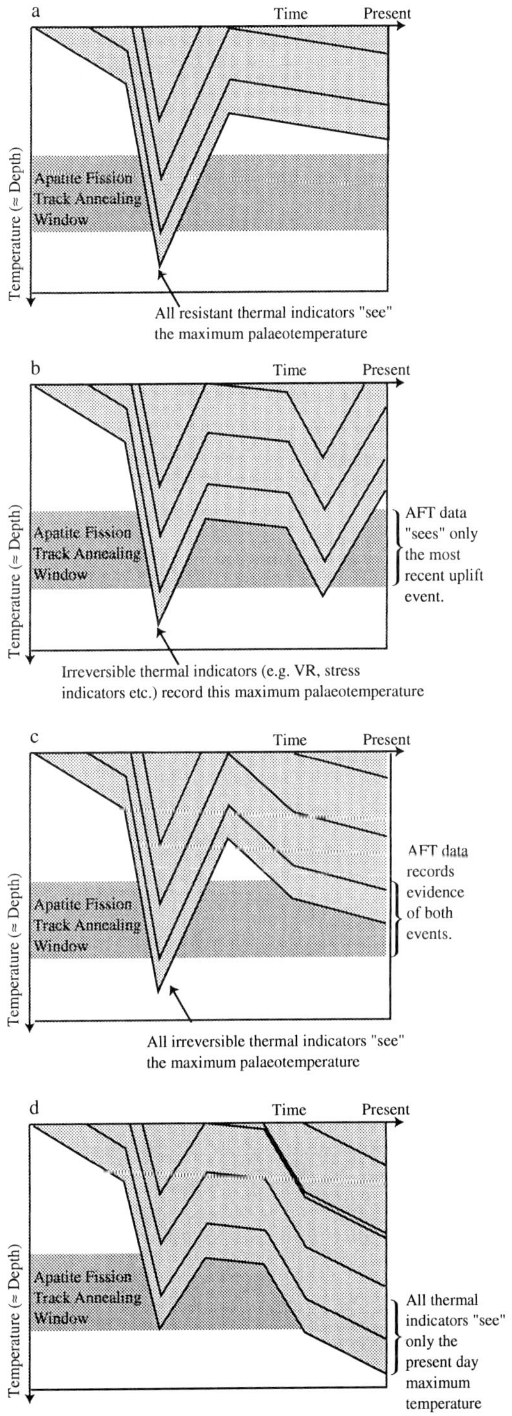
a
Time
Present
Apatite Fission
Track Annealing
Window
Temperature (≈ Depth)
All resistant thermal indicators "see"
the maximum palaeotemperature
b
Time
Present
Apatite Fission
Track Annealing
Window
Temperature (≈ Depth)
AFT data
"sees" only
the most
recent uplift
event.
Irreversible thermal indicators (e.g. VR, stress
indicators etc.) record this maximum palaeotemperature
c
Time
Present
Apatite Fission
Track Annealing
Window
Temperature (≈ Depth)
AFT data
records
evidence
of both
events.
All irreversible thermal indicators "see"
the maximum palaeotemperature
d
Time
Present
Apatite Fission
Track Annealing
Window
Temperature (≈ Depth)
All thermal
indicators "see"
only the
present day
maximum
temperature

Reverse thermal models attempt to use the indicator kinetics together with the available stratigraphic data to work backwards from the present day. The thermal history is inverted from the thermal indicator data taking into account the known uncertainties in the data. This approach is typified by the models of Lerche (1990). These models improve with the number of thermal indicators employed, the quantity, and the depth range over which the data is available. However, the accuracy, and uniqueness of the temperature history diminishes the further back in time the model is run, being seriously degraded at times earlier than the time that the maximum temperatures were attained.

So-called "forward" thermal models sequentially calculate the thermal effects of depositing a layer and compacting the underlying sequence starting from some point in the past and working toward the present. The temperature field is calculated from the heat flow equation, typically using the surface temperature and the base sediment heat flow as boundary conditions. Thermal indicator data are used to constrain the models so that those models which do not explain the current temperature and thermal indicator values are rejected. Forward thermal models are non-unique with many possible burial pathways giving a thermal history which can explain the observed indicator data.

Both reverse and forward thermal models are subject to uncertainties in the indicator data, stratigraphic data and model parameters such as thermal conductivity, heat flow etc. The inversion procedures used in reverse thermal models (see Lerche 1990) usually incorporate uncertainties in the thermal indicator data. Forward models, which form by far the most commonly available type of model do not generally incorporate uncertainties. However, uncertainties in indicator, and model parameter values can be incorporated into forward models via either stochastic finite element or Monte Carlo methodologies. The thermal model used in the Irish Sea analysis (see below) employs a Monte Carlo methodology for assessing uncertainties.

4. Reconstructing the Burial History

Typically data from a number of wells and possibly also outcrops needs to be synthesized into a consistent thermal history. The steps involved in synthesizing the data include:-

1. Data gathering and data quality control.
2. Determination of the present day heat flow.
3. Assessment of the preserved stratigraphy within the wells and or outcrops and regionally.
4. Recognition of and reconstruction of missing section at unconformities
5. Construction of thermal history

6. Testing the robustness of the thermal model and identification of alternative solutions.
7. Assessment of sensitivities and uncertainties.

Consistency is the overall aim. Models which fit the data, but in which the fundamental controlling variables change inexplicably between the modelled locations are intrinsically less likely than a consistent model in which variations in the controlling parameters are explainable. An excellent overview of the practice of thermal modelling can be found in the papers of Waples *et al* (1992a & b).

The reconstruction of the thermal history involves a number of parameters. These are:-

1. The heat flow history
2. The amount of missing section
3. Surface temperature history variations
4. Lithology and lithology variations

Regional studies of heat flow (see for instance, Fig. 7) show that variations in heat flow largely occur on a relatively long wavelength basis (10's to 100's of km). Exceptions occur, for instance over radiogenic granites such as those found in the SW of England. In these circumstances relatively strong lateral gradients in heat flow occur. For the most part, however, the heat flow does not differ greatly between closely spaced wells or outcrops.

The primary control on heat flow is provided by tectonic setting (see for instance, Allen & Allen, 1990; Pollack *et al.*, 1991). Thus tectonic setting used in conjunction with the geological history can be used as a primary guide to the heat flow history.

From a region with a uniform geological history, crust of relatively uniform thickness and composition then the heat flow will not vary greatly. Thus in fitting thermal indicator data from multiple locations it is generally inappropriate to invoke strongly varying heat flow or heat flow histories. If large differences are called on then explanations such as a change in basement composition must be sought. Low values might reflect a basic igneous or carbonate basement, while high values are usually due to granites. Similarly, when considerable differences in paleobathymetry exist across an area then the sea floor temperature can also be expected to vary considerably and these differences will propagate into the subsurface.

Lateral lithological variations within the stratigraphic column between wells result in lateral variations in thermal conductivity and therefore temperature (at constant heat flow). Such differences will translate themselves ultimately into

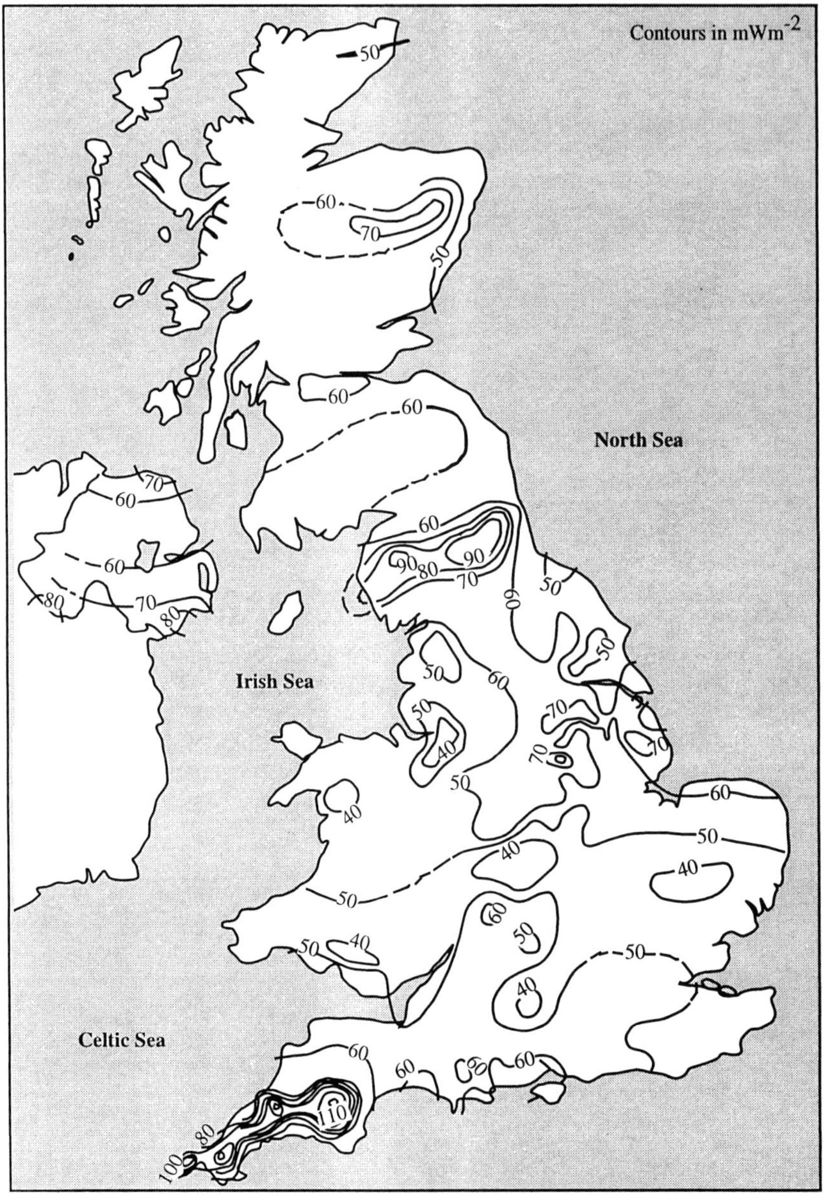

Figure 7
Heat flow map of the UK. After Rollin (1987).
Heat flow generally varies smoothly except where basement lithology controls high heat flows (e.g. high heat flows associated with granites in Cornwall and the Lake District.)

differences in thermal indicator values for identical burial histories and heat flow history.

Where a common heat flow history cannot account for the thermal indicator data from multiple locations then differences in the amount of missing section should be examined. However, the assessment of the amount of missing section must also be consistent. Large scale erosion is invariably associated with tectonic activity and thus the distribution of the eroded stratigraphy should be reflect this. In addition, the total amount of each unit eroded should ideally reflect likely or known variations in the thickness of the unit.

5. Application to the East Irish Sea and Cheshire Basins

The offshore East Irish Sea Basin (EISB) and onshore Cheshire Basin (Fig. 8) are part of a chain of Permo-Triassic basins developed in the UK. A pre-rift succession of Carboniferous shales and coal measures is overlain by a Permian and Lower Triassic syn-rift sequence. The Permian consists of an eolian unit, the Collyhurst Sandstone, which is overlain by the Manchester Marls and evaporites. The Lower Triassic Sherwood Sandstone Group consists of an arid floodplain succession which includes fluvial sandstones, sheetflood sands and minor eolian incursions. This in turn is overlain by a post-rift sequence comprising the Mercia Mudstone Group (marls and evaporites) and localised Liassic mudstone outliers (Fig. 9) at Prees in the Cheshire Basin, in the Solway Basin and near Carlisle. A seismically transparent sequence overlying the Mercia Mudstone Group in the Keys sub-basin of the EISB is also considered to be a Liassic section (Jackson *et al.*, 1987). The syn-rift sediments are exposed at surface, or subcrop a thin Quaternary cover over large portions of the Cheshire Basin and EISB. Although evidence exists for structural inversion, explicit inversion structures are not common (Knipe *et al.*, 1993). Further details of the geology of the EISB and Cheshire Basins may be found in Jackson and Mulholland (1993), Jackson *et al.*, (1987), Jones and Ambrose (1994), Knipe *et al.*, (1993), Williams and Eaton (1993), Macchci and Meadows (1987).

Both oil and gas are known from the EISB. Gas prone source rocks are provided by the Westphalian Coals, while oil is believed to be sourced from Namurian shales (Hardman *et al.*, 1993). In general the northern EISB appears gas prone, while the both oil and gas are to be found in the southern portion. Variability in the source rock facies within the Namurian is largely unconstrained. However, the rapid change from oil to gas in some portions of the southern EISB suggest the primary control is not source rock facies. The challenge for the petroleum explorationist is to predict the distribution of oil and gas with the aim of identifying the economically more attractive oil accumulations.

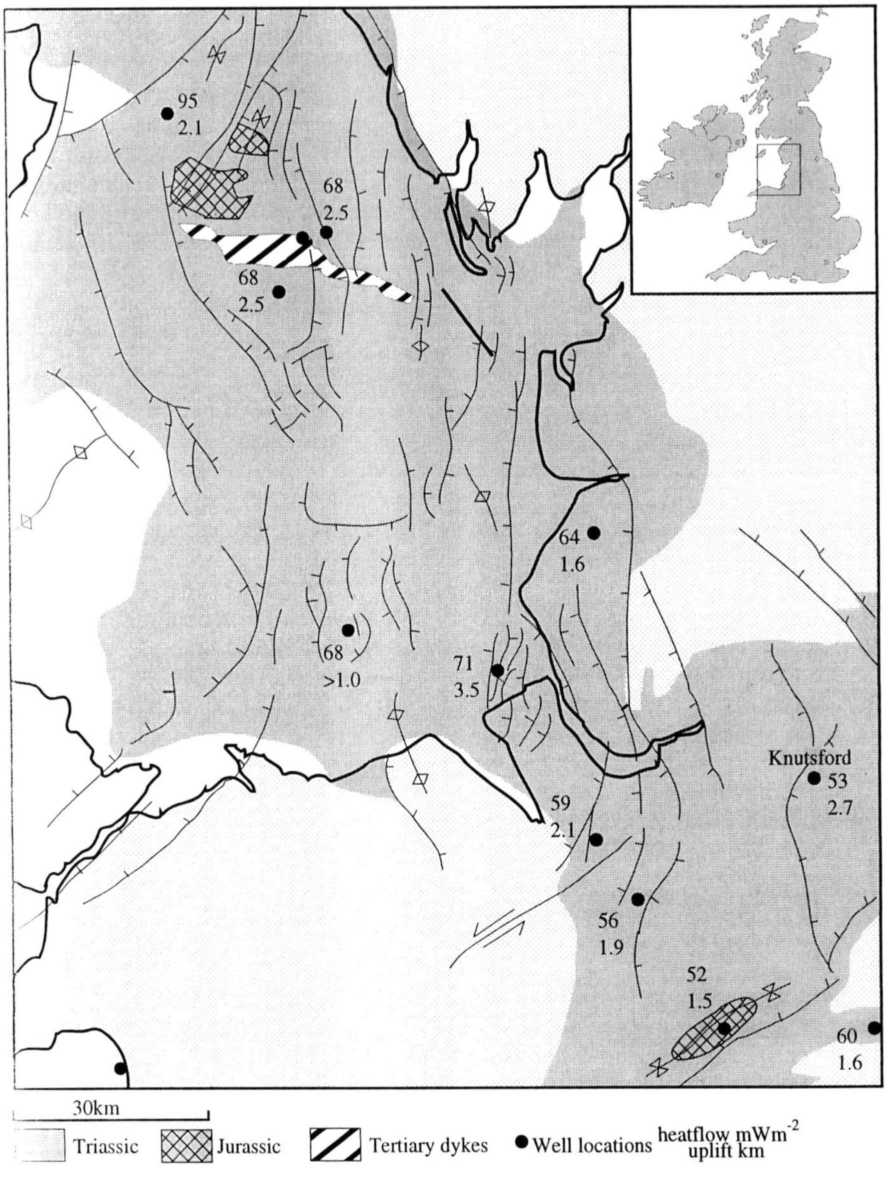

Figure 8.
Map of the Cheshire Basin and Irish Sea showing the location of the wells used in this study, and the heatflow and Early Tertiary uplift which to give the best fit to the thermal indicator data.

Vitrinite reflectance data from the Westphalian coals are variable, but generally exhibit high maturity (Fig. 10). Explanations for the high maturity of the source rocks and distribution of oil and gas include:-

1. Deep Mesozoic burial, followed by uplift either during base Cretaceous ("Cimmerian") tectonism (Hardman *et al.*, 1993) or at the beginning of the Tertiary (Holliday, 1993). Tertiary uplift is possibly related to regional doming as a result of a mantle plume associated with the British Tertiary Igneous Province (White, 1988; Cope, 1994; Brodie and White, 1994; Nadin *et al.*, 1995).

2. High Early Tertiary heat flow associated with the British Tertiary Igneous Province (Hardman *et al.*, 1993)

3. High temperatures related to fluid movement (Green *et al.*, in press; Duddy *et al.*, 1994; Green *et al.*, 1993a; McCulloch, 1994,).

The EISB, Cheshire Basin and the adjacent Lake District have been the subject of a considerable number of AFT studies (Green *et al.*, in press; Green *et al.*, 1993a; Lewis *et al.*, 1992a). Additional AFT data are also available from a number of other areas of the UK, including Northern Ireland (McCulloch, 1993, 1994), NW Scotland (Hurford, 1977a, 1977b; Lewis *et al.*, 1992b), Northern England (Green 1986; Holliday, 1993), Midlands (Green, 1989), and the Southern North Sea (Bray *et al.*, 1992). In addition vitrinite reflectance data and down-hole temperature measurements are available from exploration wells in the EISB and Cheshire basins.

The following approach is used to investigate the variation in the (generally high) maturity of the Westphalian coals:-

1. Synthesis of the regional stratigraphic framework
2. Analysis of the subsidence pattern to arrive at the expected post-rift subsidence history.
3. Uplift history indicated by the sonic velocity data.
4. Multiple thermal indicator inversion utilizing vitrinite reflectance, AFT age and length data and bottom hole temperature data.
5. Review of the inversion evidence.

We choose to order topics in this way as it forms a natural way of examining the problem, and in particular the starting point for any evaluation of the uplift history of an area is a regional synthesis of the stratigraphic data.

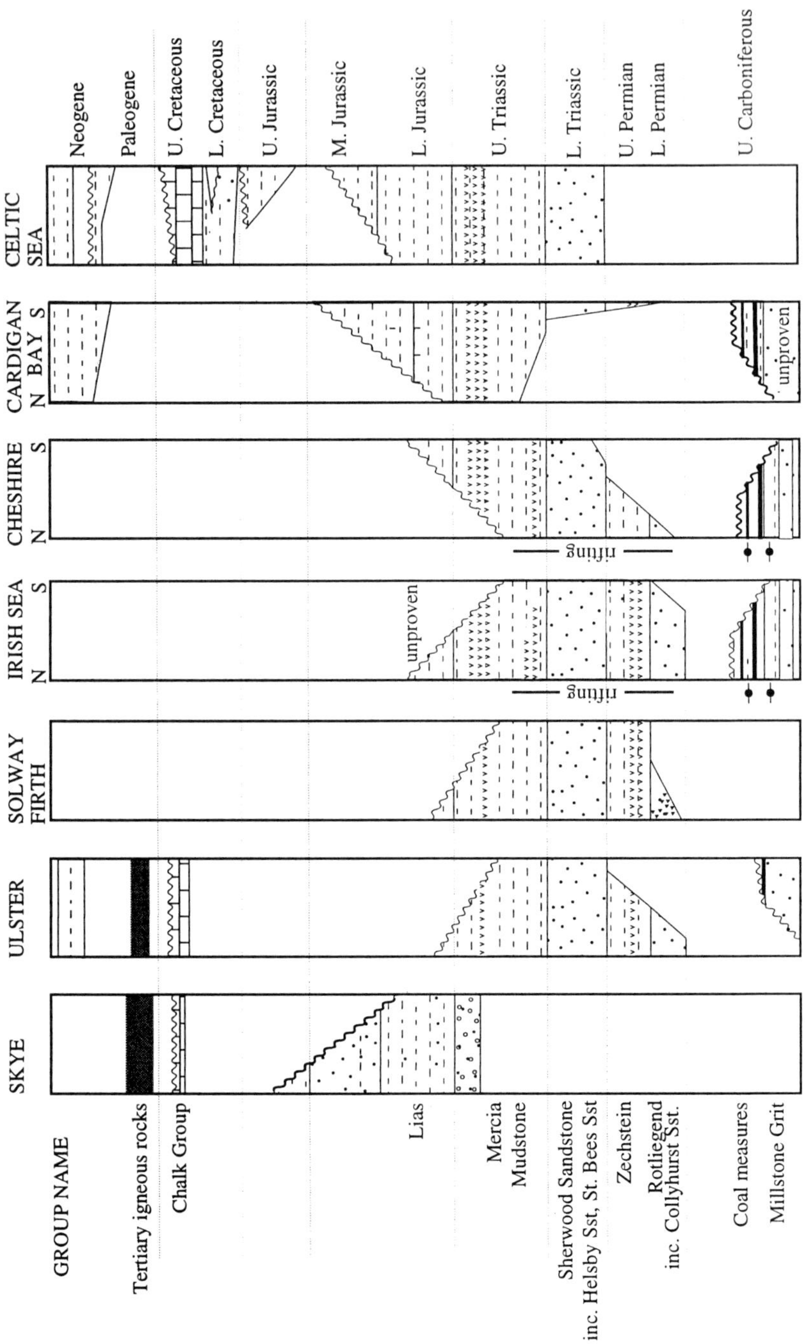
CELTIC SEA
CARDIGAN BAY
CHESHIRE
IRISH SEA
SOLWAY FIRTH
ULSTER
SKYE
N
S
Neogene
Paleogene
U. Cretaceous
L. Cretaceous
U. Jurassic
M. Jurassic
L. Jurassic
U. Triassic
L. Triassic
U. Permian
L. Permian
U. Carboniferous
unproven
rifting
GROUP NAME
Tertiary igneous rocks
Chalk Group
Lias
Mercia
Mudstone
Sherwood Sandstone
inc. Helsby Sst, St. Bees Sst
Zechstein
Rotliegend
inc. Collyhurst Sst.
Coal measures
Millstone Grit

5.1 EVIDENCE FOR MISSING SECTION

5.1.1. Regional stratigraphic trends

Figure 9 summarizes the stratigraphy of the EISB and surrounding basins. The upper part of the Triassic is partially or totally eroded throughout most of the East Irish Sea and Cheshire basins, but thickness trends in the Mercia Mudstone indicate continued extensional fault activity during the late Triassic. In the deepest part of the EISB, the Keys Basin, 3 km of Mercia Mudstone is preserved. Deposition is assumed to have continued throughout the Early Jurassic. The thick succession in the Keys sub-basin is presumed to include around 500m of Liassic sediments (Jackson *et al.*, 1987), and there are outliers of Liassic shales close to the basin margins in the Cheshire (600m) and Solway Basins, as well as in Northern Ireland, the Kish Bank Basin and Cardigan Bay. Lack of significant normal displacement affecting the Upper Triassic sediments suggests little faulting occurred during the Lower Jurassic (as would be expected in the post-rift phase) so a relatively even blanket of 600-800m of Liassic deep water sediments are assumed to have covered the entire area.

The similarity of relative sea-level curves and faunas in the Celtic Sea and Sea of Hebrides basins has been used to infer an open marine connection through the EISB during the Late Jurassic (Ziegler, 1987a). To the south, in Cardigan Bay Basin rapid subsidence continued throughout the Jurassic (Millson, 1987) with more than 2000m of Middle - Upper Jurassic sediments being deposited. Further north in the Sea of Hebrides Basin there is only a thin Jurassic succession (Brodie and White, 1994). A similar thin Jurassic sequence is present in Northern Ireland. A conservative estimate of 100-300m of middle and upper Jurassic sediments is made for the EISB based on the assumption that the East Irish Sea Basin received a greater amount of sediment than the marginal areas in Northern Ireland.

Early Cretaceous erosion has been inferred by some authors (e.g. Hardman *et al.*, 1993), but, although there is evidence for Lower Cretaceous ("Late Cimmerian") tectonism in the Celtic Sea Basin and in Northern Ireland, where the Chalk unconformably oversteps Jurassic and Triassic, it is not possible to prove a Cretaceous age for any inversion structures in the EISB. Even without late Cimmerian tectonism, there would probably have been little accommodation space in the EISB in the Lower Cretaceous, as the thermal subsidence associated with the Permo-Triassic rifting would have already slowed down significantly, and there was a eustatic low-stand. Thus there was probably little or no Early Cretaceous

---

Figure 9.
Comparison of the stratigraphy of the East Irish Sea Basin with the surrounding basins. In both the Cheshire Basin and the East Irish Sea Basin the post-rift phase is missing (i.e. there are no sediments preserved between Early Jurassic and Quaternary Periods). However, the missing stratigraphy can be inferred by comparison with the Cardigan Bay basin, Northern Ireland and Southern Scotland (see text.)

Figure 10.
Vitrinite reflectance data from the East Irish Sea Basin exhibits a number of clearly defined trends for the different wells. After Hardman *et al* (1993).

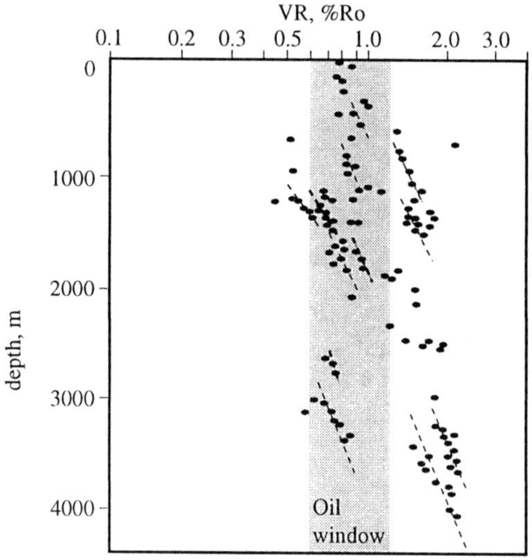

deposition. In contrast, Upper Cretaceous chalk is inferred to have been present over most, if not all, of the British Isles despite the limited outcrop at the present day, based on remnant chalk and flint clasts in central Ireland, Arran and Northumberland (Ziegler, 1987a). Widespread cover was presumably associated with the eustatic sea-level high-stand in the Late Cretaceous.

Thus at least 1 km, and possibly as much as 4 km of sediment could have been present in the East Irish Sea Basin prior to uplift and erosion to the present day level (300-2500m of Upper Triassic, 500-800m of Lower Jurassic, 1-300m of Middle Jurassic, and 100-300m of Upper Cretaceous).

5.1.2 Evidence From Subsidence Patterns

An independent estimate of the amount of material missing can be provided by considering the thickness of sediment expected to have been present in the rifted basin. Modelling the subsidence history of wells in the Irish Sea Basin suggests a Triassic stretching factor (ß) of 1.2, which is in good agreement with the 16% extension estimated from structural reconstruction (Tucker and Arter, 1987). Assuming rifting occurred on continental crust of normal thickness, extension with a ß factor of 1.16-1.2 should result in post-rift (thermal) subsidence of 1.1-1.3 km in the Jurassic. The lack of post-rift section thus implies uplift and erosion of at least 1.3 km. This is a minimum value since should there have been reactivation of the normal faults and erosion of the top part of the syn-rift sequence, the ß factor will be underestimated.

5.1.3 Evidence From Sonic Velocities/densities

Uplift estimates of around 2 km have been made from shale sonic velocity data by Colter and Ebbern (1978) and 2.2-2.3 km in the north of the EISB from sandstone densities (Chadwick *et al.*, 1994). This is a similar magnitude to the other estimates of missing section discussed above.

5.2 MULTIPLE THERMAL INDICATOR INVERSION

The results of thermal modelling of 12 wells (e.g. Fig. 11) in the EISB and Cheshire Basin using Shell proprietary pseudo-inverse thermal modelling packages show that in all cases the high heat flow associated with initial rifting had no impact on the present day maturity. The available AFT, VR and temperature data can be fitted by two scenarios:

1) A large amount of erosion (between 1.6 and 3.6 km), mostly during the Paleocene

2) A smaller amount of erosion (between 0 and 1 km) plus an externally imposed Paleocene heat flow pulse (for example a change in the mantle heat flow). Such a heat flow pulse has been suggested by some authors e.g. (Hardman *et al.*, 1993). It is generally assumed to be connected with the intrusion of dykes in the EISB which are associated with the British Tertiary Igneous Province of Northern Ireland and the Sea of Hebrides Basin, although advection of heat by fluid flow has also been suggested (McCulloch, 1994).

A third scenario - a heat flow pulse alone - can be ruled out, because there has clearly been some erosion (no post-rift sediments preserved and high sonic velocities).

By analogy with the Skye area (where thermal resetting of AFT ages is limited to areas within 8 km of the main igneous complex (Lewis *et al.*, 1992b), the presence of a Tertiary dyke complex in the Irish Sea Basin (Fig. 8) is not thought to have a significant impact on thermal maturity in the area, except immediately adjacent to the intrusions. Matching the observed AFT and VR data in the well which intersects the Tertiary dyke complex is impossible, because the shallowest part of the well is almost as mature as the deepest part due to the intrusion of 200m of Tertiary igneous rocks, whereas another well was apparently not affected by a dyke 2 km away.

We prefer solution 1 because the regional reconstruction of missing stratigraphy indicates that a large amount of section has been removed, and that uplift alone can match the observed VR and AFT data.

Thermal modelling using a transient heat flow solution indicates that uplift of around 1-2 km in a period of less than 20 Myr will produce a transient increase in

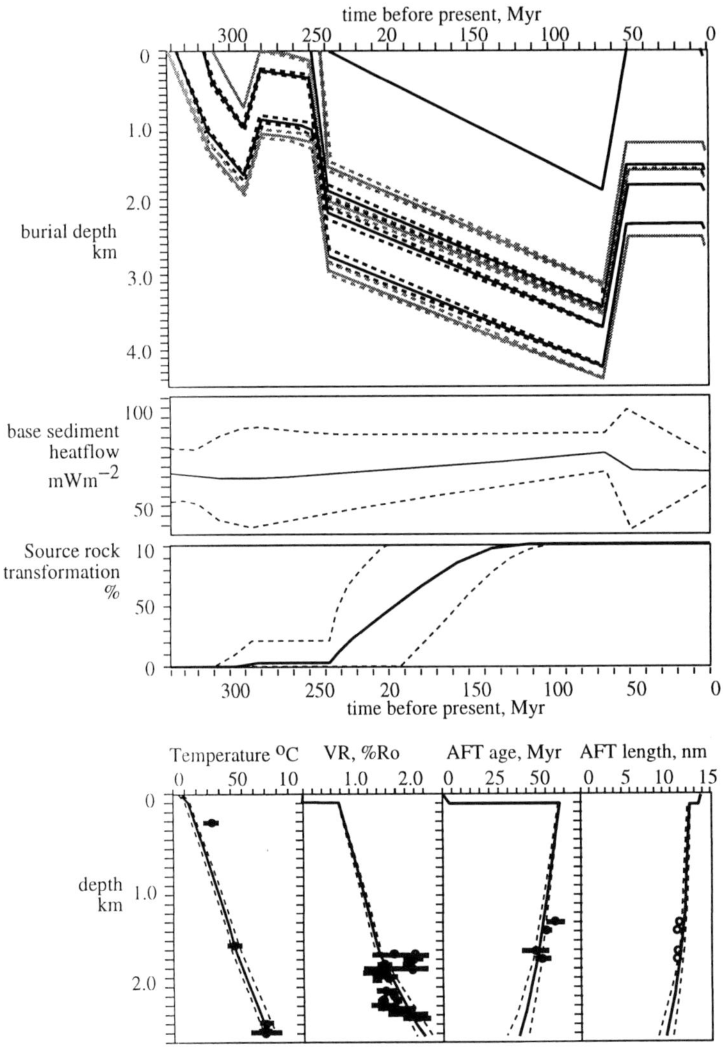

Figure 11.
Example of calculated burial history, present day temperature, vitrinite reflectance and AFT length and age for a well in the Irish Sea. Data are shown as points with associated error bars. Predictions are shown as lines with associated uncertainties (dotted lines). Burial depth is uncertain because decompacted thickness is reliant on porosity - depth relations which are themselves uncertain.
Using a fairly constant heat flow and a large amount of uplift a good match to the measured data points is obtained. Note that during the uplift phase there is a large uncertainty on the base sediment heatflow which fits the thermal indicator data. This is in line with the fact that the irreversible indicator (VR) sees only the maximum temperature (i.e. prior to uplift) and thus only provides constraint prior to uplift, whereas the AFT data pass rapidly through the annealing window.

heat flow of about 15 mW/m^2 above the background rate. Therefore usage of a steady state thermal model, will require input of a heatflow spike to mimic the transient effect of rapid uplift.

Thermal modelling using a transient heat flow solution indicates that uplift of around 1-2 km in a period of less than 20 Myr will produce a transient increase in heat flow of about 15 mW/m^2 above the background rate. Therefore usage of a steady state thermal model, will require input of a heatflow spike to mimic the transient effect of rapid uplift.

Throughout most of the EISB the heatflow required by the best - fit model is around 70 mW/m^2 (Fig. 8). In the Cheshire Basin it is around 55-60 mW/ m^2). The uplift in the northern EISB is estimated at 1.9-2.8 km with local variations over fault blocks (Fig. 8). However, at one well in the south of the basin the best fit to the AFT, VR and temperature data is around 3.6 km. The larger amount of uplift at this location probably reflects its position near the crest of a large (25 km) anticline with a relief of around 750m, although it is possible that the southern Irish Sea suffered more uplift than the north. The small amount of uplift (1.6 km) required in the onshore well location suggests that this area was outside the main rift axis, and received relatively little sediment.

Estimated uplift in the Cheshire Basin is between 1.5 km and 2.7 km. Except at Knutsford no AFT data was available. Thus although the magnitude of uplift is constrained by thermal modelling the timing is not. For example in some wells the VR data is derived from the Carboniferous section, and without AFT data, a Carboniferous maximum burial cannot be ruled out.

AFT length distributions provide further constraint on the timing and rate of uplift, because slow uplift allows annealing to continue for a longer period than if the apatite is rapidly removed from the partial annealing zone, resulting in a wider range of ages, and a younger mean age. Slow cooling allows considerable shortening of track lengths, whereas rapid cooling from temperatures of over 120 °C will preserve a narrow distribution of long tracks. Similarly, AFT ages down-hole should be very similar in the case of very rapid uplift, and should show a more distinct decrease as depth increases if uplift occurred more slowly.

Modelling of the AFT lengths for the Knutsford well in the Cheshire Basin (Figure 12) indicates rapid Paleocene uplift of around 1.5 km was followed by continued slower uplift. The similarity of annealed AFT ages throughout the EISB indicates that rapid uplift occurred throughout the basin at this time (Latest Cretaceous - Early Tertiary). Although resetting of the AFT ages is theoretically possible due to flow of hot fluids, this is considered to be highly unlikely as it would require that the fluids affected all the units from which the AFT data are derived, and in all the various fault blocks within the EISB simultaneously.

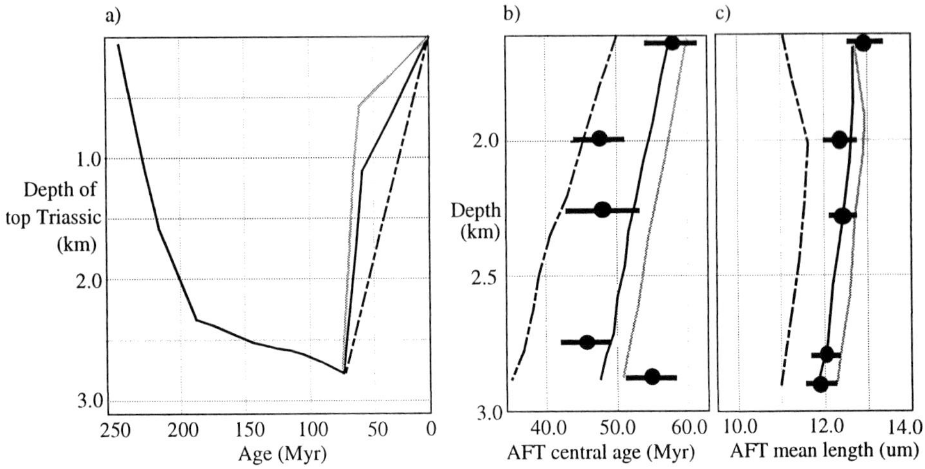

Figure 12.
Effect of different uplift histories on the fit to the AFT data at the Knutsford well in the Cheshire Basin. Three scenarios for the uplift history (a) were used to forward model the expected apatite age (b) and length (c) using the kinetic model of Lutz and Omar (1991): gradual uplift from 65 Ma to present day; rapid uplift of 1.5 km followed by gradual uplift; and rapid uplift of 2.3 km followed by gradual uplift. Comparison of the predictions (lines) with the observed AFT data shows the best fit is for the second scenario (rapid uplift of 1.5 km followed by gradual uplift after 50 Ma), although none of the models explain all the data. This may be due to variations in the composition of the apatites, which was not taken into account.

Furthermore, if uplift and erosion had occurred in the Early Cretaceous the sediments would already be at their present day depth at the time of annealing. This would require extremely high temperature fluids (> 120 °C to cause rapid annealing) at shallow levels (1.5 - 3 km) where the ambient temperature would be 40 - 90 °C. Although such temperature differences may be observed in local hydrothermal systems, such a temperatures difference over a whole basin would be highly unlikely.

Some AFT single grain ages in the EISB and Cheshire Basin are as young as 20-25 Ma, which could hint at a second uplift event during the Miocene, recorded only by those apatites with a more unusual chemistry and hence a higher susceptibility to annealing.

5.3 STRUCTURAL EVIDENCE FOR INVERSION IN THE EISB

The large variation in the amount of uplift derived from thermal modelling suggests that passive regional uplift is not the sole mechanism, and localised structural inversion must have occurred.

Analysis of seismic sections indicates that true reverse offsets are rare in the East Irish Sea Basin, although they are observed close to the basin margin in the NE of the basin (Knipe *et al.*, 1993), and at Liverpool Bay (Williams and Eaton, 1993). Other evidence for structural inversion includes folds developed parallel and adjacent to faults which have amplitudes of up to 200m, displacement reduction, such as that observed along the Tynwald fault system (Knipe *et al.*, 1993) and some geometries characteristic of inversion (such as short cut faults in the footwall block near the top of major normal faults, and low angle splays).

Structural restoration provides an estimate of 1.5 km shortening (2.5%) (Knipe *et al.*, 1993). Although in reality this is concentrated on specific structures, it represents an average tectonic uplift of 200m. This could be magnified to almost 900m by subaerial erosion of the basin fill and the accompanying isostatic uplift. This is significantly less than the observed amount of uplift in the EISB, which ranges from 1.0 - 3.6 km.

Sand-box modelling of inversion tectonics indicates that in the early stages of inversion shortening is accommodated by bulk strain in the syn-rift sequence, with reverse displacements only occurring in the post-rift section (Buchanan and McClay, 1991). General upwarping into a broad anticline is also characteristic. True reverse movement is not observed within the syn-rift sequence until more than 11% shortening has occurred. However, even at this stage, erosion of the post-rift and upper syn-rift section would result in problems in recognizing inversion, since compressional structures within the syn-rift section are limited to folds developed in the hanging wall adjacent to the reactivated fault planes. Therefore significantly more compression could have taken place than can be readily observed in the East Irish Sea Basin.

5.4 PALEOCENE UPLIFT OF THE BRITISH ISLES

Recognition of the pattern of Tertiary uplift in the British Isles provides a framework in which to place the results of the EISB and Cheshire Basins work and allows the recognition of other areas within the British Isles in which this type of approach might be valuable.

In the southern Celtic Sea, Western Approaches, Hampshire-Weald, and southern North Sea basins preservation of Chalk Group sediments, and Paleocene/Eocene deposition allow accurate constraints to be placed on the timing of uplift. Elsewhere estimates of the timing depend on fitting the AFT length data. The extra information from AFT studies is very important in constraining the timing and rate of uplift in areas where no syn-tectonic sediments are preserved, and even in areas where good control exists AFT can provide support for the interpretation and additional constraint on the heat flow history.

Apatite fission-track work on outcrop and well samples (Fig. 13) suggests that significant uplift occurred in much of Britain during the Paleocene (Lewis *et al.*, 1992a). However, there is no simple pattern of regional uplift. Instead, inversion was localised on different axes - with only the Irish Sea, Sole Pit and Cleveland areas undergoing enough erosion to exhume totally annealed apatites.

5.5 EISB AND CHESHIRE BASIN: SUMMARY AND INTERPRETATION

- The VR, AFT and sonic velocity data in the Irish Sea Basin are explained by uplift alone. The magnitude of uplift ranges from 1 km (well X of Hardman *et al.*, 1993) to 3.6 km (well 110/20-1) and is generally greatest (>2 km) near the basin margins, while the centre experienced less uplift (1 - 1.5 km).

- AFT length modelling suggests that in the East Irish Sea and Cheshire Basins only around half of the uplift (1 - 1.5 km) occurred during the Paleocene, whereas the rest occurred later.

- The magnitude of Paleocene uplift in the EISB is greater than that observed in the southern North Sea (0.3 - 1km), Inner Moray Firth (~1 km) or Celtic Sea Basins (~1 km).

- The effects of dyke intrusion on maturity are only important immediately adjacent to the intrusions, but overmaturation could be a risk in these areas.

- There is no evidence for Tertiary heat pulse, other than transient effects associated with (and hence slightly post-dating) uplift.

- The heat-flow in the EISB is higher than in the Cheshire Basin. The present day heat flow pattern is consistent with that observed for the British Isles as a whole.

- Thermal modelling does not require a phase of Lower Cretaceous ("Cimmerian") uplift, although this cannot be ruled out.

-

- Maximum paleo-temperature, and hence maximum maturity was achieved in the Late Cretaceous prior to Tertiary uplift.

- There is clear evidence that at least some of the uplift in the East Irish Sea Basin is due to structural inversion, including inversion of normal faults and development of anticlinal structures. However, few of these structures exhibit significant relief. Reverse offsets are only observed close to the basin margin.

In a regime of regional compression structural inversion in former basinal areas will occur as the result of competence contrasts between the basin filling units and the more rigid blocks between, thus it is possible that inversion in the Irish Sea could be linked to a driving force as far away as the Alps. Inversion of basins throughout the European plate has been linked to compression associated with Alpine tectonism (Ziegler, 1987b). Calculations show that the "regional" uplift

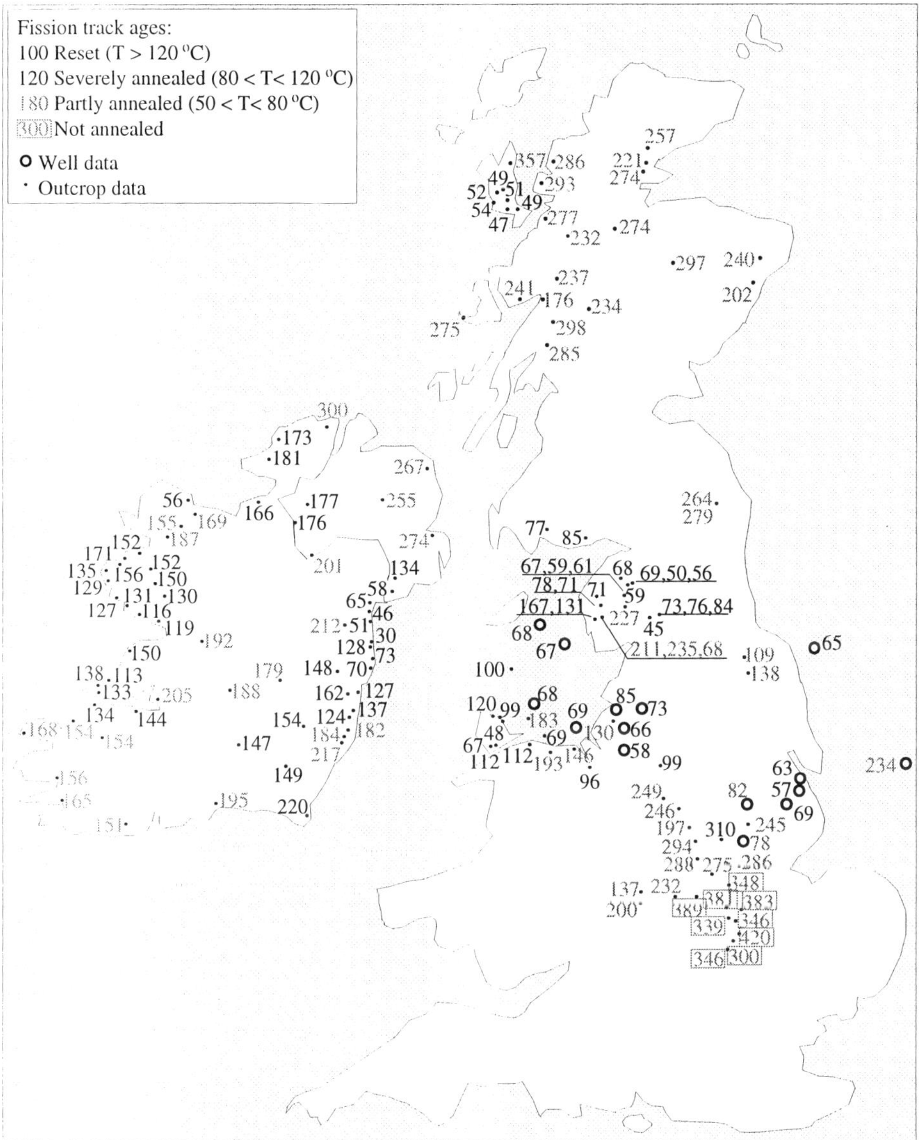

Figure 13.
Apatite fission-track data for the UK. Large areas have been subjected to partial or complete annealing, due to higher paleotemperatures. This is interpreted as indicating uplift over wide areas.

of 1 - 2 km throughout the Irish Sea could be accomplished by compression of only a few percent coupled with erosion and isostatic rebound without development of significant inversion structures except adjacent to rigid basement blocks. Furthermore, the more rigid the continental crust (in terms of its elastic thickness) the further from the inversion axis uplift due to isostatic rebound could extend.

5.6 IMPLICATIONS FOR THE UPLIFT HISTORY OF THE UK

Although there is evidence for some structural inversion in the Irish Sea the magnitude of this (relief of 1 km at most) is relatively small when compared to that observed in the Celtic Sea Basins and the southern North Sea, where structural inversion accounts for all the uplift, and little or no subaerial erosion has occurred. In contrast, further north in the Sea of Hebrides Basin and other basins off the west coast of Scotland and in the northern North Sea there is no clear evidence for structural inversion associated with uplift (and there is even some extension), although as in the EISB recognition of relatively small amounts of inversion may be hindered by the lack of post-rift section. Thus there is a trend toward less structuration in the north (which is reasonable if compression is due to stresses associated with Alpine collision), although the total amount of uplift is similar.

The large amounts of uplift in the EISB could thus result from the superposition of uplift due to two causes: regional uplift of a few hundred to a thousand metres, perhaps caused by igneous underplating, and structural inversion of basinal areas caused by regional compression. Uplift due to structural inversion of earlier basins would be magnified and prolonged by erosion and isostatic rebound. This effect would be particularly marked in those areas lying close to or above sea-level (such as those in the NW British Isles which experienced uplift due to thermal doming), and would be negligible in areas which remained within depositional regimes (such as in the Celtic Sea and North Sea basins).

The regional pattern of uplift (Fig. 14) can be summarised as follows:

- Regional uplift occurred in the Early Tertiary. The largest magnitudes of uplift apparently occur in the Triassic basins of the western British Isles. Although uplift occurred on the mainland, subsidence continued in many parts of the North Sea.

- Some Triassic and Jurassic rift basins suffered structural inversion. Development of reverse offsets, and other inversion features is most clear in the more southerly basins (Western Approaches Trough, southern North Sea). Compressional structures which are demonstrably Tertiary in age are not developed outside basinal areas.

- Uplift in Britain was not restricted to the Paleocene: in the southern North Sea there were several phases of uplift in the Late Cretaceous, and in the Sole Pit

Basin, Cleveland Basin, eastern Ireland and the Celtic Sea Basins there is evidence for Oligocene uplift. The timing of uplift in Scotland, Ireland and the Lake District is poorly constrained due to only partial annealing of AFT.

- Some basins which underwent uplift in the Early Tertiary subsequently became Tertiary depocentres (e.g. the Cardigan Bay Basin, the eastern Inner Moray Firth) while others have received no Tertiary sediments (e.g. the Cleveland axis, the Irish Sea).

- There is a trend from clearly inverted basins in the south, such as the Hampshire / Weald, Sole Pit and Celtic Sea basins, to basins which exhibit little or no evidence for structural inversion in the north, such as the Sea of Hebrides and Solway Firth basins.

Several theories have been put forward to explain regional uplift: Some authors use the Jurassic outcrop pattern (Cope, 1994), or the coincidence in timing to suggest that uplift is due to regional doming over a hot-spot associated with North Atlantic rifting and British Tertiary Igneous Province (BTIP) activity in the Paleocene (White, 1988; Brodie and White, 1994; Nadin *et al.*, 1995). However, removal of the heat source should result in subsidence back to the pre-heating level, whereas there has been little or no Tertiary subsidence in much of NW Britain. Furthermore, due to the long time constant for transmission of thermal effects through the crust, the area of the hot-spot would be expected to still exhibit a slightly higher than average heat flow. Although heat flows in northwest England are generally higher than in the south (Fig. 7), there is no clear pattern which would indicate the influence of a mantle plume. Similarly, the regional pattern of uplift (Fig. 14) is not indicative of doming.

Such problems with the original thermal dome theory led to the development of a modified theory, in which uplift is caused by isostatic effects in response to underplating of igneous rocks at the base of the continental crust and is thus permanent (Brodie and White, 1994). Since Tertiary igneous rocks outcrop in Ireland, Scotland, the Lake District, the Cheshire Basin, and as far south as Lundy this theory could explain the uplift of all of western Britain and Ireland. However, the pattern of uplift cannot be explained <u>only</u> by doming, since some areas which are uplifted are too far away from a possible hotspot to be affected. Similarly, the magnitude of uplift inferred for NW Scotland and central Ireland is not significantly different from that inferred in central England whereas tilting centred on a spot to the north or west would require over 3 km of uplift in Ireland and/or western Scotland. Furthermore although doming could explain some of the regional uplift trend, thermal doming cannot explain local variations in uplift.

Inversion of basins throughout the European plate has been linked to compression associated with Alpine tectonism (Ziegler, 1987b) and similar localisation of compression into previously rifted (and hence weaker) zones is

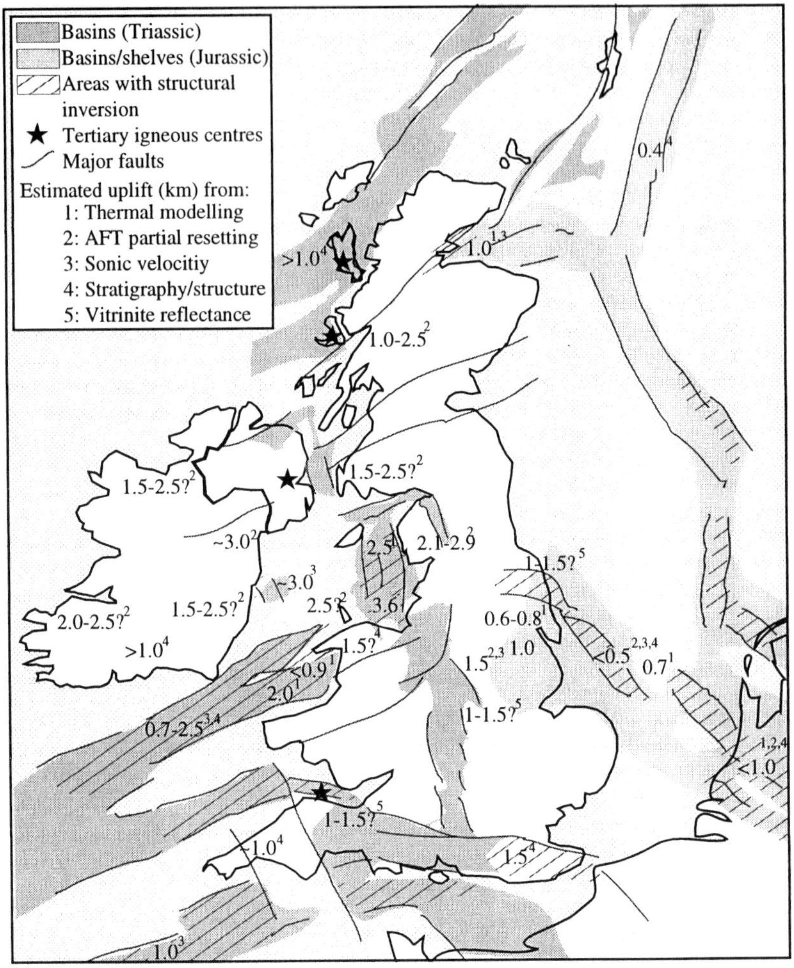

Fig. 14.
Regional pattern of Tertiary uplift observed in the UK.
Note that the estimated uplift in the southern North Sea is the figure for the Paleocene phase of inversion only, whereas in NW Britain the figures are for total Tertiary uplift. Paleocene uplift in the Cheshire and East Irish Sea Basins was around 1 - 2 km.

reported in the mid-continent rift region of the US (Cannon, 1994). However, it is not clear how Alpine stresses could be transmitted so far from the collision zone, or why different basins should apparently be inverted at different times. Possibly the timing of inversion in the various basins in Britain reflects a subtle shift in the direction of maximum compression, with the SW-NE (Caledonide) trends in the west of Britain subject to Paleocene reactivation and compression, whereas the SE-NW trend of the Sticklepath, Cleveland and Sole Pit areas became the focus of reactivation/uplift in the Oligocene, and the Cardigan Bay Basin underwent further subsidence. Inversion in the E-W oriented Wessex-Weald basin continued throughout the Early Tertiary. However, alpine compression cannot easily explain the uplift of the Lake District and western Scotland, where no inversion structures have been recorded.

6. Conclusions

Apatite fission-track thermochronometry is a powerful tool for recording cooling histories, particularly as the data record not only information about the amount of cooling but also its timing. AFT data are, however, only one of a large number of thermal and stress indicators all of which provide important independent information about the burial history of the sediments. As the arguments over the EISB and Cheshire Basins demonstrate, interpretations based solely on AFT data cannot supply a unique answer to burial history problems. The use of multiple thermal indicator data improves resolution. However, the inherent limitations on the use of indicator data and the lack of unique solutions from forward modelling ensures that it is difficult or impossible to distinguish between some hypotheses based on indicator data alone. Ultimately the additional data necessary to refine the thermal history can only be found from (regional) geological data, and fundamental geodynamic constraints.

References

Abbot G.D., Wang G.Y., Eglington T.I., Home A.K. & Petch G.S. 1990. The kinetics of sterane biological marker release and degradation during the hydrous pyrolysis of vitrinite kerogen. *Geochim. Cosmochim. Acta*, **54**, 2451-2461.

Alixant J-L. 1989. *Real-time effective stress evaluation in shales: pore pressure and permeability estimation*. PhD thesis. Louisiana State University and Agricultural and Mechanical College.

Allen, P. A., and Allen, J. R , 1990. *Basin Analysis Principles and Applications*. Blackwell Scientific Publications. 451. pp

Barnard P.C., Cooper B.A. and Fisher M.J. 1976. Organic maturation and hydrocarbon generation in the Mesozoic sediments of the Sverdrup Basin, Arctic Canada. *4th Int. Pal. Conf. Lucknow*, 581-588.

Bensley D.F, & Crelling J.C., 1994. The inherent heterogeneity within the vitrinite maceral group. *Fuel*. **73**, 1306-1316.

Brigaud, F., and Vasseur, G., 1989. Mineralogy, porosity and fluid control on thermal conductivity of sedimentary rocks. *Geophys. J.*, **98**, 525-542

Bray, R. J., Green, P. F., and Duddy, I. R., 1992. Thermal history reconstruction using apatite fission-track analysis and vitrinite reflectance: a case study from the UK East Midlands and Southern North Sea. In: *Exploration Britain: Into the next decade*. (ed Hardman, R. F. P.), Geol. Soc. Special Publ., No. 67, 3-25.

Brodie, J. and White, N., 1994, Sedimentary basin inversion caused by igneous underplating: Northwest European continental shelf. *Geology*, **22**, 147-150.

Buchanan, P. G. and McClay, K. R., 1991. Sandbox experiments of inverted listric and planar fault systems, *J. Geol. Soc. London*, **150**, 489-499

Burnham A.K. & Sweeney J.J., 1989. A chemical model of vitrinite maturation and reflectance. *Geochim. Cosmochim. Acta*, **53**, 2649-2657.

Cannon, W. F., 1994. Closing of the mid-continent rift - a far field effect of the Grenvillian compression. *Geology*, **22**, 155-158

Chadwick, R. A., Kirby, G. A., and Baily, H. E., 1994. The post-Triassic structural evolution of north-west England and adjacent parts of the East Irish Sea. *Proc. Yorks. Geol. Soc.*, **50**, 61-76

Colter, V. S., 1978. Exploration for gas in the Irish Sea. *Geologie en Mijnbouw*, 57, 503-516

Cope, J. C. W., 1994. A latest Cretaceous hotspot and southeasterly tilt of Britain. *J. Geol. Soc. London,* **151**, 905-908

Duddy, I. R., Green, P. F., Bray, R. J., and Hegarty, K. A. 1994. Recognition of the thermal effects of fluid flow in sedimentary basins. In: *Geofluids: origin, migration and evolution of fluids in sedimentary basins*. (ed Parnell, J.). Geol. Soc. Spec. Publ. No. **78**, 325-345

Ellacott M.V., Russell N.J. & Wilkins R.W.T., 1994. Trouble shooting vitrinite reflectance problems using FAMM: A Gippsland and Otway Basin case history. *APEA Journal*, **34**, 216-230.

George S.C., Smith J.W. & Jardine D.R., 1994. Vitrinite reflectance suppression in coal due to a marine transgression: Case study of the organic geochemistry of the Greta Seam, Sydney Basin. *APEA Journal*, **34**, 241-255.

Gijzel, van P., 1990. Transmittance colour (TCI) of amorphous organic matter: a new thermal maturity indicator for hydrocarbon source rocks, and it's correlation with mean vitrinite reflectance and thermal alteration index (TAI). *Bull. Am. Assoc. Petrol. Geol.*, 73, 1177

Giles, M. R., 1997, *Diagenesis: A Quantitative Perspective*. Kluwer, pp 520.

Giles, M. R., Indrelid, S. L., Beynon, G. V., and Amthor, J. , in prep. The Origin Of Large Scale Quartz Cementation: Evidence from large data-sets and coupled heat-fluid-mass transport modelling. In: IAS Special Publication on quartz cementation. (ed Worden, R.)

Gleadow, A. J. W., Duddy, I. R., Green, P. F., and Hegarty, K. A., 1986. Fission track lengths in the apatite annealing zone and interpretation of mixed ages. *Earth Sci. Planet Letts.,* **78,** 245-254.

Gleadow A.J.W., Duddy I.R. & Lovering G.M., 1983. Fission track analysis: A new tool for the evaluation of thermal histories and hydrocarbon potential. *APEA Journal* **23**, 93-102.

Green, P. F., Duddy, I. R., and Bray, R. J., In press. Variation in thermal history styles around the Irish Sea and adjacent areas: implications for hydrocarbon occurrence and tectonic evolution. Geol. Soc. Special Publication No. 124: Petroleum Geology of the Irish Sea and Adjacent Areas, pages 73 - 93.

Green, P. F., Duddy, I. R., and Bray, R. J., 1993b. Early Tertiary heating in Northwest England: fluid or burial (or both?) (extended abstract). In: *Geofluids '93: Contributions to an international Conference on fluid evolution, migration and interaction in rocks*. (eds Parnell, J., Ruffell, N. R., and Moles, N. R.), 119-123.

Green, P. F., Duddy, I. R., Bray, R. J., and Lewis, C. L. E., 1993a. Elevated palaeotemperatures prior to early Tertiary cooling throughout the UK region: implications for hydrocarbon generation. In: *Petroleum Geology of Northwest Europe: Proceedings of the 4th Conference, Barbican, London* (ed Parker, J.), 1067-1074

Green, P. F., Duddy, I. R., Laslett G.M., Hegarty K.A., Gleadow A.J.W. and Lovering J.F. 1989. Thermal annealing of fission tracks in apatite 4. Quantitative modelling techniques and extension to geological timescales. In: *Chem. Geology*, **79**, 155-182

Green, P. F., 1989, Thermal and tectonic history of the East Midlands shelf (onshore UK) and surrounding regions assessed by apatite fission-track analysis. *J. Geol. Soc. London*, **146**, 755-773.

Green, P. F., 1986. On the thermotectonic evolution of Northern England: evidence from fission-track analysis. *Geol. Mag.,* **123**, 493-506.

Green, P. F., Duddy, I. R., Gleadow, A. J. W., Tingate, P. R., and Laslett, G. M., 1986. Thermal annealing of fission-tracks in apatite. 1. A quantitative description. *Chem. Geol.*, **59**, 237-253.

Hardman, M., Buchanan, J., Herrington, P., and Carr, A., 1993. Geochemical modelling of the east Irish Sea Basin. its influence on predicting hydrocarbon type and quality. In: *Petroleum Geology of Northwest Europe: Proceedings of the 4th Conference, Barbican, London* (ed Parker, J.), 809-821

Harrison, M. T., and Be, K., 1983, $^{39}Ar/^{40}Ar$ age spectrum analysis of detrital microclines from the southern San Joaquin Basin, California: An approach to determining the thermal evolution of sedimentary basins. *Earth Science Plant. Letts.,* **64,** 2242-256

Holliday, D. W., 1993. Mesozoic cover over northern England: Interpretation of apatite fission-track data. *J. Geol. Soc. London*, 150, 657-660

Hurford, A. J., 1977a. Fission tack dates of two Galloway granites, Scotland. *Geol. Mag.* **114**, 299-304

Hurford, A. J., 1977b. A preliminary fission-track dating survey of Caledonian 'newer and last granites' from the Highlands of Scotland. *Scottish J. of Geology*, **13**, 271-284

Jackson, D. I., and Mulholland, P., 1993. Tectonic and stratigraphic aspects of the East Irish Sea Basin and adjacent area: contrasts in their post-Carboniferous structural styles. In: *Petroleum Geology of Northwest Europe: Proceedings of the 4th Conference, Barbican, London* (ed Parker, J.), 791-808

Jackson, D. I., Mulholland, P., Jones, S. M., and Warrington, G., 1987. The geological framework of the East Irish Sea Basin. In: *Petroleum Geology of Northwest Europe.* (eds Brooks, J., and Glennie, K.), Graham and Trotman, London , 191-203

Jones, N. S., and Ambrose, K., 1994. Triassic sandy braidplain and aeolian sedimentation in the Sherwood Sandstone Group of the Sellafield area, west Cumbria. *Proc. Yorks. Geol. Soc.*, 61-76

Kappelmeyer, O. and Haenel, R.,1974. *Geothermics with special reference to application*. Gebrüder Borntraeger, Stuttgart, 238. pp

Kaviany, M., 1991. *Principles of heat transfer in porous media.* Springer-Verlag. 630. pp

Knipe, R., Cowan, G., and Balendran, V. S., 1993. The tectonic history of the East Irish Sea Basin with reference to the Morecambe Fields. In: *Petroleum Geology of Northwest Europe: Proceedings of the 4th Conference, Barbican, London* (ed Parker, J.), 857-866

Landis C.R. & Castaño J.R., 1994. Solid hydrocarbons. *Organic Geochem.* **22**, 137-149.

Lang, W. H. J., 1978. The determination of prior depth of burial (uplift and erosion using interval transit time. SPWLA, 19th *Annual Log. Symp. Trans*., paper B.

Lang, W. H. J., 1994. Compaction/diagenesis of sediments and compaction gradients in relation to interval transit time, *The Log Analyst*, July-August, 54-58

Laslett, G. M., Green, P. F., Duddy, I. R., and Gleadow, A. J. W., 1987. Thermal annealing of fission-tracks in apatite. 2. A quantitative analysis. *Chem. Geol.* **65**, 1-13.

Lerche, I., 1990. *Basin Analysis, Quantitative Methods, 2 Vol.* Academic Press, Inc. 562 (Vol. 1), 570 (Vol. 2). pp

Lewis, C. L. E., Green, P. F., Carter, A., and Hurford, A. J., 1992a. Elevated K/T temperatures throughout Northwest Europe: Three kilometers of Tertiary erosion? *Earth Planet. Sci. Letts*., 112, 131-145

Lewis, C. L. E., Carter, A. and Hurford, A. J., 1992b. Low temperature effects of the Skye Tertiary Intrusions on Mesozoic sediments in the Sea of Hebrides basin. In: *Basins on the Atlantic Seaboard: Petroleum Geology, Sedimentology and Basin Evolution* (ed Parnell, J.), Geol. Soc. Special Publication. 62., 175-188.

Lopatin, N. V. ,1971. Temperature and geologic time as factors in coalification. Akade. Nauk SSSR, *Izve. Ser. Geol.*, **3**, 95-106.

Lutz T.M. and Omar G. 1991. An inverse method of modelling thermal histories from apatite fission track data. *Earth. Planet. Sci. Lett*., **104**, 181-195.

Macchei, L., and Meadows, N. S., 1987. Field excursion to the Permo-Carboniferous of Cumbria and Cheshire, Poroperm Geochem Excursion Guide 12, 140 pp.

McCulloch, A A , 1994. Low temperature thermal history of eastern Ireland: effects of fluid flow. *Mar Petrol. Geol.*, **11**, 389-399

McCulloch , A.A. 1993. Apatite fission-track results from Ireland and the Porcupine basin and their significance for the evolution of the North Atlantic. *Marine and Petrol. Geol.*, **10**, 572-589.

McKenzie A.S. & McKenzie D.P., 1983. Isomerization and aromatization of hydrocarbons in sedimentary basins formed by extension. *Geol. Mag.*, **120**, 417-528.

McKenzie, D., 1978. Some remarks on the development of sedimentary basins. *Earth Planet. Sci. Letts*., 40, 25-32.

Millson, J. A., 1987. The Jurassic evolution of the Celtic Sea Basins. In: *Petroleum Geology of Northwest Europe*, (eds Brooks, J., and Glennie, K.), Graham and Trotman, London , 599-610

Nadin, P. A., Kusznir, N. J., and Toth, J., 1995. Dynamic uplift associated with the development of the Iceland plume: Constraints from North Sea stratigraphy. *Terra Abstr.* (Oxford), 7/1, 158

Naeser, C. W., 1981. The fading of fission-tracks in the geologic environment - data from deep drill holes. *Nucl. Tracks* **5**, 248-259.

Naeser, C. W.,1979. Thermal history of sedimentary basins : Fission-track dating of subsurface rocks. In. *Aspects of diagenesis* (ed Scholle P. A. P. R) SEPM Special Publication No. **26**., 109-112.

Naeser, C. W., and Faul, H., 1969. Fission track annealing in apatite and sphene. *J. Geophys. Res.*, **74**, 705-710

Pollack, H. N., Hurter, S. J. and Johnson, J. R. 1991. *The new global heat flow compilation*. Dept. Geol. Sci., Univ. Michigan, Ann Arbor, Michigan 48109-1063, USA

Pytte, A. M., and Reynolds, R. C., 1989. The thermal transformation of smectite to illite. In: *Thermal history of sedimentary basins: Methods and case histories*. (eds Naeser, N. D., and McCulloh, T. H.), Springer-Verlag, 130-140.

Robert P. 1988 *Organic metamorhpism and geothermal history*. Elf-Aquitaine and Riedel Publishing, Dordrecht, 311pp

Rollin K.E. 1987. *Catalogue of geothermal data for the land area of the United Kingdom. Third revision: April 1987. Investigation of the geothermal potential of the UK*, British Geological Survey, Keyworth.

Sclater, J. G., and Christie, P. A. F., 1980. Continental stretching: An explanation of the post Mid-Cretaceous subsidence of the Central North Sea Basin. *J. Geophys. Research*, 85(B7), 3711-3739

Somerton, W. H.,1992. *Thermal properties and temperature related behavior of rock/fluid systems*. Elsevier, 257. pp

Staplin, F. L., 1969. Sedimentary organic matter, metamorphism and oil and gas occurrence. *Bull. Can. Petrol. Geol.*, **17**, 47-66.

Sweeney, J.J., and Burnham, A.K., 1990. Evaluation of a simple model of vitrinite reflectance based on chemical kinetics. *Bull. Am. Assoc. Petrol. Geol.*, **74**, 1559-1570.

Tagami, Carter and Hurford 1996. Natural long-term annealing of the zircon fission-track system in Vienna Basin deep borehole samples. Chem. Geol., **130**, 147-157

Tagami Ito and Nishimura 1990. Thermal annealing characteristics of spontaneous fission-tracks in zircon. *Chem. Geol.*, **80**, 159-169.

Terzaghi, K., 1923. Die Berechnung der Durchlässigkeitsziffer des Tones aus dem verlauf der hydronamischen Spannungserscheinungen, Sitz. Akad. Wissen. Wien Math-naturw. Kl. 132, 105-124

Terzaghi, K., 1936. The shearing resistance of saturated soils. In: Proceedings of the Ist International Conference on Soil Mechanics, Harvard 1,54-56

Terzaghi, K., and Peck, R. B.,1948. *Soil Mechanics In Engineering Practice*. Wiley. 566. pp

Tissot, B. P., and Welte, D. H., 1978, *Petroleum Formation and Occurrence (2nd Edition)*. Springer-Verlag, New York, pp. 699.

Tucker, R. M., and Arter, G., 1987. The tectonic evolution of the North Celtic Sea and Cardigan bay basins with special reference to basin inversion. *Tectonophysics*, 137, 291-307

Turcotte, D. L., and Schubert, G., 1982. *Geodynamics: Applications of Continuum Physics to Geological Problems*. John Wiley & Sons, 450.pp

Wangen, M., 1992. Pressure and temperature evolution in sedimentary basins. *Geophys. J. Int.*, **110**, 601-613.

Waples, D. W., Kamata, H. and Suizu, M., 1992a. The art of maturity modeling part 1 : Finding a satisfactory solution. *Bull. Am. Assoc. Petrol. Geol.*, **76(1),** 31- 46.

Waples, D. W., Suizu, M. and Kamata, H., 1992b. The art of maturity modeling part 2: Alternative models and sensitivity analysis. *Bull. Am. Assoc. Petrol Geol.*, **76**(1), 47-66

White, R. S., 1988. A hot-spot model for Early Tertiary volcanism in the North Atlantic. In: *Early Tertiary volcanism and the opening of the NE. North Atlantic*. (eds Morton, A. C. and Parson, L. M.), Geol. Soc. Spec. Publ. **39**, 3-13

Wilkins R.W.T., Wilmshurst J.R., Russell N.J. Hladky G., Ellacott M.V. & Buckingham C.,1992. Fluorescence alteration and the suppression of vitrinite reflectance. *Org. Geochem.* **18**, 629-640.

Williams, G. D., and Eaton, G. P., 1993. Stratigraphic and structural analysis of the Late Palaeozoic of NW Wales and Liverpool Bay: implications for hydrocarbon prospectivity. *J. Geol. Soc. London,* **150**, 489-499

Woodside, W., and Messmer, J. H.,1961. Thermal conductivity of porous media (Parts I and II*). Jour. Applied Phys.*, **32**(9), 1688-1707.

Yamada, Tagami, Nishimura and Ito. 1995. Annealing kinetics of fission tracks in zircon: an experimental study. *Chem. Geol.*, **119**, 293-306

Yamada, Tagami and Nishimura 1995. Confined fission-track length measurement of zircon: assessment of factors affecting the palaeotemperature estimate. *Chem. Geol.,* **122**, 249-258

Ziegler, P., 1987a. Evolution of the Western Approaches Trough, *Tectonophysics*, **137**, 341-346

Ziegler, P., 1987b. Late Cretaceous and Cenozoic intra-plate compressional deformations in the Alpine foreland - a geodynamic model, *Tectonophysics*, **137**, 389-420

APATITE FISSION TRACK THERMOCHRONOLOGY OF PALEOZOIC SANDSTONES AND THE HILL-INTRUSION, NORTHERN LINKSRHEINISCHES SCHIEFERGEBIRGE, GERMANY

U. GLASMACHER, M. ZENTILLI AND A.M. GRIST
Fission Track Res. Lab., Dept. of Earth Sciences, Dalhousie University, Halifax, Canada.

Abstract

A fission track investigation was carried out on apatite from Paleozoic sandstones along a NW-SE transect and the Hill-intrusion in the northern Linksrheinisches Schiefergebirge. The study included age determinations, confined track length and chemical analysis as well as computer modeling of the thermal history. Apatite fission track ages vary between 130±11 Ma and 239±13 Ma. Mean confined track lengths range from 12.35±0.36 μm to 13.20±0.73 μm. Fluorapatite grains with no chlorine content are the main components. Single grain ages of the Paleozoic sandstones and the tonalite decrease with increasing fluorine and decreasing OH and chlorine content. Single grain ages from wallrock sandstones of the galena-sphalerite-calcite-quartz veins show no relation with the chemical composition. Computer modeling of the thermal history using an inverse model allowed to distinguish an Upper Paleozoic to Lower Mesozoic component (up to 220 Ma), a Mesozoic component, and a post-Campanian component. An average cooling rate below 5°C/Ma was estimated for the Upper Paleozoic thermal history. The subsidence of the Graben of Malmedy led to the accumulation of 3,000m of sediments during Permian and Lower Triassic. Apatite grains of the Devonian sandstones of the Venn-Weser-Inde nappe were completely annealed after deposition and reached the PAZ at 260 Ma. With a normal geothermal gradient this temperature would imply 3,000m to 4,000m of now eroded deformed Devonian and Carboniferous units.
The thermal history of the Mesozoic component is constrained by slow cooling of 0.3°C/Ma up to 120 Ma and faster cooling of ≈1.5°C/Ma up to 80 Ma. An exhumation of ≈1,000m might have occurred if a stationary normal geothermal gradient is applied to the first time interval. The cooling between 120 Ma and 80 Ma which point to an exhumation of ≈2,000m is interpreted as the main Mesozoic exhumation of the northern Linksrheinisches Schiefergebirge. Apatite fission track data of a Namurian sandstone taken adjacent to the galena-sphalerite-calcite vein in the Venn-Weser-Inde nappe point to a mineralization age between 200 Ma and 160 Ma. The age of the mineralization in the Eastardennen-Eifel nappe revealed by the fission track data is younger (160 -140 Ma).

1. Introduction

The post-Variscan thermal and exhumation history of the foreland and fold-and-thrust belt in the northern Linksrheinisches Schiefergebirge is part of the general post-Variscan geological evolution of the Rhenish Massif. Meyer *et al.* (1983) discussed an uplift of the Rhenish Massif between Early Jurassic (Liassic: 187 Ma) and Late Cretaceous (Santonian: 88 Ma) of 200m and also argued for a fast uplift (160m) of the central and northern parts of the Massif in the Upper Maastrichtian. Murawski *et al.* (1983)

P. Van den haute and F. De Corte (eds.), Advances in Fission - Track Geochronology, 151-172.

emphasize the beginning of uplift at the Jurassic/Cretaceous boundary (144 Ma) and connect this movement with the subsidence of the Netherlands Central Graben. In the uppermost Santonian, Campanian and Maastrichtian the western part of the Rhenish Massif was an area of marine sedimentation. Albers and Felder (1979) describe a regressive tendency for the Upper Maastrichtian. During Lower Tertiary (Paleogene) sedimentation and subsidence occurred in the western and northwestern parts of the Rhenish Massif. The last uplift, which is still in progress and has caused a recent elevation of up to 690m, began in the southern margins of the western part of the Rhenish Massif in the Upper Miocene and moved towards the East and North.

Main lithological units of the northern Linksrheinisches Schiefergebirge are deformed allochthonous and autochthonous siliciclastic rocks of Cambrian to Upper Carboniferous age. Intercalated are Middle to Upper Devonian reef limestones and shelf to lacustrine limestones-dolostones of Lower Carboniferous age (Figure 1). Thick coal seams are part of the Upper Carboniferous molasse deposits in the foreland and the northern part of the fold-and-thrust belt.

From northwest to southeast the area is divided into four main tectonic units; Liége-Wurm syncline, Herve-Aachen-Imbricated zone, Venn-Weser-Inde nappe and the Eastardennen-Eifel nappe. The Cambrian and Ordovician rocks of the Venn-Weser-Inde and the Eastardennen-Eifel nappe as well as the Lower Devonian siliciclastics of the Eastardennen-Eifel nappe are folded, cleaved and partly metamorphosed to lower greenschist facies (Kramm 1982, Kramm *et al.*, 1985a, Spaeth *et al.*, 1985, Echle *et al.*, 1985). The Devonian and Carboniferous units of the Liége-Wurm syncline, the Herve-Aachen-Imbricated zone and the Venn-Weser-Inde nappe are also folded but cleavage is only developed close to thrust faults. The maturation of the organic material range from 0.9% Ro in the Upper and Lower Carboniferous sequence to 3% Ro in the Lower Devonian units (Wellens 1995).

K/Ar-dating of illite and sericite in various grain fractions (<2μm) of Paleozoic slates, phyllites and cleaved magmatic dikes as well as two middle Devonian intrusions (384 ± 16 Ma, Kramm and Buhl 1985) show that the deformation occurred at around 300 Ma (Kramm *et al.*, 1985b, Tschernoster *et al.*, 1995). Also the K/Ar date (243 ± 6 Ma) of very fine-grained K-feldspar of the Hill intrusion is interpreted to represent a lower temperature (≈ 200°C) step in the post-Variscan exhumation history of the region. In parts of the region exhumation of deformed Paleozoic siliciclastic and carbonate rocks must have occurred prior to the deposition of red bed type coarse to medium grained siliciclastic rocks of Upper Permian and Lower Triassic age (Bless *et al.*, 1990). Today the Permian terrestrial deposits (< 250m) occur in the NE-SW trending Graben of Malmedy where they are unconformably underlain by Cambrian and Lower Devonian sedimentary rocks. Further to the east Permian sedimentary rocks, which only occur in small outcrops, are unconformably overlain by middle Lower Triassic coarse clastic sedimentary rocks. Also Triassic and Jurassic terrestrial and marine sediments (≈ 530m) were deposited in an N-S trending basin at the eastern margin of the northern Linksrheinisches Schiefergebirge (Mader 1982, Walter 1992). The variable thickness of the Lower Triassic siliciclastic unit, the clastic intercalations and the dominance of dolostones in the Middle Triassic, point to a depositional environment close to the western shore of the Triassic and Jurassic sea. No sedimentary record younger than the Lower Jurassic limestone-marl-black-shale intercalation is known from the area.

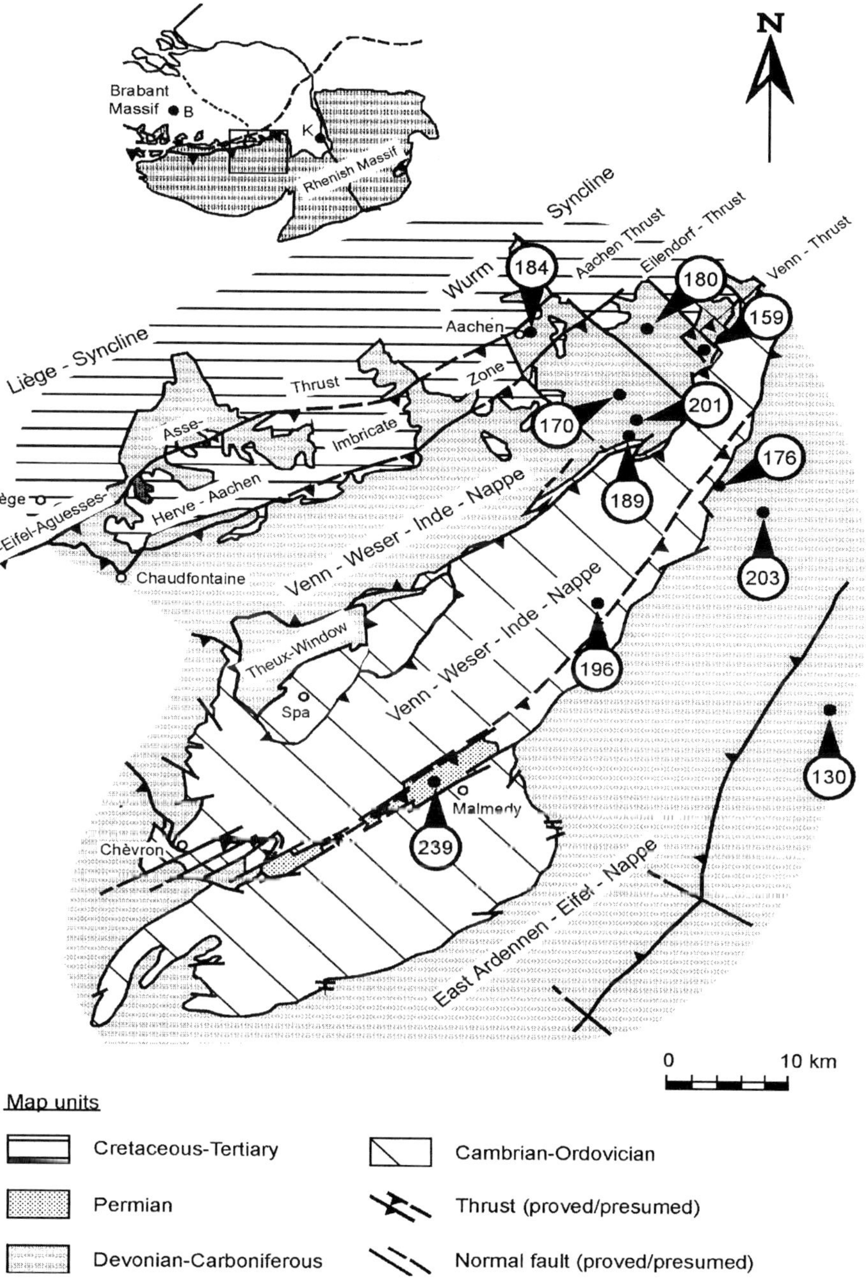

Figure 1. Geological map of the northern Linksrheinisches Schiefergebirge with fission track ages. Black circles indicate the location of the samples.

Paleozoic rocks that are under investigation were unconformably overlain by marine marls and limestones at Cretaceous (Campanian, around 80 Ma) time. Also Lower Tertiary sands (Oligocene) are described from the highest parts (690m) of the area.

After the deformation and prior to the deposition of Cretaceous carbonate beds the Paleozoic and Triassic sedimentary rocks were thermally and chemically overprinted by the advection of hydrothermal fluids (<190 °C); (Redecke 1992, Muchez *et al.*, 1995). These fluids moved along NW-SE trending steeply dipping normal faults and migrated through the Devonian and Carboniferous limestone and dolomitc beds in the Venn-Weser-Inde nappe and the Lower Triassic coarse-grained siliciclastic units in the east. The interaction of those fluids with the carbonate beds caused galena-sphalerite-carbonate veins, stockworks and stratiform mineralization which occur in an large NE-SW trending area of the Venn-Weser-Inde nappe. Galena-sphalerite-quartz veins and stockworks in siliciclastic rocks are only known from the Lower Devonian siliciclastic units of the Eastardennen-Eifel nappe and as disseminated galena mineralization in the Lower Triassic sandstones. To which extent the hydrothermal fluids have interacted with the deformed Paleozoic siliciclastic units of the Venn-Weser-Inde nappe is not known. Minor calcite veinlets crosscutting Upper Devonian siliciclastic rocks are characterized by a different stable isotopic ($\delta^{13}C$, $\delta^{18}O$) and chemical signature (Redecke 1992, Stroink 1993). At present convective heat transfer by thermal waters (up to 72°C) occurs in the Aachen region along thrust and normal faults as well as in Devonian limestone beds (Plum 1989, Pommerening 1993).
An apatite fission track study was carried out to fill the gap of the thermal history between the dated Variscan deformation and the present hydrothermal activity and to date the Mesozoic hydrothermal system. This study was also done to evaluate the thermal overprint of the Paleozoic rocks due to the convective heat transfer by the Mesozoic hydrothermal system.

2. Method and samples

The apatite fission track method is based on the measured density and etchable length distribution of linear tracks of crystal damage produced during the spontaneous fission decay of trace amounts of ^{238}U in apatite (Wagner 1968, Wagner and Van den Haute 1992). Fission track densities provide a measure of mineral age with respect to low-temperature thermal history, with complete erasure of apatite fission tracks occurring at temperatures in the range of 100-150°C over geologic time periods. The precise value is dependent on the chemical composition of apatite and the heating rate. As described by various authors (Gleadow and Duddy 1981, Gleadow *et al.*, 1986, Green *et al.*, 1985, 1986, 1989 a, b, James and Durrani 1986, Barbarand *et al.*, 1996, Carpéna 1996, Gleadow *et al.*, 1996, Green 1996, Kohn and Foster 1996) the chemical composition of apatite has a direct influence on the annealing kinetics of fission tracks. Fission tracks in apatite with dominant chlorine content ($Cl/(Cl+F) \approx 0.8$) and low fluorine content anneal slower and completely fade at higher temperatures than fission tracks in apatite with low chlorine content ($Cl/(Cl+F) \approx 0.0$) and high fluorine content. Further cooling of those sandstones results in a slower annealing of the fission tracks in Cl-dominated apatite. This would cause a higher probability that etchable fission tracks in these grains would intersect a polished surface, and therefore a higher track density and an older fission track age for chlorine dominated grains in a sample having mixed apatite compositions (Laslett *et al.*, 1988, Green 1988). If during slow cooling in the PAZ, or after reaching the stable zone, samples are reheated and again cooled, the fission tracks in low Cl-apatite grains anneal more rapidly and at lower temperatures than co-existing Cl-rich grains, and therefore the age difference increases. This process also has an influence on the length distribution of confined fission tracks. Therefore an average length distribution without screening the data in relation to the chemical composition of the apatite grains will not only represent the thermal history (T-t) but also the chemical variability of the sample.

Twelve sandstone horizons of various Paleozoic age were sampled to provide an overview on the stratigraphic distribution. They were taken along a NW-SE profile perpendicular to the main structural trend to evaluate the exhumation and/or uplift rates of the different tectonic units (Figure 1). Also sampled were the middle Devonian Hill-intrusion and two sandstone beds, which represent the wallrock of post-Variscan galena-sphalerite-calcite-quartz veins. After mineral separation was complete only eleven samples were found to contain sufficient detrital apatite (Table 1).

TABLE 1. Description of the samples (Tt.= tonalite, Hill-intrusion) used for fission track analysis, summary of the track count data and the ages of the samples. (n = no. of grains counted, N_s = no. of spont. tracks, N_i = no. of ind. tracks, ρ_s = density of spont. tracks (x10^5/cm^2), ρ_i= density of ind. tracks (x10^5/cm^2), N_d = 7,500 (no. of flux dosimeter tracks counted on CN-1). All samples pass the χ^2-test at 95% (χ^2 >0.05) and ages are calculated using the pooled statistics.

S.-No.	Strat.	Lith.	N	Alt. (m)	Sp. tracks ρ_s	Sp. tracks N_s	Ind. Tracks ρ_i	Ind. Tracks N_i	χ^2	age (Ma)±2σ
1	Permian	Sd.	20	410	15.2	865	11.5	656	0.8912	239 ± 26
2	U. Dev. (Fam.)	Sd.	23	160	24.9	343	24.7	339	0.9963	184 ± 28
3	U. Dev. (Fam.)	Sd.	29	168	20.1	1454	20.5	1477	0.8343	180 ± 16
4	U. Dev. (Fam.)	Sd.	25	240	28.1	544	30.1	583	0.9989	170 ± 22
5	L. Dev. (Sieg.)	Sd.	22	330	13.5	753	12.2	681	0.9156	201 ± 24
6	L. Dev. (Ged.)	Sd.	31	310	8.55	560	8.26	541	0.9525	189 ± 24
7	L. Dev. (Ged.)	Sd.	19	470	17.6	528	18.6	546	0.1996	176 ± 24
8	L. Dev. (Sieg.)	Sd.	12	320	12.1	197	10.9	177	0.6515	203 ± 42
9	U. Carb. (Nam.)	Sd.	27	60	20.1	506	23.1	581	0.9828	159 ± 20
10	L. Dev. (Ems.)	Sd.	24	570	15.5	272	21.9	384	0.9997	130 ± 22
11	M. Dev.	Tt.	20	520	5.98	259	5.57	241	0.9758	196 ± 36

Apatite mineral separates were obtained using standard heavy liquid and magnetic separation techniques (Grist and Ravenhurst 1992a,b). The individual separates (generally containing >100 apatite grains) were mounted in epoxy and polished to expose internal grain surfaces. Spontaneous tracks were revealed by etching in 5M HNO_3 for 20s at 25°C. Fission track ages were determined using the external detector method (Hurford and Green 1982). A thin sheet of low-uranium muscovite was placed in contact with the polished surface of each grain mount to serve as a detector of neutron-induced fission.

Samples were irradiated at the Dalhousie University Slowpoke reactor in the presence of a glass neutron dosimeter (CN1) of known uranium content. The thermal-neutron fluence obtained was approximately 9×10^{15} neutrons/cm^2. The cadmium ratio for gold in the irradiation site at the Slowpoke reactor is 4.3 (Chatt and Holzbecher 1992). Micas were etched at room temperature in 48% HF for 20 minutes. Track densities (tracks/cm^2) for both spontaneous and induced fission track populations were measured at a magnification of 1,250 (100X dry objective, 12.5X oculars). Fission track ages were calculated using a weighted mean zeta calibration factor (Fleischer *et al.*, 1975; Hurford and Green 1983), determined using the Fish Canyon Tuff apatite and Durango age calibration standards. Etchable length distributions of horizontal, confined tracks were measured at 1,562.5 magnification (100X dry objective, 12.5X oculars, 1.25X microscope tube) using a digitizing pad interfaced with a personal computer. The precision of each individual track length measured is approximately ±0.15 μm. A value of 105.4 ± 1.8 was used for the zeta factor. All samples passed the Chi-square test at the 95% confidence level.

The chemical composition of apatite grains which were previously counted was determined using a Jeol 733 Microprobe with an EDS- and WDS-analyzing system (Table 2, 3). The operating parameters similar to those recommended by Stormer et al. (1993) are 10μm beam size, 15 KV acceleration voltage and 15 nA current. Fluorine was counted for 20 sec. and Oxygen for 40 sec. The raw data were corrected using the Link ZAF matrix-correction program. Instrument calibration was performed on cobalt metal (±0.5%). The analytical program includes P, Si, S, La, Ce, Fe, Mn, Mg, Ca, Na, Sr, F, Cl and O. Durango apatite was used as a standard for O, F, Ca and P, tugtupite for Cl, sanidine for Si, garnet for Fe, MnO_2 for Mn, jadeite for Na, kaersutite for Mg, chalcopyrite for S, celestite for Sr and monazite for La, Ce and Nd. The detection limit for the major elements is 0.1 wt.% and for the trace elements 0.01 wt.%. OH has been calculated assuming 21 atoms in the unit cell of apatite, F+Cl+OH equal 1 and total O equal 12.

TABLE 2. Chemical composition of Fluoropatite from different locations described as average with 2δ-error of mean (Si, Ce, Na, Fe, Mn, Cl < lower detection limit).

S.-No.	n	P (wt.%)	Ca (wt.%)	F (wt.%)	OH (wt.%)	O (wt.%)	Total (wt.%)
1	8	18.42 ± 0.12	39.48 ± 0.20	3.42 ± 0.22	0.34 ± 0.09	38.56 ± 0.40	100.22 ± 0.54
2	11	18.21 ± 0.06	39.24 ± 0.12	3.47 ± 0.28	0.37 ± 0.11	39.64 ± 0.26	100.93 ± 0.32
3	19	18.27 ± 0.08	39.17 ± 0.16	3.03 ± 0.20	0.69 ± 0.18	38.79 ± 0.22	99.95 ± 0.26
6	14	18.52 ± 0.06	39.46 ± 0.06	2.83 ± 0.22	0.90 ± 0.22	37.47 ± 0.24	99.18 ± 0.26
9	19	18.28 ± 0.08	39.19 ± 0.12	3.12 ± 0.14	0.58 ± 0.12	38.31 ± 0.24	99.48 ± 0.30
10	15	18.32 ± 0.08	39.15 ± 0.12	3.20 ± 0.20	0.51 ± 0.18	38.20 ± 0.22	99.38 ± 0.36
11	1	18.49	39.18	3.19	0.52	38.33	99.71

TABLE 3. Chemical composition of chlorine bearing Fluorpatite from different locations described as average with 2δ-error of mean (Si, Ce, Na, Fe, Mn < below lower detection limit).

S.-No.	n	P (wt.%)	Ca (wt.%)	F (wt.%)	Cl (wt.%)	OH (wt.%)	O (wt.%)	Total (wt.%)
1	3	18.31 ± 0.30	38.93 ± 0.34	2.38 ± 0.26	0.37 ± 0.24	1.09 ± 0.30	38.92 ± 0.30	100 ± 0.66
2	4	18.20 ± 0.14	39.31 ± 0.18	2.67 ± 0.29	0.26 ± 0.16	0.92 ± 0.23	39.30 ± 0.20	100.66 ± 0.50
3	4	18.16 ± 0.04	39.14 ± 0.16	2.88 ± 0.24	0.18 ± 0.06	0.77 ± 0.40	39.32 ± 0.64	100.45 ± 0.48
6	2	18.53 ± 0.06	39.15 ± 0.12	2.89 ± 0.04	0.10 ± 0.00	0.75 ± 0.01	38.41 ± 0.18	99.83 ± 0.38
9	1	18.46	39.43	3.19	0.14	0.47	38.53	100.22
10	2	18.06 ± 0.12	38.69 ± 0.66	2.25 ± 0.01	0.64 ± 0.40	1.04 ± 0.20	38.13 ± 0.32	98.81 ± 076
11	5	18.30 ± 0.04	38.88 ± 0.14	3.28 ± 0.06	0.17 ± 0.04	0.37 ± 0.04	38.72 ± 0.32	99.72 ± 0.50

The etchable track length distribution in relation to the chemical composition of the grains was used to model the thermal history. The computer based random search inversion model of Issler (1996) calculates a range of thermal histories that provide statistically acceptable fits of calculated fission track age and length parameters to the observed data. The computer program is a modification of the inverse model of Willett (1992). The program allows to use different annealing models as the base of the calculation. In relation to the chemical composition of apatite grains the thermal history of the intrusion (11) was modeled using the annealing model of Crowley *et al.*, (1991) for the Durango apatite.

The sandstone samples were modeled with the fluorapatite annealing characteristics as described by Crowley *et al.*, (1991). Although the program allows to constrain the temperature history very much, the data have been modeled with as little constraint as possible. The age steps were set to 20 Ma for the older and to 10 Ma for the younger part of the thermal history, with 5 Ma substeps. The modeling was stopped at a 0.05 significance level. 25,000 forward models were permitted and 250 thermal solutions retained in parameter search space. The ACUR-parameter of 0.1 gives an approximate numerical error for calculated track lengths of ≤0.2%. The corresponding errors in temperature are approximately ≤0.2°C. The program does not allow to correct the length with respect to the c-axis or the a-axis. For the first run the upper temperature limits were in general set to 160 °C and changed only for those ages where temperature information was available due to the established geological history of the area. The lower temperature limits were set to zero for all runs. Further runs were conducted with the step by step adjustment of the upper temperature limit.

3. RESULTS

3.1. FISSION TRACK AGES

Apatite fission track pooled ages of deformed Devonian sandstones in the northern Linksrheinisches Schiefergebirge are in the range of 170±22 Ma (Middle Jurassic±2σ error) to 203±42 Ma (Lower Jurassic) with an average of 186 Ma (Table 1, Figure 1). Similar ages are obtained from Gedinnian and Famennian sandstones (2, 3, 4, 6, 7) whereas the Upper Siegenian sandstones (5, 8) have the oldest fission track ages of the Devonian sandstones. Sandstones of the same stratigraphic position but in different nappes have similar ages. The Hill-intrusion (11) gives a fission track pooled age of 196±36 Ma which is in the range of the Devonian sandstone ages. The Permian sandstone (1) of the Graben of Malmedy has the oldest pooled age of 239±26 Ma which is only slightly younger than the depositional age of the sandstones. The Namurian (Pb-Zn-North 9) and Emsian sandstone (Pb-Zn-South 10) taken in cm distances from the post-Variscan vein-type mineralization, show younger fission track pooled ages of 159±20 Ma and 130±22 Ma.

The pooled ages of all samples are younger than the depositional age of the sandstones or emplacement age of the intrusion. The reflectance of the organic material in associated shale and slate layers show that all Devonian sandstone samples have been heated above 160 °C during burial (Stroink 1993, Heinen 1996, Rottke 1996), which is sufficient for a complete annealing of the fission tracks in apatite (Wagner and Van den Haute 1992). Therefore the pooled ages of all pre-Permian rocks represent finite ages which are derived during the cooling of the samples from peak- to post-depositional temperatures.

In the case of the Permian sandstone the only available temperature information is from a fluid inclusion study done on quartz veinlets and breccias which occur along normal fault planes of the Graben (Glasmacher 1994, 1995). The homogenization temperature of two phase fluid inclusions are 190 °C. Even if we assume that the original total thickness of the Permian siliciclasic unit was more than 250m and that the thermal gradient was higher than normal due to a strong advective component, it seems unlikely that the fission tracks in apatite have been completely annealed. Therefore the pooled age of the Permian sandstone might represent partly reset provenance age for the sandstone. No relation exists between the present altitude and the pooled age of the samples.

The geological interpretation of the pooled age is strongly dependent on the distribution and degree of homogeneity, as well as on the chemical composition of the apatite grains. To evaluate the distribution and degree of homogeneity the standard radial plot (Galbraith 1988, Galbraith 1990, Galbraith and Green 1990), and diagram combining a composite probability density curve with a frequency distribution histogram (Hurford *et al.*, 1984, Green 1989, Brandon 1992) were applied to the single grain age data for the samples. Both statistical tools showed the same trends, therefore the radial plots were chosen to display distribution and homogeneity of the single grain age data. In general the radial plots confirm a single age population for most of the samples. Only a few sample (2, 3, 9, 5) show a spread in age data which might point towards two populations (Figure 2, 3).

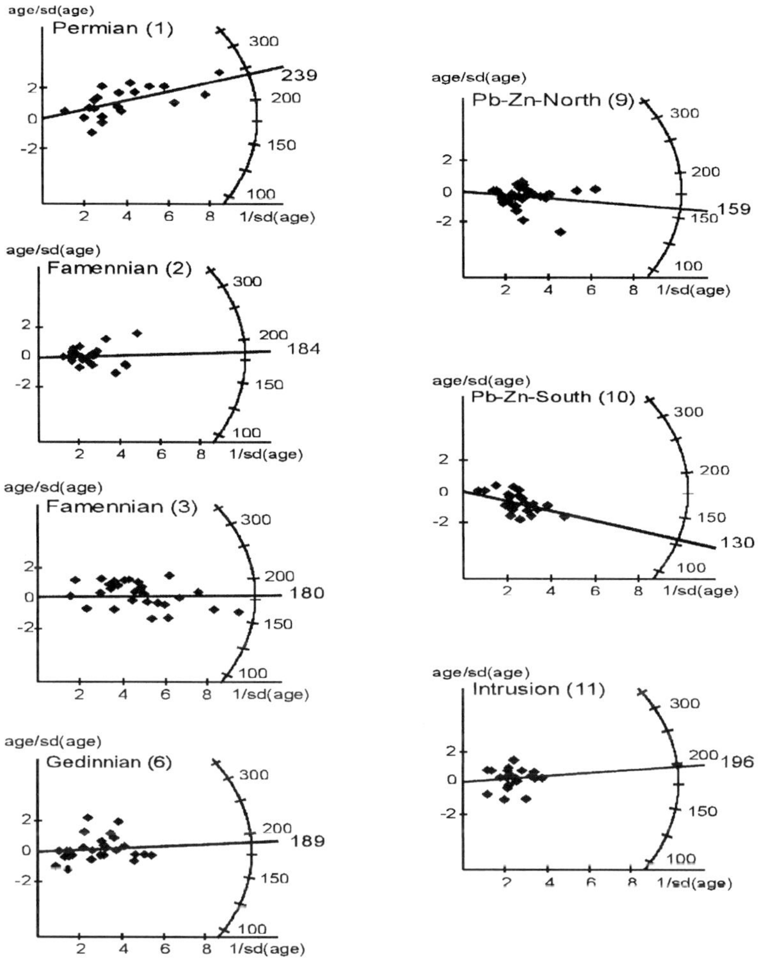

Figure 2. Radial plots of selected samples from the northern Linksrheinisches Schiefergebirge, Germany.

3.2. CHEMICAL COMPOSITION

To test the dependency of the single grain age distribution on the chemical variation, the chemical composition of dated apatites were determined for the Permian (1), the Famennian (2, 3) and the Gedinnian sandstones (6) as well as for both sandstones (9, 10) which were thermally affected by the post-Variscan hydrothermal fluids (Table 2, 3, Figure 4). A set of randomly chosen grains covering the entire age range of the dated apatites from the intrusion (11) were also analyzed (Figure 5). The chemical data were grouped into Cl-bearing and Cl-free apatites and screened by applying an age/σ-ratio of 1.8 per grain. All grains beneath the ratio were rejected. The sandstone samples in general are dominated by fluorapatite with chlorine content below the detection limit. Only a small amount of grains per sample (less than 28%) are Cl-bearing (<0.65 wt.%).

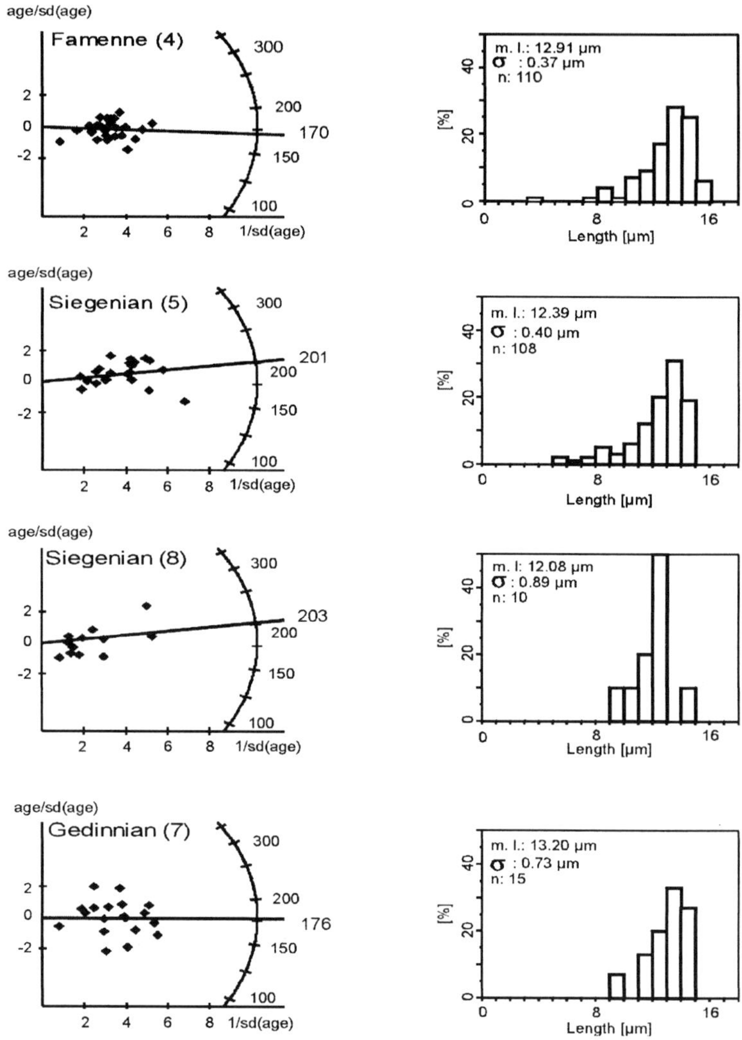

Figure 3. Radial plots and track length distribution of selected samples from the northern Linksrheinisches Schiefergebirge, Germany (m.l.: mean length, n: number of confined tracks).

The apatite grains of the intrusion are dominated by low chlorine contents in the range of 0.22 wt.% to 0.13 wt.%.

Fluorapatite grains of the Permian (1), Famennian (2, 3) and Gedinnian (6) sandstones show a well developed negative correlation of fluorine (r<0.77) and a positive correlation of OH (r<0.79) with the decrease of the single grain ages between 260 Ma and 140 Ma (Figure 4). The fluorine and OH versus age pattern of the Gedinnian sandstone exhibit a curved shape with flat slope at grain ages above 200 Ma and a step slope below 200 Ma. This pattern is possibly an analytical artifact but if not it might represent a change in the degree of cooling (around 150 Ma).

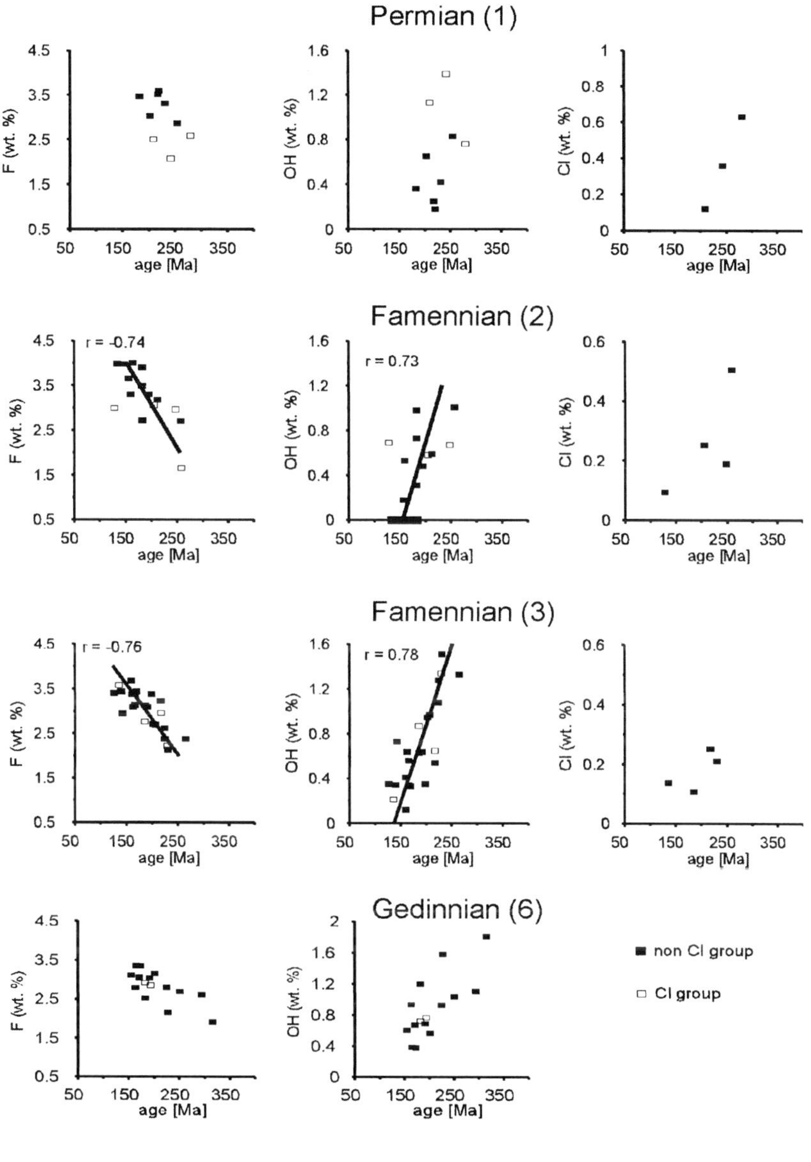

Figure 4. Fluorine, chlorine and OH variation with decreasing single grain ages.

A positive correlation can be seen between the chlorine content of Cl-bearing fluorapatite grains and the decrease in single grain age. The Cl-bearing fluorapatite grains of the middle Devonian intrusion show a similar weak positive correlation (Figure 5). In general the dependency of the distribution of single grain ages on the fluorine and chlorine content of the fluorapatite point to a single cooling history of the sandstone samples with no major thermal overprint.

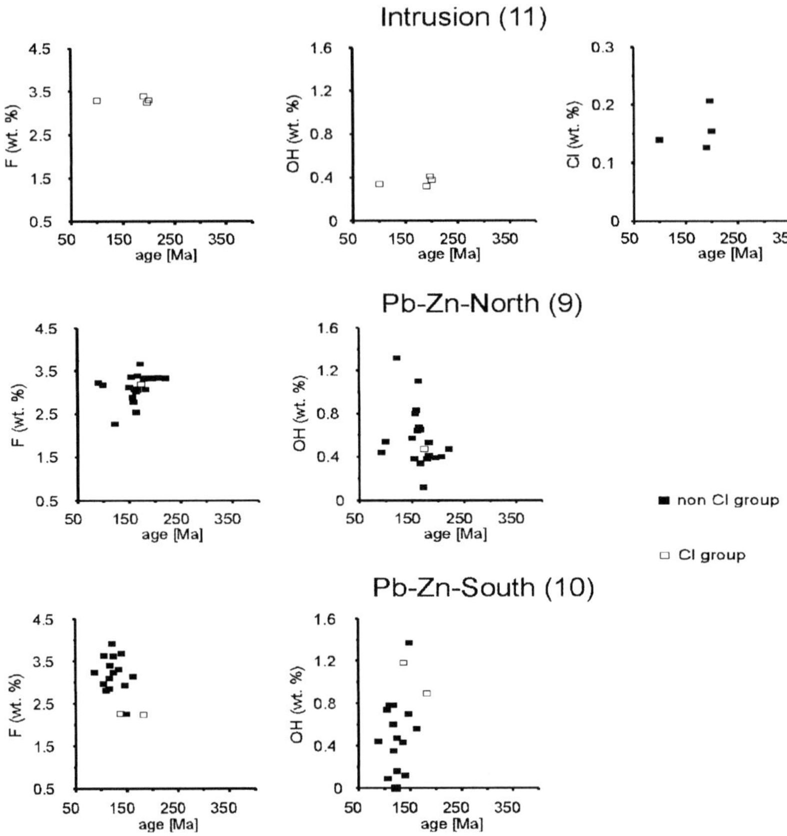

Figure 5. Fluorine, chlorine and OH variation with decreasing single grain ages of the middle Devonian intrusion and the sandstones taken adjacent to the post-Variscan galena-sphalerite-calcite-quartz veins.

Both sandstones (Namurian: 9, Emsian: 10), which have been affected by post-Variscan hydrothermal fluids, show no correlation between chemical composition and single grain ages, but in general a wide spread of fluorine and OH values between 190 Ma and 160 Ma (Pb-Zn-North, 9) and between 160 Ma and 100 Ma (Pb-Zn-South, 10). This trend might point to a fast cooling of all grains.

3.3. TRACK LENGTH DISTRIBUTION

The mean length distributions of confined tracks range from 12.1 μm to 13.2 μm with a standard deviation between 1.3 μm and 2.1 μm (Table 4, Figure 3, 6, 7). All are unimodally distributed with a negative skewness and a leptokurtic shape. According to Gleadow et al. (1986) a fission track length distribution of similar shape, mean length and standard deviation is typical for a slow cooling history.

The confined track length distribution of the Siegenian and Gedinnian sandstone (7, 8) of the Eastardennen-Eifel nappe can only be taken as an indication because only a few confined tracks could be measured.

TABLE 4: Confined fission track data.

S.- No.	No. of confined tracks	Mean length (μm)	Std.-Dev. (μm)	Std. Err. (2σ) (μm)
1	114	12.7	1.6	0.3
2	82	12.5	1.8	0.4
3	120	13.0	1.7	0.3
4	110	12.9	1.9	0.4
5	108	12.4	2.0	0.4
6	103	12.8	1.9	0.4
7	15	13.2	1.4	0.7
8	10	12.1	1.3	0.9
9	131	12.4	2.1	0.4
10	68	13.0	1.5	0.4
11	104	12.5	1.9	0.4

3.4. MODELING THE THERMAL HISTORY

The confined track length distribution together with the fission track age were used to model the thermal history of the samples. The computer based modeling depends on the annealing characteristics of apatite, which is strongly dependent on the chemical composition. In relation to the chemical composition of the apatite grains the thermal history of the sandstone samples was modeled with the fluorapatite annealing characteristics as described by Crowley *et al.* (1991). The intrusion (11) was modeled using the Durango apatite annealing model of Crowley *et al.* (1991). For the modeling procedure the geological information of the region was converted into time using the time scale of the Geological Society of America. With the exception of the Namurian sandstone (Pb-Zn-North 9) a temperature of 20°C at 80 Ma (Campanian) was used to represent the onset of marine sedimentation during Upper Cretaceous. The depositional age of the Permian sandstone (≈260 Ma) was encountered by a temperature of 20 °C. The preclosure-age temperatures of the Devonian and Carboniferous sandstones were estimated by using temperatures described by Stroink (1993). Although those temperatures were applied for the preclosure-age temperatures of the potential source region of the Permian sandstone, the preclosure-age temperature of the intrusion was constrained by the results of a K/Ar-dating of fine grained K-feldspar.

The computer program failed to find a solution if the upper temperature limit for the age range between 60 Ma and 20 Ma was set below 60 °C. The extend of the onset of partial annealing up to recent time was necessary for the Famennian sandstone from Aachen (2). According to Green (1996) the model overpredicts the amount of low temperature annealing, but in the case of the sandstone from Aachen it might be related to a thermal overprint due to the present migration of thermal fluids (up to 72 °C) in the city limits of Aachen.

The final mean exponential temperature solutions of all modeled samples are displayed as time-temperature curves (Fig. 5, Fig. 6). Also for comparison the observed (histogram) and calculated (PDF-curve) track length distribution are shown. The final exponential mean temperature solution was accepted if the following criteria were all met:

1) calculated mean apatite fission track ages were less than ±5% apart from the measured ages,
2) the calculated mean closure ages was in the range of the oldest single grain ages with an age/σ-ratio above 1.8,
3) the calculated mean track-length distribution was in good visible agreement with the measured track length-distribution,
4) the thermal history of the PAZ displayed by the F/age-ratio and the Cl/age-ratio was in visible agreement with the calculated exponential mean temperature solution.

If in the future more geological information becomes available, it might be possible to find better solutions for the samples. And with a more extensive sample set it might be possible to obtain a better temperature resolution. To compare the time-temperature curves (TTC) of the different samples the major time interval between 220 Ma and 120 Ma of the Permian sandstone (1) has been chosen for all samples. Also nearly all TTC's show a change in the cooling rate at around 120 Ma.

After deposition of the Permian sandstone at 260 Ma (assumed 20°C) the TTC display a strong increase of temperature (86°C) up to 106°C at 220 Ma. Between 220 Ma and 120 Ma the temperature decreased very slowly down to 75°C with an average cooling rate of 0.3°C/Ma. The closure temperature of 135°C was reached at an average closure age of 280 Ma. Therefore the Lower Permian cooling rate of the source region which is believed to be the Paleozoic rocks of the Variscan orogen (Maillieux 1931), was less than 6°C/Ma. A closure age older than the depositional age of 260 Ma shows also that the apatite grains were only partially reset. The lower limit of the PAZ was reached at around 110 Ma.

The intrusion which is located further to the NE but in the same nappe, also displayed a cooling rate of 0.3°C/Ma between 220 Ma (94°C) and 120 Ma (61°C). The slight difference in temperature might be due to the difference in altitude between both locations. The cooling rate between 120 Ma and 80 Ma for both samples are in a similar range (intrusion: 1.2°C/Ma, Permian: 1.5°C/Ma). At 230 Ma the intrusion cooled beneath the upper limit of the PAZ (135°C) and at 120 Ma beneath 60°C. The increase of temperature below 150 Ma should not be over interpreted also the Devonian sandstones (2, 3, 6) show a more or less similar pattern of temperature increase before the temperature decreases with a cooling rate of 1.8°C/Ma towards 80 Ma.

In general the cooling rate between 220 Ma and 120 Ma of the Devonian sandstones (2, 3, 6) is only slightly higher (0.4°C/Ma) than those of the southern area. All samples have the same boundary temperatures of around 124°C and 85°C for this time interval even that the variation of temperatures in the time interval are different.

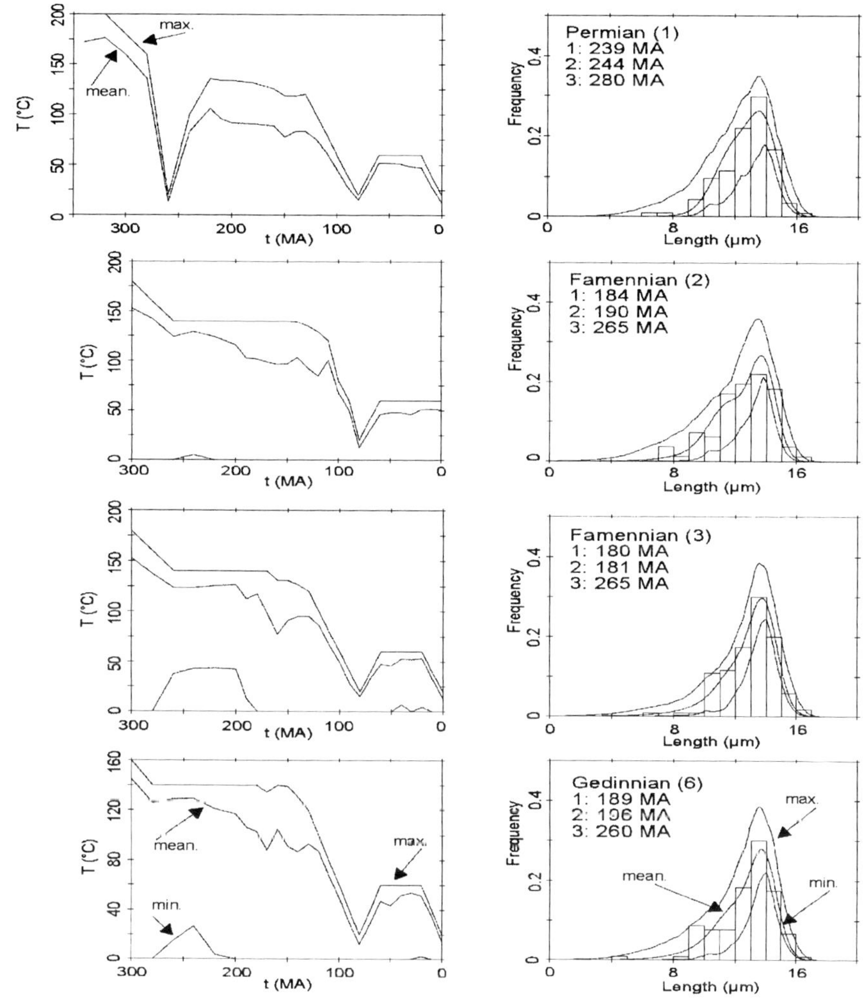

Figure 6. Exponential mean temperature solution (middle curve) displayed in a time-temperature diagram (left) and frequency distribution of measured track length data (right) overlain by the calculated PDF-curve (mean: middle curve) of the Devonian and Permian sandstone samples (1: measured fission track age, 2: calculated fission track age, 3: calculated closure age).

Whereas the Famennian sandstone (2) displays a more or less monotonous decrease of temperature with an increase between 150 Ma and 120 Ma, the other Famennian (3) sample further to the south shows a steeper decrease between 180 Ma and 150 Ma and a stronger increase of temperature between 150 Ma and 120 Ma. This sample has been taken close to the occurrence of post-Variscan Pb-Zn-mineralization in the Venn-Weser-Inde nappe. The TTC of the Gedinnian sample (6) is similar to those of the Famennian sandstone (2). A closure age of 265 Ma at a temperature of 130°C was calculated for the Upper Devonian sandstones (2, 3) and 260 Ma at the same temperature for the Lower Devonian sandstone (6). All three sandstones reached the lower limit (60°C) of the PAZ at around 100 Ma.

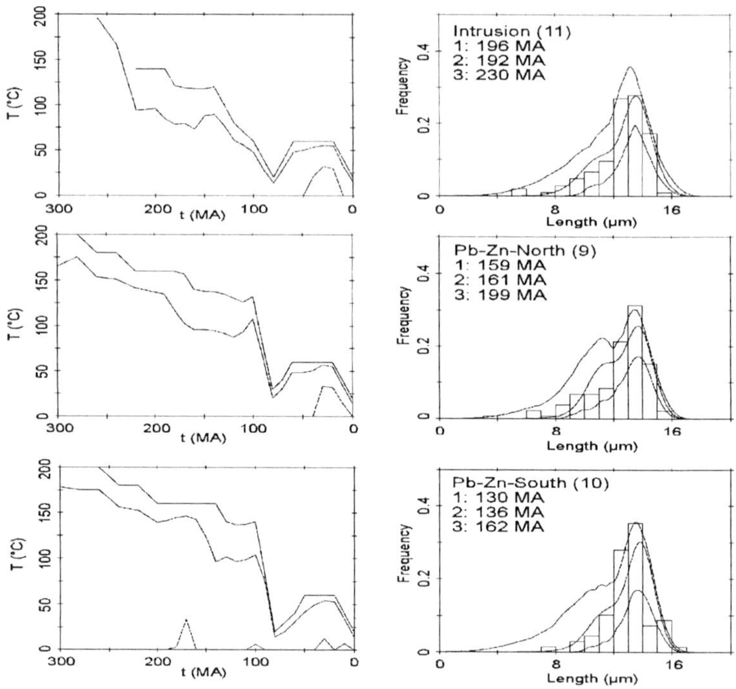

Figure 7. Exponential mean temperature solution (middle curve) displayed in a time-temperature diagram (left) and frequency distribution of measured track length data (right) overlain by the calculated PDF-curve (mean: middle curve) of the Middle Devonian Intrusion and wallrock sandstones of post-Variscan galena-sphalerite-calcite-quartz veins (1: measured fission track age, 2: calculated fission track age, 3: calculated closure age).

The Namurian sandstone (Pb-Zn-North 9) displays a TTC pattern between 190 and 160 Ma similar to the Famennian sandstone (3). In comparison with the other sandstone samples of the Venn-Weser-Inde nappe the Namurian sandstone (Pb-Zn-North 9) remained longer above the closure temperature (135°C), which was reached around 200 Ma shortly before the steep decrease in temperature occurred. The lower limit of the PAZ was reached at around 90 Ma The TTC of the Emsian sandstone (Pb-Zn-South 10) of the Eastardennen-Eifel nappe, which was also taken adjacent to a galena-sphalerite-quartz vein, displays a similar steep decrease of temperature but at the time interval of 160 Ma to 140 Ma. The closure temperature (143°C) was reached at 162 Ma and the lower limit of the PAZ at around 90 Ma. Between 140 Ma and 100 Ma the sample stayed in the PAZ.

4. **Discussion**

The pattern of the time-temperature curves revealed by the computer-based modeling of the fission track data lead to a general subdivision of the post-orogenic Variscan thermal history into three major components: 1) The oldest component was still influenced by processes related to the Variscan orogeny lasted until the late Triassic (220 Ma). It was followed by a 2) Mesozoic component up to the Upper Cretaceous (80 Ma) and 3) a post-Campanian thermal evolution of the region. Also the thermal history displayed by the TTC divides the northern Linksrheinisches Schiefergebirge during Permian and Lower Triassic into a southern and northern region. The limited number of samples does not allow to confine the boundary exactly but it seems that the southern and northern region are identical with the Eastardennen-Eifel nappe and the Venn-Weser-Inde nappe.

Information regarding the Upper Paleozoic thermal evolution of the Eastardennen-Eifel nappe are still stored in the apatite grains of the Permian sandstone (1) despite the partial thermal overprint of the grains during the Graben development (260 Ma - 220 Ma). Mailieux (1931) described the surrounding Paleozoic sedimentary units as the source rocks of the Permian sediments. If a normal constant geothermal gradient is considered as a minimum gradient for the time interval between 280 Ma and 260 Ma, the cooling rate of less than 6°C/Ma would imply an average exhumation rate of less than 0.15 mm/a. Applying the same geothermal gradient for the time interval of the Graben development would lead to an accumulation of up to 3,000m of sediments. The thermal evolution of the Middle Devonian intrusion (11) as part of the Eastardennen-Eifel nappe during the Upper Permian and the Triassic defined by a K/Ar-date of K-feldspar at 250 Ma, which was interpreted as a cooling age (Tschernoster *et al.*, 1995) and the closure age of apatite grains at 230 Ma also point to a similar average cooling rate for the Eastardennen-Eifel nappe. The different thermal histories of the intrusion, which is located outside the influence of the subsidence area and the Permian sandstone during the Graben development (260 Ma - 220 Ma), can either be explained by opposite movements of both crustal segments or as caused by the cooling of the regional thermal field due to the subsidence of cool sediments into the upper crust. If the second process has caused the cooling of the intrusion then a higher geothermal gradient has to be applied to the surrounding crust. A higher geothermal gradient would also lower the emplacement depth of the intrusion. The Devonian sandstone samples of the Venn-Weser-Inde nappe and the Aachen-Herve-Imbricated zone in the northern region reached the PAZ at around 260 Ma. The application of a similar normal geothermal gradient to estimate the sedimentary cover at the closure time (260 Ma) leads to a depth of the samples between 3,000 m and 4,000 m. Regarding the age of this cover it seems most probable to attribute them mainly to the Devonian and Carboniferous. The balanced cross section of v. Winterfeld and Walter (1993) suggests an extension of the Venn-Weser-Inde nappe and the Eastardennen-Eifel nappe further to the northwest. The equal and very low cooling rate (0.3°C/Ma) of all three sandstone samples between 260 Ma and 220 Ma point to a more stable behavior of the northwestern crustal segments.

With the beginning of the Mesozoic thermal component at 220 Ma, the TTC of all samples display in general a similar slow cooling (≈0.3°C/Ma) trend up to the early Cretaceous (120 Ma) and a rapid cooling (1.2 - 1.8°C/Ma) up to the Campanian marine transgression (≈80 Ma) for both nappes. The application of a normal but stable geothermal gradient to these data leads to an exhumation rate of 0.01mm/a up to 120 Ma and an average of 0.05mm/a afterwards. With the exception of the Permian sandstone (1) the general slow cooling trend is moderated between 150 and 120 Ma, which might have been caused either by a more or less unstable thermal field or a higher activity of tectonic movements. Especially the TTC of the Famennian sandstone (3), which is located in the main area of post-Variscan hydrothermal activitys, displays fast cooling down to 80 °C between 180 and 170 Ma and an slight increase of temperature (95 °C) up to 130 Ma. Regionally the late Jurassic and the early Cretaceous are characterized by tectonic movements (Meyer *et al.*, 1983, Murawski *et al.*, 1983), intrusion of magmas along the southern margin of the North German Basin and the opening of the North Atlantic (Ziegler 1988, Ziegler 1990, Berthelsen 1992, Franke 1992, Walter 1992). Murawski *et al.* (1983) proposed the beginning of a major uplift at the Jurassic/Cretaceous boundary (144 Ma) which is connected with the subsidence of the Netherlands Central Graben north of the Brabant Massif. Based on apatite fission track data Van den Haute and Vercoutere (1989) and Vercoutere and van den Haute (1993) proposed a major uplift in late Jurassic (150 - 140 Ma) and early Cretaceous (90 Ma) for the Brabant Massif which is located north of the area under investigation. In comparison their pooled ages are in the same range as those described in this study. The confined track length distribution has a similar pattern but the mean track length of their samples is longer which points to a faster cooling. The differences between both data sets might be interpreted as a later start of the uplift in the northern Linksrheinisches Schiefergebirge. In an area with hydrothermal mineralization spread as widely as it is in the northern Linksrheinisches Schiefergebirge, cooling or heating might not be directly related to exhumation or subsidence.

The TTC of the Namurian sandstone (Pb-Zn-North 9) taken adjacent to a galena-sphalerite-carbonate vein display similar fast cooling in the middle Jurassic as the Famennian (3) sandstone does but show also that the Namurian sandstone remained longer above the PAZ. The upper limit of the PAZ was reached at around 200 Ma shortly before the cooling of the sample occurred. As described by Redecke (1992) the hydrothermal solutions reached a maximum temperature of 190°C and cooled down to 95°C. Therefore it seems to be very likely that the TTC-pattern for that time interval (≈220 Ma -160 Ma) displays the complete annealing of the fission tracks due to the advective heat transfer. Also the lowest temperature of the hydrothermal fluids is the same as the final temperature shown by the TTC after cooling. This temperature is in agreement with the temperature displayed by the TTC of the other Devonian sandstones (2, 3,) in the surrounding area. Post-Variscan mineralization in other areas of the Rhenish Massif dated by Hagedorn *et al.* (1991) and Boneß *et al.* (1990) occurred also between 180 and 160 Ma. The fast cooling of the Famennian (3) sandstone which was not a wallrock of the mineralization, but is located in the area of mineralization, shows an imprint due to the hydrothermal solutions. The TTC of the Emsian sandstone (Pb-Zn-South 10) taken adjacent to a galena-sphalerite-quartz vein in the southern part of the area displays a similar rapid cooling but between 160 Ma and 140 Ma down to the same temperature as the TTC of the other non-mineralized samples show at 140 Ma.

The post-Campanian thermal evolution of the area displayed by the TTC is problematic. As mentioned before it is still not known if the pattern modeled by the computer program is an artifact of the inverse model as Green (1996) proposed. In the case of the northern Linksrheinisches Schiefergebirge the thermal history is constrained by sedimentation up to the Miocene followed by an uplift which is still active and has caused an elevation of up to 690m. Also the Famennian (2) sandstone from Aachen might have been thermally influenced by the recent active thermal springs.

5. Conclusions

The post-orogenic Variscan thermal history of the northern Linksrheinisches Schiefergebirge in general can be divided into three components: 1) an old component which was still influenced by processes related to the Variscan orogeny, 2) a Mesozoic component up to the Upper Cretaceous, and 3) a post-Campanian component. Information regarding the Upper Paleozoic thermal history are recorded by fission tracks in apatite grains in the Permian sandstone (1) of the Graben of Malmedy, which were not completely annealed during the 3,000 m subsidence of the Graben. An average cooling rate below 6°C/Ma can be estimated for the Late Paleozoic thermal history. Apatite grains of the Devonian sandstones (2, 3, 6) of the Venn-Weser-Inde nappe were completely annealed after deposition and reached the upper limit of the PAZ (135°C) at 260 Ma. At a normal geothermal gradient this would represent a 3,000m to 4,000m thick cover. Regarding the age of the cover, deformed Devonian and Carboniferous units are more probable candidates than Permian sedimentary rocks. Projecting the recent geological situation with the addition of the balanced cross section on top of the sampled outcrops would imply the extension of the Venn-Weser-Inde nappe and the Eastardennen-Eifel nappe further to the northwest.

The thermal history of the Mesozoic component is constrained by slow cooling of 0.3°C/Ma up to 120 Ma and faster cooling of ≈1.5°C/Ma up to 80 Ma. An exhumation of ≈1,000m might have occurred if a normal geothermal gradient is applied to the first time interval. The cooling between 120 Ma and 80 Ma, which points to an exhumation of ≈2,000m is interpreted as the main Mesozoic exhumation of the northern Linksrheinisches Schiefergebirge. Apatite fission track data of the Namurian sandstone (Pb-Zn-North 9) taken adjacent to the galena-sphalerite-calcite vein in the Venn-Weser-Inde nappe point to a mineralization age between 200 Ma and 160 Ma. The age of the mineralization in the Eastardennen-Eifel nappe revealed by the fission track data is younger (160 -140 Ma). Siliciclastic units represented by the Famennian (3) sandstone that are not mineralized, but occur in the area of mineralization, are also affected by the advective heat transfer.

The post-Campanian thermal evolution of the northern Linksrheinisches Schiefergebirge cannot be constrained by the apatite fission track data.

Acknowledgments

The authors like to thank the BMBF (Germany) and BEB-OIL and Gas GmbH (Germany) who funded this research with a cooperative grant. We also appreciated the constructive comments of Max Rohrmann and an unknown reviewer.

References

Albers H.J. and Felder W.M. (1979): Litho-, Biostratigraphie und Palökologie der Oberkreide und des Alttertiärs (Präobersanton - Dan/Paläozän) von Aachen-Südlimburg (Niederlande, Deutschland, Belgien). In: Aspekte der Kreide Europas. *Internat. Union geol. Sci.* **6**, 47-84.

Barbarand J., Pagel M., Blanc P., Braun J.-J., Chaussidon M., Vetil C. and Walgenwitz F. (1996): Combined Cathodoluminescence spectra, fission-track age and chemistry of detrital apatite grains: Towards a better understanding of fission track annealing. *Int. Workshop on Fission Track Dating* (abstr. vol.), 5.

Berthelsen A. (1992): Mobile Europe. In: *A Continent revealed, The European Geotraverse* (eds. Blundell D., Freeman R. and Mueller St.), pp. 11-32.

Bless M.J.M., Bouckaert J., Camelbeeck T., Dejonghe L., Demou-Lin A., Dupuis C., Felder P.J., Geukens F., Gullentops F., Hance L., Jagt J.W.M., Juvigne E., Kramm U., Ozer A., Pissart A., Robas-Zynski F., Schumacker R., Smolderen A., Spaeth G., Steemans Ph., Streel M., Vandenven G., Vanguestaine M., Walter R. & Wolf M. (1991): The Stavelot Massif from Cambrian to recent. A survey of the present state of knowledge. *Ann. Soc. Géol. Belg.*, 113, 53-73.

Boneß M., Haak U. & Feldmann K.H. (1990): Rb/Sr-Datierung der hydrothermalen Pb-Zn Vererzung von Bad Grund (Harz), BRD. *Chem. Erde* **50**, 1-25.

Brandon M.T. (1992): Decomposition of Fission-Track grain-age distributions. *Am. J. Sc.* **292**, 535-564.

Carpéna J. (1996): Compositional variation of natural apatites subjected to fission track analysis. *Int. Workshop on Fission Track Dating (abstr. vol.)*, 7.

Chatt A. and Holzbecher J. (1992): Geoanalytical usage of Dalhousie University Slowpoke-2 Reactor, in: *Short Course on low temperature thermochronology* (eds. Zentilli M. and Reynolds P.H.), Min. Ass. Can. Short Course **V 20**, Appendix 3, 211-224.

Crowley K.D., Cameron M. and Schaefer R.L. (1991): Experimental studies of annealing of etched fission tracks in fluorapatite. *Geoch. Cosmoch. Acta* **55**, 1449-1465.

Donelick R.A. (1988): Etchable Fission Track Length Reduction in Apatite: Experimental Observations, Theory and Geological Application. *Unpub. Ph.D. thesis*, Renssalaer polytechnic Institute, Troy, New York, USA, 414p.

Echle W., Plüger W.L., Zielinski J., Frank B. and Scheps V. (1985): Petrography, Mineralogy, and Geochemistry of the Salmian Rocks from Research Borehole Konzen, Hohes Venn (West Germany). *N. Jb. Geol. Paläont. Abh.* **171**, 31-50.

Franke W. (1992): Phanerozoic structures and events in central Europe. In: *A Continent revealed - The European Geotraverse* (eds. Blundell D., Freeman R. and Mueller St.), 164-179.

Fleischer R.L., Price P.B. and Walker R.M. (1975): Nuclear tracks in Solids: Principles and Applications. University of California Press, Berkeley, California.

Galbraith R.F. (1988): Graphical display of estimates having differing standard errors. *Technometrics* **30**, 271-281.

Galbraith R.F. (1990): The radial plot: Graphical Assessment of spread in ages. *Nucl. Tracks. Radiat. Meas.* **17**, 207-214.

Galbraith R.F. and Green P.F. (1990): Estimating the component ages in a finite mixture. *Nucl. Tracks. Radiat. Meas.* **17**, 197-206.

Glasmacher U. (1994): Variscan and Postvariscan fluid systems and thermal history in the northern part of the Linksrheinisches Schiefergebirge. *J. of the Czech Geol. Soc. (abst. vol.)*, **39**, 37-38.

Glasmacher U. (1995): Variszische und postvariszische Fluidsysteme, In: *KW-relevante Eigenschaften potentieller Mutter- und Speichergesteine am Nordrand des Linksrheinischen Schiefergebirges*: (eds. Walter R., Glasmacher U. & Wolf M.). **V**, pp. 1-41.

Gleadow A.J.W. & Duddy I.R. (1981): A natural long term track annealing experiment for apatite. *Nucl. Tracks Radiat. Meas.* **5**, 169-174.

Gleadow A.J.W., Duddy I.R., Green P.F. & Lovering J.F. (1986): Confined fission tracks lengths in apatite: a diagnostic tool for thermal history analysis. *Contrib. Mineral. Petrol.* **94**, 405-415.

Gleadow A.J.W., Foster D.A. and Galbraith R.F. (1996): Fission track dating of apatites from a mid-oceanic continental fragment: The Seychelles Islands in the Indian Ocean. *Int. Workshop on Fission Track Dating (abstr. vol.)*, 41.

Green P.F. (1989): Thermal and tectonic history of the East Midlands shelf (onshore UK) and surrounding regions assessed by apatite fission track analysis. *J. Geol. Soc. Lon.* **146**, 755-773.

Green P.F. (1996): The importance of compositional influence on fission track annealing in apatite, *Int. Workshop on Fission Track Dating (abstr. vol.)*, 43.

Green P.F., Duddy I.R., Gleadow A.J.W., Tingate P.R. & Laslett G.M. (1986): Thermal anneling of fission tracks in apatite: 1. A quantitative description. *Chem. Geol. (Isot. Geosci. Sect.)* **59**, 237-253.

Green P.F., Duddy I.R., Lovering J.F. (1988): Can fission track annealing in apatite be described by first-order kinetics? *Earth Planet. Sci. Lett.* **87**, 216-228.

Green P.F., Duddy I.R., Gleadow A.J.W. & Lovering J.F. (1989a): Apatite fission-track analysis exploration. In: *Thermal History of Sedimentary Basins: methods and case histories* (eds. Naeser N.D. & McCulloh Th.H.), pp. 181-195.

Green P.F., Duddy I.R., Laslett G.M., Hegarty K.A., Gleadow A.J.W. & Lovering J.F. (1989b): Thermal annealing of fission tracks in apatite: 4. Quantitative modelling techniques and extension to geological timescales. *Chem. Geol. (Iso. Geos. Sec.)* **79**, 155-182.

Grist A.M. and Ravenhurst C.E. (1992a): Mineral Separation Techniques used at Dalhousie University. In: *Short Course on low temperature thermochronology* (Zentilli M. and Reynolds P.H.), Min. Ass. Can. Short Course **V 20**, Appendix 2, pp. 203-209.

Grist A.M. and Ravenhurst C.E. (1992b): A Step-By-Step Laboratory guide to fission track thermochronology at Dalhousie University. In: *Short Course on low temperature thermochronology* (Zentilli M. and Reynolds P.H.), Min. Ass. Can. Short Course **V 20**, Appendix 1, pp. 190-201.

Hagedorn B., Lippolt H.J. & Wernicke R. (1991): Untersuchungen zur Isotopensystematik von Hämatit und Adular - Geochronologische Fallstudie Harz, In: *DFG-Schwerpunktprogramm Intraformationale Lagerstättenbildung*, (ed. Friedrich G.), Report **2**, pp. 355-366.

Heinen V. (1996): Simulation der präorogenen devonisch-unterkarbonischen Beckenentwicklung und Krustenstruktur im Linksrheinischen Schiefergebirge, *Aachener Geowiss. Beitr.*, **15**, 161p.

Hurford A.J. & Green P.F. (1982): A user's guide to fission track dating calibration. *Earth Planet. Sci. Lett.* **59**: 343-354.

Hurford A.J. & Green P.F. (1983): The zeta age calibration of fission track dating. *Chem. Geol.* **1**, 285-317.

Hurford A.J., Fitch F.J. and Clarke A. (1984): Resolution of the age structure of the detrital zircon populations of two Lower Cretaceous sandstones from the Weald of England by fission track dating. *Geol. Mag.* **121**, 269-277.

Issler A.R. (1996): An inverse model for extracting thermal histories from apatite fission track data: Instructions and software for the Windows 95 Environment. *GSC Open File* **2325**, 84p.

James K. and Durrani S.A. (1986): The effect of crystal composition on Fission-Track annealing and closure temperatures in geological minerals: Implications for the cooling rates of terrestrial and extraterrestrial rocks. *Nucl. Tracks. Radiat. Meas.* **11**, 277-282.

Kohn B.P. and Foster D.A. (1996): Exceptional apatite chlorine variation in the Stillwater complex, Montana: Thermochronological consequences. *Int. Workshop on Fission Track Dating (abstr. vol.)*, 67.

Kramm U. (1982): Die Metamorphose des Stavelot-Venn Massivs, nordwestliches Rheinisches Schiefergebirge: Grad, Alter und Ursachen, *Decheniana* **135**, 121-178.

Kramm U. and Buhl D. (1985): U-Pb Zircon Dating of the Hill Tonalite, Venn-Stavelot Massif, Ardennes, *N. Jb. Geol. Paläont. Abh.* **171**, 329-337.

Kramm U., Spaeth G. and Wolf M. (1985a): Variscan Metamorphism in the NE Stavelot-Venn Massif, Ardennes: A New Approach to the Question of Regional Dynamothermal or Contact Metamorphism, *N. Jb. Geol. Paläont. Abh.* **171**, 311-327.

Kramm U., Buhl D. and Chernyshev I.V. (1985b): Caledonian or Variscan Metamorphism in the Venn-Stavelot Massif, Ardennes? Arguments from a K-Ar and Rb-Sr Study, *N. Jb. Geol. Paläont. Abh.* **171**, 339-349.

Laslett G.M., Green P.F., Duddy I.R. and Gleadow A.J.W. (1987): Thermal annealing of fission tracks in apatite 2. A quantitative analysis, *Chem. Geol. (Iso. Geos. Sec.)* **65**, 1-13.

Mader D. (1982): Sedimentologie und Genese des Buntsandsteins in der Eifel, *Z. dt. geol. Ges.*, **133**, 257-307.

Maillieux E. (1931): Remarques sur les galets dévoniens du Poudingue de Malmédy, *Bull. Mus. Roy. Hist. Nat. Belg.* **7**, 1-15.

Meyer W., Albers H.J., Berners H.P., v. Gehlen K., Glatthaar D., Löhnertz W., Pfeffer K.H., Schnütgen A., Wienecke K. and Zakosek H. (1983): Pre-Quaternary Uplift in the Central Part of the Rhenish Massif, In: *Plateau Uplift* (eds. Fuchs K. et al.), pp. 39-46.

Muchez Ph., Slobodnik M., Viane W.A. and Keppens E. (1995): Geochemical constraints on the origin and migration of paleofluids at the northern margin of the Variscan foreland, southern Belgium, *Sed. Geol.* **96**, 191-200.

Murawski H., Albers H.J., Bender P., Berners H.-P., Dürr St., Huckriede R., Kauffmann G., Kowalczyk G., Meiburg P., Müller R., Muller A., Ritzkowski S., Schwab K., Semmel A., Stapf K., Walter R., Winter K.-P. and Zankl H. (1983): Regional tectonic setting and geological structure of the Rhenish Massif. In: *Plateau Uplift* (eds. Fuchs K. et al.), pp. 9-38.

Pommerening J. (1993): Hydrogeologie, Hydrogeochemie und Genese der Aachener Thermalquellen, *Mitt. Ing.- u. Hydrogeol.* **50**, 168p.

Plum H. (1989): Genetische Klassifikation und geochemische Interpretation der Mineral- und Thermalwässer der Eifel und Ardennen, *Mitt. Ing.- u. Hydrogeol.* **34**, 170p.

Redecke P. (1992): Zur Geochemie und Genese variszischer und postvariszischer Buntmetallmineralisation in der Nordeifel und der Niederrheinischen Bucht, *Mitt. Min. Lag.* **41**, 152 p.

Rottke W. (1996): Zur Diagenese devonischer und unterkarbonischer Karbonatgesteine am Nordrand des Linksrheinischen Schiefergebirges, *Aachener Geowiss. Beitr.* **12**, 152p.

Stormer J.C., Pierson M.L. and Tacker R.C. (1993): Variation of F and Cl X-ray intensity due to anisotropic diffusion in apatite during electron microprobe analysis. *Am. Min.* **78**, 641-648.

Stroink L. (1993): Zur Diagenese paläozoischer Sandsteine am Nordrand des Linksrheinisch-Ardennischen Schiefergebirges, *Aachener Geowiss. Beitr.* **1**, 206p.

Spaeth G., Fielitz W. and Frank B. (1985): Caledonian Deformation and Very Low-Grade Metamorphism in the Northeastern Part of the Stavelot-Venn Massif, *N. Jb. Geol. Paläont. Abh.* **171**, 297-310.

Tschernoster R., Glasmacher U., Spaeth G. & Clauer N. (1995): K-Ar-Datierungen zur Abkühlungsgeschichte ausgewählter Magmatite und Metapelite aus dem Stavelot Venn Massiv, In: *KW-relevante Eigenschaften potentieller Mutter- und Speichergesteine am Nordrand des Linksrheinischen Schiefergebirges*: (eds. Walter R. Glasmacher U. & Wolf M.) **IV**, pp. 1-20.

Vercoutere C. & Haute v. d. P. (1993): Post-Palaeozoic cooling and uplift of the Brabant Massif as revealed by apatite fission track analysis, *Geol. Mag.* **130**, 639-646.

Van Den Haute P. and Vercoutere C. (1990): Apatite fission-track evidence for a mesozoic uplift of the Brabant Massif: Preliminary results, *Ann. Soc. Géol. Belg.* **112**, 443-452.

von Winterfeld C. & Walter R. (1993): Die variszische Deformationsfront des nordwestlichen Rheinischen Schiefergebirges - Ein bilanziertes geologisches Tiefenprofil über die Nordeifel, *N. Jb. Geol. Paläont. Mh.* **5**, 305-320.

Wagner G.A. (1968): Fission track dating of apatites, *Earth and Planetary Letters* **4**, 411-415.

Wagner G.A. and Van den Haute (1992): Fission-track dating. Kluwer Academic Publishers, Dordrecht, The Netherlands.

Walter R. (1992): Geologie von Mitteleuropa, E. Schweizerbart´sche Verlagsbuchhandlung, Stuttgart.

Wellens M. (1995): Charakterisierung der organischen Substanz in den potentiellen Muttergesteinen, In: *KW-relevante Eigenschaften potentieller Mutter- und Speichergesteine am Nordrand des Linksrheinischen Schiefergebirges*: (eds. Walter R., Glasmacher U. & Wolf M.). **II**, pp. 1-64.

Willett S.D. (1992): Modelling thermal annealing of fission tracks in apatite, In: *Short Course on low temperature thermochronology* (Zentilli M. and Reynolds P.H.), Min. Ass. Can. Short Course **V 20**, Appendix 1, pp. 190-201.

Ziegler P.A. (1988): Evolution of the Arctic-North Atlantic and the Western Tethys. *APPG Mem.* **43**, 1-198.

Ziegler PA. (1990): Geological Atlas of Western and Central Europe-2nd ed., Shell Internationale Petroleum Maatschappij B.V., *Geol. Soc. Lond.*, Elsevier, Amsterdam.

EVIDENCE FOR THE UNROOFING HISTORY OF HERCYNIAN GRANITOIDS IN CENTRAL PORTUGAL DERIVED FROM LATE PALAEOZOIC AND MESOZOIC SEDIMENTARY ZIRCONS

A.J.S.C. PEREIRA[1], A. CARTER[2], A.J. HURFORD[2], L.J.P.F. NEVES[1] and M.M. GODINHO[1]

[1] *Departamento de Ciências da Terra, Universidade de Coimbra, 3049 Coimbra Codex, Portugal.*

[2] *London Fission Track Research Group, Department of Geological Sciences, Birkbeck and University College London, Gower Street, London WC1E 6BT, United Kingdom.*

The Lusitanian Basin on the western Iberian margin comprises Mesozoic and Cenozoic sediments derived from the surrounding Precambrian to Lower Palaeozoic and Hercynian basement. Apatite fission-track data from Hercynian granitoids to the east of the basin show that the intrusives were being actively eroded throughout the Jurassic and Cretaceous; however, exactly when these rocks first became exposed and eroded is not clear. In this study we have attempted to reconstruct the erosional history of the granitoids of Central Portugal by examining the detrital zircon fission-track signatures contained within post-granite emplacement sediments.

Results show that the fission-track method can discriminate between zircon sources within the Mesozoic sediments, although it cannot identify every zircon source because of the manner in which some grains etch. The combined apatite and zircon fission-track data set suggest that syn-tectonic granites experienced considerable unroofing before emplacement of the younger post-tectonic granites. Fission-track data show that denudation rates during the Upper Carboniferous were ~0.8-1.0 mm/yr, and then decreased during the Mesozoic and Tertiary to an average rate of ~0.02-0.05 mm/yr. The number of identified zircon age modes present in the Mesozoic sediments peaked (six modes) during the Cretaceous and is associated with material derived from Mesozoic igneous sources.

1. Introduction

The basement rocks of central Portugal belong to part of the Hesperic Massif, which form the largest exposure of granitoids within the European Hercynian belt; timing of emplacement of these granitoids is well established, mostly between 330 - 280 Ma (e.g. Priem *et al.*, 1970; Pinto *et al.*, 1987; Macedo, 1988; Beetsma, 1995), in contrast to their unroofing history which is poorly constrained. Reconnaissance apatite and zircon fission-track data (Pereira and Hurford, 1994) and petrographic observations on Upper Carboniferous sediments suggest that some of the Hercynian granites were first exposed by the Late Carboniferous (Neiva, 1943). This would require substantial and rapid erosion of the cover rocks which, on the basis of estimated granite emplacement depths, were originally 7-12 km thick (Godinho *et al.* 1997). Evidence for further, more recent

P. Van den haute and F. De Corte (eds.), Advances in Fission - Track Geochronology, 173-186.

erosion of the granites can be found in the large volume of clastic sediments within the Mesozoic Lusitanian Basin. Thus, exactly when the Hercynian granitoids became exposed *en masse* is unclear. Significant levels of unroofing occurred either during the Late Palaeozoic/Early Mesozoic or throughout the mid- to Late Mesozoic.

A key objective of this study is to use detrital zircon fission-track (FT) ages to identify granite sources within Late Paleozoic - Mesozoic sediments of Central Portugal, thereby helping to constrain the unroofing histories of the Hercynian granites.

Portuguese granites show a varied and complex range in geochemistry (*e.g.* Neves, 1991), and their classification depends on a combination of geochemistry, petrology, field relationships and radiometric dating (*e. g.* Ferreira *et al.,* 1987). Two main groups have been recognised, the syn- to late tectonic granites, emplaced between 330-300 Ma, and the post-tectonic granites, emplaced between 290-280 Ma. In provenance studies where most of this evidence has been lost or diagenetically altered, identification of specific granite groups becomes difficult, so the addition of a new method such as zircon FT could be important in discriminating between granite sources.

2. Geological Framework

The Hesperic Massif represents the south-western extension of the European Hercynides. The region has been subdivided into five principal tectono-metamorphic domains, of which the Central Iberian Zone (CIZ) is the focus of this study. Hercynian granitoids occupy a large area in the CIZ and have been emplaced within a basement of Late Precambrian to Palaeozoic rocks; in these, metapelites and metagreywackes are the main lithologies, whilst quartzites, black shales and sandstones are less frequent. Pre-Hercynian igneous rocks, usually in relatively small outcrops, are also found within Late Precambrian and Palaeozoic strata. By the end of the Hercynian orogeny, conglomerates, in some cases with granitoid pebbles, sandstones and shales were deposited in intramontane limnic basins; the age of the deposit ranges from the Upper Carboniferous to Permian (Wagner and Sousa, 1983).

The evolution of the Portuguese western Iberian margin after the Hercynian orogeny was influenced by the formation of sedimentary basins created by the extensional collapse of Hercynian thrust sheets associated with opening of the Atlantic Ocean. This led to the formation of the Lusitanian Basin, which contains a series of unconformity bounded sequences that record the extensional events. Late Triassic and Late Jurassic to Early Cretaceous fluvial sandstones are related with the two main rift sequences, which also represent significant denudation of the Hercynian basement. Dykes, sills and stocks, mostly of basic to intermediate composition, were emplaced during the Mesozoic in the Massif and in the marginal Basin. K-Ar dating of some of these indicates four main groups of ages at 230-210 Ma, 190-180 Ma, 160-130 Ma and 80-90 Ma. (Ferreira and Macedo 1979; Willis, 1988).

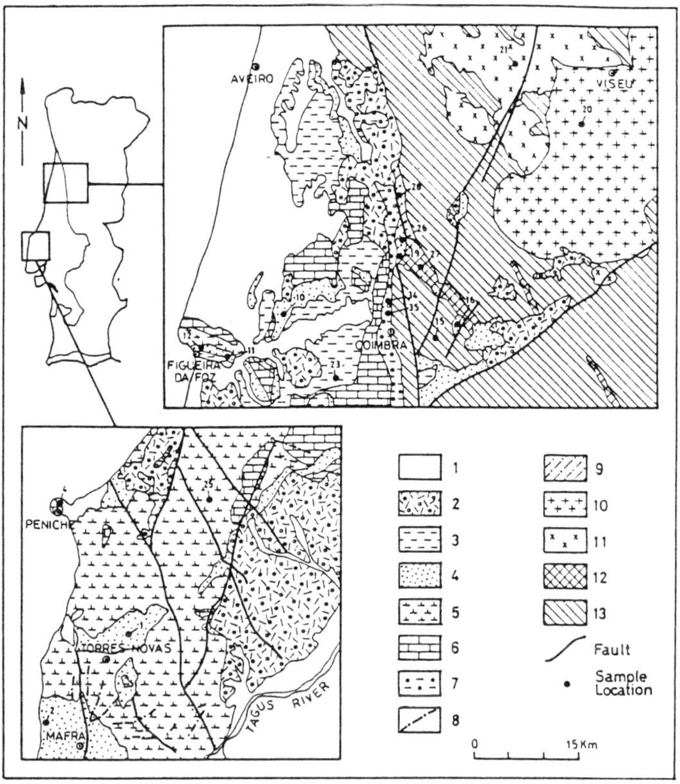

Figure 1. Map of regional geology and sample localities. 1. Quaternary; 2. Tertiary; 3. Upper Cretaceous; 4. Lower Cretaceous; 5. Upper Jurassic; 6. Lower and Middle Jurassic; 7. Triassic and Lower Jurassic; 8. Mesozoic igneous rocks, 9. Upper Carboniferous; 10. Hercynian post-tectonic granites; 11. Hercynian syn- to late-tectonic granites; 12. Ordovician - Silurian; 13. Precambrian - Cambrian.

3. Sampling Strategies and Fission-track Methodology

Zircon was selected as the principal mineral for this study, as it is chemically stable, mechanically robust, and has a partial annealing range (~200-350°C, Yamada *et al.*, 1995; Tagami *et al.*, 1996) that offers higher levels of thermal stability than apatite. Also, assessment of heavy mineral yields indicated that detrital zircon is generally abundant in the clastic sediments of this region whilst apatite is less common. Samples were taken in a broad NE-SW traverse extending from the Palaeozoic-Hercynian basement to the Upper Cretaceous, covering two different sectors of the Lusitanian Basin; Viseu, Figueira da Foz and Mafra are located at the limits of the sampling area (Figure 1). Samples were also collected from granites and surrounding basement rocks, as these are likely sources for the Late Paleozoic and Mesozoic clastic sediments. Thickness of the Mesozoic is locally highly variable, but regionally increases westwards. Three samples were collected from one of the Late Paleozoic intramontane deposits that occur in Portugal (Buçaco Basin, located North of Coimbra). Their age has been subject to some discussion: Upper Carboniferous or Permian (Wagner and Sousa, 1983), and in the latest 1992 edition of the 1:500,000 geological map of Portugal these deposits are labelled as Upper Carboniferous (Stephanian-C). Detrital zircon data may provide a maximum age for these sediments.

Data from detrital apatites collected from representative samples can be used to provide independent confirmation that zircons have not been to resetting temperatures since deposition. The cumulative thickness of Mesozoic - Tertiary sediments suggest that samples taken from these rocks have probably not experienced apatite total resetting temperatures. The original thickness of the Upper Carboniferous/Permian sedimentary rocks is unclear as the record is incomplete. Accordingly, increased levels of post-depositional heating cannot be ruled out for Upper Palaeozoic sediments, especially if a link is established between the granitic clasts of some of these rocks and the Hercynian granites, which would imply high rates of denudation and deposition. Apatite FT data is therefore important in assessing the likelihood of post-depositional annealing of fission tracks in zircon.

Zircon and apatite were separated from each sample using standard crushing, heavy liquid and magnetic methods; typical weight of samples was 3-5 kg. Preparation and analysis followed procedures given in Storey *et al.* (1996). Detrital zircons were analysed using several mounts of differing etch times (4-60 hours) in order to detect multiple age/uranium populations. For high track density (semi-metamict) grains, minimum ages were calculated on the basis of uranium content and an estimated spontaneous track density. Samples were irradiated in the thermal facility of the Risø Reactor, National Research Centre, Rosklide, Denmark, (cadmium ratio for Au > 400), using Corning glasses CN-5 for apatite and CN-2 for zircon as neutron dosimeters. Counting and track length measurements used a microscope total magnification of 1250x with 100x objectives, dry for apatite and oil for zircon. Central ages (Galbraith 1992) were calculated using the IUGS-recommended zeta calibration approach (Hurford, 1990) using personal (AC) values of $\zeta = 339\pm5$ for glass CN5 (apatite) and $\zeta = 127\pm5$ for glass CN2 (zircon).

Extraction of age components from pooled data has been the subject of a number of provenance studies (*e.g.* Hurford and Carter, 1991 and references therein; Naeser *et al.* 1987, Corrigan and Crowley, 1992; Brandon and Vance, 1992; Garver and Brandon, 1994). Methods used include simple identification of peaks within grain-age probability distributions (Hurford *et al.,* 1984); more complex peak fitting methods used to decompose grain age distributions into component Gaussian distributions (Brandon, 1992); mathematical analysis based on Poissonian models and goodness of fit (Galbraith, 1988; Galbraith and Green, 1990); and mixture modelling using several methods, one of which is based on Galbraith and Green (op. cit.). The difficulties of identifying modes and estimating standard errors is a major problem in histogram based methods. In this study we have used the approach of Sambridge and Compston (1994), which is based on several competing numerical methods, each of which has recognised strengths and limitations. Consistent results from several methods offers a greater degree of confidence than data from any single method.

4. Results and Interpretation

Apatite results given in Table 1 and Table 2 show that data quality is generally good apart from the Jurassic and Cretaceous samples which have low numbers of measured confined track lengths. The Upper Carboniferous (OM28), Triassic (OM35) and Cretaceous (OM12) samples each contain a single population of grain ages, whilst the Jurassic sample (OM4) has a mixed population of grain ages between ~50 and ~150 Ma (see radial plot, Figure 2), although this is due to only one grain and may not be statistically significant. The relationship between central age, individual grain ages and depositional age (shaded region on the radial plots in Figure 2) is a primary indicator of post-depositional annealing. Central ages for Upper Carboniferous, Triassic and Jurassic samples are reduced to below their stratigraphic age, suggesting that some annealing has occurred. No evidence for annealing is present in the Cretaceous sample as the central age is older than the depositional age.

Insufficient track lengths were measured for the Jurassic and Cretaceous samples to permit both a qualitative interpretation and modelling. Good quality data (100 measurements) exists for the Upper Carboniferous and Triassic samples (OM28 and OM35). These have similar negatively skewed distributions (Figure 2), indicative of uninterrupted cooling and prolonged residence within the apatite partial annealing zone an interpretation which can be refined by modelling. For this the method of Gallagher (1995) was used, the results being summarised in Figure 3. The Upper Carboniferous sample (OM28) reached maximum palaeotemperatures during latest Triassic - Early Jurassic times. Similar levels of annealing are observed in the Triassic sample (OM35) although the onset of cooling was earlier, albeit at a similar rate. A probable explanation is the higher stratigraphic position of OM35. Resetting for both samples is effectively total although the oldest preserved tracks are concordant with sample depositional ages and suggests that maximum temperatures have probably not been much above 110°C. Apatite data from the Upper Carboniferous and Triassic samples suggest that sedimentation was not excessively thick, reaching levels between 2-3 km assuming an average geothermal gradient of ~30°C/km.

Attempts at obtaining zircon FT data directly from the granites (sampled in this study) failed due to preferential etching along growth zones interfering with track revelation, a feature which may serve as a source indicator (identical grains were observed in Triassic sample OM34). Although no zircon data was obtained in this study, evidence for rapid early cooling of the granites can be found in apatite and zircon data from the Tondela-Oliveira do Hospital region (Pereira and Hurford, 1994) and apatite data from the Viseu granites sampled for this study. The apatite granite results (Table 1) have central ages of 206±9 Ma (syn-tectonic granite, OM21) and 122±6 Ma (post-tectonic granite, OM20), with near-identical negatively-skewed length distributions and standard deviations indicative of similar cooling histories (Figure 3b). The relatively old apatite ages imply early cooling within the apatite..

Sixteen sedimentary samples ranging in age from Late Precambrian / Lower Palaeozoic to Cretaceous underwent zircon fission-track analysis (Table 1). Overall,

TABLE 1 - Fission track data for rocks of Central Portugal

Sample no.	Stratigraphic. age/ emplacement age and lithology	Mineral and no. of crystals	ρ_d (n_d)	ρ_s (n_s)	ρ_i (n_i)	$P(\chi^2)$	RE (%)	Age ± 1σ (Ma)
OM16	Upper Cretaceous sandstone	Zircon 18	0.446 (3086)	22.53 (3778)	2.546 (427)	5	15	244 ± 16
OM23	Upper Cretaceous sandstone	Zircon 31	0.446 (3086)	12.87 (6047)	2.514 (1181)	<1	54	142 ± 15
OM1	Lower Cretaceous sandstone	Zircon 8	0.425 (2942)	18.67 (1501)	2.301 (185)	<1	36	251 ± 39
OM2	Lower Cretaceous sandstone	Zircon 40	0.425 (2942)	15.41 (6964)	2.260 (1021)	<1	40	186 ± 14
OM10	Lower Cretaceous sandstone	Zircon 21	0.425 (2942)	18.75 (2989)	1.983 (316)	<1	39	251 ± 27
OM11	Lower Cretaceous sandstone	Zircon 19	0.425 (2942)	15.70 (3351)	1.382 (295)	54	5	299 ± 19
OM12	Lower Cretaceous sandstone	Zircon 26	0.425 (2942)	14.52 (4186)	1.880 (542)	<1	47	208 ± 22
OM12	Lower Cretaceous sandstone	Apatite 21	1.246 (6910)	1.532 (380)	2.173 (539)	63	8	147 ± 10
OM 25	Upper Jurassic sandstone	Zircon 26	0.439 (3086)	19.31 (5540)	1.805 (518)	<1	23	293 ± 20
OM4	Middle Jurassic sandstone	Apatite 17	1.246 (6910)	1.424 (521)	2.159 (790)	<1	36	132 ± 16
OM34	Triassic sandstone	Zircon 16	0.421 (2334)	16.47 (2534)	1.762 (271)	85	0.1	240 ± 16
OM35	Triassic sandstone	Zircon 11	0.421 (2334)	13.36 (1759)	1.481 (195)	25	8	231 ± 19
OM35	Triassic sandstone	Apatite 20	1.332 (3691)	5.111 (1778)	5.568 (1937)	7	10	204 ± 9
OM19	Upper Carboniferous sandstone	Zircon 12	0.444 (3086)	21.77 (2233)	1.199 (123)	10	19	489 ± 53
OM26	Upper Carboniferous sandstone	Zircon 17	0.421 (2334)	14.78 (2903)	1.260 (248)	60	0.4	299 ± 21
OM28	Upper Carboniferous sandstone	Zircon 18	0.421 (2334)	16.52 (2313)	1.393 (195)	25	19	309 ± 28
OM28	Upper Carboniferous sandstone	Apatite 26	1.332 (3691)	2.567 (1103)	3.389 (1456)	7	9	169 ± 8
OM20	granite ~280 Ma	Apatite 20	1.399 (7756)	1.685 (974)	3.270 (1890)	8	12	122 ± 6
OM21	granite ~310 Ma	Apatite 17	1.399 (7756)	4.291 (1594)	4.851 (1802)	3	11	206 ± 9
OM27	Ordovician sandstone	Zircon 3	0.421 (2334)	26.83 (595)	1.218 (27)	3	32	504 ± 138
OM15	Precambrian greywacke	Zircon 12	0.448 (3000)	16.96 (1692)	1.092 (109)	16	18	430 ± 49

Dosimeter (ρ_d) and induced (ρ_i) track densities (x10^6 cm^{-2}) measured on external mica detectors (g=0.5). Spontaneous (ρ_s) track densities (x10^6 cm^{-2}) measured on internal mineral surfaces; numbers of tracks counted (*n*) shown in brackets.

Central ages calculated using the external detector method by Carter using ζCN5=339±5 (apatite) and ζCN2=127±5 (zircon).

P (χ^2) indicates the probability of the single grain ages representing a homogeneous population.

TABLE 2. Apatite track length data for some samples from from Central Portugal

Sample no.	Track length ± 1σ (μm)	SD (no. of tracks)
OM12	11.96 ± 0.51	2.64 (28)
OM4	12.95 ± 0.24	1.15 (22)
OM35	13.20 ± 0.17	1.73 (108)
OM28	12.66 ± 0.22	2.31 (108)
OM20	12.62 ± 0.17	1.68 (100)
OM21	12.54 ± 0.17	1.67 (100)

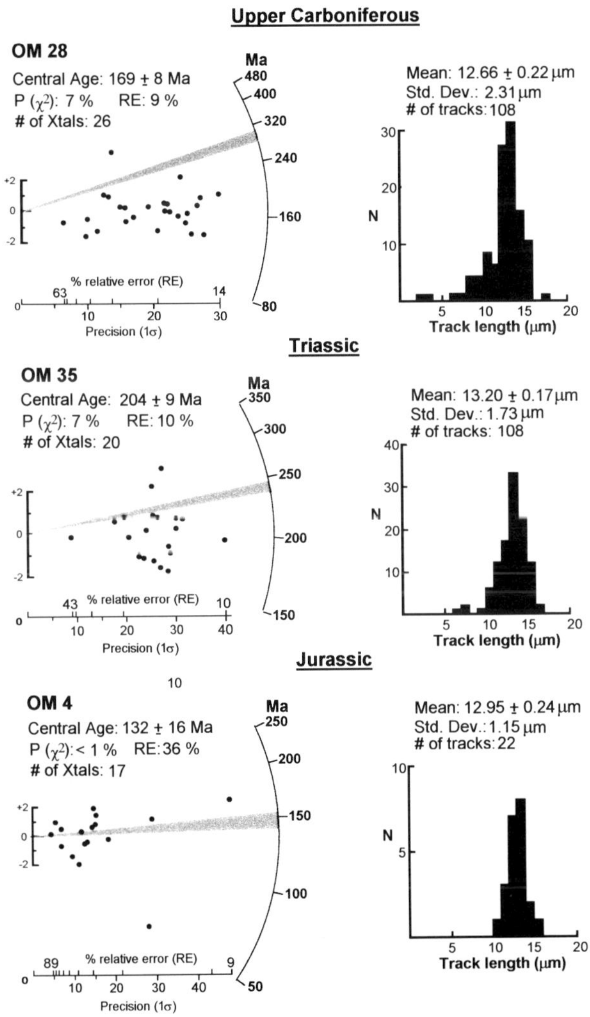

Figure 2. Radial and track length plots of representative apatite data. Shaded region represents sample depositional age.

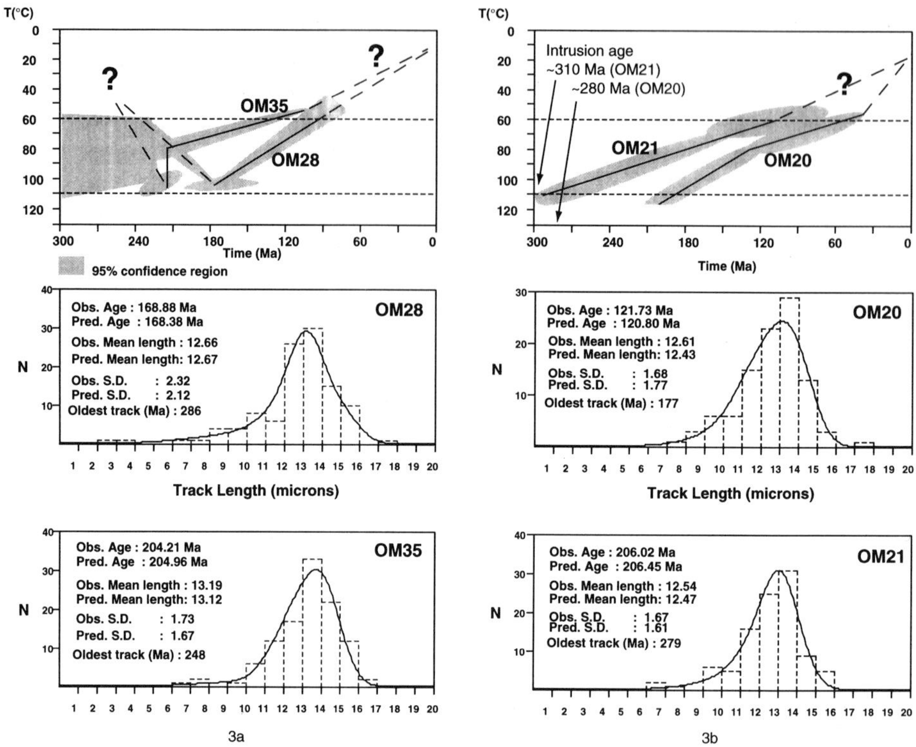

Figure 3. Cooling histories derived from modeling apatite age and length data. Shaded area represents 95% confidence region showing which parts of the cooling history are constrained by the FT data.
a) Charts of the cooling path for samples OM28 and OM35
b) Cooling paths for syn-tectonic and post-tectonic granites from the Viseu region.

the number of grains analysed was low, <20 grains in ~70% of samples, despite making several mounts per sample. Lack of dateable grains is attributed to a combination of preferential etching of growth zones, uranium zoning, metamict grains and, in some samples, small grain size. Low numbers of grains can restrict interpretation, since decomposition of pooled data relies on an adequate number of grains within the data-set. Optimum numbers of grains is not fixed and depends on: the degree of separation between modes, the differing precision inherent in each mode and the relative proportions of grains in each population. In the simplest scenario, a two component mixture, ~20 grains are generally adequate, providing that the modes are suitably different in age. Detrital zircon FT data are generally biased towards better etched grains and thus, whilst some populations may be omitted, others may be relatively over represented. Accordingly no attempt is made to estimate the relative proportions of a source within a sediment. Consideration of the regional geology suggests potential zircon sources to be restricted: Hercynian granite sources, a Palaeozoic metamorphic or igneous sources (>400 Ma) and additional sources from localised Mesozoic igneous rocks.

The oldest samples analysed come from the Precambrian and Ordovician metasediments. The number of grain ages obtained for both samples (OM27 and OM15) is low due to an abundance of metamict grains and thus only rarer, low uranium (semi-metamict) zircons could be counted, hence data quality is poor and of high uncertainty. The measured ages (430±49 Ma - OM15 and 504±138 Ma - OM 27) are comparable with the oldest modes present in the Upper Carboniferous and Cretaceous sediments. Three samples were collected from Upper Carboniferous intramontane deposits, rocks important as they could contain the first evidence for unroofing of the Hercynian granites (granitic boulders with a petrography similar to some of the older granites). The zircon FT data for these samples all comprise single populations of grain ages, two are indistinguishable at 299±21 Ma and 309±28 Ma, the third is older at 489±53 Ma and derives from the pre-Hercynian source. In the two samples of identical age no older metamict zircons were observed, the population clearly being derived from a single source. Combining the grain age data for the two samples produces a central age of 302±17 Ma. Whole rock Rb/Sr and K-Ar dating of representative "Older syn-tectonic Granites" and "Younger post-tectonic Granites" produced ages of 330-315 Ma and 290-280 Ma respectively (Oen 1970; Priem *et al.*, 1970; Pinto, 1987; Neves, 1991; Pereira and Hurford, 1994, Beetsma 1995) and therefore the most likely source for the Upper Carboniferous sandstones are the older granites.

Two Triassic samples yielded similar single age populations (age dispersion about the central age is <10%) at 231±19 Ma and 240±16 Ma which, when combined, produce a central age of 236±12 Ma. The occurrence of large K-feldspar megacrysts similar to those observed in younger granites may indicate that these granites were the source of zircons at 236±12 Ma. Unlike the Upper Carboniferous samples, other undated zircon sources were present in these samples, including highly-zoned euhedral grains identical to zircons from syn-tectonic granites, an implication that these granites could also be sources for the zircons observed in the Triassic sediments.

A single sample collected from the Upper Jurassic (which consists mainly of siltstones and limestones) gave a central age of 293±20 Ma and a mixed population of grain ages (relative error 24%, χ^2 <5%) - see radial plot in Figure 4. Analysis of the principal age components reveal two modes, a dominant one at 302±19 Ma (21 grains) and a minor younger mode at 182±18 Ma (5 grains). The older age is consistent with the syn-tectonic granite source found in Upper Carboniferous samples, whilst the younger mode may derive either from Mesozoic igneous rocks, or from a granite which entered the zircon partial annealing zone (~200-350°C) close to 180 Ma, from where it was rapidly cooled and eroded by ~140 Ma (sediment depositional age). The latter hypothesis is inconsistent with available granite apatite data (this study and in Pereira and Hurford, 1994), which show that by ~180 Ma intrusions were at temperatures <100°C. However, local variations cannot be totally ruled out. Staple *et al.* (1996) note an increased influx of coarse clastics into the Lusitanian Basin during the Late Jurassic, which implies significant denudation of the source terrain.

The dominance of clastic sediments during the Early Cretaceous show that extensive regional denudation occurred at this time. Seven samples were collected for analysis, and in general the number of dateable grains is higher than for previous samples primarily due to changes in provenance. Central ages for the six samples range from 142±15 Ma to 299±19 Ma, with all but one sample (OM11, 299±19 Ma) comprising mixed populations of grain ages. The radial plots for some of the mixed age samples are shown in Figure 4, together with their modelled principal component ages. Approximately 20% of grains in these samples were metamict, having minimum ages ~400 Ma, and therefore it is probable that these derive from the pre-Hercynian source. The next mode, at 300±19 Ma, is consistent with a syn-tectonic granite source, whilst post-tectonic sources are suggested by modes at 268±16 Ma and possibly 202±19 Ma (identical to zircon FT data, 267±23 Ma and 204±15 Ma, obtained on granites from the Tondela-Oliveira do Hospital region, Pereira and Hurford, 1994 and unpublished data). The remaining modes at 122±5 Ma and 78±4 Ma do not compare with any existing data, and Mesozoic intrusive rocks are a possible source. The youngest mode at 78±4 Ma is remarkably close to the age of the Sintra Granite (82 Ma, Macintyre and Berger, 1982), north-west of Lisbon.

5. Discussion

Apatite FT data show that post-depositional heating of the sediments has not been to levels sufficient to cause partial track annealing in zircon and, therefore, all measured zircon FT ages record original cooling of the source rock. Following analysis and extraction of dateable age components within the zircon data-set a number of distinct sources can be recognized as summarized in Table 3. Pre-Hercynian rocks have a distinct FT Palaeozoic signature ~400-450 Ma, which confirms results found by K-Ar, Ar-Ar and Rb/Sr dating of amphiboles and micas (*e.g.* Peucat *et al.*, 1990; Dallmeyer and Ibarguchi 1990). The metamict or near-metamict grains in Upper Carboniferous, Triassic and Cretaceous sediments therefore derive from the Lower Palaeozoic and Precambrian rocks.

A major problem encountered in this study was obtaining zircon data directly from the Hercynian granites. Because no zircons were dated, the identification of a particular mode as granitic may only be carried out indirectly using other sources of information such as published radiometric data (Oen, 1970; Pinto *et al.*, 1987; Macedo, 1988; Neves, 1991; Pereira and Hurford, 1994) and apatite FT cooling histories. The apatite data from the granites obtained in this study record regional cooling as a consequence of denudation during the Late Palaeozoic and Mesozoic, in this case probably in response to rifting associated with opening of the Atlantic Ocean. This cooling history shows that the granites reached shallow crustal levels soon after their emplacement, and therefore zircon FT ages should be close to other radiometric data: thus the mode at ~300 Ma, found in the majority of Mesozoic sediments, most probably derives from the syn-tectonic granites. Evidence for zircons from the younger post-tectonic granites is already detected in the Triassic sediments, but their presence is much more important during the Cretaceous.

Upper Carboniferous, Triassic and Jurassic samples all have simple zircon age spectra with grains deriving from either the Precambrian-Palaeozoic metamorphic basement or the syn-tectonic granites. The 300 Ma mode provides a constraint on the maximum age of the Upper Carboniferous samples, restricting the depositional age of the sediments to a maximum of 302±17 Ma (Stephanian-C).

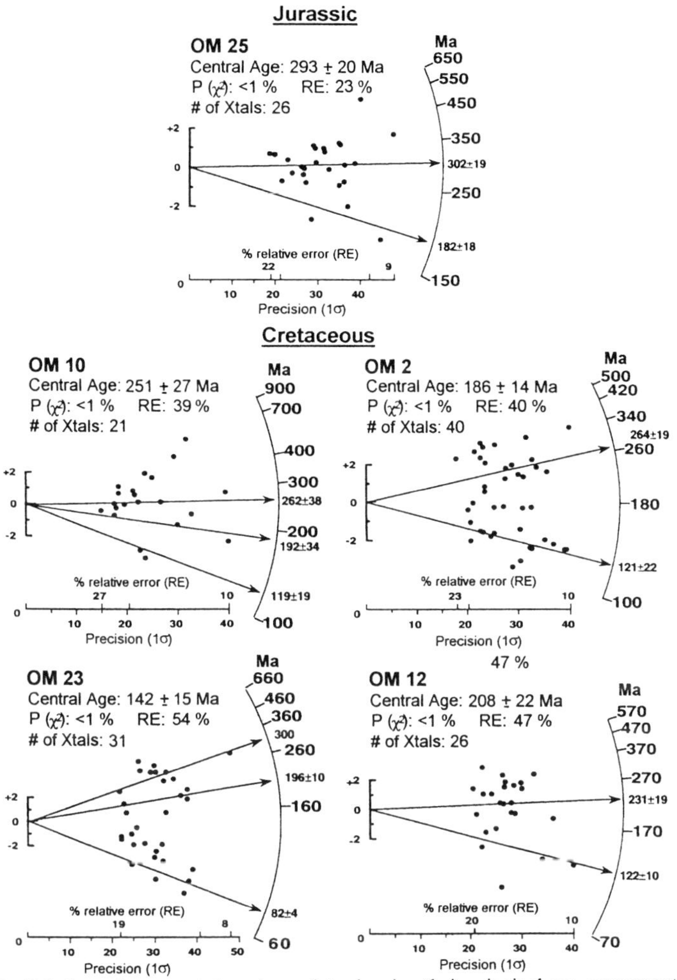

Figure 4. Radial plots of representative zircon data showing their principal age components, secondary older modes which are poorly constrained are not included.

By Cretaceous times a dramatic increase in the number of zircon sources occurs; up to six age modes can be identified in the combined data-set. Whilst Palaeozoic metamorphic basement and syn-tectonic granite sources are present, post-tectonic granite sources occur together with material possibly derived from Mesozoic intrusives, which occur as four main groups at 230-210 Ma, 190-180 Ma, 160-130 Ma and 80-90 Ma. (Ferreira and Macedo 1979, Willis, 1988).

The FT data have established that sediments from the Upper Carboniferous contain zircons derived from both syn-tectonic Hercynian granites and pre-Hercynian basement. Evidence for zircons from the younger post-tectonic granites can be found in the Triassic sediments, but it is only during Cretaceous times that late Hercynian zircons occur more widely. In the Cretaceous there is also an accompanying increase in

TABLE 3. Principal age components (Ma).

Cretaceous	~400 (m)	300±19 (26)	268±16 (61)	202±19 (27)	122±5 (36)	78±4 (17)
Jurassic		302±19 (21)		182±18 (5)		
Triassic		growth zone	236±12 (27)			
Upper Carboniferous	489±53 (12)	302±18 (35)				
Ordovician-Silurian	504±138 (3)					
Pre-Cambrian	430±49 (12)					

number of grain ages comprising mode written in brackets.
m = estimated minimum age of semi-metamict grains.
growth zone = grains present with a recognized characteristic of preferentially etched growth zones, typical of syn-tectonic granites in the Viseu region.

zircon sources of different age, which probably include the Mesozoic intrusives. Using these data, together with the apatite FT results, the rate of granite unroofing may be estimated. Exposure and erosion of some Hercynian granites occurred as early as Upper Carboniferous/Permian, as indicated by the presence of granitic clasts and zircon FT data. To bring the granites to the surface in a relatively short period of time, substantial denudation is required. Emplacement depths estimated on the basis of thermobarometric data on surface samples are 11.1±0.7 km for the syn-tectonic and 7.4±0.7 km for the post-tectonic granites (Godinho *et al.* 1997). Between ~300 Ma and the present-day this amount of cover has been removed. How much of this cover was originally granite is not known.

The mean age estimated for the syn-tectonic granites outcropping in Viseu region is 313±3 Ma. Taking in account an emplacement depth of ~11 km (Godinho *et al.*, 1997), the presence of zircons from those granites in the Upper Carboniferous sediments (Stephanian-C, ~300 Ma) suggests denudation rates of 0.8-1.0 mm/year (using an assumed standard geothermal gradient of 30°C/km). The amount of crust removed by erosion in the Upper Carboniferous compares with (or excedes) the amount removed in the last ~300 Ma (Pereira, 1991; Pereira and Hurford, 1994; Godinho *et al.*, 1997). This implies an average decrease of denudation rate since Upper Carboniferous, and could explain the dominance of zircons from the earlier Hercynian granites in the Upper Carboniferous and early Mesozoic sediments. Previous estimations of the average Early Mesozoic denudation suggests values ~0.05 mm/year (Pereira and Hurford, 1995; Godinho *et al.*, 1997) decreasing to a maximum of 0.02 mm/year to the present. Similar conclusions are reached from the results of modelling the apatite FT data obtained in this study.

6. Conclusions

Zircon FT data discriminates between different zircon sources in the Early Mesozoic sediments and those in the Lusitanian Basin (Central Portugal). Combined apatite and zircon FT data suggest older, syn-tectonic Hercynian granites were unroofed before the younger post-tectonic granites were emplaced. Consequently Early Mesozoic clastic sequences are dominated by detritus from the syn-tectonic granites. Denudation rates for the older granites in the Viseu region during the Upper Carboniferous were ~0.8-1.0 mm/yr, decreasing during the Mesozoic and Tertiary to an average rate of ~0.05-0.02 mm/yr. Consequently amounts of denudation during the Upper Carboniferous alone are greater or, at least, similar to those experienced over the following 300 Ma. Limitations imposed by zircon etching from some granites restricts the specific identification of granite sources. Additional studies, in particular of Cretaceous clastic sequences, should refine further the unroofing history of the area.

Acknowledgements

Professor A.F. Soares is thanked for his helpful comments on an earlier version of the manuscript. The authors are grateful to the British Council, JNICT and NERC Grant GR3/8291 for financial support. Dr. L.V.P. Duarte and Dr. J.F. Marques provided field guidance.

References

Beetsma, J.J. (1995): The Late proterozoic and Hercynian crustal evolution of the Iberian Massif, N. Portugal. Doctoral thesis, University of Amsterdam.
Brandon M. T. (1992): Decomposition of fission-track grain age distributions. *Amer.J. Sci.* **292**, 535-564.
Brandon M. T. and Vance J. A. (1992): Tectonic evolution of the Cenozoic Olypmic subduction complex, Washington State, as decuced from fission track ages for detrital zircons. *Amer. J. of Sci.* **292**, 565-636.7
Corrigan J. D. and Crowley K. D. (1992): Unroofing of the Himalayas: a view from apatite fission-track analysis of Bengal Fan sediments: *Geoph Res Lett.* **19**, 2345-2348.
Dallmeyer R.D. and Ibarguchi J.I.G. (1990): Age of amphibolitic metamorphism in the ophiolitic unit of the Morais Allochthon (Portugal): implications for early Hercynian orogenesis in the Iberian Massif. *J. Geol. Soc. Lond.* **147**, 873-878.
Ferreira M.R.V.P. and Macedo C.A.R. (1979): Actividade magmática durante o Mesozóico: I. Achega para a datação K-Ar das rochas filonianas básicas intrusivas na Zona Centro-Ibérica (Portugal). *Memórias e Notícias*, Publ. Mus. e Lab. Mineral. e Geol., Univ. Coimbra **87**, 29-49.
Ferreira N., Iglesias M., Noronha F., Pereira E., Ribeiro A. and Ribeiro M.L. (1987): Granitóides da Zona Centro Ibérica e seu enquadramento geodinâmico. . In: *Geologia de los granitoides y rocas asociadas del Macizo Hespérico* (eds. Bea F., Carnicero A., Gonzalo J, Plaza M.L. and Alonso M.D.), Editorial Rueda, Madrid, pp. 37-51.
Galbraith R.F. (1988): Graphical display of estimates havering different standard errors. *Technometrics* **30**, 271-281.
Galbraith R.F. and Green P.F. (1990): Estimating the component ages in a finite mixture. *Nucl. Tracks Radiat. Meas.* **17**, 197-206.
Gallagher, K. (1995): Evolving temperature histories from apatite fission-track data. *Earth Planet. Sci. Letts.* **136**, 421-435
Garver J.I. and Brandon M.T. (1994): Fission-track ages of detrital zircons from Cretaceous strata, southern British Columbia: Implications for the Baja BC hypothesis. *Tectonics* **13**, 401-420.
Godinho M.M., Pereira A.J.S.C. and Neves L.J.P.F. (1997) - Denudation rate of the western Hesperian Massif since Upper Carboniferous - a first approach. EUG9, Terra Nova, **9**, Abstract Supplement No. 1, pp. 104.
Hurford A. J. (1990): Standardization of fission track dating calibration: Recommendation by the Fission Track Working Group of the I.U.G.S. Subcommission on Geochronology. *Chem. Geol. (Isotope Geosci. Sect.)* **80**, 171-178.
Hurford A.J., Fitch F.J. and Clarke A. (1984): Resolution of the age structure of the detrital zircon populations of two Lower Cetaceous sandstones from the Weald of England by fission-track dating. *Geol. Mag.* **121**, 269-277.

Hurford A. J. and Carter A. (1991): The role of fission track dating in discrimination of provenance. In: *Developments in Sedimentary Provenance studies* (eds. Morton A.C. and Haughton, P.D.W.), Geological Society Special Publication No. 57, pp 67-78.

Macedo C.A.R. (1988): Granitóides, complexo xisto-grauváquico e Ordovícico entre Trancoso e Pinhel (Portugal Central). Geologia, petrologia, geocronologia. Doctoral thesis, University of Coimbra.

Macintyre R.M. and Berger G.W. (1982): A note on the geochronology of the Iberian Alkaline Province. *Lithos* **15**, 133-136.

Naeser N.D., Zeitler P.K., Naeser C.W. and Cerverny P.F. (1987): Provenance studies of fission track dating of zircon - etching and counting procedures. *Nucl Tracks Radiat. Meas.* **13**, 121-126.

Neiva J.M.C. (1943): A idade dos granitos portugueses. *Bol. Soc. geol. Port.*, **3**, 5-25.

Neves, L.J.P.F. (1991): Transferências de matéria e energia na interface de granitóides biotíticos porfiróides - granitóides moscovítico-biotíticos na região de Torredeita (Viseu, Portugal Central). Doctoral thesis, University of Coimbra.

Oen Ing Soen (1970): Granite intrusion, folding and metamorphism in central-northern Portugal. *Bol. Geol. min.* **81**, 271-298.

Pereira A.J.S.C. (1991): Transferências de calor e ascensão crustal no segmento Tondela-Oliveira do Hospital (Portugal Central) após a implantação dos granitos hercínicos sin a tardi-orogénicos. Doctoral thesis, University of Coimbra.

Pereira A.J.S.C. and Ponte M.J.B. (1992): Geologia da região de Tondela-Oliveira do Hospital (Portugal Central). *Memórias e Notícias*, Publ. Mus. Lab. Mineral. Geol., Univ. Coimbra **114**, 1-17.

Pereira A.J.S.C. and Hurford A.J. (1994): Time of emplacement and cooling history of the late hercynian granites from Tondela-Oliveira do Hospital region (Central Portugal). *Memórias e Notícias*, Publ. Dep. Ciências da Terra and Mus. Lab. Mineral. Geol., Univ. Coimbra **118**, 63-79.

Peucat J.J, Griffiths J.B., Ibarguchi J.I.G., Dallmeyer R. D., Menot R.P., Cornichet J. and Leon M.I.P.L. (1990): Geochemical and geochronological cross section of the deep Variscan crust; the Cabo Ortegal high-pressure nappe (northwestern Spain). *Tectonophysics* **177**, 263-292.

Pinto M.C.S., Casquet C., Ibarrola E., Corretgé L.G and Ferreira M.P. (1987): Síntese geocronológica dos granitóides do Maciço Hespérico. In: *Geologia de los granitoides y rocas asociadas del Macizo Hespérico* (eds. Bea F., Carnicero A., Gonzalo J, Plaza M.L. and Alonso M.D.), Editorial Rueda, Madrid, pp. 69-86.

Priem H.N.A., Boelrijk N.A.I.M., Verschure R.H., Hebeda E.H. and Verdurmen E.A.T.H. (1970): Dating events of acid plutonism through the Paleozoic of the western Iberian Peninsula. *Eclogae Geologicae Helvetiae* **63**, 255-274.

Sambridge M.S. and Compston W. (1994): Mixture modelling of multi-component data sets with application to ion probe zircon ages. *Earth Plan. Sci. Letts* **128**, 373-390.

Staple G., Pereira A.J.S.C., Andreissen P., Neves L.J.P.F., Cloetingh S. and Godinho, M.M. (1996): Exhumation associated with the N-Atlantic opening and Alpine tectonics in W-Iberia deduced from fission track analysis. *Proceedings of International Workshop on Fission Track Dating*, Ghent, Belgium, August 1996, *(refereed abstract)*, pp. 109.

Storey B.C., Brown R., Carter A., Doubleday P. Hurford A.J., Macdonald D.I.M. and Nell P.A.R. (1996): Fission-track evidence for the thermotectonic evolution of a Mesozoic-Cenozoic fore-arc region, Antartica. *J.Geol. Soc. Lond.* **153**, 65-82.

Tagami T., Carter A. and Hurford A.J. (1996): Natural long-term annealing of the zircon fission-track system in Vienna Basin deep borehole samples: constraints upon the partial annealing zone and closure temperature. *Chemical Geology* **130**, 147-157.

Wagner R.H. and Sousa M.J.L. (1983): The Carboniferous Megafloras of Portugal - a revision of identifications and discussion of stratigraphic ages. *Mem. Serv. geol. Port.* **29**, 127-152.

Willis M.G. (1988): The tectonic history of the Lusitanian basin of Portugal. Doctoral thesis, The Open University, Milton Keynes, UK.

Yamada R., Tagami T., Nishimura S. and Ito H. (1995): Annealing kinetics of fission tracks in zircon: an experimental study. *Chem. Geol. (Isot. Geosci. Sect.)* **122**, 249-258.

APATITE FISSION-TRACK THERMOCHRONOLOGY OF THE UPPERMOST TECTONIC UNIT OF CRETE, GREECE: IMPLICATIONS FOR THE POST-EOCENE TECTONIC EVOLUTION OF THE HELLENIC SUBDUCTION SYSTEM

S.N. THOMSON[1], B. STÖCKHERT[1], H. RAUCHE[2] & M.R. BRIX[1]
1. Institut für Geologie, Ruhr-Universität Bochum, D-44780 Bochum, Germany
2. Institut für Geowissenschaften, Friedrich-Schiller-Universität Jena, Burgweg 11, D-07749 Jena, Germany

Abstract

Apatite fission-track thermochronology is applied to 31 samples from various components of the uppermost tectonic unit of the Island of Crete, Greece. This unit is one of several that presently lie in the hanging wall to a major extensional detachment fault that juxtaposes rocks metamorphosed at high pressure and low temperature during the Late Oligocene - Early Miocene below against rocks that lack any Oligo-Miocene metamorphism. The data reveal a phase of accelerated Middle Miocene denudation that can be linked to erosion between ca. 17 Ma and 11 Ma. The discovery of the base of a denuded apatite partial annealing zone limits the total amount of denudation during this time period to ca. 4km. This implies a mean denudation rate of ca. 650 m/m.y.

Apatite fission-track ages of up to ca. 30 Ma and shortened confined tracks from high-grade metamorphic rocks of the uppermost tectonic unit require that no significant denudation of these rocks occurred during accretion of the other Cretan tectonic units in the Oligocene / Early Miocene. This information means that significant re-evaluation of previous tectonic models of the post-Eocene tectonic development of the Cretan segment of the Hellenic Subduction Zone is required. It is proposed that the Hellenic Subduction Zone has acted as a retreating plate boundary since at least the Late Eocene.

1. Introduction

The Island of Crete consists of several tectonic units accreted during northward directed subduction and collision of the continental margin of the Apulian Microplate with Europe during the Oligocene (Creutzburg & Seidel, 1975; Seidel et al., 1982; Bonneau, 1984; Hall et al., 1984; Fassoulas et al., 1994; Jolivet et al., 1994). These units are unconformably overlain by Late Miocene and younger sedimentary rocks (Postma et al., 1993). Subsequent 'roll-back' of the subduction zone since ca. 12 Ma has resulted in crustal thinning beneath the Cretan Sea and the disruption of pre-Neogene tectonic units by normal faulting into horsts and basins (Meulenkamp et al., 1988). The present day tectonic setting of the eastern Mediterranean is shown in Figure 1. Today Crete represents a structural high within the

P. Van den haute and F. De Corte (eds.), Advances in Fission - Track Geochronology, 187-205.

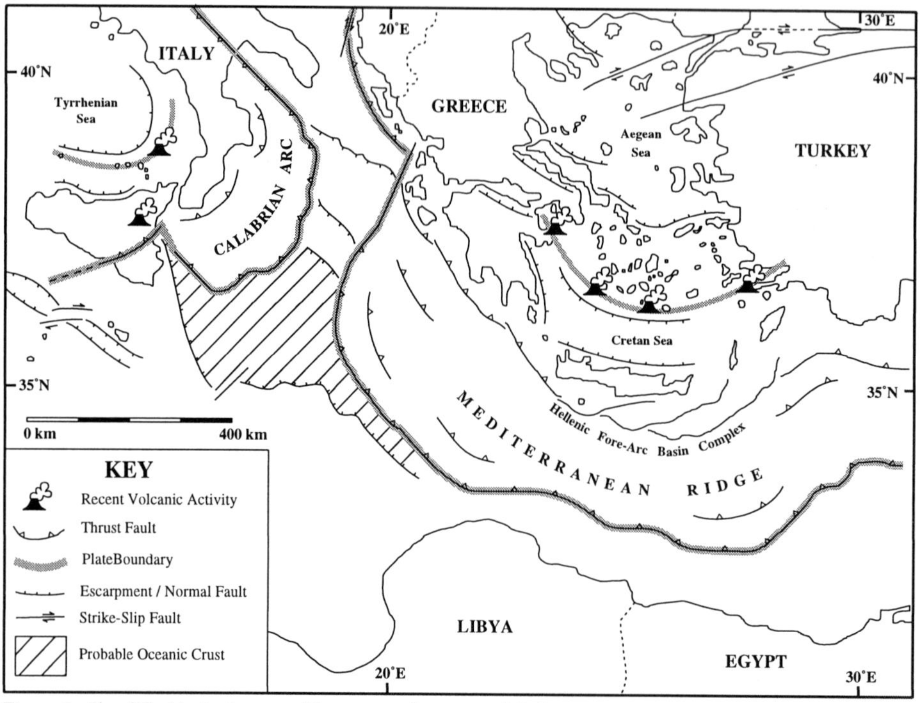

Figure 1. Simplified tectonic map of the present day eastern Mediterranean area summarised from Jolivet et al. (1994), Meulenkamp et al. (1988) & Van Dijk & Scheepers (1995).

'fore-arc' of the active northward directed subduction of the African Plate beneath the Aegean Sea.

There are five major tectonic units, commonly referred to as nappes, exposed on Crete (Figure 2 & 3) that, based on metamorphism, belong to two main groups. The two lowermost units, the mainly carbonate Plattenkalk Unit and tectonically above, the largely clastic, partly carbonate Phyllite-Quartzite Unit, show high-pressure-low temperature (HP-LT) metamorphism. This has been shown by K-Ar dating to be of Oligo-Miocene Age (ca. 20 to 25 Ma - Seidel et al., 1982). Above are three tectonic units that lack any Oligo-Miocene HP-LT metamorphism. Recent interpretations of the nappe pile (e.g. Kilias et al., 1994; Fassoulas et al., 1994 and Jolivet et al., 1994) propose that the HP-LT metamorphic nappes were brought to the surface by extensional tectonism and separated from the unmetamorphic 'upper plate' nappes above by a low-angle detachment fault. Directly above this fault lies the Tripolitza Unit. This consists largely of massive Mesozoic-Early Tertiary platform carbonate rocks grading into Tertiary flysch. Above this is the Pindos Unit comprised of deeper water Mesozoic-Early Tertiary carbonates and cherts and also overlain by Tertiary flysch. The subject of this study, the so-called uppermost tectonic unit (UM Unit) of the nappe pile in Crete (Seidel et al., 1976; 1981), is a mélange that contains serpentinitised mantle fragments, oceanic pillow basalt, various carbonate rocks, 70 Ma old high-grade metamorphic rocks, and other Mesozoic metamorphic slices. This unit is described in more detail in the next section.

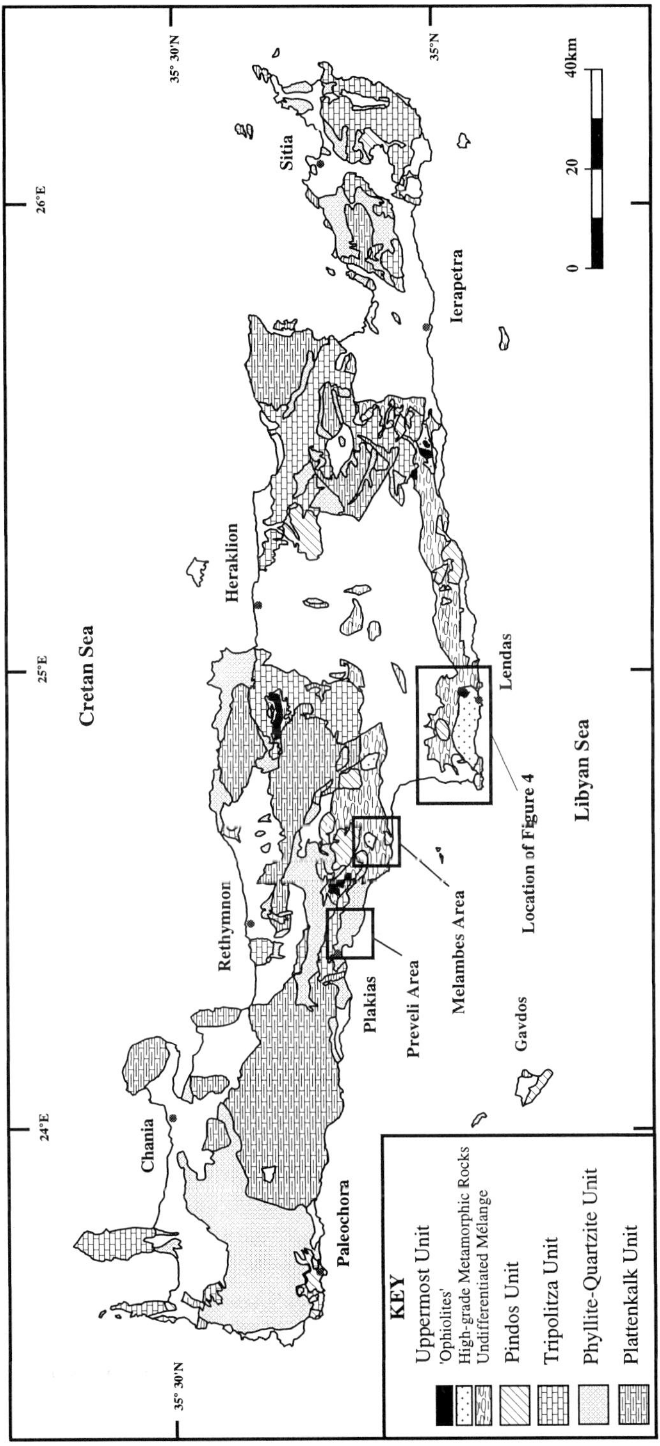

Figure 2. A sketch map of the geology of Crete based on Creutzburg et al. (1977). The sampling localities of Melambes and Preveli are shown along with the location of Figure 4 (the Lendas area).

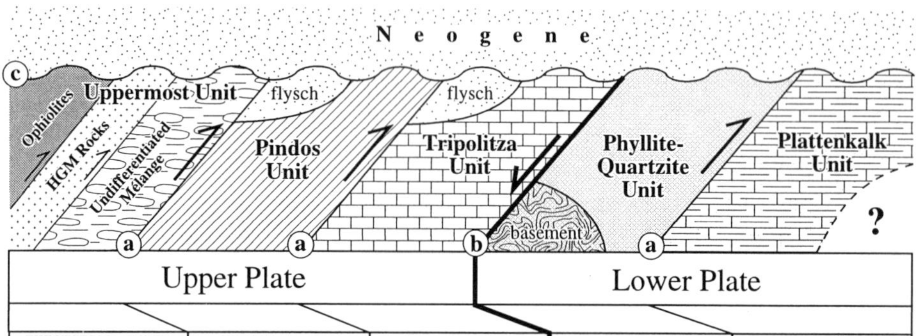

THE UPPERMOST UNIT (see text for details)	THE PINDOS UNIT	THE TRIPOLITZA UNIT	THE PHYLLITE-QUARTZITE UNIT	THE PLATTENKALK UNIT
Ophiolites - L. Jurassic to E. Cretaceous, mainly serpentinites. **HGM Subunit** - HT amphibolites, marble, migmatites and leucogranites (K-Ar : ~70Ma - Seidel et al., 1981). **Undifferentiated Mélange** - includes the Preveli Subunit Unit and the 'Very-Low Grade' Metamorphic Subunit	Unmetamorphosed Triassic - Palaeocene pelagic sequence including cherty limestones, radiolarites, calc-turbidites and calc-breccias. These pass upwards into Eocene turbidite sandstones and shales with intercalated carbonate conglomerates.	Unmetamorphosed Triassic - Eocene platform carbonate rocks overlain by late Eocene calcareous turbidites and later flysch-like sandstone / shale turbidites.	Deformed Permian - Triassic shallow marine clastic rocks with some carbonate, volcanic, evaporite, and Hercynian basement rocks. This Unit shows HP/LT metamorphism (peak conditions: 300-400°C, 8-10 kbar - Theye et al., 1992). K-Ar data suggest an age of ~20 to 25Ma for this metamorphism (Seidel et al., 1982).	Permian - Eocene recrystallised mainly pelagic cherty limestones and shallow marine carbonate rocks overlain by several metres of L. Eocene-Oligocene marls and shales. Metabauxites reveal HP/LT metamorphism (~350°C, ~10kbar - Theye et al., 1993).

Figure 3. Simplified tectono-stratigraphy of the five main tectonic units described from Crete (loosely based on Seidel et al., 1982). The main contacts shown are (a) Thrusts / accretion of Oligocene age (Bonneau, 1984), (b) ?Miocene extensional detachment fault (Fassoulas et al., 1994; Jolivet et al., 1994; Kilias et al., 1994) separating HP-LT metamorphic rocks below (lower plate) from rocks lacking Tertiary metamorphism (upper plate) above, and (c) Neogene unconformity, rocks immediately above which are of ?Langhian to Tortonian in age (Postma et al., 1993; Meulenkamp et al., 1994).

This study applies apatite fission-track (FT) thermochronology to the UM Unit of the Cretan Nappe pile. Apatite FT thermochronology uses both qualitative interpretation, such as the concept of a denuded partial annealing zone (Gleadow & Fitzgerald, 1987; Fitzgerald et al., 1995) and recently developed quantitative techniques, such as the inverse thermal modelling of apatite fission-track data outlined by Gallagher (1995). In compressional tectonic environments these can be used to resolve low-temperature cooling histories resulting from both erosional and tectonic denudation processes. Here it is used to help constrain the tectonic history of Crete between HP-LT metamorphism of the 'lower plate' units at ca. 25Ma and post-denudation Neogene sedimentation at ca. 12Ma and in addition, to obtain more information on the age and development of the various components of the still poorly understood UM Unit. This work forms a part of a more substantial study that aims to construct low temperature thermal histories for the various tectonic units exposed on Crete. This will eventually allow a detailed reconstruction of Late Tertiary tectonic development, and in particular allow assessment of the importance of extension to the overall denudation that brought the HP-LT Plattenkalk and Phyllite-Quartzite Unit rocks from depths of up to 35 km at ca. 25 Ma (Late Oligocene) to the surface by Middle to Late Miocene times. Although the orogenic belt exposed on Crete has been disturbed by much subsequent Neotectonism, the lack of significant post-tectonic erosion means that many shallow level crustal orogenic features are still preserved. This includes syn-orogenic sedimentary rocks, late-stage extensional tectonic features and the 'upper plate' rocks in the hanging wall to the low angle detachment fault. Such features are often only poorly preserved in other more eroded orogenic belts such as the European Alps. The UM Unit rocks of the 'upper plate' comprise the most suitable lithologies (sandstones and metamorphic rocks as opposed to mainly carbonate rocks within the Tripolitza and Pindos Units) from which to obtain apatite for fission-track analysis from above the 'Cretan Detachment Fault'. Thus apatite FT thermochronology from the UM Unit rocks should also allow the low temperature cooling history in response to denudational processes above the major detachment fault to be assessed.

2. The Uppermost Tectonic Unit of Crete

The uppermost tectonic unit (UM Unit) of the Cretan nappe pile (or uppermost tectonic nappe of Seidel et al., 1976) comprises a number of differing components. For the purposes of this study these are divided into five separate sub-units.

2.1. UNMETAMORPHOSED FLYSCH SUBUNIT (PINDOS UNIT *SENSU STRICTU*)

Directly underlying the UM Unit proper is a sequence of Lower to Middle Eocene (Aubouin et al., 1965) sandstones, shales and calc-turbidites with a flysch-like character. These pass downward into a typical Pindos facies of basinal sedimentary rocks of Late Triassic to Palaeocene age that include pelagic limestones, radiolarites, calc-turbidites and calc-breccias (Seidel, 1971). In places (e.g. Asteroussia Mountains of southern Crete) the sequence passes up into a 'Blocky Flysch' that contains large (100's meter scale) shelf limestone blocks of Mesozoic to Eocene age as well as abundant clasts of granite, metamorphic rocks and serpentinite (Bonneau, 1984).

2.2. THE VERY LOW GRADE (VLG) METAMORPHIC SUBUNIT

This subunit, parts of which have previously been described as the 'Arvi Nappe' (Bonneau, 1972), includes pillow-basalts, tuffs, dolerite dykes, turbiditic psammites and sedimentary 'mass-flows' and red pelagic limestones of Late Cretaceous age. Petrologic studies (Robert & Bonneau, 1982; Hall, 1987) show these rocks have been altered at low temperatures, with formation of pumpellyite, epidote, chlorite, calcite and muscovite, commonly in amygdales and veins. This has been considered to represent sub-sea floor metamorphism due to hydrothermal circulation (Bonneau, 1984). This subunit tectonically overlies, but is always found in close association with the Pindos Unit.

2.3. THE PREVELI SUBUNIT

Rocks of this subunit have also been described as the Vatos Nappe (Bonneau & Lys, 1978), the Kalypso Unit of Gavdos Island (Vicente, 1970) and the Miamou Unit of central Crete (Bonneau et al., 1974). It includes meta-basalts, meta-rhyolites, meta-sediments and pegmatitic gabbro dykes that show a greenschist to HP/LT glaucophane metamorphism. It has been dated by the K-Ar method on phengite and hornblende from the rocks of Gavdos as ca. 148 Ma (Seidel et al., 1977). These pass up into a flysch-like formation containing Jurassic-Cretaceous limestone clasts. The whole sequence is overlain by pelagic red marls of Late Cretaceous age. This subunit is thrust upon rocks of the Pindos Unit on the Island of Gavdos (Vicente, 1970), and appears to lie tectonically below the high-grade metamorphic rocks near Miamou (Bonneau, 1984).

2.4. THE HIGH GRADE METAMORPHIC (HGM) SUBUNIT

This subunit, often referred to as the 'Asteroussia Nappe' (Bonneau, 1972), consists of high-grade metamorphic rocks that include amphibolites, marbles, mica schists, migmatites and leucogranites (Creutzburg & Seidel, 1975). Seidel et al. (1976; 1981) have demonstrated high temperature-low pressure conditions of metamorphism and K-Ar ages from hornblende and biotite of ca. 70-75Ma (Late Cretaceous). Granitic intrusions in the Kalo Chorio region of northeastern Crete also give Late Cretaceous K-Ar ages (Baranyi et al., 1975). The much larger former extent of this subunit is indicated by the widespread presence of gneissic pebbles within Neogene sediments, particularly in eastern Crete (Fortuin, 1977; Postma et al., 1993).

2.5. THE 'OPHIOLITE' SUBUNIT

The highest rocks of the Cretan nappe pile comprise mainly serpentinite with, in places, associated basic dykes and Jurassic-Cretaceous sedimentary rocks. These are found as relict nappes thrust upon the rocks of the HGM Subunit. K-Ar age determinations yield ages of 140 to 156 Ma (Seidel et al., 1981). These authors also propose that these rocks were formed as ophiolites in either an island arc or continental margin environment.

Various interpretations of the UM Unit rocks are proposed in the literature. Creutzburg and Seidel (1975) and Seidel et al. (1981) regard the rocks above the unmetamorphosed Eocene Pindos flysch as a 'serpentinite-amphibolite association' forming a composite nappe consisting of ophiolites at the top, and an ophiolitic mélange, comprising the VLG, Preveli and HGM Subunits, at the base. Hall et al. (1984) interpret the various rocks of the UM Unit as exotic blocks of varying size within a 'blocky flysch' that forms a major olistostrome in the upper part of the Pindos Unit. In contrast, Bonneau (1984) proposes that the rocks of the VLG Subunit form an 'ophiolitic olistostrome' related to the rocks of the Pindos Unit, whereas the Preveli, HGM and Ophiolite Subunits are 'mappable units with a different geodynamic significance' that form a series of individual nappes.

3. Apatite Fission-track Thermochronology

3.1. ANALYTICAL DETAILS

Separated apatites were mounted, polished and etched according to the techniques outlined by Hurford et al. (1991). All the samples were analysed using the external detector method. The samples were irradiated in the graphite reflector of the Risø reactor at the National

Research Centre, Roskilde, Denmark, the neutron fluence being monitored using Corning CN-5 glass.

Ages were calculated using the IUGS recommended Zeta-Calibration approach (Hurford & Green, 1983). The apatite CN-5 zeta was obtained using the Fish Canyon, Durango and Mount Dromedary age standards (see Hurford 1990).

Spontaneous and induced fission-track densities were determined using a Zeiss Axioplan microscope at ×1250 magnification with the help of an automated Märzhäuser stage. Apatite fission-track lengths were measured using an attached drawing tube and digitising tablet calibrated against a stage micrometer.

3.2. DATA PRESENTATION

31 apatite FT ages and 20 apatite confined FT length measurements have been obtained from the rocks of the UM Unit in the areas of Lendas, Melambes and Preveli (see Figure 2). Table 1 presents the data in IUGS standard format (Hurford 1990). The ages are quoted as 'central ages' as recommended by Galbraith & Laslett (1993). The Central Age is a modal age that is weighted to allow for differing precision of individual grain ages. It contains two uncertainties. The main age error is a measure of analytical error (quoted here at the 1σ level), whereas the 'age dispersion' is a measure of the spread in individual grain ages. A low age dispersion (<20%) is indicative of a single grain age population.

3.3. QUALITATIVE DATA INTERPRETATION

For the purposes of interpretation the data are split into three groups. The first group comprises samples collected from unmetamorphosed Paleocene-Eocene flysch-like sedimentary rocks (UMF - UnMetamorphosed Flysch), the second group comprises rocks from the 'Very Low Grade Metamorphic Subunit' (VLG) and the final group are rocks from the 'High Grade Metamorphic Subunit' (HGM) that tectonically overlie the VLG and UMF Subunits.

The majority of data were obtained from samples collected from the area around Lendas, southern Crete. These are shown on a geological sketch map of the area in Figure 4. Other data were obtained from samples in the Melambes area (TH71 to 74) and the Preveli area (TH121 & 122). The locations of these latter two areas are shown in Figure 2.

In the Lendas area the apatite FT ages show a relatively wide spread from 14.0±1.7 Ma to 32.0±8.8 Ma. The apatite confined track length measurements show a similar spread from 12.31μm to 14.84μm with standard deviations between 0.68μm and 2.70μm.

The data from the UMF Subunit give apatite FT ages of between 14.0±1.7 Ma and 17.7±1.0 Ma. The mean track length measurements are all over 14μm (14.27μm to 14.84μm) with correspondingly low standard deviations of between 0.68μm and 1.25μm. Qualitative interpretation of these results indicates that all the UMF rocks of the UM Unit cooled quickly through the apatite partial annealing zone between ca. 19 and 13 Ma.

The apatite FT ages from the VLG rocks of the Lendas area are generally slightly older, being between 15.1±2.1 Ma and 20.0±2.5Ma (this excludes imprecise ages from samples Ra93030, TH97 & TH 105). The corresponding track lengths distributions are

TABLE 1. Apatite fission-track data from the uppermost tectonic unit of Crete, Greece.
Notes: (i). analyses by external detector method using 0.5 for the $4\pi/2\pi$ geometry correction factor; (ii). ages calculated using dosimeter glasses: CN5 with ζ_{CN5} =368.9±8; (iii). $P\chi^2$ is the probability of obtaining a χ^2 value for v degrees of freedom where v = no. of crystals - 1; (iv). VLG = Very Low Grade Metamorphic Unit, HGM = High-grade Metamorphic Unit, UMF = Unmetamorphosed Flysch;

Sample Number Rock Type (Geological Unit)	Location Latitude, Longitude	Height	No. of Crystals	Track Density (x 10^6 tr cm^{-2}) Spontaneous ρ_s (N_s)	Induced ρ_i (N_i)	Dosimeter ρ_d (N_d)	Age Dispersion ($P\chi^2$)	Apatite FT Central Age (±1σ)	Apatite Mean Track Length (μm ± 1 s.e.) (no. of tracks)	Standard Deviation (μm)
Lendas Area										
Ra 93030	Acri Tracheli	5m	6	0.0777	0.8089	1.288	0.1%	**22.8±5.1**	-	
Quartzite (VLG)	34°56'01"N, 24°52'51"E			(22)	(229)	(8899)	(49%)			
TH 82	Kali Limenes	250m	20	0.7423	5.732	1.262	0%	**30.1±1.5**	**12.76±0.22**	**2.17**
Mica Schist (HGM)	34°57'55"N, 24°49'13"E			(489)	(3776)	(8717)	(96%)		(100)	
TH 83	Antiskari	140m	11	0.0738	0.5413	1.159	0%	**29.1±6.8**	-	
Volcanic (VLG)	34°57'33"N, 24°52'09"E			(21)	(154)	(8003)	(99%)			
TH 84	Platio Peramata	60m	16	0.2044	2.181	1.160	0%	**20.0±2.5**	**14.09±0.48**	**1.19**
Amphibolite (VLG)	34°56'16"N, 24°52'16"E			(70)	(747)	(8014)	(99%)		(7)	
TH 88	Lendas	310m	20	1.053	9.935	1.160	0%	**22.6±1.2**	**13.46±0.19**	**1.87**
Leucogranite (HGM)	34°56'50"N, 24°54'20"E			(394)	(3718)	(8012)	(97%)		(100)	
TH 90	Kali Limenes	145m	20	0.3999	3.076	1.161	0%	**27.8±1.9**	**13.08±0.25**	**1.28**
Bi. Gneiss (HGM)	34°56'41"N, 24°49'25"E			(243)	(1869)	(8020)	(99%)		(27)	
TH 92	Apessokari	210m	20	0.3161	4.205	1.161	0%	**16.1±1.2**	**14.84±0.09**	**0.82**
Arenite (UMF)	35°00'09"N, 24°57'12"E			(213)	(2834)	(8015)	(>99%)		(78)	
TH 93	Apessokari	380m	20	0.4727	5.718	1.159	3%	**17.7±1.0**	**14.77±0.10**	**0.83**
Greywacke (UMF)	34°59'25"N, 24°57'01"E			(393)	(4754)	(8002)	(48%)		(76)	
TH 94	Miamou	400m	5	0.0424	0.6075	1.278	0%	**16.4±4.9**	-	
Greywacke (UMF)	34°59'10"N, 24°56'51"E			(12)	(172)	(8825)	(65%)			
TH 97	Krotos/Lutra	190m	4	0.0690	0.6207	1.159	0%	**23.7±7.2**	-	
Calc-arenite (VLG)	34°56'20"N, 24°57'44"E			(12)	(108)	(7999)	(65%)			
TH 101	Krotos/Lutra	350m	20	1.326	12.43	1.159	0%	**22.8±0.9**	**13.17±0.20**	**2.03**
Pegmatite (HGM)	34°56'34"N, 24°57'43"E			(793)	(7431)	(8000)	(99%)		(100)	
TH 102	Krotos/Lutra	350m	20	0.6852	6.188	1.161	0%	**23.7±1.1**	**12.99±0.19**	**1.93**
Migmatite (HGM)	34°56'34"N, 24°57'43"E			(564)	(5093)	(8017)	(98%)		(100)	
TH 103	Kali Limenes	5m	8	0.1853	2.615	1.160	0%	**15.1±2.1**	-	
Quartzite (VLG)	34°55'26"N, 24°47'48"E			(54)	(762)	(8008)	(98%)			
TH 104	Miamou	460m	4	0.0312	0.2310	1.286	0%	**32.0±8.8**	-	
Granodiorite (HGM)	34°57'57"N, 24°56'16"E			(15)	(111)	(8883)	(>99%)			
TH 105	Ag. Kirillos	340m	20	0.1993	2.704	1.291	0%	**17.5±1.9**	**14.40±0.38**	**1.00**
'Phyllite' (VLG)	34°59'14"N, 24°55'38"E			(90)	(1221)	(8916)	(>99%)		(8)	
TH 106	Ag. Kirillos	340m	9	0.0203	0.2301	1.294	0%	**21.0±9.0**	**14.65±0.25**	**0.86**
'Phyllite' (VLG)	34°59'14"N, 24°55'38"E			(6)	(68)	(8941)	(96%)		(13)	
TH 108	Plagia Peramata	60m	20	0.4324	5.448	1.290	0%	**18.9±1.0**	**14.54±0.11**	**1.08**
Greywacke (UMF)	34°56'33"N, 24°50'48"E			(369)	(4650)	(8908)	(92%)		(100)	
TH 109	Pombia	300m	20	0.1143	1.938	1.287	0%	**14.0±1.7**	-	
Sandstone (UMF)	35°00'07"N, 24°51'54"E			(74)	(1255)	(8891)	(99%)			
TH 110	Pombia	420m	20	0.1460	2.184	1.280	0%	**15.8±1.9**	**14.70±0.23**	**0.68**
Sandstone (?UMF)	34°59'33"N, 24°51'54"E			(74)	(1107)	(8841)	(>99%)		(10)	
TH 112	Moni Ampezarion	450m	20	0.4412	5.951	1.275	0%	**17.4±0.9**	**14.68±0.14**	**0.98**
Quartzite (UMF)	34°56'20"N, 24°52'57"E			(418)	(5638)	(8808)	(99%)		(52)	
TH 113	Antiskari	400m	20	0.1318	1.944	1.274	0%	**15.9±1.5**	**14.27±0.44**	**1.25**
Greywacke (UMF)	34°58'15"N, 24°53'35"E			(128)	(1887)	(8800)	(>99%)		(9)	
TH 114	Antiskari	420m	3	0.2102	3.340	1.276	0%	**14.8±3.6**	-	
Sandstone (UMF)	34°58'18"N, 24°54'34"E			(18)	(286)	(8816)	(83%)			
TH 115	Antiskari	420m	17	0.2849	4.103	1.293	0%	**16.5±1.4**	-	
Sandstone (?VLG)	34°58'18"N, 24°54'34"E			(157)	(2261)	(8933)	(>99%)			
TH 116	Plagia Peramata	10m	20	0.4723	5.683	1.292	0%	**19.8±1.1**	**14.68±0.10**	**1.01**
Quartzite (VLG)	34°56'24"N, 24°50'25"E			(392)	(4717)	(8924)	(67%)		(100)	
TH 118	Krotos	470m	4	0.1317	1.834	1.279	0%	**16.9±4.7**	-	
Greywacke (UMF)	34°57'45"N, 24°57'33"E			(14)	(195)	(8833)	(99%)			
Melambes Area										
TH 71	Melambes	190m	40	0.0956	1.298	1.160	0%	**15.7±1.7**	**14.50±0.41**	**0.91**
Conglom. (UMF)	35°08'30"N, 24°40'22"E			(91)	(1239)	(8011)	(99%)		(6)	
TH 72	Melambes	210m	20	0.3358	4.601	1.160	0%	**15.6±1.0**	**14.05±0.25**	**0.97**
Greywacke (UMF)	35°06'27"N, 24°41'11"E			(269)	(3685)	(8009)	(98%)		(16)	
TH 73	Melambes	510m	9	0.1448	2.470	1.161	0%	**12.6±3.0**	-	
Gneiss (HGM)	35°07'56"N, 24°39'38"E			(19)	(324)	(8018)	(>99%)			
TH 74	Saktouria	500m	20	0.4063	5.526	1.159	0%	**15.7±0.9**	**14.37±0.13**	**0.93**
Greywacke (UMF)	35°07'29"N, 24°37'03"E			(335)	(4556)	(8005)	(99%)		(51)	
Plakias Area										
TH 121	Moni Preveli	200m	20	0.3434	2.632	1.283	0%	**30.8±2.5**	**12.31±0.49**	**1.97**
Phyllite (Preveli)	35°09'26"N, 24°27'45"E			(175)	(1341)	(8860)	(>99%)		(17)	
TH 122	Kato Moni Preveli	90m	20	0.3103	2.445	1.275	0%	**29.8±2.2**	**12.44±0.39**	**2.70**
Phyllite (Preveli)	35°10'00"N, 24°28'09"E			(207)	(1631)	(8806)	(90%)		(50)	

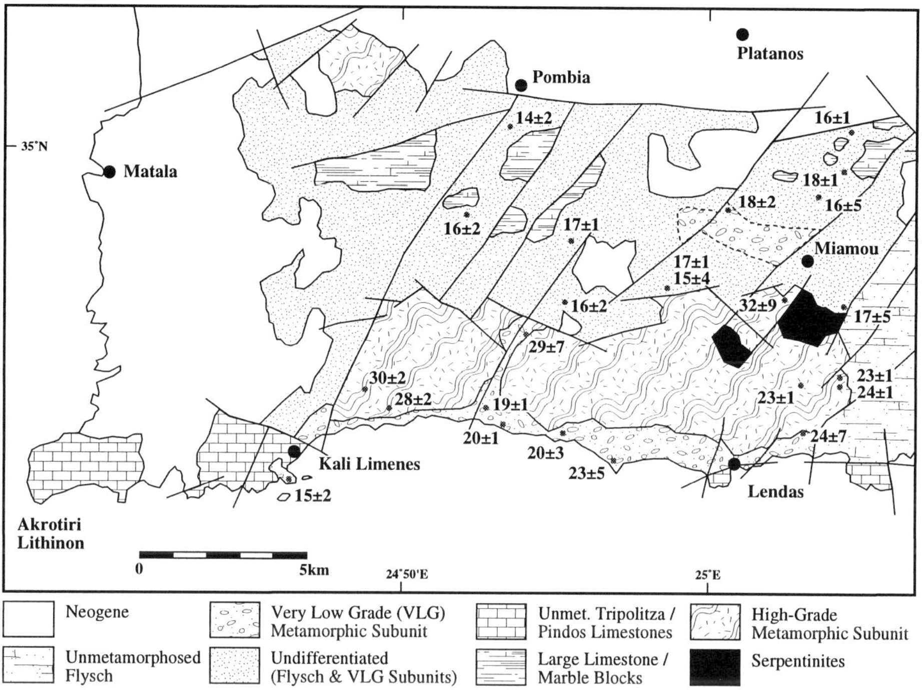

Figure 4. Simplified geological and location map showing the apatite fission-track central ages obtained from the various components of the uppermost tectonic unit exposed in the region of Lendas, southern Crete.

similar to those from the unmetamorphosed sedimentary rocks, with mean track lengths between 14.09μm and 14.68μm and standard deviations between 0.86μm and 1.19μm.

The HGM Subunit rocks of the UM Unit give significantly older apatite FT ages, ranging from 20.0±2.5 Ma to 30.8±2.5 Ma (excluding the imprecise age from sample TH104). The apatite confined track length distributions all show a significant proportion of short tracks (< 14μm). This results in low mean track lengths of between 12.31μm and 13.46μm, and large standard deviations of between 1.28μm and 2.70μm (this excludes sample TH84, where only 7 tracks were measured). Several of the track length distributions (see Figure 6) suggest the presence of two groups of track lengths (a bimodal distribution), especially samples TH82, TH88 and TH102. Such a bimodal distribution is indicative of a sample that has spent some time within the apatite partial annealing zone (between ca. 60°C and 110±10°C) to produce the shorter tracks, then later cooled to below 60°C to allow the accumulation of the longer tracks (Gleadow & Fitzgerald, 1987).

The apatite FT ages of all the units discussed are plotted against both their mean track length and standard deviation of the track length distribution in Figure 5, to produce so-called 'boomerang plots' (Brown et al., 1994; Gallagher & Brown, 1997). The plots both indicate the presence of a thermal event at ca. 17Ma, probably linked to the initiation of a phase of accelerated denudation, that affected all of the rocks of the UM Unit.

This pattern of ages is indicative of the presence of a denuded fossil apatite partial annealing zone (PAZ) where different levels of a former apatite fission-track 'stratigraphy'

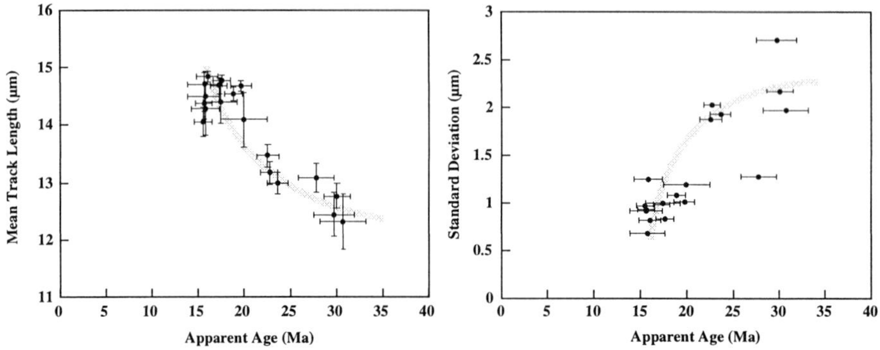

Figure 5. Apatite age data from the Uppermost Unit plotted against mean track length and the standard deviation of the track length distribution to produce 'boomerang plots'. Both plots indicate a thermal event at ca. 17Ma.

are now exposed at the surface (Fitzgerald, 1994; Gallagher & Brown, 1997). However, Neogene fault block tilting and extensional faulting on Crete (Meulenkamp et al., 1994) means that plotting an apatite age vs. elevation profile (e.g. Fitzgerald et al., 1995) for most profiles or regions on Crete, including the UM Unit in the area around Lendas, are of no use when looking for the presence of a 'break-in-slope', which can be used to estimate the timing of initiation of an accelerated denudation event.

To try to avoid this we attempted to plot the apatite FT age data against the perpendicular distance from a reference plane that existed prior to Neogene deformation. A similar approach has also been used by Lihou et al. (1995). For the Lendas area the ?Oligocene (Bonneau, 1984) thrust contact at the base of the HGM Subunit rocks is used. The result is shown in Figure 6. To aid interpretation the apatite track length distributions and corresponding radial plots are also shown. What emerges is a relatively well-defined 'break-in-slope' representing the base of an uplifted fossil apatite PAZ with the break-in-slope marking the onset of rapid cooling at ca. 17 Ma. This is in good agreement with the boomerang plot shown in Figure 5. The interpretation of such plots is dealt with in detail by Fitzgerald et al. (1995). In the profile here, some spread in the data occurs due to the regional nature of the data set, the use of a reference plane rather than a true vertical profile (the HGM Subunit thrust contact was unlikely to have been totally horizontal and planar over the area of study during cooling or denudation) and the relatively small vertical profile (<1000m). It should thus not be treated as a 'vertical profile' *sensu strictu* .

However it allows some interpretation to be made. First, the onset of rapid cooling may have occurred slightly earlier in places, probably the western part of the Lendas area, as here the older apatite ages are found, or second and more likely, the former apatite fission-track stratigraphy has been slighly more eroded since the onset of accelerated denudation in the eastern part of the Lendas area.

The confined track lengths distributions below the break-in-slope have long track lengths (>14.5µm) and narrow standard deviations (<1.08µm), and confirm rapid cooling through the apatite PAZ after ca. 17 Ma. The profile above the break-in-slope represents rocks that lay within the apatite PAZ prior to the onset of rapid cooling. This slope does not quantify a denudation rate, but instead reflects the shape of the fossil PAZ or apatite fission-

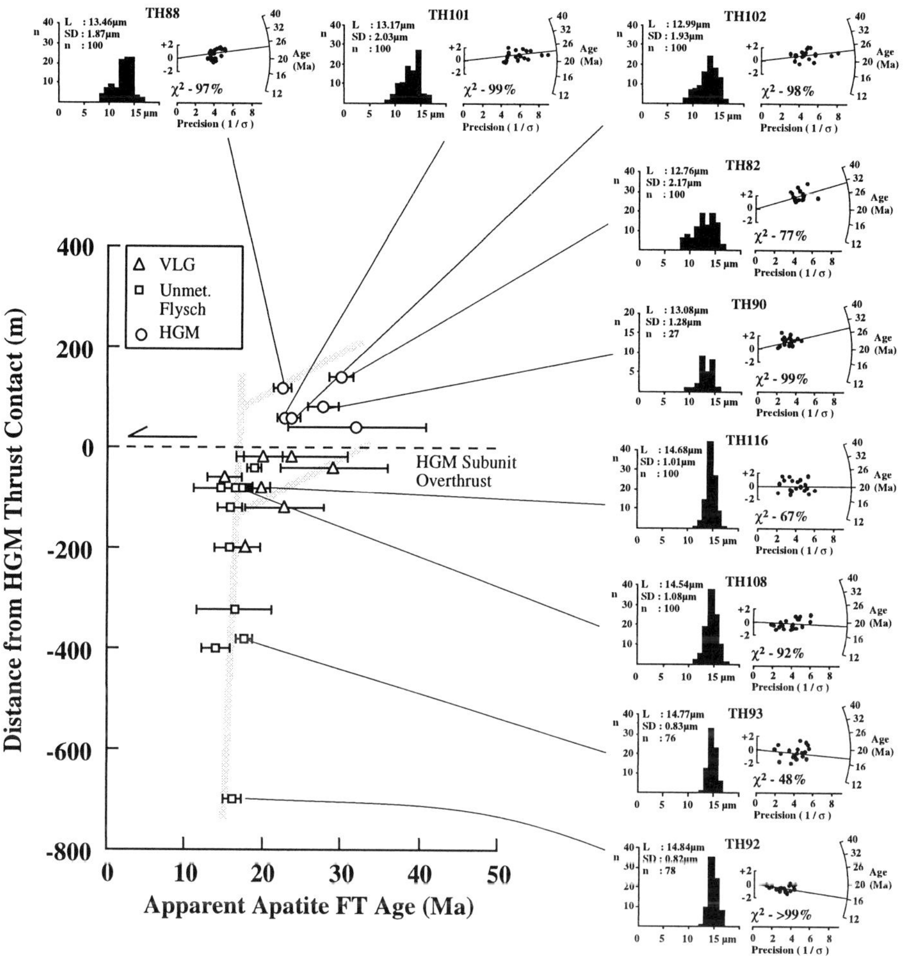

Figure 6. Apatite age profile from the samples of the Lendas area including apatite confined track length distributions and radial plots (Galbraith, 1990), showing apparent age (± 1 σ) against perpendicular distance from the High-grade Metamorphic Unit overthrust. Although the data show some scatter, a distinct break-in-slope is seen at ca. 17Ma, marking the onset of accelerated cooling.

track stratigraphy. The shape of this fossil PAZ is dependent on the original age of the apatites, the duration of land surface stability and the geothermal gradient. The shape of the fossil PAZ is also relatively sensitive to low denudation rates. The effects of varying these parameters have been modelled by Brown et al. (1994). The age profile from the Lendas area indicates that before the onset of rapid cooling at ca. 17 Ma the land surface and geotherm of the UM Unit must have been relatively stable for at least 15 Ma, with cooling rates not exceeding 5°C/m.y. This indicates that the overthrusting of the HGM Subunit rocks of the Lendas area upon the VLG Subunit and the unmetamorphosed Eocene flysch probably occurred very soon after the deposition of the sedimentary rocks themselves, probably in the Late Eocene or Early Oligocene.

One other factor that can influence the shape of the age profile above the break-in-slope is apatite chemistry, and in particular the Cl / Cl+F ratio. O'Sullivan & Parrish (1995) have demonstrated that when a spread in apatite composition exists between grains within a single sample, the apparent age of fluorine-rich grains at temperatures within the apatite PAZ (110±10°C to ca. 60°C) will be reduced relative to the chlorine-rich grains. This results in a spread of ages on a radial plot and lowers the χ^2 probability. However, although the single grain apatite ages in the samples from above the break-in-slope in Figure 6 are older, the spread in single grain ages is not significant indicating that the apatite composition within these samples is similar.

The apatite FT data in the Melambes area (Figure 2) all have apatite FT ages and track length distributions all indicating rapid cooling at ca. 15-17 Ma. It is likely that here the fossil apatite PAZ has been totally removed, implying that the UM Unit in this region is more denuded than that in the Lendas area.

Apatite FT data have also been obtained from two samples of the Preveli Subunit near to the Preveli Monastery (Figure 2). They give ages of ca. 30Ma and the form of the confined track length distributions, with mean lengths <13μm and large standard deviations >1.97μm, indicate that these samples lie within a denuded fossil apatite PAZ as is also seen in the UM Unit rocks of the Lendas area.

3.4. QUANTITATIVE THERMAL HISTORY MODELLING

Detailed thermal history analyses have been carried out on three samples from the UMF, VLG and HGM subunits of the UM Unit. The 'MonteTrax' thermal modelling method used in this study has been outlined in detail by Gallagher (1995). This method uses a stochastic approach, whereby the bounds of possible time-temperature points are specified and then thermal histories are selected from within these bounds. A genetic algorithm is used to search the parameter space and define time-temperature histories that show a good-fit to the observed data with a maximum likelihood approach used to define the confidence limits of predicted versus observed fitting of the data.

Figure 7 shows the results of thermal history modelling of the apatite FT data from the UM Unit. Several time-temperature constraints confine the time-temperature space in which the models are run. The UM Unit rocks must have been at the surface (taken here as 20°C) when they were unconformably overlain by the onset of Neogene sedimentation between ca. 15 Ma and 12 Ma (Postma et al., 1993). The majority of the unmetamorphosed flysch was deposited during the Eocene, thus these samples are constrained by their depositional age. However as these rock samples have been totally annealed during the Miocene, the older 'burial' or heating of these samples cannot be predicted by thermal history modelling. The depositional age is included only to illustrate that these rocks must have been significantly buried after deposition. The VLG metamorphic rocks of probable late Mesozoic age (Seidel et al., 1977), were probably not buried during the late Eocene / early Oligocene. Hence for these samples the T-t history starts from below the apatite annealing zone. The thermal history of the HGM rocks are constrained by ca. 70 Ma K-Ar hornblende, muscovite and biotite ages (Seidel et al., 1981) that indicate rapid late

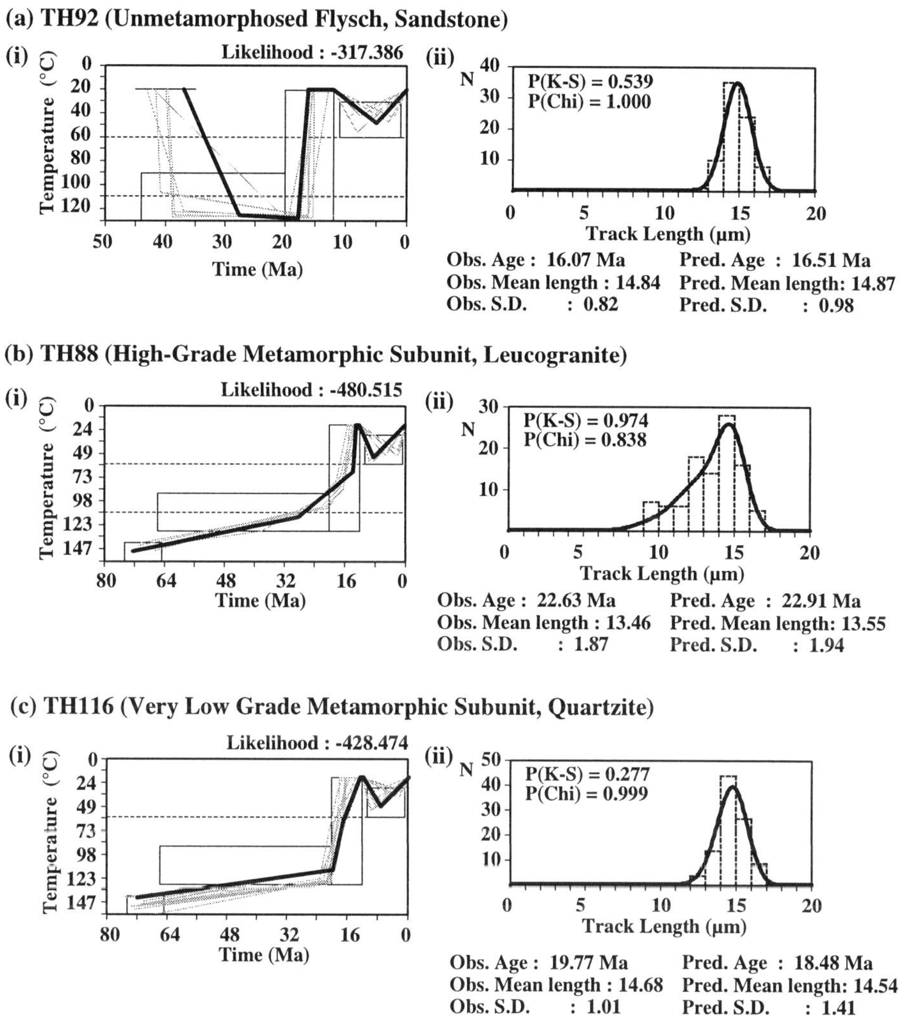

Figure 7. 'Best-fit' thermal histories, obtained from modelling apatite fission-track data using the approach of Gallagher (1995), from three different components of the uppermost tectonic unit in the Lendas area. Parts (i) show the 'best fit' thermal history in bold, that produces the predicted results in (ii), with other thermal histories that fit the observed apatite age and length data in grey. Parts (ii) show the predicted track-length distribution in bold, and the observed track length data in grey. The results of comparing the two distributions using the χ^2 (Chi) and the Kolmogorov-Smirnoff (K-S) statistical fitting methods are also shown (see Gallagher, 1995 for details).

Cretaceous cooling to below the closure temperature of Ar in biotite (ca. 320°C - Harrison et al., 1985). A confining box at 70±5 Ma and 150±10°C (an arbitrary temperature between the closure temperature of Ar in biotite and the base of the apatite PAZ) is applied to the model of the HGM rocks. Zircon fission-track data from the UM Unit rocks are in the process of being acquired to better constrain the cooling history and hence the T-t model between ca. 230°C and 310°C.

The modelling applied to the apatite FT data from the UMF rocks confirms the interpretation made from the data as an apatite FT age profile (Figure 6) with relatively fast rates cooling (15-25°C / Ma) between the base of the apatite annealing zone (110±10°C) and the surface (ca. 20°C) between ca. 19 Ma and 13 Ma.

The modelled thermal histories from the HGM rocks of the UM Unit require that these rocks entered the apatite PAZ (i.e. cooled to below 110±10°C) at a relatively slow cooling rate (<1°C / Ma) between ca. 40 Ma and ca. 30 Ma with continued slow cooling within the apatite PAZ until the onset of a period of accelerated cooling between ca. 18-14 Ma. The slow cooling within the apatite annealing zone allowed the development of the apatite age-depth profile between ca. 40 Ma and 20Ma that was later denuded, and is represented today by the break-in-slope discovered within the UM Unit rocks of the Lendas area. The accelerated cooling of these rocks was initiated when they were at a temperature of about 70-80°C.

The apatite FT data from the VLG rocks require a thermal history similar to that experienced by the HGM rocks. One difference is that the accelerated cooling of these rocks needs to have been initiated slightly earlier (between 18 and 20Ma) and from slightly higher temperatures, very close to the base of the apatite PAZ (ca. 110±10°C).

Some of the apparent Neogene heating of the samples between 12Ma and present is probably an artefact of the Laslett et al. (1987) annealing model used. This model tends to overestimate apatite fission-track annealing over geological time at temperatures <60°C (P. Green pers. comm.).

4. Discussion

A proposed late Tertiary cooling history for the components of the UM Unit in the area around Lendas, southern Crete is illustrated in Figure 8. According to Bonneau (1984) the VLG Subunit rocks and the HGM Subunit rocks were emplaced upon the Eocene flysch during the early Oligocene. This can be correlated with the results of apatite FT thermochronology that reveal the cooling of the HGM rocks into the apatite PAZ (to below 110±20°C) sometime between ca. 40 Ma and ca. 30 Ma. This must have caused the consequent burial and hence heating of the Eocene flysch-like sedimentary rocks to above 110±10°C, shortly after their deposition. The VLG Subunit rocks, that lie tectonically below the HGM Subunit rocks must have remained above 110±10°C during this time. To allow the development of an apatite FT age stratigraphy, all the rocks of the UM Unit must have then cooled very slowly until the onset of accelerated cooling between ca. 18 and 16Ma. This is confirmed by the thermal history modelling of the apatite data from the HGM Subunit rocks. Quantitative and qualitative interpretation of the apatite FT data reveal that all the rocks of the UM Unit underwent a period of rapid accelerated cooling from ca. 17 Ma, until they were unconformably overlain by coarse grained clastic sedimentary rocks possibly as early as the Langhian (ca. 15Ma) (the Mithi Formation of Fortuin, 1977). Some later Neogene burial of these rocks has been revealed by Meulenkamp et al. (1994), deduced from an examination of the Neogene sedimentary rocks of the Iraklion Basin of central Crete.

The presence of a break-in-slope or base of a fossil apatite PAZ (Figure 6) allows the

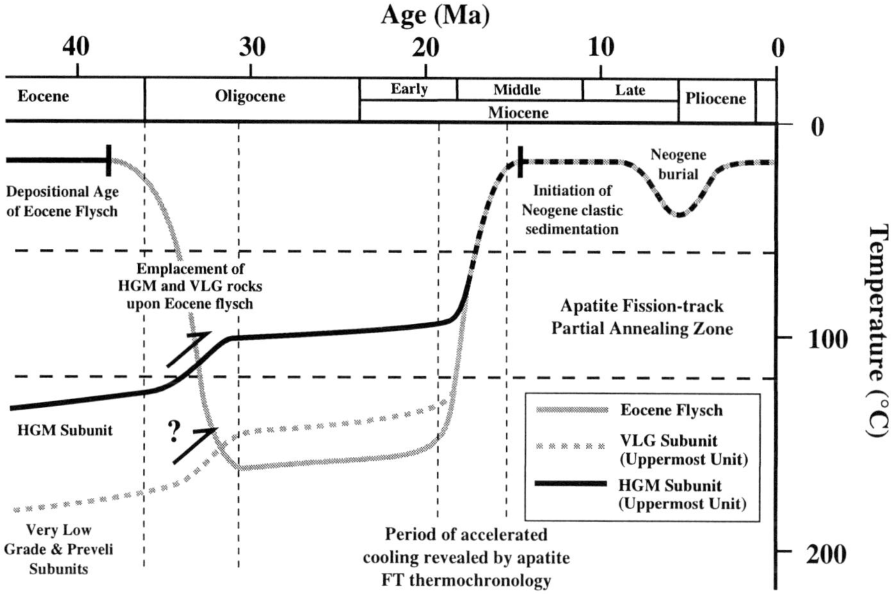

Figure 8. Proposed late Tertiary cooling history of the components of the uppermost tectonic unit in the area around Lendas, southern Crete deduced from apatite fission-track thermochronology, with other structural and stratigraphic constraints.

total amount of denudation since the onset of accelerated cooling in the UM Unit rocks at ca. 18-16Ma to be estimated. This is done using the equation given in Brown (1991):

$$\text{Total Denudation} = ((T_{PAZ} - T_S) / G) + d \qquad (1)$$

where T_{PAZ} is the temperature at the base of the fossil apatite PAZ, taken as 110±10°C, T_S is the palaeo-mean surface temperature, G is the palaeo-geothermal gradient, and d is the elevation difference between the base of the fossil PAZ and the present day mean surface elevation. The value for palaeo-mean surface temperature is taken to be ca. 15°C. Work on fluid inclusions from the rocks of the Phyllite-Quartzite Unit of Crete (Küster & Stöckhert, 1995) indicates a palaeo-geothermal gradient at ca. 17 Ma of ca. 25°C / km, and the value for d in the Lendas area is ca. 0 km. Putting these values into (1), means that 3.8 ± 0.4 km of overburden has been removed from the rocks of the UM Unit in the Lendas area since ca. 17 Ma. Meulenkamp et al. (1994) show that by the Tortonian (ca. 11 Ma) the Asteroussia region was undergoing subsidence. Thus the phase of accelerated denudation responsible for the removal of ca. 4km of overburden took place during a relatively short time period of ca. 6 Ma. This requires a mean denudation rate of ca. 650m / Ma.

This period of accelerated denudation revealed by apatite FT thermochronology is contemporaneous with the initiation of large scale clastic sedimentation in the Ierapetra Basin of eastern Crete (Fortuin, 1977; Fortuin & Peters, 1984; Postma et al., 1993). Here the onset of sedimentation is marked by probable Langhian (ca. 15Ma) continental red-bed sedimentary rocks (the Mithi Formation) derived from the UM Unit. It is therefore most probable that the mid-Miocene phase of accelerated denudation was a result of increased erosion.

The results of apatite FT thermochronology from the UM Unit rocks of Crete have several important implications for the tectonic development of the Hellenic Subduction System. The most significant is the discovery that the rocks in the so-called 'upper plate' of the Cretan nappe pile underwent no significant denudation between the Late Eocene - Early Oligocene (ca. 35Ma) and the Middle Miocene (ca. 15Ma). This implies that during ongoing 'accretion' associated with deformation and high-pressure metamorphism of the 'lower plate' rocks, little or no erosion (or denudation) took place at the surface. This implies a lack of topographic elevation at this time that contradicts the tectonic model of Jolivet et al. (1994) which proposes that during convergence in the Oligocene, collision with the Adria or 'Ionian' microplate formed an accretionary wedge. Then as a result of underplating, with coeval HP-LT metamorphism, an overall 'cold crust' with a deep brittle-ductile transition was developed. This resulted in a thickened crustal wedge that supported a high topography with subsequent unroofing of the HP-LT rocks occurring due to collapse of this thick crust when it entered the warmer back-arc region.

However the lack of significant erosion during accretion of the tectonic units of Crete leads us to propose that the Hellenic Subduction Zone was acting as a 'retreating plate boundary' (in the sense of Royden, 1993a) as early as the Late Eocene. With this model, the main driving force of subduction is caused by the negative buoyancy of the subducted plate. Royden (1993a) states that orogenic belts formed at retreating plate boundaries are commonly typified by topographically low mountains, the development of regional extension in the overriding plate, little or no erosion or denudation, low-grade metamorphism, little involvement of the crystalline basement in thrusting and flysch deposition within adjacent foredeep basins. These features fit remarkably well with what is observed in the rocks of Crete. A lack of erosion (or denudation) and hence relief during the Oligocene and early Miocene is confirmed by apatite FT analysis, with the uppermost tectonic unit of Crete having remained in the upper ca. 4km of the upper crust through the entire Oligo-Miocene accretionary process. During this accretion the lower plate Phyllite-Quartzite and Plattenkalk Units were buried to depths of up to 35 km associated with HP-LT metamorphism and subsequently brought back to the surface by regional extensional processes (Kilias et al., 1994; Fassoulas et al., 1994; Jolivet et al., 1994). In addition, almost no crystalline basement is seen within the tectonic units of Crete (see Creutzburg & Seidel, 1975). Finally the present day Mediterranean Ridge accretionary complex (Kastens, 1991) to the south of Crete contains significant amounts of largely undated sedimentary rocks that were probably deposited in the fore-deep basin adjacent to the orogenic belt.

The second discovery made from the results of apatite FT thermochronology on the rocks of the UM Unit is the period of accelerated denudation related to ca. 4km of erosion between ca. 17Ma and ca. 11Ma, likely to be related to an increase in topography and hence relief. The cause of this is probably related to isostatic rebound caused by a change in the tectonic regime during the Middle Miocene, after the accretion of the Adria microplate continental fragment upon which the Pindos, Tripolitza and HP-LT Units were deposited.

5. Conclusions

Apatite FT thermochronology carried out on the rocks of the uppermost tectonic unit of Crete reveals that the rocks in the 'upper plate' to extension underwent a phase of accelerated Middle Miocene denudation linked to erosion between ca. 17-11Ma. The presence of the base a denuded apatite partial annealing zone allows the total amount of erosion (or denudation) during this time to be estimated as ca. 4km. This implies a mean denudation rate of ca. 650m/m.y.

In addition, the presence of older apatite FT ages with corresponding bi-modal track length distributions from the high-grade metamorphic rocks of the uppermost tectonic unit, require that no significant denudation occurred at the surface during the accretion of the Cretan tectonic units, including the burial to ca. 35km and subsequent denudation of the HP-LT Phyllite-Quartzite and Plattenkalk Units, between the Oligocene and Early Miocene.

In light of this information a new tectonic model for the evolution of the Cretan segment of the Hellenic Subduction Zone is proposed. This advocates that the subduction zone has acted continuously as a retreating plate boundary since at least the Late Eocene. The subsequent onset of accelerated denudation at ca. 17 Ma is attributed to a change in the plate tectonic regime subsequent to the accretion of the Adria microplate continental fragment upon which the Pindos, Tripolitza and HP-LT Units were deposited.

6. Acknowledgements

This work has been supported by the Deutsche Forschungsgemeinschaft grants STO 196/6-1 and STO 196/8-1. Sample preparation was carried out under the supervision of Frank Hansen. Thanks to the IGME in Athens and Rethymnon for permission to collect samples from Crete, Nikos Fytrolakis for his interest and support, Eberhard Seidel for discussions in the field and Kerry Gallagher for providing his modelling program 'MonteTrax'.

7. References

Aubouin, J., Dercourt, J., Neumann, M. & Sigal, J. (1965): Un élément externe de la zone du Pinde: la série d'Ethia (Crète, Grèce). *Bull. Soc. géol. Fr.*, **7**, 753-757.

Baranyi, I., Lippolt, H.J. & Todt, W. (1975): Kalium-Argon-Datierungen an zwei Magmatiten von Kalo Chorio, Nord-Ost-Kreta. *N. Jb. Geol. Paläont., Mh.*, **1975**, 257-262.

Bonneau, M. (1982): Evolution dynamique de l'Arc Egéen depuis le Jurassique supérieur jusqu'au Miocène. *Bull. Soc. géol. Fr.*, **24**, 229-242.

Bonneau, M. (1984): Correlation of the Hellenic Nappes in the south-east Aegean and their tectonic reconstruction. *Geol. Soc. Lond., Spec. Pub.*, **17**, 517-527.

Bonneau, M. & Lys, M. (1978): Sur la présence de Permien fossilifère dans l'unité de Vatos (Crète): sa nature interne et l'ampleur des charriages dans l'arc égéen. *C. R. Acad. Sci. Paris.*, **287**, 423-426.

Bonneau, M., Beauvais, L. & Middlemiss, F-A. (1974): L'unité de Miamou (Crète, Grèce) et sa macrofaune d'age Jurassique supérieur (Brachipodes, Madréporaires). *Ann. Soc. Géol. Nord.*, **94**, 71-85.

Brown, R.W. (1991): Backstacking apatite fission-track "stratigraphy" : A method for resolving the erosional and isostatic rebound components of tectonic uplift histories. *Geology*, **19**, 74-77.

Brown, R.W., Summerfield, M.A. & Gleadow, A.J.W. (1994): Apatite fission track analysis: Its potential for the estimation of denudation rates and implications for models of long-term landscape development. In: *Process Models and Theoretical Geomorphology.* (ed. Kirkby, M.J.), J. Wiley & Sons Ltd. pp. 23-53.

Creutzburg, N. & Seidel, E. (1975): Zum Stand der Geologie des Präneogens auf Kreta. *N. Jb. Geol. Paläont., Abh.*, **141**, 259-285.

Creutzburg, N., Drooger, C.-W., Meulenkamp, J.-E., Papastamatiou, J., Sanneman, W., Seidel, E. & Tataris, A. (1977): General geological map of Crete (scale 1: 200 000). Institute of Geological and Mining Research, Athens.

Fassoulas, C., Kilias, A. & Mountrakis, D. (1994): Postnappe stacking extension and exhumation of high-pressure/ low temperature rocks in the island of Crete, Greece. *Tectonics*, **13**, 127-138.

Fitzgerald, P.G. (1994): Thermochronological constraints on post-Paleozoic tectonic evolution of the central Transantarctic Mountains, Antartica. *Tectonics*, **13**, 818-836.

Fitzgerald, P.G., Sorkhabi, R.B., Redfield, T.F. & Stump, E. (1995). Uplift and denudation of the central Alaska Range: A case study in the use of apatite fission track thermochronology to determine absolute uplift parameters. *J. Geophys. Res.*, **100**, 20,175-20,191.

Fortuin, A.R. (1977). Stratigraphy and sedimentary history of the Neogene deposits in the Ierapetra region, eastern Crete. *GUA Papers of Geology*, **No. 8**. 164pp.

Fortuin, A.R. & Peters, J.M. (1984): The Prina Complex in eastern Crete and its relationship to possible Miocene strike-slip tectonics. *J. Struct. Geol.*, **5**, 459-476.

Galbraith, R.F. (1990): The radial plot: graphical assessment of spread in ages. *Nucl. Tracks*, **17**, 207-214.

Galbraith, R.F. & Laslett, G.M. (1993): Statistical models for mixed fission track ages. *Nucl. Tracks*, **21**, 459-470.

Gallagher, K. (1995): Evolving temperature histories from apatite fission-track data. *Earth Planet. Sci. Lett.*, **136**, 421-435.

Gallagher, K. & Brown, R. (1997): The onshore record of passive margin evolution. *J. Geol. Soc. Lond.*, **154**, 451-457.

Gleadow, A.J.W. & Fitzgerald, P.G. (1987): Uplift history and structure of the Transantartic Mountains: new evidence from fission track dating of basement apatites in the Dry Valleys area, southern Victoria Land. *Earth Planet. Sci. Lett.*, **82**, 1-14.

Hall, R. (1987): Geochemistry and origin of Cretaceous igneous rocks of the southern Aegean. *Ofioliti*, **12**, 343-355.

Hall, R., Audley-Charles, M.G. & Carter, D.J. (1984): The significance of Crete for the evolution of the Eastern Mediterranean. *Geol. Soc. Lond., Spec. Pub.*, **17**, 499-516.

Harrison, T.M., Duncan, I. & McDougall, I. (1985): Diffusion of ^{40}Ar in biotite: temperature, pressure and compositional effects. *Geochim. Cosmochim. Acta*, **49**, 2461-2468.

Hurford, A.J. (1990): Standardization of fission track dating calibration: recommended by the Fission Track Working Group of the I.U.G.S. Subcommision on Geochronology. *Chem. Geol. Isot. Geosci. Sect.*, **80**, 171-178.

Hurford, A.J. & Green, P.F. (1983): The zeta age calibration of fission track dating. *Isot. Geosci.*, **1**, 285-317.

Hurford, A.J., Hunziker, J.C. & Stöckhert, B. (1991): Constraints on the late thermotectonic evolution of the western Alps: Evidence for episodic rapid uplift. *Tectonics*, **10**, 758-769.

Jolivet, L., Daniel, J.M., Truffert, C. & Goffé, B. (1994): Exhumation of deep crustal metamorphic rocks and crustal extension in arc and back-arc regions. *Lithos*, **33**, 3-30.

Kastens, K.A. (1991): Rate of outward growth of the Mediterranean Ridge accretionary complex. *Tectonophysics*, **199**, 25-50.

Kilias, A., Fassoulas, C. & Mountrakis, D. (1994): Tertiary extension of continental crust and uplift of Psiloritis metamorphic core complex in the central part of the Hellenic Arc (Crete, Greece). *Geol. Rundschau*, **83**, 417-430

Küster, M. & Stöckhert, B. (1995): The thermobarometric potential of fluid inclusions in high pressure metamorphic rocks - constraints on the exhumation path. *Bochumer geol. u. geotech. Arb.*, **44**, 91-98

Laslett, G.M., Green, P.F., Duddy, I.R. & Gleadow, A.J.W. (1987): Thermal annealing of fission tracks in apatite: 2. A quantitative analysis. *Chem. Geol. Isot. Geosci. Sect.*, **65**, 1-13.

Lihou, J., Hurford, A.J. & Carter, A. (1995): Preliminary fission-track ages on zircons and apatites from the Sardona unit, Glarus Alps, eastern Switzerland: late Miocene-Pliocene exhumation rates. *Schweiz. Mineral. Petrogr. Mitt.*, **75**, 177-186.

Meulenkamp, J.E., Van Der Zwaan, G.J. & Van Wamel, W.A. (1994): On Late Miocene to recent vertical motions in the Cretan segment of the Hellenic arc. *Tectonophysics*, **234**, 53-72.

Meulenkamp, J.E., Wortel, M.J.R., Van Wamel, W.A., Spakman, W. & Hoogerduyn Strating, E. (1988): On the Hellenic subduction zone and the geodynamic evolution of Crete since the late Middle Miocene. *Tectonophysics*, **146**, 203-215.

O'Sullivan, P.B. & Parrish, R.R. (1995): The importance of apatite composition and single-grain ages when interpreting fission track data from plutonic rocks: a case study from the Coast ranges, British Columbia. *Earth Planet. Sci. Lett.*, **132**, 213-224.

Postma, G., Fortuin, A.R., & Van Wamel, W.A. (1993): Basin-fill patterns controlled by tectonics and climate: the Neogene 'fore-arc' basins of eastern Crete as a case history. *Spec. Publ. Int. Assoc. Sediment.*, **20**, 335-362.
Robert, U. & Bonneau, M. (1982): Les basaltes des nappes du Pinde et d'Arvi (Crète) et leur signification dans l'évolution géodynamique de la Mediterranée orientale. *Ann. Géol. Pays Hellén.*, **31**, 373-408.
Royden, L.H. (1993a): The tectonic expression of slab pull at continental convergent boundaries. *Tectonics*, **12**, 303-325.
Royden, L.H. (1993b): Evolution of retreating subduction boundaries formed during continental collision. *Tectonics*, **12**, 629-638.
Seidel, E. (1971): Die Pindos-Serie in West-Kreta, auf der Insel Gavdos und im Kedros-Gebiet (Mittel-Kreta). *N. Jb. Geol. Paläont. Abh.*, **137**, 443-460.
Seidel, E., Kreuzer, H. & Harre, W. (1982): A Late Oligocene/Early Miocene High Pressure Belt in the External Hellenides. *Geol. Jb.*, **E23**, 165-206.
Seidel, E., Okrusch, M., Kreuzer, H., Raschka, H. & Harre, W. (1976): Eo-alpine metamorphism in the uppermost unit of the Cretan nappe system - petrology and geochronology - Part 1. The Léndas area (Asteroussia Mountains). *Contrib. Mineral. Petrol.*, **57**, 259-275.
Seidel, E., Okrusch, M., Kreuzer, H., Raschka, H. & Harre, W. (1981): Eo-alpine metamorphism in the uppermost unit of the Cretan nappe system - petrology and geochronology - Part 2. Synopsis of High-Temperature metamorphics and associated ophiolites . *Contrib. Mineral. Petrol.*, **76**, 351-361.
Seidel, E., Schliestedt, M., Kreuzer, H. & Harre, W. (1977): Metamorphic rocks of Late Jurassic age as components of the ophiolitic mélange on Gavdos and Crete (Greece). *Geol. Jb.*, **B28**, 3-21.
Theye T. & Seidel E. (1993): Uplift-related retrogression history of aragonite marble in Western Crete (Greece). *Contrib. Mineral. Petrol.*, **114**, 349-356.
Theye, T., Seidel, E. & Vidal, O. (1992): Carpholite, sudoite and chloritoid in low-grade high pressure metapelites from Crete and the Peloponnese, Greece. *Eur. J. Min.*, **4**, 487-507.
Van Dijk, J.P. & Scheepers, P.J.J. (1995): Neotectonic rotations in the Calabrian Arc; implications for a Pliocene-Recent geodynamic scenario for the Central Mediterranean. *Earth Sci. Rev.*, **39**, 207-246.
Vicente, J.C. (1970): Étude géologique de l'île de Gavdos (Grèce), la plus méridionale de l'Europe. *Bull. Soc. géol. France.*, **12**, 481-495.

POST-OROGENIC COOLING HISTORY OF EASTERN SOUTH AUSTRALIA FROM APATITE FT THERMOCHRONOLOGY

M.M. MITCHELL, B.P. KOHN AND D.A. FOSTER

Australian Geodynamics Cooperative Research Centre
School of Earth Sciences, La Trobe University, Bundoora, Victoria 3083, Australia

Abstract

An apatite fission track (FT) study of shallow borehole and outcrop samples from the Cambrian Arrowie Basin and Late Proterozoic Flinders Ranges sedimentary rocks, and the underlying Early to Mid Proterozoic basement rocks of the Curnamona Craton of eastern South Australia, yields ages of 200-350 Ma and mean track lengths of 11-14 μm. Forward modelling of the data suggests at least two periods of significant cooling subsequent to the extensive Early Palaeozoic Delamerian Orogeny. Regionally, these represent cooling since ~400 Ma from temperatures >110°C via an estimated minimum of ~2.5 km denudation, although local denudation may have attained 4-5 kms. Although the time-temperature cooling paths of individual samples may be poorly constrained and variations in the quality of data set exists, some regional coherence is apparent. Late Palaeozoic cooling, probably related to the Alice Springs Orogeny, is preserved ubiquitously throughout the region. The cooling affects of denudation related to this episode have not been reported in southern Australia and were previously believed to have been predominantly confined to central Australia and eastern Australia, where it is known as the Kanimblan Orogeny. A second, less well constrained period of regional cooling is suggested for the mid Cretaceous to Tertiary, probably in response to kilometre scale denudation. This cooling is of smaller magnitude and less well constrained in its timing, and as such is difficult to resolve in detail. This episode is believed to be associated with intraplate stresses related to the reconfiguration of the Australian plate and its margins during this time.

Keywords

Apatite FT thermochronology, Flinders Ranges, Arrowie Basin, Curnamona Craton, Australia.

1. Introduction

The Flinders Ranges are the most northerly part of an extensive arcuate-shaped province of deformed Late Proterozoic to Cambrian rocks collectively known as the Adelaide Fold Belt (Figs 1 and 2; Preiss, 1987). The Fold Belt was intensely deformed during the extensive Cambrian to early Ordovician Delamerian Orogeny as were the adjoining regions to the east; the Early to Mid Proterozoic Curnamona Craton, and the Cambrian Arrowie Basin (Fig. 2). The structural architecture of the region is interpreted as entirely Late Proterozoic to Ordovician with the major phase of tectonic development during the Delamerian Orogeny when metamorphic grades attained

P. Van den haute and F. De Corte (eds.), Advances in Fission - Track Geochronology, 207-224.

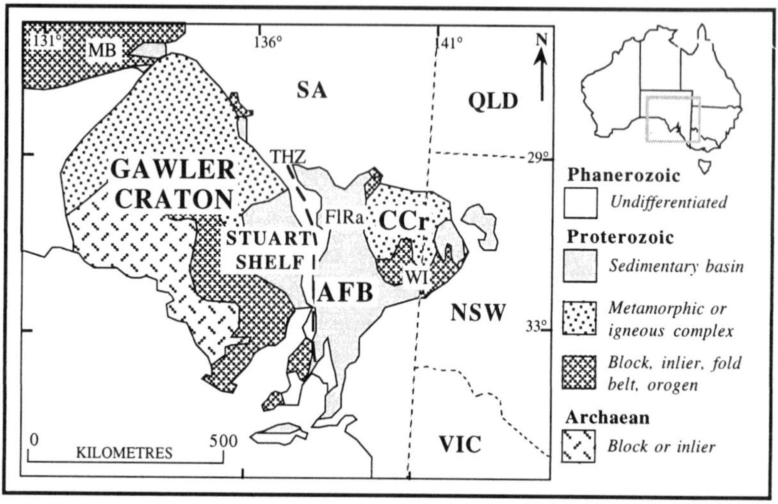

Figure 1. Map of Precambrian orogens and basins in South Australia (SA) and far western New South Wales (NSW) (modified after Parker, 1993). Abbreviations are as follows: AFB— Adelaide Fold Belt; CCr— Curnamona Craton; FlRa— Flinders Ranges; WI— Willyama Inliers; MB— Musgrave Block; THZ— Torrens Hinge Zone (dashed line); VIC—Victoria; QLD—Queensland.

greenschist facies and locally up to migmatite grade (eg. Sandiford et al., 1995). This is with the exception of minor Mesozoic and Cainozoic rejuvenation, suggested to account for stratigraphic and structural relationships, and the present-day elevated topography of the region (Preiss, 1987; Bourman and Lindsay, 1989; Flottmann and Sandiford, 1995; Teasdale et al., 1995), the area has been described as essentially intracratonic and 'stable' with no significant uplift and erosion since the Delamerian Orogeny (eg. Preiss, 1987; Parker and Rankin, 1993).

In the study area, thin and sporadic Phanerozoic cover rocks provide little stratigraphic constraint on the tectonic history of the region. In such cases where large time gaps exist in the geological record, apatite FT analysis has proved a useful tool for the elucidation of 'missing', otherwise undetectable, cooling or denudation events (eg. Fitzgerald, 1994; Foster and Gleadow, 1996; O'Sullivan et al., 1996). The purpose of this study is to investigate the Phanerozoic cooling history of eastern South Australia, which is largely unconstrained.

2. Apatite FT Thermochronology

The damage trail of spontaneous decay of ^{238}U may be chemically etched to reveal a 'fission track' (Fleischer et al., 1975). In typical compositions of apatite, fission tracks are preserved at temperatures below ~125°C over geological time periods, thus providing an ideal technique for the examination of the thermal history of upper crustal rocks (eg. Naeser, 1979; Gleadow et al., 1986; Green et al., 1989; Brown et al., 1994). The reduction in length of fission tracks (annealing) subjected to temperatures of ~70 to 125°C over geological time periods has been examined in both natural borehole samples, and quantitative laboratory studies (Gleadow and Duddy, 1981; Gleadow et al., 1983; Green et al., 1986, 1989).

In addition to temperature and time, apatite chlorine composition has been found to be an important factor controlling the thermal annealing characteristics of fission tracks (eg. Green et al, 1985, 1989). Green et al., (1985) showed that in a sedimentary borehole sample currently at ~90°C, individual grains displayed a range of FT age from the stratigraphic age to zero (total fission track annealing). The FT age variation in response to temperature was found to correlate with chlorine content of the individual apatite crystals. Chemical analysis of single grain ages is therefore a useful tool in determining maximum paleotemperatures and complex thermal histories for partially annealed samples (eg. O'Sullivan and Parrish, 1995; Mitchell, in press). The Durango apatite standard on which much of the modelling of FT annealing is based, has a chlorine content of ~0.4 wt% (Young et al., 1969). Values greater than this are generally uncommon in granitic basement terrains (eg. Foster and Gleadow, 1996).

Thermal history analysis based on forward modelling of fission track parameters (confined track length and fission track age) ie. the thermal history below ~110°C, was mathematically described by Laslett et al. (1987) based on experimental analysis of Durango apatite. A computer model based on the work developed by Gallagher (1995), Monte Trax, which employs a Monte Carlo type approach with a genetic algorithm, has been used throughout this study.

3. Regional Setting

3.1. CURNAMONA CRATON

The Curnamona Craton comprises the outcropping Willyama, Mt Painter and Mt Babbage Inliers and subsurface cratonic rocks forming an extensive province of Early to Mid Proterozoic rocks to the east of the Adelaide Fold Belt (Fig. 2). The craton was also intersected in shallow boreholes below a mantle of Cambrian to Cainozoic sediments. The craton is believed to have formed a relatively stable region during the Late Proterozoic and Cambrian, and appears to have remained relatively undeformed, perhaps acting as a rigid crustal block, during the Delamerian Orogeny (Parker, 1993).

3.2. ADELAIDE FOLD BELT—FLINDERS RANGES

The Late Proterozoic sedimentary sequence (locally known as the Adelaidean sequence; Preiss, 1987) was deposited in a complex system of partly intracratonic rift and subsident basins with minor associated igneous activity (Preiss, 1993). The Adelaide Fold Belt consists largely of folded and faulted Adelaidean rocks, formed in a zone of west to northwest convergent deformation against the Australian Craton (Jenkins and Sandiford, 1992). The Fold Belt is separated from the Gawler Craton to the west, by the Torrens Hinge Zone and marks a transition from flat-lying thin sediments of the Stuart Shelf, to the intensely deformed sediments of the Fold Belt (Fig.

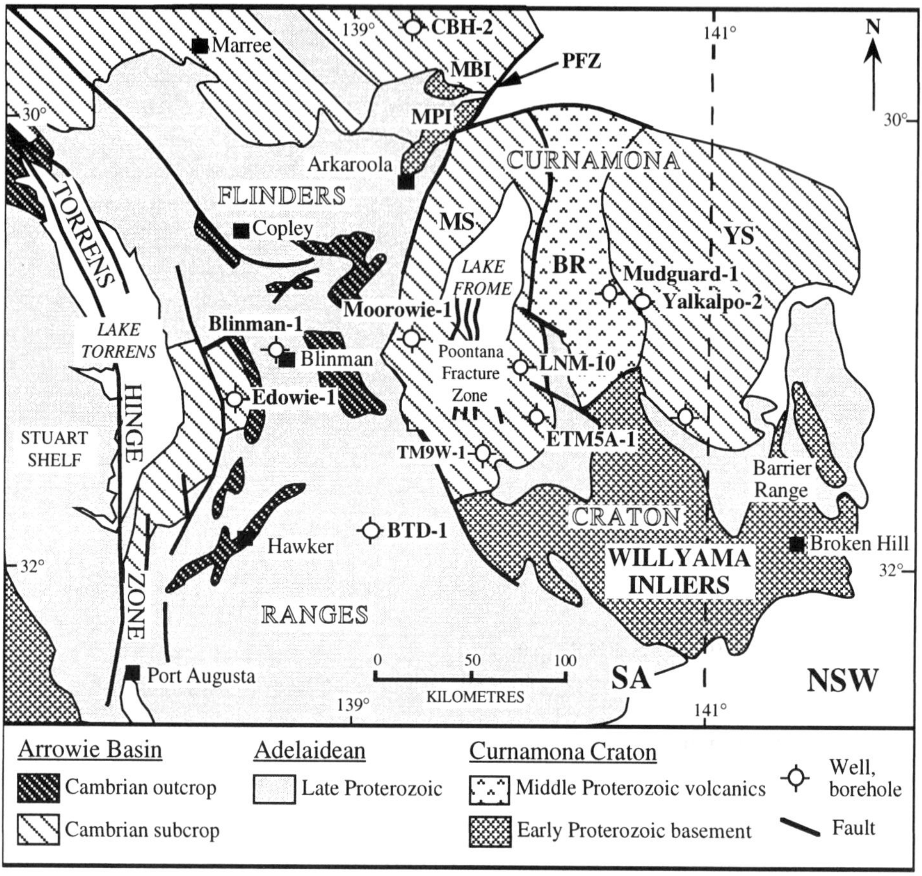

Figure 2. Geologic sketch map of the northern Adelaide Fold Belt, Arrowie Basin and Curnamona Craton (modified after Gravestock and Hibburt, 1991). The stratigraphic divisions of the Precambrian, Cambrian Arrowie Basin and structural elements are shown, as well as the boreholes examined. Abbreviations as follows: MS— Moorowie Syncline; BR— Benagerie Ridge; YS— Yalkalpo Syncline; MPI— Mount Painter Inlier; MBI— Mount Babbage Inlier; PFZ— Paralana Fault Zone.

2). The Torrens Hinge Zone is a major crustal feature which has controlled deformation and sedimentation at various times during the Late Proterozoic to Cainozoic (eg. Thomson, 1969; Parker, 1993; Flottmann and Sandiford, 1995).

A pre-Cretaceous erosion surface in the northern Flinders Ranges (Sheard and Flint, 1992; Twidale and Campbell, 1993), sedimentary depositional patterns (Alley and Benbow, 1995) and structural relationships in adjoining and overlying Cainozoic basins (Bourman and Lindsay, 1989; Flottmann and Sandiford, 1995; Flottmann and Cockshell, 1996) indicate minor Phanerozoic tectonic activity in the region, despite its intracratonic location. However, neither the timing nor the amount of erosion during the post-orogenic period of the Phanerozoic can be accurately constrained from the previous evidence.

3.3. ARROWIE BASIN

Sediments of the Cambrian Arrowie Basin were deposited initially on a shallow marine shelf, however, tectonic activity restricted open marine access and resulted in redbed and evaporite deposition (Gravestock and Hibburt, 1991). The present basin margins result from the effects of the Delamerian Orogeny and subsequent denudation, and differ considerably from the original basin geometry (Fig. 2; Gravestock and Hibburt, 1991). The Arrowie Basin can be separated into four structural and geographic provinces: 1) thin flat-lying deposits of the Stuart Shelf, 2) structurally complex fault slices preserved in the Torrens Hinge Zone, 3) thick and dispersed outcrops in the Flinders Ranges, and 4) large areas of preserved Cambrian sediments in the depocentres (Moorowie and Yalkalpo Synclines) overlying the Curnamona Craton (Fig. 2; Gravestock and Hibburt, 1991). Although these depocentres were probably connected during the Early Cambrian, they are now separated by the Benagerie Ridge.

4. Analytical Methods

Samples were prepared following the procedures outlined in Moore et al. (1986) and Green et al. (1986) and analysed using the external detector method (eg. Gleadow, 1981) with the Corning 5 standard glass dosimeter (Bellemans et al., 1995). Samples were irradiated in the X-7 facility of the HIFAR reactor, Lucas Heights, NSW, Australia. Fission track ages were calculated using the zeta calibration method and errors using the method of Green (1981) and Hurford and Green (1982). The $P\chi^2$ statistic measures the probability of the age grains belonging to a single population (Galbraith, 1981). Where a significant spread in age is found ($P\chi^2 <5\%$) more than one age population may be present, therefore the conventional analysis (Green, 1981) based on a Poissonian variation is not valid, and the central age is reported (Galbraith and Laslett, 1993). Multiple age populations may be related to either variations in detrital provenance for sedimentary rocks, and/or variable annealing characteristics within a single sample due to compositional populations. In such cases, quantitative modelling using Monte Trax cannot be employed as this approach assumes a single age population. However, where a sample fails the $P\chi^2$ test because of the presence of a relatively small number of outliers, useful information may still be obtained by modelling the data.

Samples were collected over an extensive geographic area as well as shallow boreholes. However, lithologies are predominantly quartzite resulting in poor apatite yields. A total of 81 samples were analysed, however, only 37 samples yielded data of sufficient quality (Tables 1 and 2).

5. Apatite FT Results

Two adjoining regions were examined in this study: the Flinders Ranges section of the Adelaide Fold Belt, and the Curnamona Craton (Figs 2 and 3). The Arrowie Basin

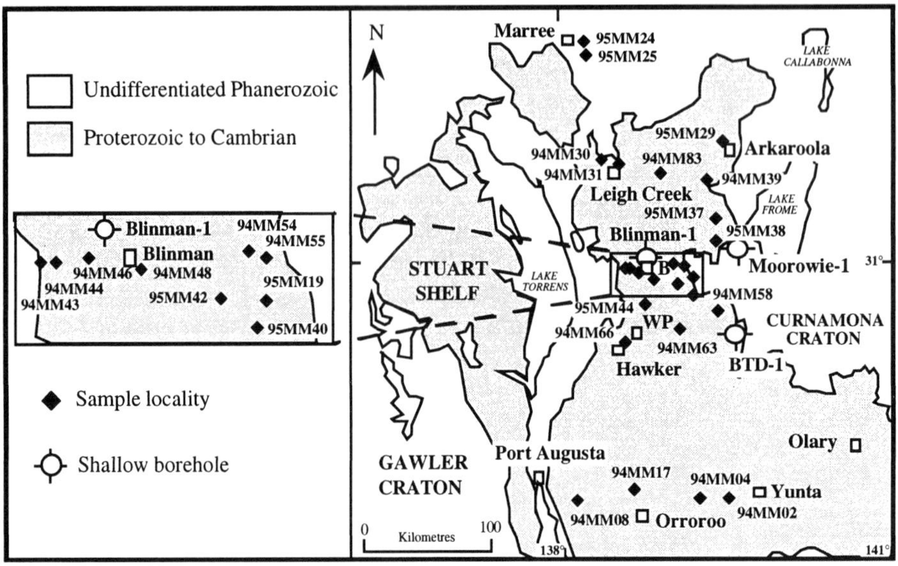

Figure 3. Regional map of the northern Adelaide Fold Belt showing apatite FT sample localities for outcrop and shallow boreholes (modified after Preiss, 1987). The inset shows the Blinman area where sample density is higher.

overlies each of these regions, but was primarily examined as part of the Phanerozoic cover of the Curnamona Craton and therefore the results for these two sections are discussed together prior to the Adelaide Fold Belt results. Ages and mean lengths are quoted at $\pm 1\sigma$ throughout the text. Apatite compositions were found to be predominantly Fl-apatites with a range of values between 0-0.3 wt% Cl with the exception of one sample (95MM19 discussed in detail below).

5.1. CURNAMONA CRATON AND ARROWIE BASIN

Seven boreholes (Fig. 4) were examined, but only Moorowie-1 (Fig. 5) reached a depth >800 m. The present-day geothermal gradient at Moorowie-1 is ~32°C/km calculated from bottom hole temperature (BHT) data (Delhi Petroleum, 1984). Limited bore hole temperature data is available for the other 6 boreholes. However, based on geothermal gradients of 20-30°C/km observed regionally, all samples other than those from Moorowie-1, are considered to be at present-day temperatures significantly <70°C. Samples were collected from the Cambrian and Adelaidean metasedimentary units, as well as the underlying Early to Mid Proterozoic Curnamona Craton basement (Fig. 2).

A plot of age versus mean track length for shallow borehole samples from the Arrowie Basin and Curnamona Craton shows a broad, poorly defined trend of decreasing age and length from ~250 to 330 Ma and mean lengths of ~13 μm to ages <250 Ma with shorter lengths (Fig. 4). The spread in data is consistent with a protracted cooling history with samples having the observed range of apatite Cl compositions (although this is generally not a broad range) having spent significant time at temperatures where partial track annealing is optimised ie. ~60-90°C. Only one FT

Table 1. Apatite FT results from shallow borehole samples for the Curnamona Craton and Arrowie Basin.

Borehole Sample #	Depth (m KB) Stratigraphic Age	Locality Lat (S) Long (E)	Standard track density ($x10^6cm^{-2}$)	Fossil track density ($x10^6cm^{-2}$)	Induced track density ($x10^6cm^{-2}$)	$P\chi^2$ (%)	Apatite FT Age* (Ma)	Mean Track Length (μm)	SD (μm)
CBH-2									
95MM02	727	29°36'36.0"	1.349	2.535	2.496	<1	249±20	11.97±0.13	1.27
	Cambrian	139°28'48.0"	(4212)	(1691)	(1665)		(20)	(101)	
Mudguard-1									
95MM03	191-196	30°27'39.6"	1.349	1.160	0.978	5.8	296±16	13.33±0.13	1.28
	Mid Prot.	140°16'55.2"	(4212)	(942)	(794)		(23)	(100)	
Yalkalpo-2									
95MM05	507	30°30'36.0"	1.349	1.112	1.069	6.4	260±14	13.18±0.15	1.67
	Cambrian	140°22'48.0"	(4212)	(890)	(856)		(27)	(123)	
ETM5A-1									
95MM10	483	31°25'39.0"	1.349	1.405	1.606	48.8	220±43	13.17±0.53	1.68
	Mid Prot.	140°14'27.9"	(4212)	(49)	(56)		(12)	(10)	
TM9W-1									
95MM11	796-802	31°41'10.7"	1.349	2.160	2.757	<1	201±12	12.60±0.14	1.45
	Early -Mid Prot.	139°55'10.3"	(4212)	(1999)	(2552)		(20)	(106)	
Edowie-1									
95MM13	800	31°10'28.2"	1.349	1.853	1.410	1.2	329±20	12.93±0.13	1.32
	Cambrian	138°15'18.0"	(4212)	(1268)	(965)		(20)	(101)	
LNM-10									
95MM14	229	31°07'06.0"	1.349	1.533	1.494	0.7	258±15	13.17±0.15	1.52
	Adelaidean	140°10'05.9"	(4212)	(1287)	(1254)		(20)	(100)	
Moorowie-1									
95MM17	1852	30°58'07.6"	1.349	0.729	0.742	<1	212±120	-	-
	Cambrian	139°23'30.0"	(4212)	(54)	(55)		(2)		
95MM18B	3030		1.349	0.118	2.885	<1	19±11	8.42±0.95	2.52
	Adelaidean		(4212)	(33)	(806)		(21)	(7)	

All elevations are above mean sea level (ASL) and depths are converted to elevations using the elevation of the ground surface at borehole site and the sample depth with respect to KB (Kelly Bushing-drilling rig floor). Numbers of tracks counted, grains analysed and confined lengths measured are shown in brackets.

*Central age used where pooled data fail $P\chi^2$ test at 5%. Fission track data is quoted with ±1σ error. Ages determined using CN5 dosimeter glass (Bellemans et al., 1995) and calculated using a zeta of 378.5±5.0

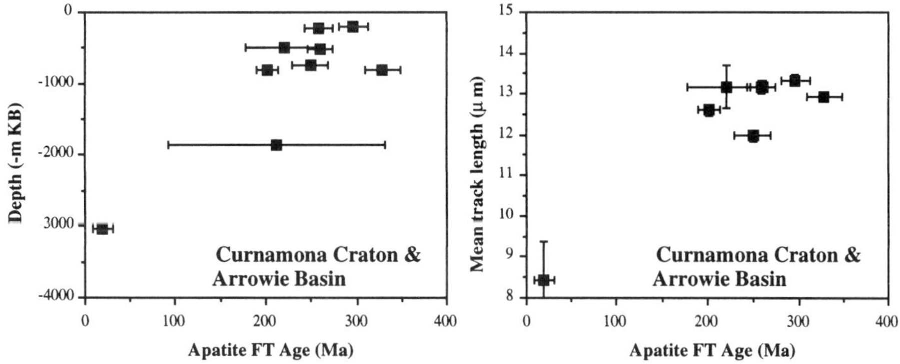

Figure 4. Plots of apatite FT age (Ma) (±1σ) against depth (-m KB), and mean track length (μm) (±1σ) for shallow borehole samples from the Curnamona Craton and Arrowie Basin. See Table 1 for individual sample details. Note that sample 95MM18B from Moorowie-1 with a mean track length of ~8.5 μm, is severely annealed by the present-day down-hole temperatures (see Figure 5).

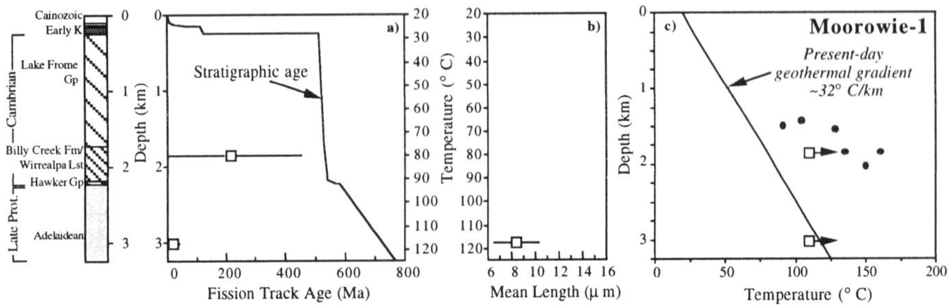

Figure 5. Plots of apatite FT parameters and maximum paleotemperatures from two samples from Moorowie-1. A representative stratigraphic column is shown at the left. Apatite FT parameters (a) FT age, and (b) mean track length are plotted against depth and present-day temperature. Plot (c) shows the present-day geothermal gradient in the well and paleotemperature estimates for apatite FT (open squares) and VR data (black dots; Delhi Petroleum, 1984) plotted against depth.

sample age in excess of ~300 Ma was observed in the borehole study, that from Edowie-1 (95MM13) in the Flinders Ranges near Blinman (Fig. 2). This sample shows a mixture of single grain ages (Fig. 7) and has a moderately reduced mean length (~12.9±0.1 μm).

The sample from well CBH-2 (95MM02), to the north of the Mt Painter Inlier, has an age of 249±20 Ma and a considerably reduced mean track length of ~12 μm (Table 1). The Mudguard-1 sample (95MM03) from the Benagerie Ridge, has the oldest age (296±16 Ma) and longest length (13.3±0.1 μm) for the Curnamona Craton samples. Results from Yalkalpo-2 (95MM05) on the Yalkalpo Syncline, are similar to CBH-2, but with a longer mean length of ~13.2 μm. TM9W-1 (95MM11), located towards the south of the Moorowie Syncline, shows the youngest FT age from the Curnamona Craton (201±12 Ma), however, the mean track length is intermediate between that of the other samples (~12.6 μm; Table 1).

Moorowie-1 is located to the east of the Flinders Ranges and intersected a thick section of the Cambrian Arrowie Basin sequence (Fig. 5a). Vitrinite reflectance data ranged from 0.59 to 1.34 Ro% over the depth interval 1424 to 2024 m (Delhi Petroleum, 1984). The paucity of VR measurements (26 analyses from 6 samples) places an uncertainty on the reliability of the results. However, when examined together, the results clearly show that this section of the well has experienced elevated paleotemperatures. Maximum paleotemperature estimates based on the VR results (Burnham and Sweeney, 1989; Duddy et al., 1991; Bray et al., 1992) suggest that this well section has experienced temperatures of ~90 to 160°C (Fig. 5c). Unfortunately, the apatite FT results from this well are poor and inconclusive due to low apatite yields (95MM17) and the relatively high present-day temperatures (95MM18B; Fig. 5).

5.2. FLINDERS RANGES

Twenty six outcrop samples are reported from throughout the Flinders Ranges and provide a sparse coverage of an extensive area from Orroroo, in the south, to near Marree, in the north (Fig. 3; Table 2). Results from two shallow borehole samples

Table 2. Apatite FT results from outcrop and shallow borehole samples from the Flinders Ranges.

Borehole Sample #	Elevation (mASL) or Depth (-m KB) Stratigraphic age	Locality Lat (S) Long (E)	Standard track density ($x10^6 cm^{-2}$)	Fossil track density ($x10^6 cm^{-2}$)	Induced track density ($x10^6 cm^{-2}$)	$P\chi^2$ (%)	Apatite FT age (Ma)	Mean track length (μm)	SD (μm)
94MM02	~375	32°43'47.0"	1.269	3.417	2.771	1.5	292±25	12.52±0.14	1.42
	Adelaidean	139°08'54.9"	(4686)	(608)	(493)		(20)	(100)	
94MM04	~400	32°43'37.1"	1.287	3.066	2.145	14.7	339±21	11.9±0.28	1.75
	Adelaidean	138°58'22.0"	(4686)	(689)	(482)		(28)	(39)	
94MM08	300	32°37'10.5"	1.407	3.158	3.679	0.2	228±18	11.94±0.22	1.92
	Adelaidean	138°1'7.4"	(5491)	(764)	(890)		(23)	(79)	
94MM17	560	32°34'01.8"	1.407	2.508	2.766	0.1	237±18	12.78±0.16	1.59
	Adelaidean	138°33'49.2"	(5491)	(779)	(859)		(22)	(100)	
94MM30	300	30°34'28.8"	1.377	2.833	2.063	89.8	348±25	12.76±0.23	1.61
	Adelaidean	138°22'59.8"	(4686)	(504)	(367)		(20)	(48)	
94MM31	220	30°33'31.7"	1.407	2.879	3.592	<1	219±20	13.15±0.13	1.35
	E. Cretaceous	138°25'50.2"	(5491)	(1047)	(1306)		(21)	(109)	
94MM39	280	30°33'23.7"	1.395	2.049	1.637	63.2	322±17	13.05±0.13	1.28
	Adelaidean	139°10'6.9"	(4686)	(1004)	(802)		(17)	(102)	
94MM43	240	31°8'7.3"	1.431	3.361	3.354	0.2	273±21	11.43±0.16	1.66
	Adelaidean	138°31'56.5"	(4686)	(876)	(874)		(22)	(103)	
94MM44	340	31°7'27.9"	1.407	2.589	3.209	<1	200±18	12.02±0.16	1.71
	Adelaidean	138°33'58.7"	(5491)	(1917)	(2376)		(22)	(111)	
94MM46	610	31°06'40.8"	1.248	3.648	2.607	5.8	322±14	12.63±0.16	1.58
	Adelaidean	138°39'32.9"	(4559)	(1577)	(1127)		(20)	(100)	
94MM48	560	31°06'56.0"	1.407	2.842	2.651	1.9	283±20	12.64±0.16	1.59
	Adelaidean	138°41'26.1"	(5491)	(745)	(695)		(20)	(100)	
94MM54	330	31°05'48.3"	1.407	2.288	2.733	4.9	219±13	12.51±0.22	1.59
	Cambrian	138°53'33.9"	(5491)	(591)	(706)		(20)	(54)	
94MM55	260	31°07'31.9"	1.281	2.603	2.021	<1	309±24	12.95±0.14	1.36
	Cambrian	138°57'5.7"	(4559)	(1377)	(1069)		(20)	(101)	
94MM58	200	31°28'54.1"	1.314	2.831	2.237	<1	308±24	12.41±0.16	1.64
	Adelaidean	139°08'10.4"	(4559)	(963)	(761)		(20)	(100)	
94MM63	400	31°39'27.4"	1.364	2.029	1.805	99.4	284±31	12.78±0.23	1.53
	Adelaidean	138°47'8.6"	(4559)	(190)	(169)		(12)	(44)	
94MM66	420	31°40'15.9"	1.381	3.330	2.588	95	328±19	12.86±0.18	1.58
	Adelaidean	138°35'48.4"	(4559)	(795)	(618)		(25)	(75)	
94MM83	550	30°33'18.6"	1.407	2.118	3.086	<1	215±25	12.93±0.21	1.63
	Adelaidean	138°49'18.7"	(5491)	(823)	(1199)		(20)	(53)	
95MM19	~260	31°16'52.1"	1.267	0.965	0.624	1.4	351±30	14.03±0.15	1.5
	Cambrian	138°56'54.2"	(3955)	(706)	(457)		(21)	(107)	
95MM24	35	29°43'13.9"	1.267	2.879	2.400	19.5	281±15	10.24±0.19	1.88
	Adelaidean	138°07'26.9"	(3955)	(865)	(721)		(21)	(102)	
95MM25	40	29°46'21.1"	1.267	1.398	1.340	99.5	246±21	12.53±0.17	1.63
	Adelaidean	138°08'38.1"	(3955)	(288)	(276)		(21)	(90)	
95MM29	380	30°18'01.8"	1.267	1.204	1.206	0.1	236±14	12.3±0.2	1.75
	Adelaidean	139°18'40.8"	(3955)	(1582)	(1585)		(20)	(118)	
95MM37	130	30°51'42.6"	1.267	2.087	2.203	19.5	223±14	12.02±0.16	1.58
	Adelaidean	139°14'22.9"	(3955)	(557)	(588)		(29)	(100)	
95MM38	100	30°57'17.6"	1.267	2.674	2.474	7.0	254±13	12.6±0.14	1.41
	Adelaidean	139°14'03.2"	(3955)	(962)	(890)		(27)	(101)	
95MM40	270	31°20'33.4"	1.267	0.664	0.477	40	326±39	12.11±0.51	2.1
	Adelaidean	138°54'26.3"	(3955)	(170)	(122)		(19)	(17)	
95MM42	320	31°16'00"	1.267	2.852	2.252	29.9	297±17	11.84±0.18	1.8
	Adelaidean	138°52'37.8"	(3955)	(779)	(615)		(27)	(100)	
95MM44	345	31°28'37.9"	0.789	3.778	1.873	9.0	294±18	12.17±0.16	1.56
	Adelaidean	138°41'12.7"	(3078)	(964)	(478)		(24)	(100)	
BTD-1	-289	32°48'41.2"	1.466	1.143	1.053	1.8	293±16	12.89±0.12	1.27
95MM09	(KB -540 m)	139°12'36.4"	(2859)	(1626)	(1498)		(20)	(104)	
	Adelaidean								
Blinman-1	472	31°05'11.0"	1.349	1.912	1.570	4.8	311±23	11.02±0.19	1.86
95MM15	(KB -178 m)	138°38'36.0"	(4212)	(795)	(653)		(19)	(101)	
	Adelaidean								

All elevations are above mean sea level (ASL) and depths are converted to elevations using the elevation of the ground surface at borehole site and the sample depth with respect to KB (Kelly Bushing-drilling rig floor). Numbers of tracks counted, grains analysed and confined lengths measured are shown in brackets.

*Central age used where pooled data fail $P\chi^2$ test at 5%. Fission track data is quoted with $\pm 1\sigma$ error. Ages determined using CN5 dosimeter glass (Bellemans et al., 1995) and calculated using a zeta of 378.5±5.0

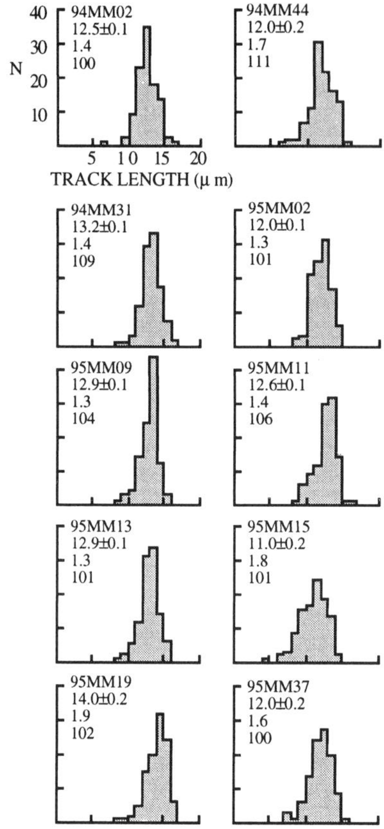

Figure 6. Confined track length histograms for several representative samples. Sample number, mean track length (μm) and standard error, standard deviation, and number of measurements are shown in the upper left corner of each plot.

(Blinman-1 and BTD-1) which penetrated the Adelaidean sequence are also reported (Table 2).

The majority of outcrop samples analysed were from the Adelaidean sequence, however, several Cambrian samples were analysed from the area to the east of Blinman (Fig. 3; Table 2). Apatite FT ages range from 200±18 to 351±30 Ma and mean lengths from 10.2±0.2 to 14.0±0.2 μm, but generally fall between 12 to 13 μm (Table 2). The younger FT ages are 200-250 Ma (eg. 94MM44) and several of these samples also contain Cretaceous to Palaeogene single grain ages (Fig. 7).

Well BTD-1 was spudded on an outcrop of Adelaidean metasediments east of Hawker (Fig. 3); one sample (95MM09) from this well yielded results within the range of values observed elsewhere in outcrop throughout the region (Table 2). The Blinman-1 sample (95MM15) is consistent with the FT ages of outcrop samples from Blinman which are at some of the highest elevations (~600-650 m ASL), and yield FT ages of 283±20 to 322±14 and lengths of ~12.2±0.2 μm. However, the borehole sample has a significantly reduced mean length of 11.0±0.2 μm.

In a plot of FT age versus mean length, all but two samples cluster in one region (Fig. 8). The sample from the Cambrian Ten Mile Creek Tuff (95MM19) records an apatite FT age of 351±30 Ma with a long mean length of 14.0±0.2 μm (Table 2). While the FT age of this sample is within error of many of the other samples, the mean track length is significantly longer than any other sample from the region. Further, this sample has an atypical apatite Cl content with values of 0.5 to 0.9 wt% Cl.Together with the long mean track length this indicates that the sample preserves a single episode of cooling with minimal subsequent annealing. The age of 351 Ma therefore indicates a Late Palaeozoic time of cooling for this sample and probably all other samples as well. The other atypical sample (Fig. 8; 95MM24) has a reduced mean length of 10.2±0.2 μm and is from an isolated outcrop to the south of Marree (Fig. 3). The reduced track length is consistent with thermal annealing resulting from localised hydrothermal activity indicated by the presence of quartz veining at the sample locality (Table 2). However, the age is older than

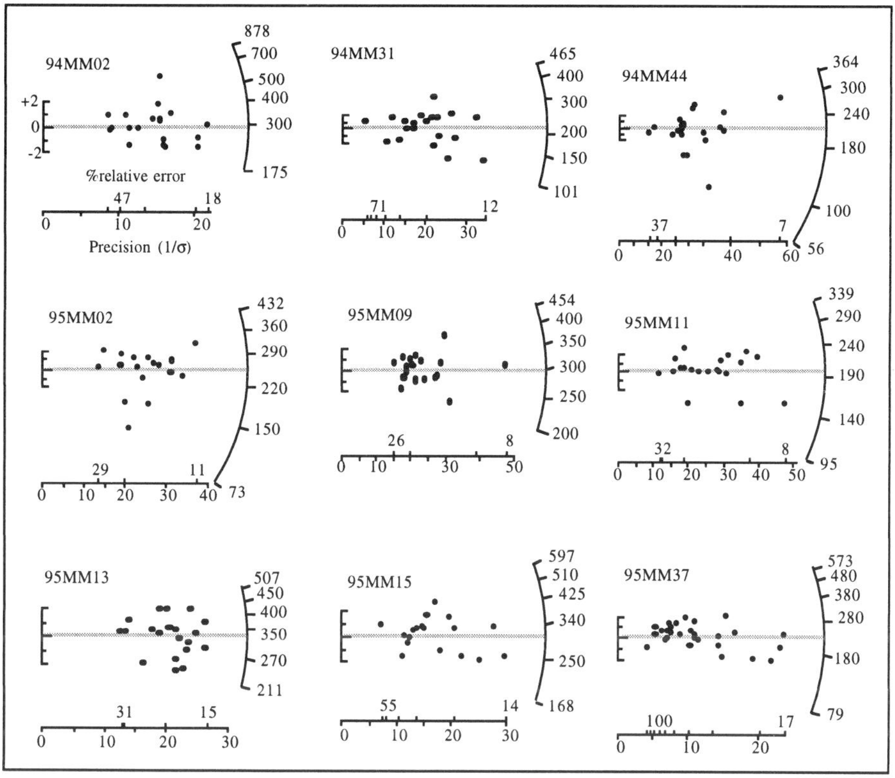

Figure 7. Representative plots of single grain ages on radial plots showing the precision in age on the X-axis ie. the most precise grains plot the greatest distance towards the right (Galbraith, 1990). Each estimate has a unit standard error on the Y-axis and the age may be determined by extrapolating a line from the origin of this axis through the point plotted to the arc of time (Ma). The central age is shown as a horizontal grey line from the Y-axis origin with $\pm 2\sigma$ error shown (values given in Tables 1 and 2).

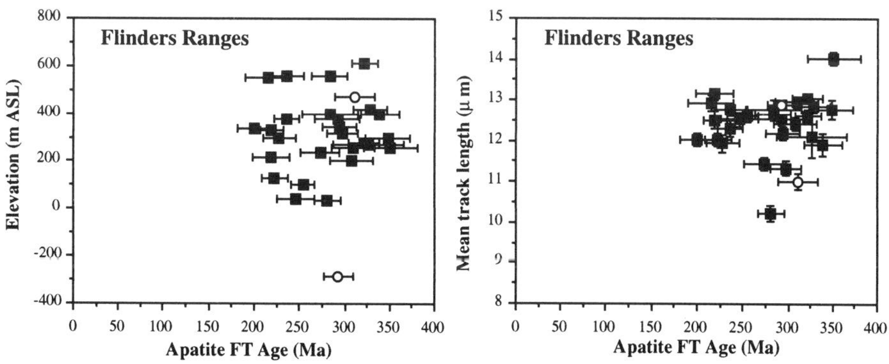

Figure 8. Plots of apatite FT age (Ma) against elevation (m above sea level-ASL), and mean track length (μm) for outcrop samples (black squares) and shallow borehole samples (open circles) from the Flinders Ranges. See Table 2 for individual sample details.

anticipated given the track length reduction and the significance of the results from this locality are unclear.

There appears to be no correlation between apatite FT parameters and elevation; samples from various elevations show (within error) concordant apatite FT ages and lengths (Fig. 8). Further, the shallow borehole sample from BTD-1 (95MM09) whilst lying at a distinctly lower elevation (-289 m ASL), yielded FT parameters consistent with those observed in the outcrop samples.

6. Discussion

Qualitative interpretation and modelling of apatite FT results (discussed below) from the Flinders Ranges and Curnamona Craton suggest that both regions have experienced two periods of post-Delamerian Orogeny cooling (Fig. 9):

1) in the Late Palaeozoic from temperatures >110°C, and

2) less well constrained mid Cretaceous through Tertiary cooling.

Widely dispersed geographic localities (ie. Marree to Orroroo) display the effects of both cooling episodes. Several best fit models are shown for individual samples in Figure 9. The results show cooling from temperatures >110°C in the Late Palaeozoic, followed by a protracted period at moderate to low temperatures until cooling below ~60°C in the mid Cretaceous through Tertiary. Local variations exist in both the degree and magnitude of cooling and are subject to a degree of uncertainty. The timing of cooling for both episodes is valid only for well constrained samples and some results are consistent with thermal history models of slow protracted cooling from the mid Palaeozoic to Recent. It is likely that some of the variations observed are related to a range of apatite Cl contents (Mitchell, submitted).

6.1. LATE PALAEOZOIC COOLING

Modelling of the apatite FT results indicates a regionally significant episode of Late Palaeozoic cooling in addition to any Early Palaeozoic post-Delamerian Orogeny cooling that occurred from temperatures higher than apatite FT annealing. The timing of cooling in the Late Palaeozoic is variable throughout the region with some samples indicating relatively rapid cooling of 30-50°C between 380 to 270 Ma while other samples suggest a more protracted cooling initiated in Devonian time with final cooling below ~60°C in the mid Cretaceous. The nature of the apatite FT data and the relative consistency of the spread of data in Figures 4 and 8 suggests that the entire region has experienced a similar period of cooling and that the samples that record clear evidence of cooling in the Late Palaeozoic are perhaps more representative. In particular sample 95MM19 shows definitive evidence of Late Palaeozoic cooling from elevated temperatures.

The amount of rock uplift and denudation during Late Palaeozoic cooling cannot be accurately constrained as poor apatite yields from samples collected in vertical profile means that a paleogeothermal gradient immediately prior to Late Palaeozoic

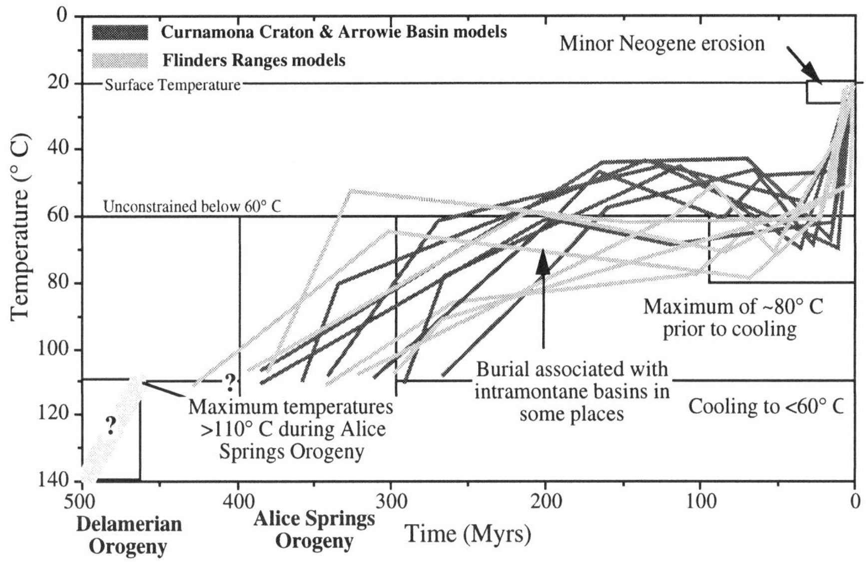

Figure 9. Modelled apatite FT thermal histories from the Flinders Ranges, Arrowie Basin and Curnamona Craton. The boxed areas are regions in which modelling suggests that FT samples have resided, and several of the best fit thermal histories (shaded lines) predicted by modelling for individual samples are shown. The stippled area represents the apatite fission track partial annealing zone for Durango apatite (Green et al., 1986, 1989). The following assumptions have been made in deriving the t-T paths shown: 1) all samples have cooled below ~110°C during the Late Palaeozoic cooling episode, 2) all samples resided at temperatures >~50°C prior to the second late Mesozoic to Tertiary cooling episode, and 3) a mean annual surface temperature of ~20°C has been assumed throughout.

cooling cannot be estimated using the FT data. However, cooling is believed to have been facilitated through regional rock uplift and denudation, with a minimum of ~3 to 4.5 km (assuming a minimum of ~90°C cooling, and a paleogeothermal gradient of ~20-30°C/km) of section being eroded throughout the Late Palaeozoic to Recent. Although minor variations in the degree of cooling are observed throughout the study area and not every sample records the same thermal history, cooling is apparently not related to either mapped structures, or topographic elevation, and the region appears to have behaved fairly coherently with no significant variation from north to south.

The Late Palaeozoic cooling is believed to be related to denudation following the Alice Springs Orogeny which has also been recognised in central Australia (Wells et al., 1970; Shaw et al., 1991; Tingate, 1990; Gravestock and Flint, 1995; Dunlap and Teyssier, 1995; Spikings et al., 1996). The effects of this orogenic event have not been previously documented in the Adelaide Fold Belt or Curnamona Craton, however, a palaeomagnetic study of hydrothermal breccias and apatite FT results from the Mt Painter Inlier suggests that deformation and tectonic activity occurred at this time (Idnurm and Heinrich, 1993; Foster et al., 1994; Mitchell et al., 1996).

6.2. MID CRETACEOUS THROUGH TERTIARY COOLING

Forward modelling also suggests that a second phase of cooling is recorded by the FT data throughout the region. However, it is of lesser magnitude than the Late Palaeozoic cooling and thus poorly constrained. Most samples have few tracks >14 μm, indicating that final cooling to temperatures <60°C was relatively recent. Some Cretaceous to Palaeogene apatite FT single grain ages from samples throughout the study area suggest that in some places final cooling below ~60°C may have occurred in the Late Cretaceous to Palaeogene, or more recently (Fig. 7; eg. 94MM44, 95MM02 and 95MM11). Modelling of the FT parameters, specifically the track length distribution, from TM9W-1 (95MM11; Fig. 6) suggests that this sample cooled from paleotemperatures of ~60-70°C some time in the Palaeogene to early Neogene. This borehole lies directly to the south of the Poontana Fracture Zone (Fig. 2), a northerly trending wrench fault complex (Gravestock and Hibburt, 1991) which may have provided a site for preferential denudation, or a conduit for hot fluids as has been inferred along the Paralana Fault zone, which marks the eastern margin of the Mt Painter Inlier (Fig. 2). Furthermore, along the Paralana Fault zone, apatite FT results suggested Palaeogene to early Neogene cooling as a result of erosion of the upper section of a hydrothermal system (Foster et al., 1994; Mitchell et al., 1996). Extremely variable temperature gradients in this area, both vertically and laterally (Mildren and Sandiford, 1995), perhaps driven by the unusually high radiogenic component of the locally shallow Early to Mid Proterozoic basement (Idnurm and Heinrich, 1993; Sandiford, 1996), may have also influenced the variable cooling signatures observed.

Late Cretaceous to mid Tertiary cooling was also reported for the Eromanga Basin, located inland to the northeast of the Flinders Ranges and partly overlying the Curnamona Craton (Gallagher et al., 1994; Tingate et al., 1996), and from the Broken Hill Block of the Willyama Inliers (O'Sullivan et al., in prep). There appears to be some variation in the amount of cooling at this time in the Flinders Ranges with many samples showing minor cooling from moderate to low temperatures. The magnitude of cooling is poorly constrained due to the low temperatures inferred, but is probably ~40-60°C, indicating either ~1.3 to 3 km of section removed, or a significant decrease in geothermal gradient, or a combination of both, at some time since the mid Cretaceous. Once again the cooling appears to be of regional extent, but with variable magnitude.

Other stratigraphic information, such as the distribution of Palaeogene sediments in valleys and creeks of the Flinders Ranges and depositional patterns within the adjacent Eyre sub-Basin (which in part overlies the Curnamona Craton), support the suggestion that some basement areas were exposed or eroding during this time (Callen et al., 1995; Twidale and Campbell, 1993). A small isolated outcrop of (?)Permian sediment to the south of Blinman (Coats, 1964), and several Triassic intramontane basins with a total thickness of <1.5 km (Kwitko, 1995) indicate that sections of the Flinders Ranges were near surface prior to deposition in the Late Palaeozoic to Early Mesozoic. Furthermore, the preservation of these sediments and the low rank sub-bituminous nature of the coal deposits contained therein (Kwitko, 1995) indicates that local areas have not undergone deep burial or elevated temperatures. Modelling of the FT results from a Cretaceous sample near Copley (94MM31) concurs with this thermal history and indicates that some areas have not undergone significant burial in the Mesozoic, but have remained at temperatures <60°C. The FT results constrain the amount of cooling to <40°, or a maximum of 1.3 to 3.0 km denudation in the local vicinity of this sample (palaeogeothermal gradient 20-30°C/km and paleosurface temperature ~20°C), but both the degree of cooling and timing may be somewhat variable throughout the region.

7. Summary and Conclusions

The results presented here show that the eastern section of South Australia encompassing the Flinders Ranges, Arrowie Basin and Curnamona Craton experienced a significant amount of post-Delamerian Orogeny cooling during two main episodes:

1) regional cooling from temperatures >110°C was initiated in the Late Palaeozoic. If geothermal gradients of ~20-30°C/km are applied, then >3 km of post-orogenic erosion has occurred. The timing of Late Palaeozoic cooling is relatively poorly constrained due to the variable effects of subsequent cooling with only one FT sample recording a significantly long mean track length (95MM19) corresponding to a FT age of ~350 Ma. Thermal modelling suggests regional cooling occurred in the Late Devonian to Carboniferous believed to be related to the Alice Springs Orogeny, the effects of which have not been previously reported in the study area. The possibility of early Permian cooling cannot be excluded for several samples.

2) although more poorly constrained, a second period of cooling is suggested by the occurrence of Cretaceous to Palaeogene single grain ages in several samples, the paucity of track lengths >14 μm, and thermal history modelling of track length distributions and by comparison of this data to previous work. The magnitude of cooling is very poorly constrained, but is possibly ~40-60°C at some time since the mid Cretaceous. This cooling appears to be of regional extent, but with variable magnitude.

The overall variation in FT parameters throughout the study region is consistent with regional cooling rather than local reactivation of pre-existing basement structures.

Acknowledgments

Sample irradiations were supported by an Australian Institute of Nuclear Science and Engineering grant. Financial support was provided by the Australian Geodynamics Cooperative Research Centre. Sample material was provided by Department of Mines and Energy, South Australia, with the assistance of David Gravestock (MESA). Outcrop samples were collected in the Flinders Ranges National Park with the permission of the Department of Environment and Natural Resources, Permit W23479-01. Constructive comments on an earlier draft of this paper were provided by C.Johnson and P.G. Fitzgerald. This paper is published with the permission of the Director of the Australian Geodynamics Cooperative Research Centre.

References

Alley N.F. and Benbow M.C. (1995): Torrens Basin. In: *The Geology of South Australia, Volume 2 The Phanerozoic*, Drexel, J.F. and Preiss, W.V. (editors), *South Australia Geological Survey, Bulletin* **54**, p.

187-188.

Bellemans F., De Corte F. and van den Haute P. (1995): Composition of SRM and CN U-doped glasses: Significance for their use as thermal neutron fluence monitors in fission track dating. *Radiation Measurements* **24**, 153-160.

Bourman R.P and Lindsay J.M. (1989): Timing, extent and character of late Cainozoic faulting on the eastern margin of the Mt Lofty Ranges, South Australia. *Transaction of the Royal Society of South Australia*, **113**, 63-67.

Bray R.J., Green P.F. and Duddy I.R. (1992): Thermal history reconstruction in sedimentary basins using AFTA and Vitrinite Reflectance: A case study from the East Midlands of England and the Southern North Sea. In: Hardman R.S.P. (editor) Exploration Britain: Geological Insights For The Next Decade. *Geological Society of London Special Publications* **67**, 3-25.

Brown R.W., Summerfield M.A. and Gleadow A.J.W. (1994): Apatite Fission Track Analysis: Its potential for the estimation of denudation rates and implications for models of long-term landscape development. *In* Kirkby M.J. (ed) *Process Models and Theoretical Geomorphology*, pp. 23-53.

Burnham A.K. and Sweeney J.J. (1989): A chemical kinetic model of vitrinite maturation and reflectance. *Geochimica et Cosmochimica Acta* **53**, 2649-2657.

Callen R.A., Alley N.F. and Greenwood D.R. (1995): Lake Eyre Basin. In: The Geology of South Australia, Volume 2 The Phanerozoic, Drexel, J.F. and Preiss, W.V. (editors), *South Australia Geological Survey, Bulletin* **54**, pp. 188-194.

Coats R.P. (1964): The geology and mineralisation of the Blinman Dome Diapir. *Report of Investigation No. 26. South Australian Geological Survey.*

Delhi Petroleum (1984): Moorowie-1 well completion report. Unpublished report for Delhi Petroleum.

Duddy I.R., Green P.F., Hegarty K.A. and Bray R. (1991): Reconstruction of thermal history in basin modelling using Apatite Fission Track Analysis: what is really possible. *Offshore Australia Conference Proceedings* (Melbourne) **1**, 49-61.

Dunlap J.W. and Teyssier C. (1995): Paleozoic deformation and isotopic disturbance in the southeastern Arunta Block, central Australia. *Precambrian Research* **71**, 229-250.

Fitzgerald P.G. (1994): Thermochronologic constraints on post-Paleozoic tectonic evolution of the central Transantarctic Mountains, Antarctica. *Tectonics* **13**, 818-836.

Fleischer R.L., Price P.B. and Walker R.M. (1975): *Nuclear tracks in solids: Principles and applications.* University of California Press, Berkeley, 605 p.

Flottmann T. and Cockshell D. (1996): Palaeozoic basins of southern South Australia: new insights into their structural history from regional seismic data. *Australian J. of Earth Science* **43**, 45-55.

Flottmann T. and Sandiford M. (1995): Tertiary Compressional Tectonics of the Mount Lofty Ranges and Gulf of St Vincent Region. Clare Valley Conference, 25-29 September 1995. Specialist group in Tectonics and Structural geology, *Geology Society of Australia.* Abstracts No. 40, pp. 44.

Foster D.A. and Gleadow A.J.W. (1996): Structural framework and denudation history of the flanks of the Kenya and Anza Rifts, East Africa. *Tectonics* **15**, 258-271.

Foster D.A., Murphy J.M. and Gleadow A.J.W. (1994): Middle Tertiary hydrothermal activity and uplift of the northern Flinders Ranges, South Australia: Insights from apatite fission-track thermochronology. *Aust. J. of Earth Sciences* **41**, 11-17.

Galbraith R.F. (1981): On statistical models for fission track counts. *Mathematical Geology* **13**, 471-488.

Galbraith R.F. (1990): The radial plot: graphical assessment of the spread in ages. *Nuclear Tracks and Radiation Measurements* **17**, 207-214.

Galbraith R.F. and Laslett G.M. (1993): Statistical models for mixed fission track ages. *Nuclear Tracks and Radiation Measurements* **21**, 459-470.

Gallagher K. (1995): Evolving temperature histories from apatite fission-track data. *EPSL* **136**, 421-435.

Gallagher K., Dumitru T.A. and Gleadow A.J.W. (1994): Constraints on the vertical motion of eastern Australia during the Mesozoic. *Basin Research* **6**, 77-94.

Gleadow A.J.W. (1981): Fission track dating methods: what are the real alternatives?. *Nuclear Tracks* **5**, 3-14.

Gleadow A.J.W. and Duddy I.R. (1981): A natural long term annealing experiment for apatite. *Nuclear Tracks* **5**, 169-174.

Gleadow A.J.W., Duddy I.R., and Lovering J.F. (1983): Fission track analysis: a new tool for the evaluation of thermal histories and hydrocarbon potential. *Petroleum Exploration Association of Australia Journal* **23**, 93-102.

Gleadow A.J.W., Duddy I.R., Green P.F. and Lovering J.F. (1986): Confined fission track lengths in apatite: a diagnostic tool for thermal history analysis. *Contributions to Mineral Petrology* **94**, 405-415.

Gravestock D.I. and Hibburt J.E. (1991): Sequence stratigraphy of the eastern Officer and Arrowie Basins: a framework for Cambrian oil search. *APEA J* **31**, 177-190.

Gravestock D.I. and Flint R.B. (1995): Post-Delamerian compressive deformation. *In* Drexel, J.F. and Preiss, W.V. (eds) *The Geology of South Australia, Volume 2, The Phanerozoic.* South Australia Geological Survey, Bulletin **54**, pp. 60-61.

Green P.F. (1981): A new look at statistics in fission track dating. *Nuclear Tracks* **5**, 77-86.

Green P.F., Duddy I.R., Gleadow A.J.W., Tingate P.T. and Laslett G.M. (1985): Fission-track annealing in apatite: track length measurements and the form of the Arrhenius plot. *Nuclear Tracks* **10**, 323-328.

Green P.F., Duddy I.R., Gleadow A.J.W., Tingate P.R. and Laslett G.M. (1986): Thermal annealing of fission tracks in apatite, 1. A qualitative description. *Chemical Geology (Isotope Geoscience Section)* **59**, 237-253.

Green P.F., Duddy I.R., Laslett G.M, Hegarty K.A, Gleadow A.J.W. and Lovering J.F. (1989): Thermal annealing of fission-tracks in apatite 4: Quantitative modelling techniques and extension to geological time scales. *Chemical Geology. (Isotope Geoscience Section)* **79**, 155-182.

Hurford A.J. & Green P.F. (1982): A users' guide to fission-track dating calibration. *Earth and Planetary Science Letters* **59**, 343-354.

Idnurm M. and Heinrich C. (1993): A palaeomagnetic study of the hydrothermal activity and uranium mineralisation of Mt Painter, South Australia. *Aust. J. of Earth Sciences* **40**, 87-101.

Jenkins R.J.F. and Sandiford M. (1992): Observations on the tectonic evolution of the southern Adelaide Fold Belt. *Tectonophysics* **214**, 27-36.

Kwitko G. (1995): Triassic Intramontane Basins. In: Drexel J.F. and Preiss W.V. (eds) *The Geology of South Australia, Volume 2: The Phanerozoic.* South Australia Geological Survey, Bulletin **54**, pp. 98-101.

Laslett G.M., Green P.F., Duddy I.R. and Gleadow A.J.W. (1987): Thermal annealing of fission tracks in apatite, 2. A Quantitative Analysis. *Chemical Geology (Isotope Geoscience Section)* **65**, 1-13.

Mildren S.D. and Sandiford M. (1995): Heat refraction and low-pressure metamorphism in the northern Flinders Ranges, South Australia. *Aust. J.of Earth Sciences* **42**, 241-247.

Mitchell M.M. (in press): Identification of multiple detrital sources for Otway Super Group sediments-implications for basin models and chronostratigraphic correlations. *Aust. J.of Earth Sciences* **44**.

Mitchell M.M. (submitted): The thermal and tectonic development of the western Otway Basin and Flinders Ranges region, Australia, from apatite fission track analysis. Unpublished PhD thesis, La Trobe University, Melbourne.

Mitchell M.M., Kohn B.P. and Foster D.A. (1996): Apatite Fission track thermochronology of the Mt Painter Inlier and the adjacent Adelaide Geosyncline. unpublished report to Department of Mines and Energy, South Australia. May 1996, Report No. R52, 18 p.

Moore M.E., Gleadow A.J.W. and Lovering J.F. (1986): Thermal evolution of rifted continental margins: new evidence from fission tracks in basement apatites from southeastern Australia. *Earth and Planetary Sciences Letters* **78**, 255-270.

Naeser C.W. (1979): Fission track dating and geologic annealing of fission tracks. In: Jager E. and Hunziker J.C. (eds) *Lectures in Isotope Geology.* Springer Verlag, New York, pp. 154-169.

O'Sullivan P.B. and Parrish R.P. (1995): The importance of apatite composition and single-grain ages when interpreting fission track data from plutonic rocks: a case study from the Coast Ranges, British Columbia. *Earth and Planetary Science Letters* **132**, 213-224.

O'Sullivan P.B., Foster D.A., Kohn B.P. and Gleadow A.J.W. (1996): Multiple post orogenic denudation events: An example from the eastern Lachlan Fold Belt, Australia. *Geology* **24**, 563-566.

O'Sullivan P.B., Kohn B.P. and Mitchell M.M. (in prep.): Documentation of Phanerozoic reactivation along a Proterozoic crustal fault: An example from the Darling River Lineament, Australia.

Parker A.J. (1993): Chapter 2: Geological Framework. In: Drexel J.F., Preiss W.V. and Parker A.J. (eds) *The Geology of South Australia, Volume 1: The Precambrian.* South Australia Geological Survey Bulletin **54**, pp. 9-31.

Parker A.J. and Rankin L.R. (1993): Phanerozoic Basins. In: Drexel J.F., Preiss W.V. and Parker A.J. (eds) *The Geology of South Australia, Volume 1, The Precambrian.* South Australia Geological Survey, Bulletin **54**, pp. 24-31.

Preiss W.V. (compiler) (1987): The Adelaide Geosyncline–late Proterozoic stratigraphy, sedimentation, palaeontology and tectonics. *South Australia Geological Survey, Bulletin* **53**.

Preiss W.V. (1993):. Chapter 6 Neoproterozoic. In: Drexel, J.F., Preiss, W.V. and Parker A.J. (eds) *The Geology of South Australia, Volume 1, The Precambrian.* South Australia Geological Survey, Bulletin **54**, pp. 171-203.

Sandiford M., Fraser G., Arnold J., Foden J. and Farrow T. (1995): Some causes and consequences of high-temperature, low-pressure metamorphism in the eastern Mount Lofty Ranges, South Australia. *Aust. J. of Earth Science* **42**, 233-240.

Sandiford M. (1996): The problem with Mt Painter. *In* Evolution of Metamorphic Belts, Abstracts. *Geological Society of Australia Special Publication* **No. 42**, p. 42.

Shaw R.D., Etheridge M.A. and Lambeck K. (1991):. Development of the Late Proterozoic to mid-Paleozoic, intracratonic Amadeus Basin in central Australia: A key to understanding tectonic forces in plate interiors. *Tectonics* **10**, 688-721.

Sheard M.J. and Flint R.B. (1992): Cretaceous wave-polished granite surfaces, northern Mount Babbage Inlier, South Australia. *Quarterly Notes of the Geological Society of South Australia*, **122**, 19-23.

Spikings R.A., Foster D.A., Gleadow A.J.W. and Kohn B.P. (1996): Phanerozoic faulting and erosion history of the Mt Isa and Murphy Inliers, Australia. *International workshop on fission-track dating, Gent, Belgium.* Abstracts, pp. 108.

Teasdale J., Schaefer B.F. and Sandiford M. (1995): Post-Delamerian thrusting along the Paralana Fault, northern Flinders Ranges, S.A. Clare Valley Conference 25-29 th Sept. 1995, *Geological Society of Australia* specialist group in Tectonics and structural geology, Abstracts **40**, 161.

Tingate P.R. (1990): Apatite Fission track studies from the Amadeus Basin, central Australia. unpublished PhD thesis, The University of Melbourne.

Tingate P.R., Alexander E.M., Michaelsen B.H., Duddy I.R. and Griffiths C.M. (1996): Tectonic events affecting the thermal history of the western margin of the Eromanga Basin. 13th Australian Geological Convention, Canberra. *Geological Society of Australia*, Abstracts **No.41**, p. 442.

Thomson B.P. (1969): The Kanmantoo Group and Early Palaeozoic tectonics. *In* Parkin L.W. (editor) *Handbook of South Australian Geology.* Geological Survey of South Australia, pp. 97-108.

Twidale C.R. and Campbell E.M. (1993): *Australian Landforms; Structure, Process and Time.* Gleneagles Publishing, Adelaide, 560 p.

Wells A.T., Forman D.J., Ranford L.C. and Cook P.J. (1970): Geology of the Amadeus Basin, central Australia. *Bulletin of the BMR Australia* **100**, 222 p.

Young E.J., Myers A.T., Munson E.L., Conklin N.M. (1969): Mineralogy and Geochemistry of Fluorapatite from Cerro De Mercado, Durango, Mexico. U.S. Geol. Survey Prof. Paper 650-D, D84-93.

COOLING AND DENUDATION HISTORY OF WESTERN MARIE BYRD LAND, ANTARCTICA, BASED ON APATITE FISSION-TRACKS

F. LISKER and M. OLESCH
Fachbereich Geowissenschaften, Universität Bremen, Germany

Abstract: The Ford Ranges and the Edward VII Peninsula, western Marie Byrd Land (MBL), form the flat eastern rift flank of the West Antarctic Rift System (WARS). Both areas were intruded by middle and late Cretaceous Byrd Coast Granites and experienced three distinct periods of low temperature cooling: from 100 to 85 Ma, from 75 to 60 Ma and since the Oligocene. The different cooling stages are related to a pulsing mantle plume, slowly wandering below MBL. They can be linked to the beginning of Gondwana break-up in the Pacific sector of Antarctica, the rifting of the New Zealand/ Campbell Plateau block from MBL and the onset of the Tertiary alkali volcanism in West Antarctica. The cumulative amount of uplift/ denudation since the late Cretaceous is ~3300 m and thus, significantly smaller than those of the western rift shoulder of the WARS (Transantarctic Mountains: up to 10 km).

1. Introduction

MBL is regarded as one of five allochthonous crustal blocks that compose West Antarctica (Storey *et al.*, 1988). These crustal provinces are all Palaeozoic to Mesozoic fore-arc and magmatic-arc terranes which were tectonically and magmatically accreted onto the Pacific margin of Gondwana during the Phanerozoic (Dalziel and Elliot, 1982). At least since Cretaceous times MBL forms the eastern part of the WARS. The WARS extends over 3000 × 750 km trending perpendicular to the Antarctic margin. It is comparable in area and dimension to the Basin and Range and the East African Rift System (e.g. Behrendt *et al.*, 1991). The main physiographic feature of the WARS is its structural asymmetry. The western rift shoulder, extending from the Transantarctic Mountains, northern and southern Victoria Land, to the Queen Maud Mountains and the Ellsworth-Whitmore-Horlick Mountains, forms a steep escarpment with maximum elevations of 4-5 km (Fig. 1). The eastern, Amundsen-Bellinghausen rift flank of MBL lacks such morphological expression.

The development of the WARS commenced with Gondwana break-up and is accompanied by late Cretaceous and Cenozoic extension and late Cenozoic volcanism (LeMasurier, 1978). Since the Cretaceous a three-stage tectonic evolution has been recognised for this rift (e.g. Tessensohn, 1994). A first amagmatic phase of diffuse crustal attenuation related to the Australia-Antarctica separation at about 100 Ma (continental shear) was followed by basin subsidence and shoulder uplift of the Transantarctic Mountains at about 60 Ma (continental separation) and rift-type magmatism from 40 Ma to present (Ross Sea rifting).

P. Van den haute and F. De Corte (eds.), Advances in Fission - Track Geochronology, 225-240.

The uplift/ denudation history of the Transantarctic Mountains and the Ellsworth-Whitmore Mountains is well documented by numerous AFT studies (e.g. Fitzgerald *et al.*, 1986; 1992; Gleadow and Fitzgerald, 1984). On the other hand, there are almost no information about the thermal and tectonic history of the Amundsen-Bellinghausen rift flank of MBL. Only a few fission-track data from the Fosdick Mountains and the Edward VII Peninsula have been recently reported by Richard *et al.* (1994) and Adams *et al.* (1995). In this study we present new AFT data from the Edward VII Peninsula and the Ford Ranges. Our aims are a) to unravel the Cretaceous and Cenozoic cooling and uplift history of MBL, b) to estimate the rate of uplift/ denudation to a first approximation, c) to determine the relationship between uplift and Gondwana break-up, plume ore rift related geological events, and d) to correlate the post-break-up processes with the history of the exposed rift shoulder of northern Victoria Land.

2. Geological Outline

Geographically the scattered bedrock outcrops of MBL are grouped into the Edward VII Peninsula, the Ford Ranges and the Kohler Ranges (Fig. 1).

The oldest rocks which form the basement of MBL are metasedimentary rocks of the Swanson Formation and their high-grade equivalents, the Fosdick (Ford Ranges) and Alexandra Metamorphic Complexes (Edward VII Peninsula). The Swanson Formation is a lower Palaeozoic turbidite sequence, which was folded and mildly metamorphosed in the Late Ordovician, at 432 - 421 Ma (Adams, 1986). Bradshaw *et al.* (1983) noted sedimentological similarities with the Robertson Bay Group of northern Victoria Land and Adams (1986) correlated the Swanson Formation with the Greenland Group of western South Island, New Zealand.

Since the Devonian two generations of granitoids intruded in the metamorphic basement of MBL: the Ford Granodiorite and the Byrd Coast Granite (Wade *et al.*, 1977a, b) (Fig. 1). Late Devonian-Carboniferous plutonism laid to the emplacement of the Ford Granodiorite (375 - 324 Ma/ Adams, 1987) in the Swanson Formation of the Ford Ranges. Those I-type granites are granodiorites to monzonites, petrologically similar to the Admiralty Intrusives of northern Victoria Land. Their intrusion is interpreted as caused by posttectonic magmatism during the uplift and adiabatic melting of crustal material (Vetter and Tessensohn, 1987) or subduction related magmatism and accretion of exotic terranes (Grindley and Davey, 1982; Borg *et al.*, 1987).

Both the Swanson Formation and Ford Granodiorite are intruded in two pulses by upper Jurassic/ lower Cretaceous (~140 Ma) and middle Cretaceous (~100 Ma) Byrd Coast Granites (K/Ar biotite and Rb/Sr whole rock isochrons: Weaver *et al.*, 1991; Adams, 1987; Adams *et al.*, 1995). The first pulse of emplacement, related to the subduction of the Pacific-Phoenix Ridge (J.D. Bradshaw, 1989) and contemporaneous with the opening of the Weddell Sea, produced the I-type granites of the Ford and Kohler Ranges in central and eastern MBL.

The intrusion of the younger Byrd Coast Granite generation between 108 and 92 Ma is associated with the initiation of continental rifting along the MBL - New Zealand/ Campbell Plateau margin. The epizonal A-type granites are coarse grained

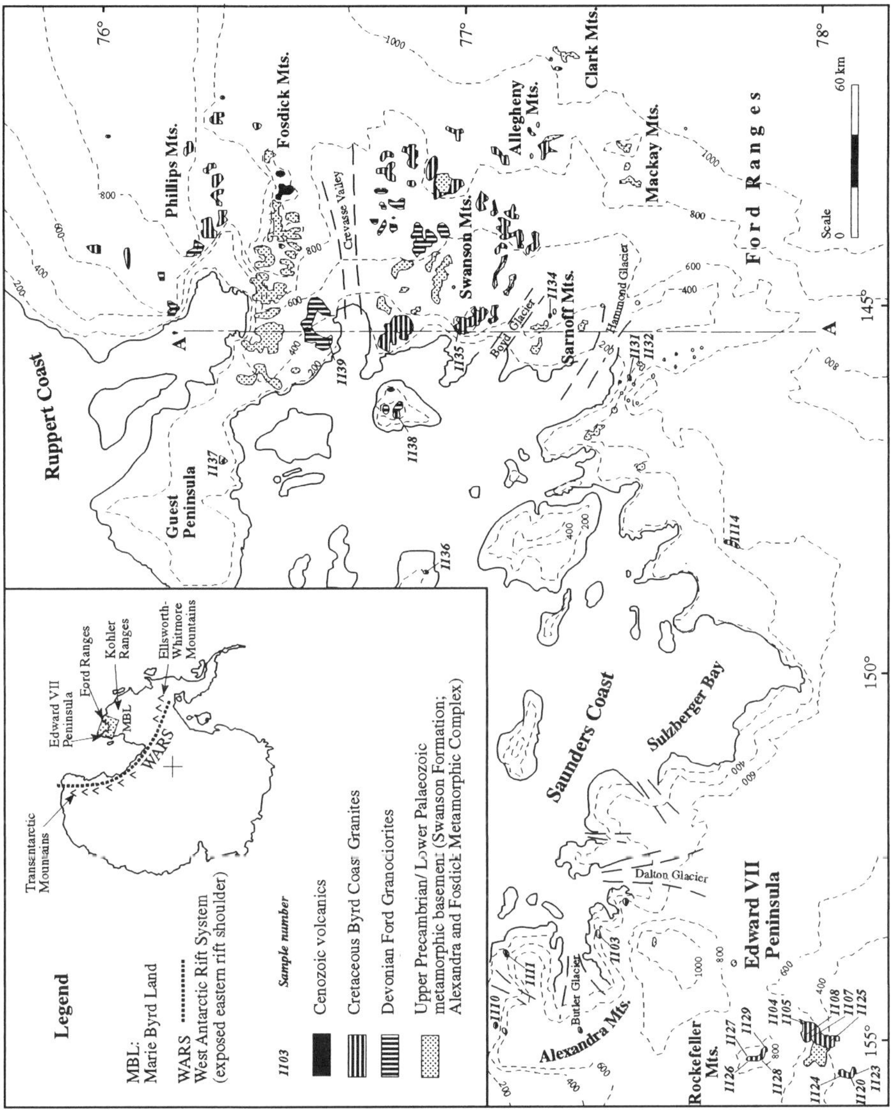

Figure 1. Geologic map of the Ford Ranges and the Edward VII Peninsula, western Marie Byrd Land, after Wade and Couch (1982). The dotted line represents the profile which is analysed in Fig. 3. The numbers in italics indicate sample localities according to Table 1.

monzogranites and syenogranites and form plutons of several kilometres in diametre as well as small sheet-like bodies and dikes. They usually show sharp intrusive contacts with the thermally metamorphosed Swanson Formation rocks.

The granite outcrops of the Ford Ranges show Byrd Coast Granites of both intrusion phases. Small plutons or dikes of the older Late Jurassic-Early Cretaceous phase predominantly occur along the eastern margin of the Ford Ranges. In contrast, the younger phase is concentrated around the Fosdick Mountains area and forms the granitic plutons and sheets of the Edward VII Peninsula.

The intrusions are coeval with the Mesozoic high-grade metamorphism and deformation of the Fosdick and Alexandra metamorphic core complexes. K/Ar and $^{40}Ar/^{39}Ar$ studies on hornblende, muscovite, biotite, and K-feldspar reported from the gneisses and migmatites of the Chester, the Fosdick and the Phillips Mountains (Richard at al., 1992; 1994) and the Alexandra Mountains (Adams *et al.*, 1995) indicate rapid exhumation of the northern Fosdick Mountains between ~110 and 95 Ma. Rapid uplift and cooling of high-grade metamorphic rocks is documented along this margin of Gondwana between 120 and 100 Ma in New Zealand (J.Y. Bradshaw, 1989; Mattinson *et al.*, 1986; Tulloch and Kimbrough, 1988) and between ~85 and 65 Ma in southern South America (Mukasa *et al.*, 1988; Dalziel and Brown, 1989).

The break-up of MBL and the New Zealand/ Campbell Plateau block (~85 - 70 Ma) was accompanied by gradual subsidence of the rifted margins, aggredation of coastal valleys and marine planation. Thus, the rocks of MBL and parts of Ellsworth Land were truncated by the planar, regionally extensive West Antarctic Erosion Surface (LeMasurier and Rex, 1983) which has an erosional relief of only 15 - 50 m. Remnants of this erosion peneplain are exposed in isolated horst blocks at elevations between 600 and 2700 m.

After a phase of prolonged tectonic stability since the early Cenozoic increasing alkaline volcanism associated with domal uplift and block faulting starting at 30 - 28 Ma is assumed to be the inception of mantle plume activity in West Antarctica. LeMasurier and Rex (1989) suggested that the development of the WARS and volcanism in MBL is related to a mantle plume or hotspot with about 600 km in diameter. The recent morphology of MBL is characterised by graben-and-horst topography (LeMasurier and Rex, 1983) and by pre-Cenozoic basement elevations up to 2800 m. The few scattered outcrops of the Alexandra and Rockefeller Mountains and the Ford Ranges in western MBL confirm this Mesozoic-Cenozoic tectonic activity.

3. Method

A set of 24 apatite samples from the western Ford Ranges and the Alexandra and Rockefeller Mountains (Fig. 1) was analysed using the external detector method. Elevations of sample localities were measured barometrically with reference to the topographic and geologic maps of Wade *et al.* (1977a; b; c; 1978). Typical sample weights were 2 - 4 kg. Apatites were separated from the host rock using standard magnetic and heavy liquid techniques as recommended by Gleadow (1981). The polished apatites were etched for 60s in 5% HNO_3. After mounting samples were irradiated in the well-thermalised graphite reflector facility of Risø/ Denmark. Thermal neutron fluence was monitored by recording track density in the reference glass SRM 612 (U.S. National Bureau of Standards). The fossil and spontaneous tracks were counted and horizontal confined track lengths were measured (Laslett *et al.*, 1984) at 1000 × magnification (air objective). For dating we used the FT-Stage program of Dumitru (1993). Ages were determined using the zeta calibration method following Hurford and Green (1983) and Green (1985) with $\zeta = 350 \pm 12$ (Lisker, 1996). Errors were calculated using the conventional method (Green, 1981) and quoted as $\pm 1\sigma$. The counting parameters used are shown at the top of Table 1. Forward modelling (MONTE TRAX / Gallagher, 1995 (Fig. 4) was used to constrain the low-temperature cooling history consistent with all measured track length distributions and ages. As a basis for the modelling the composition of the selected apatites was determined as predominantly Durango-type (Green *et al.*, 1986) semiquantitatively by EDX and WDX analysis.

4. AFT results

The AFT data are summarised in Table 1. The AFT ages of MBL range from 97 ± 5 to 72 ± 5 Ma. The analyses from the Ford Ranges and the Edward VII Peninsula are presented seperately because the different correlation of the AFT ages with the sample elevations.

4.1. FORD RANGES

All Ford Ranges samples were taken from Byrd Coast Granite with the exception of one sample (1139/ West Chester Mountains: Ford Granodiorite). The apparent AFT ages range from 97 ± 5 to 76 ± 4 Ma (Table 1, Fig. 2). Samples were taken between elevations of 950 and 230 m. Apparent AFT ages of similar altitudes increase from N to S (Fig. 3). This trend is approximately perpendicular to the general strike of the Ford Ranges and to the flanks of the West Antarctic Rift System. A similar trend has also been observed by Smith (1992) and Richard *et al.* (1994) for U/Pb and ^{40}Ar/^{39}Ar ages.
The mean track lengths vary between 13.6 and 14.2 μm with small standard deviations between 1.2 and 1.6 μm. All AFT ages are younger than the minimum Rb/Sr whole-rock isochron and ^{40}Ar/^{39}Ar potassium feldspar data of comparable samples (112 - 95 Ma) given by Weaver *et al.* (1991) or Richard *et al.* (1994). Our Ford Ranges data are in very close agreement with the AFT results from Richard *et al.* (1994) who investigated the northern Ford Ranges. Richard *et al.* (1994) report AFT ages between 95 ± 6 and 67 ± 5 Ma and mean track lengths between 13.5 and 14.6 μm with standard deviations between 1.1 and 1.6 μm for six samples from the Phillips, Fosdick and Chester Mountains.
The Ford Ranges samples can be divided into two groups, samples above and below an elevation of ~500 m. The five samples above ~500 m show no clear correlation with elevation. Especially the four AFT ages at ~600 m spread broadly between 97 ± 5 and 82 ± 4 Ma. They have mean lengths between 13.6 and 14.2 μm. The track length distributions are relatively narrow, with standard deviations between 1.2 and 1.6 μm. These narrow distribution result from a conspicuous lack of very short (<11 μm) and

TABLE 1. AFT analytical results of the Ford Ranges and the Alexandra and Rockefeller Mountains, western Marie Byrd Land. Values in parantheses are numbers of tracks counted. Ages calculated using $\zeta = 350.5 \pm 12.3$ for dosimeter glass SRM612 (U.S. National Bureau of Standards; Hurford & Green, 1983). Errors are given as $\pm 1\sigma$ (conventional method (Green, 1981). Standard: Durango apatite. * Mean age, used where pooled data fail χ^2 test at 5%. $P(\chi^2)$: Chi square probability. Correl. Coeff.: Correlation coefficient, Lat.: Latitude, Long.: Longitude, St.D.: Standard deviation. AM: Alexandra Mountains, RM: Rockefeller Mountains, FR: Ford Ranges. Kbc: Byrd Coast Granite, Mdf: Ford Granodiorite. Intrusion ages: [1] Adams et al., 1995 (K/Ar), [2] Weaver et al., 1992 (Rb/Sr), [3] Richard et al., 1994 ($^{40}Ar/^{39}Ar$)

Sample	Locality/ Petrology (Intrusion Age) [Ma]	Lat. S Long. W	Elevation [m]	Number of Grains	Standard Track Density [10^6cm^{-2}]	Fossil Track Density [10^6cm^{-2}]	Induced Track Density [10^6cm^{-2}]	Correl. Coeff.	$P(\chi^2)$ [%]	Age ($\pm 1\sigma$) [Ma]	Mean Track Length [μm]	St.D. [μm]	Number of Tracks
1103	Bowmans Pk [1] AM/ Kbc (98±2)	74°28' 153°30'	300	20	1.28 (7759)	0.103 (161)	0.297 (490)	0.62	45	73 ± 7	-	-	-
1110	Prestrup Rock AM/ Kbc	77°13' 154°31'	620	20	1.28 (7759)	2.664 (3222)	6.548 (7884)	0.91	17	91 ± 4	13.1 ± 0.2	1.6	100
1111	Mt Swadener *[1] AM/ Kbc (103±2)	77°16' 153°45'	420	8	1.28 (7759)	3.573 (2172)	9.281 (5522)	0.89	< 5	88 ± 4 86 ± 9*	14.0 ± 0.2	1.3	64
1104	Mt Paterson [2] RM/ Kbc (101±4)	78°02' 155°	540	32	1.28 (7759)	0.621 (915)	1.633 (2442)	0.93	83	83 ± 4	12.5 ± 0.2	1.8	90
1105	Mt Paterson [1] RM/ Kbc (102±2)	78°03' 155°02'	450	34	1.28 (7759)	0.931 (1125)	2.451 (3012)	0.87	27	83 ± 4	13.1 ± 0.2	1.7	100
1107	Mt Nilson [1] RM/ Kbc (102±2)	78°03' 155°20'	700	24	1.28 (7759)	1.158 (2076)	3.061 (5418)	0.60	9	85 ± 4	13.3 ± 0.2	1.8	90
1108	Strider Rock RM/ Kbc	78°02' 155°24'	620	28	1.28 (7759)	0.622 (1001)	1.678 (2740)	0.81	9	81 ± 4	13.2 ± 0.1	1.4	100
1120	Tennant Pk [2] RM/ Kbc (100±2)	78°09' 155°44'	285	32	1.28 (7759)	0.499 (817)	1.533 (2499)	0.94	82	73 ± 4	13.7 ± 0.1	1.4	100
1123	Mt Butler *[2] RM/ Kbc (97±3)	78°09' 155°47'	210	20	1.28 (7759)	0.451 (458)	1.399 (1481)	0.55	< 5	69 ± 4 72 ± 5*	-	-	-
1124	Gould Pk [1] RM/ Kbc (100±2)	78°08' 155°40'	450	12	1.28 (7759)	0.699 (660)	1.833 (1714)	0.99	86	86 ± 5	-	-	-
1125	Breckinridge Pk [1] RM/ Kbc (102±2)	78°05' 155°27'	555	53	1.27 (7730)	0.801 (1840)	2.203 (5080)	0.91	22	80 ± 4	12.8 ± 0.2	1.5	61
1126	Mt Frazier [2] RM/ Kbc (97±4)	77°52' 155°27'	870	20	1.28 (7759)	0.434 (567)	1.113 (1456)	0.94	80	87 ± 5	-	-	-
1127	Mt Jackling RM/ Kbc	77°54' 155°19'	840	48	1.27 (7730)	0.773 (2334)	2.068 (6371)	0.94	42	81 ± 4	13.2 ± 0.2	1.6	75
1128	Mt Fitzsommons [1] RM/ Kbc (100±2)	77°54' 155°16'	850	24	1.27 (7730)	0.812 (1345)	1.956 (3224)	0.94	25	92 ± 4	13.0 ± 0.2	1.5	90
1129	Mt Shideler [1] RM/ Kbc (101±2)	77°55' 155°13'	810	16	1.28 (7759)	0.37 (483)	0.968 (1220)	0.93	11	88 ± 6	-	-	-
1114	Mc Kinley Pk FR/ Kbc	77°54' 148°19'	620	25	1.28 (7759)	0.633 (574)	1.454 (1365)	0.93	22	94 ± 6	-	-	-
1131	Mt Ronne FR/ Kbc	77°34' 146°04'	575	50	1.27 (7730)	0.725 (2173)	1.813 (5547)	0.89	94	87 ± 4	14.0 ± 0.2	1.4	56
1132	Mt Ronne FR/ Kbc	77°33' 146°01'	480	24	1.27 (7730)	1.592 (734)	4.631 (2133)	0.93	5	76 ± 4	13.6 ± 0.2	1.6	50
1134	Mt Douglas FR/ Kbc	77°20' 145°20'	600	21	1.27 (7730)	0.846 (828)	2.001 (1892)	0.94	91	97 ± 5	13.7 ± 0.2	1.6	43
1135	Bilboard FR/ Kbc	77°04' 145°46'	950	20	1.27 (7730)	0.718 (451)	1.685 (1108)	0.91	69	91 ± 6	-	-	-
1136	Barela Rock FR/ Kbc	77°09' 145°53'	250	32	1.27 (7730)	0.437 (904)	1.201 (2585)	0.91	86	78 ± 4	13.9 ± 0.2	1.2	38
1137	Mitchell Pk * FR/ Kbc	76°24' 147°26'	610	23	1.27 (7730)	2.494 (2314)	6.628 (6274)	0.90	< 5	88 ± 4 83 ± 7*	14.2 ± 0.1	1.2	100
1138	Clark Knoll FR/ Kbc	76°53' 146°59'	400	26	1.27 (7730)	0.549 (1041)	1.505 (2884)	0.97	97	80 ± 4	13.7 ± 0.3	1.5	28
1139	W Chester Mts [3] FR/ Mdf (105)	76°41' 145°49'	230	20	1.27 (7730)	0.637 (678)	1.789 (1936)	0.93	48	78 ± 4	-	-	-

long (>15 μm) tracks. Track lengths of ~14 μm in connection with standard deviations ≤1.6 μm indicate rapid cooling almost entirely through the PAZ. The samples from >500 m cooled down to the base of the PAZ immediately after emplacement of the Byrd Coast Granite at the cool environment of a shallow crustal level. The spread in AFT ages of the samples at elevation of ~600 m exclude a common cooling of the entire Ford Ranges at the same time. The AFT data probably represent the individual cooling time of the single intrusive bodies. This interpretation is supported by the Rb/Sr and $^{40}Ar/^{39}Ar$ age distribution of the Ford Ranges (112 - 95 Ma), which shows a similar pattern. The data are evidence for the varied emplacement of the Byrd Coast Granite in time and intrusion (stratigraphic?) level. Richard *et al.* (1994) report an average cooling rate of ~70°C/Ma for granitoids of the northern Ford Ranges between 105 and ~95 Ma. This high cooling rate was determined by U/Pb data for monazite, $^{40}Ar/^{39}Ar$ data for hornblende, muscovite, biotite, and potassium feldspar and apatite thermochronology. Even taking in consideration slightly different intrusion ages, the elevation-age gradient between the uppermost sample (1135/ The Bilboard: 950 m) and the samples at ~600 m would coarsely support such a high cooling rate during this time. The amount of uplift accompanying or following the intrusion, however, is hardly to estimate due to the interaction of magmatic cooling and coeval erosion.

As modelling indicates, between ~90 and ~80 Ma the cooling rate rapidly decreased. The samples below 500 m rest on the very top of the PAZ at 70 - 60°C (Fig. 4). Assuming a geothermal gradient of ~30°/km in the upper crust as adopted for the Transantarctic Mountains (Gleadow *et al.*, 1984; Wagner *et al.*, 1989) or the Ellsworth Whitmore Mountains (Fitzgerald and Stump, 1991) the top of the PAZ is situated at depth of ~2000 m.

The apparent AFT ages of the four samples below 500 m range from 80 ± 4 to 76 ± 4 Ma and correlate positively with the sample elevations. The mean track lengths are slightly shorter than those of the samples from higher elevations. They vary between 13.7 and 13.9 μm, and standard deviations between 1.2 and 1.6 μm. Beginning at ~80 Ma renewed cooling started which probably represents an uplift/ denudation stage with an apparent denudation rate >100m/ Ma (Figs 2, 5). The amount of uplift/ denudation since the Late Cretaceous is estimated as ~2000 m.

4.2 EDWARD VII PENINSULA

In the Edward VII Peninsula only Byrd Coast Granites were sampled. The 15 apatite samples from the Alexandra and Rockefeller Mountains cover the maximum, albeit limited vertical elevation between 900 and 220 m above sea level. The AFT ages given in Table 1 range from 92 ± 4 to 72 ± 5 Ma. There is no obvious differentiation of cooling between the various Byrd Coast Granite plutons. All AFT ages of the Edward VII Peninsula are younger than the Rb/Sr and K/Ar data of Weaver *et al.* (1992) and Adams *et al.* (1995). They display a good correlation with the sample elevations (Fig. 2). The regular spread of the AFT ages throughout the entire Edward VII Peninsula and the absence of major tectonic faults within the region indicate a uniform multi-stage thermal history for the Alexandra and the Rockefeller Mountains. The samples of the Edward VII Peninsula can be divided into two groups, with ages ranging from ~90 to 80

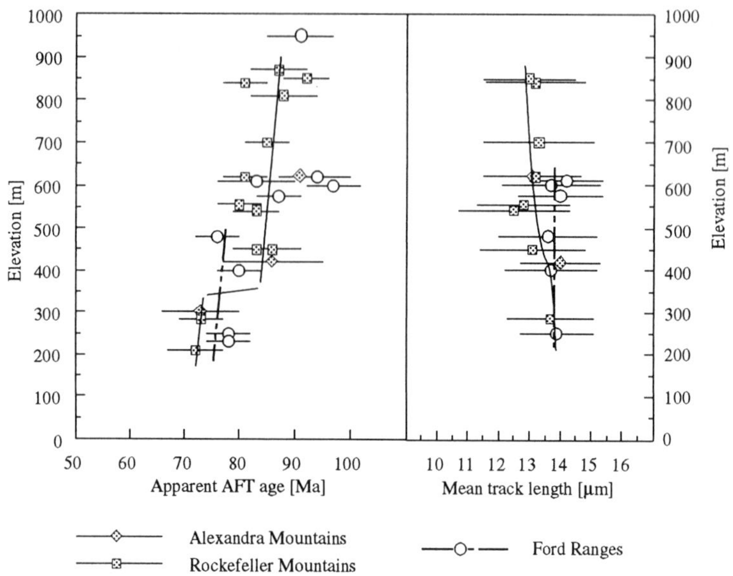

Figure 2. AFT ages and mean track length distribution plotted versus sample elevations. The apparent uplift rates are given by the slopes of the AFT age profile, the offset in slope defines the stagnation of uplift. Whereas the Ford Ranges samples show almost the same lengths and relatively small standard deviations, the samples of the Edward VII Peninsula underwent partial annealing.

Ma above 420 m (twelve samples) and ~70 Ma (three samples) below 300 m. The AFT age - elevation plot of both cooling stages is interrupted by a break in slope (Fig. 2).

The AFT ages of the samples above 420 m lie between 92 ± 4 to 80 ± 4 Ma, showing a positive correlation with elevation. The samples have mean lengths between 12.5 and 14.0 µm and standard deviations between 1.3 and 1.8 µm. The AFT ages are practically less than 20 Ma (8 -22 Ma) younger than the crystallisation age of the granites (Table 1), representing the first stage of cooling immediately after granite emplacement to temperatures of 100 - 90°C. An average cooling rate of ~15°/ Ma agrees with the shortened mean lengths and our model (Fig. 4). Radial plots and age histograms of the samples immediately above the break in slope show a bimodal age distribution (Fig. 4). The amount of accompanying uplift/ denudation cannot be determined unambiguously by the AFT data measured.

After this first maximum of AFT accumulation the cooling process stagnated for the next ~10 Ma. The investigated recent basement level remained in the isotherm of 100 - 90°C (Fig. 4). Between ~85 and ~75 Ma the rocks in the PAZ were uplifted merely by 120 m, i.e. the cooling rate of the samples *in the PAZ* slowed down to ~1°/ Ma. This interruption of the cooling process is indicated by the break in slope in Fig. 2. The change of cooling rate may represent abandoning uplift/ denudation or an uplift of isotherms due to a partial exhumation of the PAZ.

The three samples collected below 300 m have younger ages between 73 ± 7 and 72 ± 5 Ma. Due to poor quality of the apatite separates the mean track length of only one sample (1120: 13.7 µm, standard deviation 1.4 µm) could be determined. Radial plots and age histograms show that the three samples have been uplifted within the PAZ and younger tracks have been accumulated at 70 - 60 Ma (Fig. 5). For both the Alexandra and the Rockefeller Mountains the data express a phase of accelerated cooling/ denudation with an apparent denudation rate of about 70 m/Ma. The amount of uplift during this time can be calculated as 700 - 1000 m. Assuming a geothermal gradient of ~30°/km the maximum amount of uplift since about 70 Ma is estimated as much as 3300 m. The mean track length of 13.7 μm with a standard deviation of 1.6 μm exclude a very fast rock exhumation through the PAZ. The samples probably spent a significant time at the top of the PAZ. With the beginning of the Cenozoic the uplift/ denudation process adjourned again and the rocks cooled further down during a ~30 Ma period by about 10°C. Final exhumation by 2300 - 2600 m of the Alexandra and Rockefeller Mountains is caused by the onset of Cenozoic rifting.

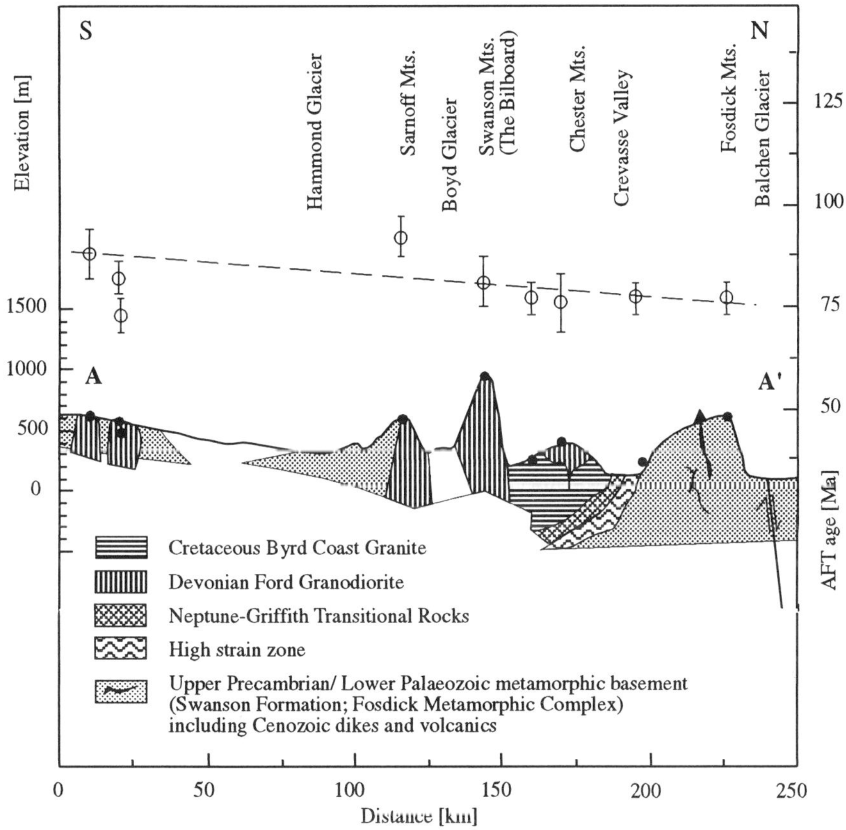

Figure 3. The profile as shown in Fig. 1 after Wade et al. (1977b; c) and Luyendyk *et al.* (1992). The dots display the real sample elevations. The circles represent apparent AFT ages in the Ford Ranges, normalised on 200 m altitude a.s.l., indicating a decreasing N-S trend.

5. Different cooling and exhumation in MBL - result of mantle plume activity?

Due to logistic limitations the nine samples of the Ford Ranges are regionally very widely scattered (Fig. 1). They do not represent an in situ profile. Otherwise, the investigated area is characterised by a limited topographic altitude of <1000 m, a large topographic wavelength not less than 20 km (Figs. 1, 3), and there are no indications for high erosion rates >500 m/Ma. Thus, the effect of isotherm perturbation should be negligible (Stüwe *et al.*, 1994) and a one-dimensional interpretation model can be applied for the MBL samples.

The tectonic regime of the Pacific Phoenix Ridge plate and the Antarctic plate is characterised by a sudden transition from subduction to extension in the Middle Cretaceous (105 ± 5 Ma). Luyendyk (1995) explains this change of the tectonic regime with the concept of subduced slab capture. This transition is accompanied by the beginning formation of the passive margin of MBL and by the intrusion of granitic plutons at shallow crustal levels followed by rapid exhumation of mid-crustal metamorphic rocks (J.D. Bradshaw, 1989; Luyendyk *et al.*, 1992). The postmagmatic/ -metamorphic history and the cooling is illustrated by our AFT data. On the first view the data seem to indicate different thermo-tectonic processes because the cooling below the apatite PAZ starts and ends earlier in the Ford Ranges than in the Edward VII Peninsula. However, the Rb/Sr, K/Ar and $^{40}Ar/^{39}Ar$ indicate a longer duration of magmatic activity in the Ford Ranges (112 - 95 Ma) than in the Edward VII Peninsula (103 - 97 Ma). The Byrd Coast Granite of the Ford Ranges intruded into a shallow crustal level, i.e. in a cool enviroment, associated with differentiated uplift of discrete blocks. This would indicate relatively fast subsidence of the uplifted isotherms, as the AFT ages between 97 and 82 Ma suggest. In the Edward VII Peninsula the granites intruded less differentiated and, obviously, some hundred metres deeper in the crust. They got stack in a slightly warmer enviroment than the granites of the Ford Ranges and cooled down slowly resulting in AFT ages between 92 and 80 Ma. This scenario explains the different ages and -particularly- track lengths of the first cooling stage in western MBL. Weaver *et al.* (1994) suggest that a mantle plume is responsible for the voluminous lithospheric magmatism in MBL. Variations in the heat flux of that mid-Cretaceous plume lead to magmatic peaks at ~100 and 30 Ma up to present. The position of the plume head centred adjacent to the Ruppert Hobbs Coast in the mid-Cretaceous is closer to the Ford Ranges area than to the Edward VII Peninsula. That might explain the intrusion of the Ford Ranges granitoids into a higher crustal level and thus, the faster cooling and more rapid uplift of the samples. Increasing distance from the plume centre may be responsible for the decrease of contact metamorphic temperature and the vague trend of increasing AFT ages in the Ford Ranges from N to S (Fig. 3). Richard *et al.* (1994) suggested that during this first cooling stage some additional tilting of the basement may have occurred.

The Gondwana break-up and the beginning of ocean-floor spreading in the south-west Pacific at about 85 Ma (Mayes *et al.*, 1990) caused the rifting of the New Zealand -

Campbell Plateau continental block from MBL accompanied by crustal extension and thinning of the Ross Embayment (Fitzgerald *et al.*, 1986). This is manifested by gradual subsidence of the topographic surface and marine transgression (Fig. 5). An erosional peneplain with a flat erosional relief of merely 15 - 50 m, the West Antarctic Erosion Surface, has been formed between 85 and 70 Ma (LeMasurier and Rex, 1983, 1994). Remnants of the West Antarctic Erosion Surface have been reported in MBL and along the Ellsworth Mountains and correspond with the Waipounamu surface of New Zealand (LeMasurier and Landis, 1996). They crop out at elevations between 500 and 2700 m and suggest differentiated Late Cretaceous and/ or Cenozoic uplift with horst-and-graben tectonics. The West Antarctic Erosion Surface may be the surface expression of the top of a fossil PAZ represented by the break in slope at elevation of ~600 m (Ford Ranges) and ~400 m (Alexandra and Rockefeller Mountains). In that case it represents the interruption of cooling in the AFT age - elevation profile (Fig. 2).

At ~75 Ma an additional stage of cooling commenced, probably lasting up to the earliest Tertiary. The beginning of this cooling phase should be equivalent to the base of the West Antarctic Erosion Surface and finish the surface subsidence in western MBL. Again the two parts of western MBL are affected in different way. While the Ford Ranges samples left the PAZ at this stage the samples from the Alexandra and Rockefeller Mountains were uplifted up to the top of the PAZ and rest there up to the final exhumation. The apparent denudation rate is estimated to about 70 m/Ma, the amount of uplift is ~700 - 1000 m. The cooling represents accelerated uplift/ denudation affected by lithospheric doming above the mantle plume in central MBL. At the same time alkaline volcanic rocks of the Chatham Islands extruded and at ~70 Ma the Darwin Mountains/ southern Chile were uplifted (Dalziel and Brown, 1989).

During the Early and Middle Cenozoic phase of relative tectonic quiescence the flat topographic relief of MBL was formed (Van der Wateren *et al.*, 1994). Alkaline basalt volcanism apparently began in MBL around 30 - 28 Ma and has episodically continued since then, associated with tectonic activity and block faulting (LeMasurier and Rex, 1983). These magmatic and tectonic processes are not a product of extensive active rifting, because the Antarctic plate has hardly moved since 85 Ma (Lawver *et al.*, 1991) and plate reconstructions will admit only minimal amounts of Cenozoic extension (Lawver and Gahagan, 1994). Lithospheric doming of the WARS over the head of the slowly westward moving (Weaver *et al.*, 1994) or a new generated mantle plume (Futa and LeMasurier, 1983; LeMasurier and Rex, 1989), or a plume swarm (Rocholl *et al.*, 1995) led to the finite cooling and an exhumation of MBL up to 2500 m (Figs. 5 and 6). This amount of uplift/ denudation agrees with the average vertical displacement rate of about 100 m/ Ma determined by LeMasurier and Rex (1994). Behrendt and Cooper (1991) and LeMasurier and Rex (1994) even suggested Pliocene uplift rates of up to 1000 m/Ma based on seismic investigations and the occurrence of microfossil-bearing glacial lake sediments of inferred Pliocene age exposed at Mount Murphy 1300 m above sea level. Minor differences between the uplift/ denudation of the Ford Ranges and the Edward VII Peninsula may arise due to the young graben-and-horst tectonic indicated by negative Bouguer (Behrendt *et al.*, 1991) and isostatic (Bentley and Robertson, 1982) anomalies between both areas and the anisotropy of magnetic susceptibility (Luyendyk *et al.*, 1992).

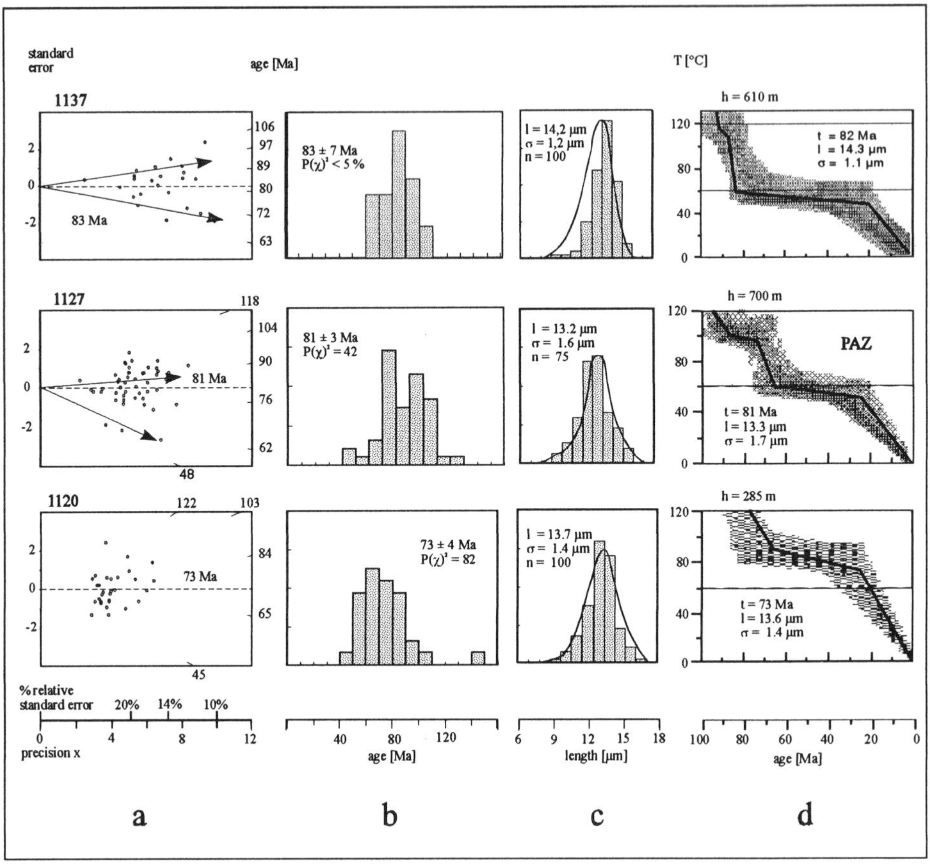

Figure 4. Plots of representative Byrd Coast Granite samples of the Ford Ranges and the Alexandra and Rockefeller Mountains, Marie Byrd Land. **a**) Radial plots: Sample number is given on left top side of plots. AFT ages (pooled ages) are given in Ma. The two arrows in the samples 1137 and 1127 show the estimation of two component ages (both Cretaceous), whereas in sample 1120 only the younger uplift/ denudation event is represented. **b**) Age histograms: All data are normalised to 100 single grain ages. The bimodal distribution in the histogram of the upper sample indicate an at least two-stage T-t-path. AFT ages are given with ± 1σ errors, χ^2 test is shown. **c**) Length histograms: All data are normalised to 100 confined tracks. The measured mean track lengths of confined tracks are shown as histograms and compared with the curve of the modelled track length distribution. **d**) T-t-path: The T-t-paths were modelled using the program MONTE TRAX (Gallagher, 1995). On top of the plots: sample altitude. t (age), l (mean track length) and σ (standard deviation) are the modelled data consistent with the T-t-paths.

A comparison of the AFT data of northern Victoria Land and MBL reveals a contrasting denudation history of the opposing rift shoulders of the WARS. Whereas in MBL the Gondwana break-up was accompanied by the intrusion of the Byrd Coast Granite and characteristic two-stage uplift, northern Victoria Land only experienced a failed rift event ("Rennick thermal event"/ Tessensohn, 1994) including some basement reheating at ~100 Ma. With the beginning of the Cenozoic rifting since about 60 Ma the Transantarctic Mountains were uplifted at episodic rates up to 10 km (e.g. Fitzgerald *et al.*, 1986; 1992; Behrendt et al., 1991). In contrast, renewed uplift/ denudation of MBL up to ~2500 m associated with volcanic activity started in the Oligocene.

Our line of evidences is highly consistent with the cooling path based on U-Pb, $^{40}Ar/^{39}Ar$ and AFT data from the Ford Ranges of Richard *et al.* (1994). Adams *et al.* (1995) determined similar apatite fission track ages on six samples from the Edward VII Peninsula but concluded that the Alexandra and Rockefeller Mountains have no common uplift history. However the error quotes, the scattering of the track data and the concentration of the Alexandra Mountains samples in lower elevations would also agree with our interpretation. However, the track lengths of ~14 μm of the above authors are difficult to explain. Those lengths would imply an earlier final uplift than our data presented here.

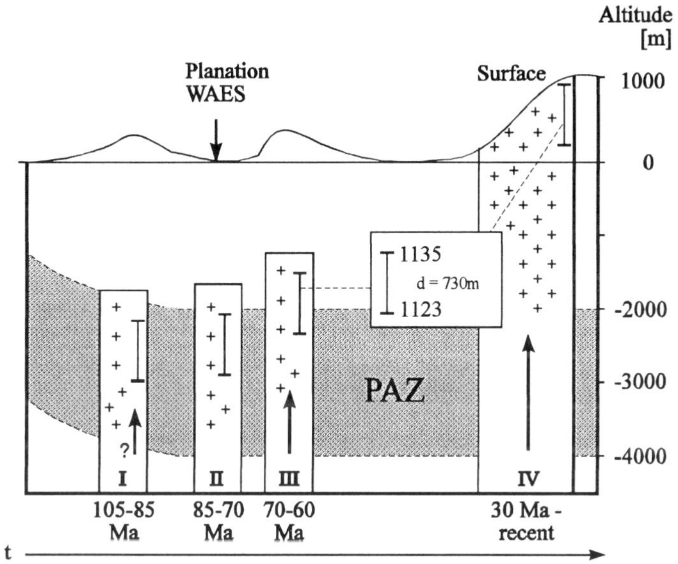

Figure 5. Schematic sequence of the cooling and denudation processes of western Marie Byrd Land. The topography of the surface and the amount of exhumation are indicated. Bars in the rock columns show the position of the separate samples (d: difference of elevation). WAES: West Antarctic Erosion Surface (LeMasurier and Rex, 1983). **I** Cooling immediately after intrusion of the Byrd Coast Granites. **II** Stagnation of denudation, followed by subsidence of the surface. The WAES has been formed. **III** Denudation in the late Cretaceous. The amount of denudation is estimated to be 500 - 1000 m. **IV** Final exhumation since the Oligocene (2000 - 2500 m).

6. Summary and Conclusions

The middle Cretaceous to Cenozoic history of MBL is different from the conjugate rift shoulder (northern and southern Victoria Land). Our data show a multi stage cooling history of the granitoids of western MBL. The unroofing took place at three discrete periods of accelerated cooling. These complex uplift/ denudation process confirms the relative sequence of events in the WARS as postulated by Tessensohn (1994): a) continental shear phase (Cretaceous, ±100 Ma), b) continental separation phase (Cretaceous/ Tertiary boundary, ±60 Ma) and c) phase of Ross Sea Rifting (Cenozoic, ±40 Ma to recent). For the first time our data prove and verify tectonic activity during the second rift stage in the Ford Ranges and the Edward VII Peninsula (cf. Fig. 5, stage III).

The sequence of events is as follows:

(1) Post intrusion cooling following the emplacement of the Byrd Coast Granites. The amount of erosion between 100 and 85 Ma can not be determined. The formation of the West Antarctic Erosion Surface was contemporaneous with tectonic quiescence between 85 - 70 Ma.

(2) Discrete increase in cooling rate between 75 - 60 Ma, involving ~1000 m of denudation (assuming a gradient of 30 °C/km). We interpret this stage as accelerated uplift/ denudation effected by lithospheric doming above a moving mantle plume beneath the MBL and Ross Sea region.

(3) Latest rifting phase since the Oligocene. From about 30 - 28 Ma until recent cooling increased involving denudation of up to 2500 m, associated with Late Tertiary volcanism and tectonism. Minor differences between the uplift/ denudation of the Ford Ranges and the Alexandra and Rockefeller Mountains are caused by horst-and-graben tectonics. This latest stage of uplift/ denudation is consistent with the uplift/ denudation and tectonic history described for the Transantarctic Mountains in northern and southern Victoria Land.

Acknowledgements

The authors are very indebted to the Bundesanstalt für Geowissenschaften und Rohstoffe, Hannover, for the invitation to participate in GANOVEX VII. They especially thank N. Roland for his support during the field work. This project was founded by a scholarship of the Universität Bremen to F. Lisker. M. Rohrmann and an anonymous reviewer critically reviewed the manuscript. We are grateful for their comments and suggestions.

References

Adams C.J. (1986): Geochronological studies of the Swanson Formation of Marie Byrd Land, West Antarctica, and correlation with northern Victoria Land, East Antarctica, and South Island, New Zealand. *New Zealand Journal of Geology and Geophysics* **29**, 345-358.

Adams C.J. (1987): Geochronology of granite terranes in the Ford Ranges, Marie Byrd Land, West Antarctica. *New Zealand Journal of Geology and Geophysics* **30**, 51-72.

Adams C.J., Seward D. and Weaver S.D. (1995): Geochronology of Cretaceous granites and metasedimentary basement on Edward VII Peninsula, Marie Byrd Land, West Antarctica. *Antarctic Science* **7**, 265-277.

Behrendt J.C. and Cooper A. (1991): Evidence of rapid Cenozoic uplift of the shoulder scarpment of the Cenozoic West Antarctic rift system and a speculation on possible climate forcing. *Geology* **19**, 315-319.

Behrendt J.C., LeMasurier W.E., Cooper A.K., Tessensohn F., Trehu A. and Damaske, D. (1991): The West Antarctic Rift System: A review of geophysical investigations. Contributions to Antarctic Research II. *Antarctic Research Series* **53**, 67-112.

Bentley C.R. and Robertson, J.D. (1982): Isostatic gravity anomalies in West Antarctica. In: *Antarctic Geoscience* (ed Craddock C.), University of Wisconsin Press, Madison, pp. 949-954.

Borg S.G., Stump E., Chappell B.W., McCulloch M.T., Wyborn D., Armstrong R.L. and Holloway J.R. (1987): Granitoids of Northern Victoria Land, Antarctica: Implications of chemical and isotopic variations to regional crustal structure and tectonics. *American Journal of Science* **287**, 127-169.

Bradshaw J.D. (1989): Cretaceous geotectonic patterns in New Zealand region. *Tectonics* **8**, 803-820.

Bradshaw J.D., Andrews P.B. and Field, B.D. (1983): Swanson Formation and related rocks of Marie Byrd Land and a comparison with the Robertson Bay Group of Northern Victoria Land. In: *Antarctic Earth Science* (eds Oliver R.L., James P.R. and Jago J.B.), Australian Academy of Science, Canberra, pp. 663-670.

Bradshaw J.Y. (1989): Origin and metamorphic history of an Early Cretaceous polybaric granulite terrane, Fiordland, southwest New Zealand. *Contrib. Mineral. Petrol.* **103**, 346-360.

Dalziel I.W.D. and Elliot D.H. (1982): West Antarctica: Problem child of Gondwanaland. *Tectonics* **1**, 3-19.

Dalziel J.W.D. and Brown R.L. (1989): Tectonic denudation of the Darwin metamorphic core complex in the Andes of Tierra del Fuego, southernmost Chile: Implications for Cordilleran orogenesis. *Geology* **17**, 699-703.

Dumitru, T. (1993) FT Stage. Stanford University.

Fitzgerald P.G. and Stump E. (1991): Early Cretaceous uplift in the Ellsworth Mountains of West Antarctica. *Science* **254**, 92-94.

Fitzgerald P.G. (1992): The Transantarctic Mountains of southern Victoria Land: the application of apatite fission track analysis to a rift shoulder uplift. *Tectonics* **11**, 634-662.

Fitzgerald P.G., Sandiford M., Barrett P.J. and Gleadow A.J.W. (1986): Asymmetric extension associated with uplift and subsidence in the Transantarctic Mountains and Ross Embayment. *Earth and Planetary Science Letters* **81**, 67-78.

Fitzgerald P.G., Sandiford M., Barrett P.J. and Gleadow A.J.W. (1986): Asymmetric extension associated with uplift and subsidence in the Transantarctic Mountains and Ross Embayment. *Earth and Planetary Science Letters* **81**, 67-78.

Futa K. and LeMasurier W.E. (1983): Nd and Sr isotopic studies on Cenozoic mafic lavas from West Antarctica: Another source for continental alkali basalts. *Contrib. Mineral. Petrol.* **83**, 38-44.

Gallagher K. (1995): Evolving temperature histories from apatite fission-track data. *Earth and Planetary Science Letters* **136**, 421-435.

Gleadow A.J.W. and Fitzgerald P.G. (1987): Uplift history and structure of the Transantarctic Mountains: New evidence from fission track dating of basement apatites in the Dry Valleys area, southern Victoria Land. *Earth and Planetary Science Letters* **82**, 1-14.

Gleadow A.J.W. (1981): Fission track dating methods: What are the real alternatives? *Nuclear Tracks* **5**, 3-14.

Gleadow A.J.W., McKelvey B.C. and Ferguson K.U. (1984): Uplift history of the Transantarctic Mountains in the Dry Valleys area, southern Victoria Land, Antarctica, from apatite fission track ages. *New Zealand Journal of Geology and Geophysics* **27**, 457-464.

Green P.F. (1981): A new look at statistics in fission-track dating. *Nuclear Tracks* **5**, 77-86.

Green P.F. (1985): Comparison of zeta calibration baselines for fission-track dating of apatite, zircon and sphene. *Chemical Geology (Isotope Geoscience Section)* **58**, 1-22.

Green P.F., Duddy I.R., Gleadow A.J.W., Tingate P.R. and Laslett G.M. (1986): Thermal annealing of fission tracks in apatite 1. A Qualitative Description. *Chemical Geology (Isotope Geoscience Section)* **59**, 237-253.

Grindley G.W. and Davey F.J. (1982): The reconstruction of New Zealand, Australia, and Antarctica. In: *Antarctic Geoscience* (ed Craddock C.), University of Wisconsin Press, Madison, pp. 15-29.

Hurford A.J. and Green P.F. (1983): The zeta age calibration of fission-track dating. *Chemical Geology (Isotope Geoscience Section)* **1**, 285-317.

Laslett G.M., Gleadow A.J.W. and Duddy, I.R. (1984): The relationship between fission track length and track density in apatite. *Nuclear Tracks* **9**, 29-38.

Lawver L.A. and Gahagan L.M. (1994): Constraints on timing of extension in the Ross Sea Region. *Terra Antartica* **1**, 545-552.

Lawver L.A., Royer J.Y., Sandwell D.T. and Scotese, C.R. (1991): Evolution of the Antarctic continental margins. In: *Geological Evolution of Antarctica* (eds Thomson M.R.A., Crame J.A. and Thomson J.W.), Cambridge University Press, Cambridge, pp. 533-540.

LeMasurier W.E. and Rex D.C. (1983): Rates of uplift and the scale of ice level instabilities recorded by volcanic rocks in Marie Byrd Land, West Antarctica. In: *Antarctic Earth Science* (eds Oliver R.L., James P.R. and Jago J.B.), Australian Academy of Science, Canberra, 663-670.

LeMasurier W.E. and Rex D.C. (1989): Evolution of linear volcanic ranges in Marie Byrd Land, West Antarctica. *Journal of Geophysical Research* **94**, 7223-7236.

LeMasurier W.E. and Rex D.C. (1994): Geologic events of the past 100 million years revealed by K-Ar dating of volcanic rocks from Marie Byrd Land: Implications for the offshore sedimentary record. *Terra Antarctica* **1**, 449-451.

LeMasurier W.E. (1978): The Cenozoic West Antarctic Rift System and its associated volcanic and structural features. *Geol. Soc. Am. Abstr. Programs* **10**, 443.

LeMasurier W.E. and Landis C.A. (1996): Mantle plume activity recorded by low relief erosion surface in West Antarctica and New Zealand. *Geol. Soc. Amer. Bull.*, in press.

Luyendyk B.P. (1995): Hypothesis for Cretaceous rifting of east Gondwana caused by subduced slab capture. *Geology* **23**, 373-376.

Luyendyk B.P., Richard S.M., Smith C.H. and Kimbrough D.L. (1992): Geological and geophysical exploration in the northern Ford Ranges, Marie Byrd Land, Antarctica. In: *Recent Progress in Antarctic Earth Science* (eds Yoshida Y., Kaminuma K. and Shiraishi K.), Terrapub, Tokyo, pp. 279-288.

Mattinson J.M., Kimbrough D.L. and Bradshaw J.Y. (1986): Western Fiordland orthogneiss: Early Cretaceous arc-magmatism and granulite facies metamorphism, New Zealand. *Contrib. Mineral. Petrol.* **92**, 383-392.

Mayes C.L., Lawver L.A. and Sandwell D.T. (1990): Tectonic history and new isochron chart of the South Pacific. *Journal of Geophysical Research* **95**, 8543-8567.
Mueller R.D., Royer J.-Y. and Lawver L.A. (1993): Revised plate motions relative to the hotspots from combined Atlantic and Indian Ocean hot spot tracks. *Geology* **21**, 555-558.
Mukasa S.B., Dalziel W.D. and Brueckner H.K. (1988): Zircon U-Pb constraints on the kinematic evolution of the northern Scotia Arc (abstract). *Geol. Soc. Am. Abstr. Programs* **20**, A12.
Richard S.M. (1992): Structure and cooling history of the Fosdick Metamorphic Complex, Marie Byrd Land, West Antarctica. In: *Recent Progress in Antarctic Earth Science* (eds Yoshida Y., Kaminuma K. and Shiraishi K.), Terrapub, Tokyo, pp. 289-294.
Richard S.M., Smith C.H., Kimbrough D.L., Fitzgerald P.G., Luyendyk B.P. and McWilliams M.O. (1994): Cooling history of the northern Ford Ranges, Marie Byrd Land, West Antarctica. *Tectonics* **13**, 837-857.
Rocholl A., Stein M., Molzahn M., Hart S.R. and Wörner G. (1995): Geochemical evolution of rift magmas by progressive trapping of stratified mantle source beneath the Ross Sea Rift, Northern Victoria Land, Antarctica. *Earth and Planetary Science Letters* **131**, 207-224.
Storey B.C., Dalziel I.W.D., Garrett S.W., Grunow A.M., Pankhurst R.J. and Vennum W.R. (1988): West Antarctica in Gondwanaland: Crustal blocks, reconstruction and break-up processes. *Tectonophysics* **155**, 381-390.
Stüwe K., White L. and Brown, R. (1994): The influence of eroding topography on steady-state isotherms. application to fission track analysis. *Earth and Planetary Science Letters* **124**, 63-74.
Tessensohn F. (1994): The Ross Sea region, Antarctica: Structural interpretation in relation to the evolution of the southern ocean. *Terra Antartica* **1**, 553-558.
Tulloch A.J. and Kimbrough D.L. (1989): The Paparoa metamorphic core complex, New Zealand: Cretaceous extension associated with fragmentation of the Pacific margin of Gondwana. *Tectonics* **8**, 1217-1234.
Van den haute P. (1986): Apatite fission-track dating applied to Precambrian terranes. *Chemical Geology (Isotope Geoscience Section)* **57**, 155-165.
Van der Wateren F.M., Luyendyk B.P., Verbers A.L.L.M. and Smith C.H. (1994): Landscape evolution model of the West Antarctic Rift System relating tectonic and climatic of the rift margins. *Terra Antartica* **1**, 453-456.
Vetter U. and Tessensohn F. (1987): S- and I-type granitoids of North Victoria Land, Antarctica, and their inferred geotectonic setting. *Geologische Rundschau* **76**, 233-243.
Wade F.A., Cathey C.A. and Oldham J.B. (1977a) Reconnaissance geologic map of the Alexandra Mountains quadrangle, Marie Byrd Land, Antarctica. Washington D.C., U.S. Geological Survey/ National Science Foundation. A-5.
Wade F.A., Cathey C.A. and Oldham J.B. (1977b) Reconnaissance geologic map of the Guest Peninsula quadrangle, Marie Byrd Land, Antarctica. Washington D.C., U.S. Geological Survey/ National Science Foundation. A-7.
Wade F.A., Cathey C.A. and Oldham J.B. (1977c) Reconnaissance geologic map of the Boyd Glacier quadrangle, Marie Byrd Land, Antarctica. Washington D.C., U.S. Geological Survey/ National Science Foundation. A-6.
Wade F.A., Cathey C.A. and Oldham J.B. (1978) Reconnaissance geologic map of the Gutenko Nunataks quadrangle, Marie Byrd Land, Antarctica. Washington D.C., U.S. Geological Survey/ National Science Foundation. A-11.
Wade F.A. and Couch D.R. (1982): The Swanson Formation, Ford Ranges, Marie Byrd Land. Evidence for and against a direct relationship with the Robertson Bay Group, northern Victoria Land. In: *Antarctic Geoscience* (ed Craddock C.), University of Wisconsin Press, Madison, pp. 609-618.
Wagner G.A., Gleadow A.J.W. and Fitzgerald P.G. (1989): The significance of the partial annealing zone in apatite fission-track analysis: projected track length measurements and uplift chronology of the Transantarctic Mountains. *Chemical Geology (Isotope Geoscience Section)* **79** 295-305.
Weaver S.D., Adams C.J., Pankhurst R.J. and Gibson I.L. (1992): Granites of Edward VII Peninsula, Marie Byrd Land: anorogenic magmatism related to Antarctic-New Zealand rifting. *Transactions, Royal Society of Edinburgh: Earth Sciences* **83**(1-2), 281-290.
Weaver S.D., Bradshaw J.D. and Adams, C.J. (1991): Granitoids of the Ford Ranges, Marie Byrd Land, Antarctica. In: *Geological Evolution of Antarctica* (eds Thomson M.R.A., Crame J.A. and Thomson J.W.), Cambridge University Press, Cambridge, pp. 345-352.
Weaver S.D., Storey B.C., Pankhurst R.J., Mukasa S.B., DiVenere V.J. and Bradshaw J.D. (1994): Antarctica-New Zealand rifting and Marie Byrd Land lithospheric magmatism linked to ridge subduction and mantle plume activity. *Geology* **22**, 811-814.

MULTIPLE THERMAL EVOLUTION OF OATES LAND (NORTHERN VICTORIA LAND, ANTARCTICA): EVIDENCE FROM APATITE FISSION TRACK ANALYSIS

T. Schäfer and M. Olesch
Fachbereich Geowissenschaften, Universität Bremen, Germany

Abstract: Results of an AFT study on basement rocks of Oates Land (northern Victoria Land: NVL, Antarctica) indicate a three stage thermal evolution since the Jurassic. The first episode, during the Late Jurassic (~180 ± 10 Ma), was probably associated with the Ferrar Event which marks the initial break-up of Gondwana. These thermal event caused different annealing in the basement of Oates Land. The second thermal overprint is related to the initiation of separation of Antarctica and Australia/Tasmania starting in mid Cretaceous times (~110 Ma). Since the Cenozoic the final denudation/uplift of the Transantarctic Mountains (TAM) along the area of the Rennick Graben has occurred with an apparent amount of ~4 up to 5 km.

1. Introduction and geologic setting

Three tectonostratigraphic terranes form the crystalline basement of NVL. The Robertson Bay terrane (Fig. 1) comprises a series of open-folded Cambrian-Ordovician flysch-type sedimentary rocks presumably deposited on continental crust (Gair et al., 1969; Sturm & Carryer, 1969; Tessensohn et al., 1981; Vetter & Tessensohn, 1987). The central Bowers terrane consists of greenstones derived from Cambrian submarine volcanic rocks overlain by Cambrian-Ordovician fluviomarine sedimentary rocks (Weaver et al., 1984). High-angle faults mark the boundary to the west against the Wilson terrane and to the east against the Robertson Bay terrane. The Wilson terrane shows relics of amphibolite (schists and gneisses of the Wilson Group) to granulite facies metamorphism (Carmigiani et al., 1989; Armienti et al., 1989).
Basement rocks of the Wilson terrane are extensively intruded by syn- to post-orogen Granite Harbour Intrusives (GHI) (Babcock et al., 1986; Borg et al., 1987; Vetter & Tessensohn, 1987; Olesch et al., 1996) caused by the early Paleozoic Ross orogeny. Adams (1986) obtained Rb-Sr intrusion ages between 490 and 470 Ma and gave an Rb-Sr isochron age of 490 Ma and monazite U-Pb ages between 488 and 484 Ma for the high-grade metamorphism in the Wilson terrane gneisses. Younger granite intrusions (390-360 Ma; Admiralty Intrusives) occur exclusively in the Bowers and Robertson Bay terranes (Borg et al., 1987).
Flöttmann & Kleinschmidt (1993) explain the geological similarities of the basement areas in the southern Wilson terrane (Daniels Range) with the basement of Oates Land

P. Van den haute and F. De Corte (eds.), Advances in Fission - Track Geochronology, 241-253.

as an continuation of the Wilson terrane to the north. The Wilson terrane of Oates Land and NVL is transacted by two major NNW-SSE trending thrust systems (Fig. 2) with contrasting senses of displacement (Flöttmann & Kleinschmidt, 1991a, 1991b, 1993). Flöttmann & Kleinschmidt (1993) interpreted the Wilson and Exiles thrusts as a structural expression of the lower Paleozoic cratonward (westward) directed subduction of oceanic crust and the accretion of the Robertson Bay and Bowers terranes at the Wilson terrane of NVL. The thrusting is interpreted to have been roughly contemporaneous with the granitic intrusions (Flöttmann & Kleinschmidt, 1993), dated at 490-470 Ma (Adams, 1986).

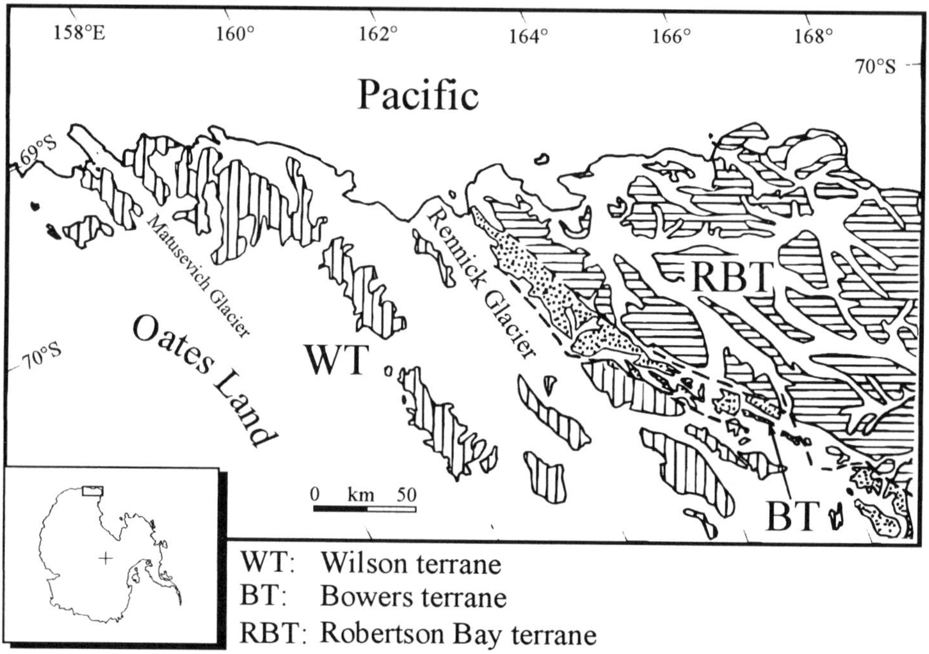

Figure 1. Geological units of northern Victoria Land and Oates Land, Antarctica. WT: Wilson Terrane, BT: Bowers Terrane, RBT: Robertson Bay Terrane.

The Jurassic (180 Ma) is characterised by strong volcanism, associated with dyke and sill intrusions in NVL. These intrusions (Ferrar Dolerite) and Kirkpatrick Basalt are evidence of the Antarctic-wide Ferrar Event, caused by the initial break-up of Gondwana (Kyle et al., 1981; Elliot, 1992). The separation of Gondwana continued in the Cretaceous (105-85 Ma) between East and West Anarctica. Since the Cenozoic the final denudation/uplift of the Transantarctic Mountains (TAM) occured. This mountain belt extends from NVL to the Weddell Sea, with elevations over 4500 m. Several explanations have been proposed for the uplift of the TAM including a "simple shear model" (Fitzgerald et al., 1986) with asymmetric passive rifting, a "flexural uplift model" (Stern & ten Brink, 1989) and a "delayed phase change model" as described by Smith & Drewry (1984).

In various sectors of the TAM several authors have applied AFT analyses to model the T-t-uplift path of the TAM (Fitzgerald et al., 1986; Gleadow & Fitzgerald, 1987;

Fitzgerald, 1992, 1994; Fitzgerald & Gleadow, 1988; Wagner et al., 1989; Fitzgerald & Stump, 1991; Stump & Fitzgerald, 1992; Balestrieri et al., 1994a, 1994b; Lisker, 1996). These authors described an episodic uplift including early phases in the Early Cretaceous and Late Cretaceous, but with a main phase of denudation/uplift during the Cenozoic.

This work presents the result of an AFT study using samples from granitoid rocks exposed in a wide area in Oates Land. The AFT analysis provides new information of the cooling history of Oates Land and adds to previous knowledge of the Mesozoic-Cenozoic evolution of NVL.

2. Method

27 apatite samples from granitoid rocks of Oates Land were analysed using the external detector method. Apatite concentrates were separated using conventional magnetic and heavy liquid techniques (Gleadow, 1981). The apatites were etched for 60 s in 5% HNO_3 and prepared for irradiation at the graphite reflector facility of Risø (Roskilde, Denmark). The thermal neutron fluence was monitored by recording the track density in a mica detector using the reference glasses SRM 612 (U.S. National Bureau of Standards) and CN 5 (Corning Museum). Fission track ages were calculated using the zeta calibration method and standard fission track age equation (Neaser, 1979; Hurford & Green, 1983; Green, 1986). The χ^2 statistic is used to detect the probability that all grains analyzed belong to a single population of ages (Galbraith, 1981). A probability of <5% is evidence of an asymmetric spread of single grain ages, in which case the "central age" (Galbraith & Laslett, 1993) is reported. Errors were calculated by the conventional method of Green (1981) and quoted as $\pm 1\sigma$. The track length distribution was recognised by measuring confined track lengths (Laslett et al., 1984). Data have been interpreted using genetic algorithms of Gallagher et al. (1991) and Gallagher (1995) based on an empirical kinetic description of laboratory annealing data in Durango apatite (Laslett et al., 1987).

3. Results

Apatite results are given in Table 1 for 27 localities and shown in Fig. 2 according to the sample number, the fission-track age and the mean track length. Apparent AFT ages range between 275 ± 14 and 43 ± 4 Ma. They clearly define geographically and thus, geologically distinct areas:

(1) Samples with apparent AFT ages >190 Ma are located in a greater distance of the Wilson and Exiles fault (Fig. 2). The mean confined track lengths of the samples range between 11.1 and 13.1 μm with standard deviations between 1.2 and 1.5 μm. These AFT ages from the samples western of the Rennick Graben are all significantly older than most of any other AFT age determined from Victoria Land. However, Fitzgerald & Gleadow (1988) as well as Lisker (1996) reported some similar ages west of the Rennick Graben. They are clearly pre-dating the Jurassic magmatic Ferrar Event. Furthermore, the AFT ages do not define a positive correlation with the sample elevation, as observed by Fitzgerald & Gleadow (1988) for samples taken also west of the Rennick Graben but in the southern Wilson terrane.

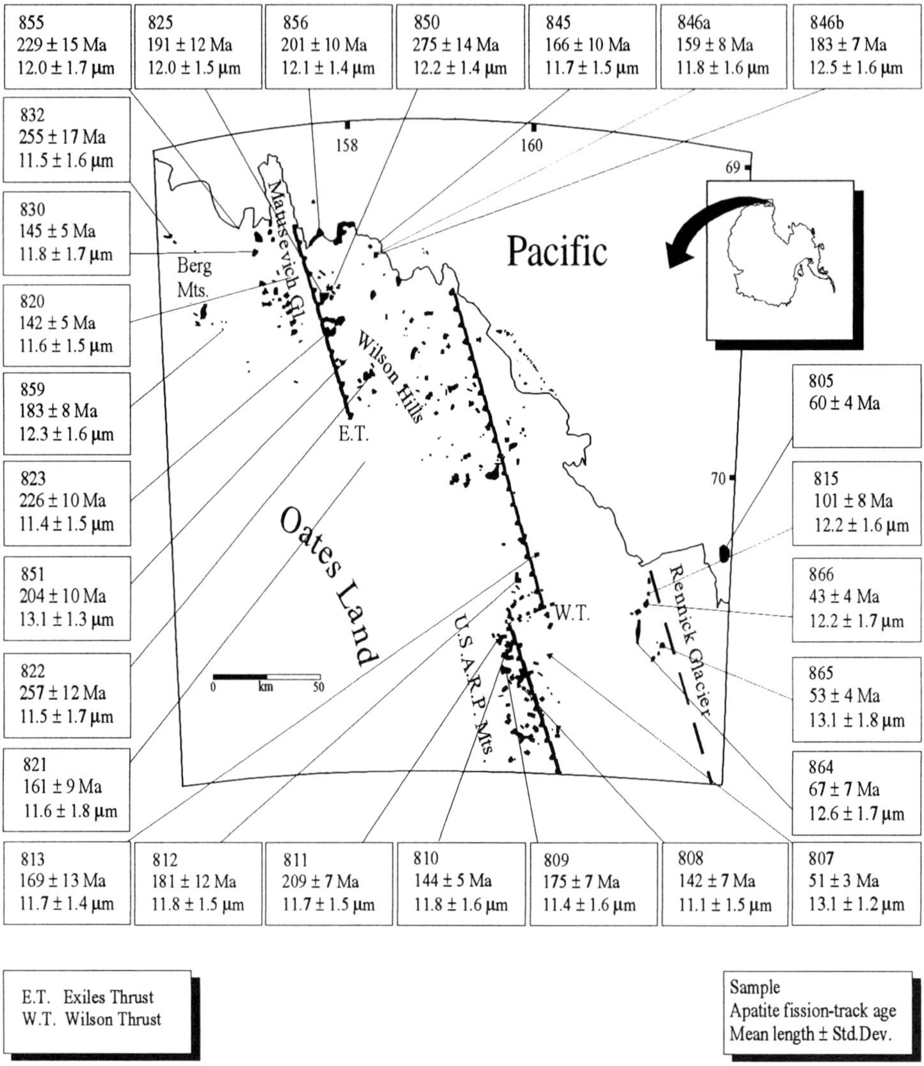

Figure 2. Map of Oates Land with the main Wilson terrane structures and the apatite fission track data. The apparent apatite fission track ages range between 43 ± 4 and 275 ± 14 Ma. The youngest samples were located along the Rennick Glacier. Intermediate ages correlate with the thrusts of Oates Land, whereas samples with ages >190 Ma more located in a greater distance to the faults.

(2) Intermediate AFT ages between ~140 and ~190 Ma correlate with the Wilson and Exiles thrust (Fig. 2). Their mean track length range between 11.1 and 12.0 μm with standard deviations between 1.5 and 1.7 μm. The majority of the AFT ages plotted against sample elevation define a positive correlation with a steep gradient similar to that in the Rennick area (Fig. 3).

(3) The youngest samples (apparent ages of ~50-60 Ma) are directly located along the western flank of the Rennick Graben. These samples have mean track lengths between

13.1 and 12.2 μm with standard deviations between 1.2 and 1.8 μm. In a plot of apparent AFT ages versus sample elevation (Fig. 3) the ages do not change significantly over large elevation intervals. These pattern is remarkably similar to that established for the southern coastal region of NVL (Mt Murchison and Lower Tucker Glacier) by Fitzgerald & Gladow (1988).

Table 1: Sample details and apatite fission track data, Oates Land (NVL, Antarctica). Analysis performed at the University of Bremen following the methods described by Green (1986). Standard and induced track densities measured on mica external detectors. Ages calculated using the zeta method (TS = 303.3 ± 9.6) for the dosimeter glass SRM 612 and (TS = 234 ± 8.6) for the dosimeter glass CN 5. Central age used when pooled age data failed the Chi square test at 5%. $P(\chi^2)$: Chi square probability. Nd, Ns, Ni: number of counted tracks. N-#: number of counted cristals. Lat.: Latitude, Long.: Longitude, Std.Dev.: Standard deviation. Brackets show the track densities. (n-L.): number of measured lengths.

Sample (N-#)	Location [Lat. S / Long. E]	Elev. [m]	Nd ($x10^6$ cm^{-2})	Ns ($x10^6$ cm^{-2})	Ni ($x10^6$ cm^{-2})	$P(\chi^2)$ [%]	Age [Ma±1σ]	Length [μm±1σ] (n-L.)	Std. dev. [μm]
805 (29)	70°14′ / 161°51′	250	4110 (1.36)	608 (0.54)	2306 (2.05)	0.1	60 ± 4	-	-
807 (20)	70°31′ / 160°09′	1400	4502 (0.99)	2289 (1.92)	5237 (4.40)	0.0	51 ± 3	13.1 ± 0.2 (100)	1.2
808 (18)	70°34′ / 159°50′	2010	4502 (0.99)	4049 (3.57)	3275 (2.89)	53.9	142 ± 7	11.1 ± 0.2 (69)	1.5
809 (17)	70°31′ / 159°43′	2420	4502 (0.99)	2693 (3.519)	1772 (2.31)	2.9	175 ± 7	11.4 ± 0.2 (94)	1.6
810 (20)	70°27′ / 159°41′	2040	4502 (0.99)	3983 (5.25)	3183 (4.20)	0.7	144 ± 5	11.8 ± 0.2 (102)	1.6
811 (25)	70°22′ / 159°47′	1670	4110 (1.36)	3920 (3.21)	3819 (3.13)	0.6	209 ± 7	11.7 ± 0.2 (65)	1.5
812 (17)	70°17′ / 159°51′	1420	4110 (1.36)	912 (1.801)	1027 (2.03)	7.1	180 ± 10	11.8 ± 0.2 (100)	1.5
813 (16)	70°14′ / 160°02′	900	4110 (1.36)	722 (1.81)	885 (2.22)	0.6	169 ± 13	11.7 ± 0.2 (30)	1.4
815 (20)	70°19′ / 161°05′	130	4110 (1.36)	1187 (1.96)	2467 (4.07)	0.0	101 ± 8	12.2 ± 0.2 (100)	1.6
820 (20)	69°30′ / 157°07′	990	4502 (0.99)	3231 (5.80)	2631 (4.73)	0.1	142 ± 5	11.6 ± 0.2 (102)	1.5
821 (20)	69°58′ / 158°01′	1500	4502 (0.99)	2076 (2.32)	(1477 (1.65)	42.5	161 ± 9	11.6 ± 0.2 (100)	1.8
822 (20)	69°40′ / 158°13′	1300	4110 (1.36)	2309 (3.33)	1813 (2.62)	67.5	257 ± 12	11.5 ± 0.2 (100)	1.7
823 (20)	69°29′ / 157°54′	1070	4110 (1.36)	3178 (5.93)	2840 (5.30)	37.4	226 ± 10	11.4 ± 0.2 (100)	1.5
825 (17)	69°19′ 157°38′	760	4110 (1.36)	771 (3.42)	821 (3.64)	97.2	191 ± 12	12.0 ± 0.2 (100)	1.5
830 (17)	69°14′ / 156°57′	470	4502 (0.99)	3844 (7.56)	3071 (6.04)	2.7	145 ± 5	11.8 ± 0.2 (100)	1.7
832 (16)	69°12′ / 156°02′	410	4110 (1.36)	684 (2.12)	541 (1.68)	42.8	255 ± 17	11.5 ± 0.2 (100)	1.6
845 (20)	69°14′ / 158°11′	750	4502 (0.99)	1410 (2.21)	975 (1.53)	92.7	166 ± 10	11.7 ± 0.2 (100)	1.5
846a (25)	69°14′ / 158°14′	100	4502 (0.99)	2257 81.85)	1634 (1.34)	28.7	159 ± 8	11.8 ± 0.2 (101)	1.6
846b (20)	69°14′ / 158°14′	100	4502 (0.99)	3286 (5.39)	2053 (3.37)	3.9	183 ± 7	12.5 ± 0.2 (100)	1.6
850 (20)	69°20′ / 157°55′	1020	4110 (1.36)	1804 (1.96)	1322 (1.44)	55.9	275 ± 14	12.2 ± 0.2 (100)	1.4
851 (20)	69°35′ / 158°01′	1000	4502 (0.99)	3512 (5.59)	1971 (3.14)	22.9	204 ± 10	13.1 ± 0.2 (100)	1.3
855 (12)	69°11′ / 156°47′	150	4110 (1.36)	663 (2.90)	587 (2.57)	70.3	229 ± 15	12.0 ± 0.2 (100)	1.7
856 (20)	69°12′ / 158°02′	470	4502 (0.99)	2567 (4.14)	1469 (2.37)	0.1	201 ± 10	12.1 ± 0.2 (100)	1.4
859 (20)	69°27′ / 156°25′	1060	4502 (0.99)	2904 (5.01)	1837 (3.17)	0.9	183 ± 8	12.3 ± 0.2 (100)	1.6
864 (15)	70°27′ / 161°01′	890	4110 (1.36)	458 (1.41)	1489 (4.58)	0.0	67 ± 7	12.6 ± 0.2 (105)	1.7
865 (18)	70°29′ / 161°15′	140	4502 (0.99)	293 (0.76)	635 (1.64)	58.4	53 ± 4	13.1 ± 0.2 (69)	1.8
866 (15)	70°19′ / 161°05′	120	4110 (1.36)	166 (0.52)	791 (2.49)	68.8	43 ± 4	12.2 ± 0.2 (37)	1.7

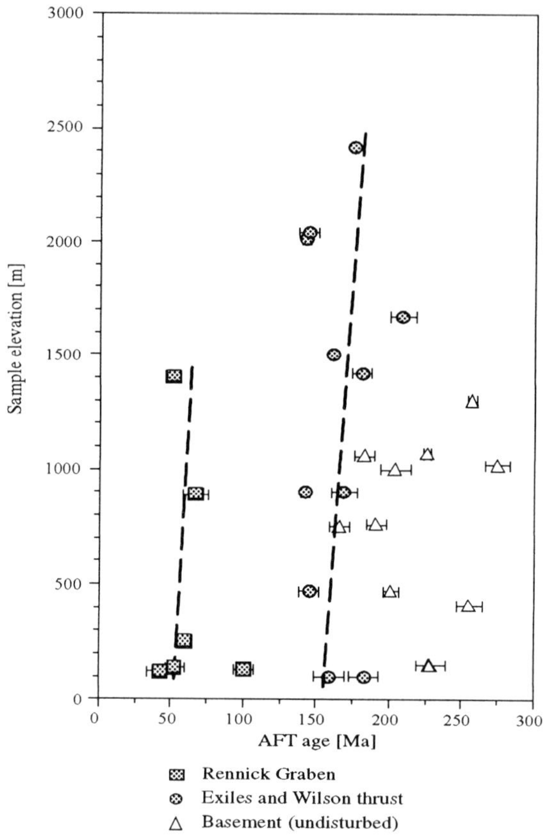

Figure 3. AFT ages for samples from Oates Land plotted against sample elevation. Note the young ages from the Rennick Graben with no change in age over the elevation intervals. Also AFT ages along the Wilson and Exiles thrust plot weakly positive against the sample elevation. Samples with greater distance to the faults do not correlate with the sample elevation.

4. Interpretation and Discussion

We determined for the area western of the Rennick Graben apparent AFT ages which similar to or predate the thermal Jurassic Ferrar Event (Fig. 4). The samples have experienced, at least partially, complete thermal overprinting at this time. The track length distributions appears to follows a regional trend. Length distributions (Fig. 5) for the samples with a greater distance to the Wilson and Exiles thrust tend from almost bimodal distributions (eg. 811, 855) to a narrowst spread (eg. 850, 856, 859). Samples which correlate with the thrust systems have a reduced shorter track component (eg. 809, 808, 823).

Late Precambrian to Palaeozoic rocks in southeastern Australia were correlated with the Wilson, Bowers, and Robertsn Bay terranes, prior to rifting (eg. Oliver, 1972; Burret & Findlay, 1984, Stump et al., 1986; Findlay, 1987). AFT ages from granitic rocks in the Dundas Tableland of the Delamarian terrane, which was correlated with the Wilson terrane (Stump et al., 1986), range from 316 ± 14 to 384 ± 10 Ma and were interpreted as slow cooling ages following Ordovician-Devonian emplacement (Gleadow & Lovering, 1978). In contrast to NVL, the Delamerian was not affected by major Jurassic magmatism, nor has it undergone significant Cenozoic uplift. Fitzgerald & Gleadow (1988) suggested a similar post-Palaeozoic thermal history for comparable AFT ages from the Lichen Hills in the southern extension of the Wilson terrane. However, Fitzgerald & Gleadow (1988) pointed out that the observed slightly younger AFT ages probably reflects very slight overprinting during Jurassic magmatism in the Lichen Hills. These trend could be also observed in Oates Land. Our younger AFT ages probably reflect a stronger thermal overprint as in the southern Wilson terrane. Furthermore, it seems that samples along the Wilson and Exiles thrust were affected much stronger by the Jurassic thermal event. The AFT ages dating the Jurassic thermal event or the samples were probably almost completely overprinted by a second post-Jurassic thermal event, reducing the AFT ages from older than 190 Ma to ages between ~140 and <190 Ma.

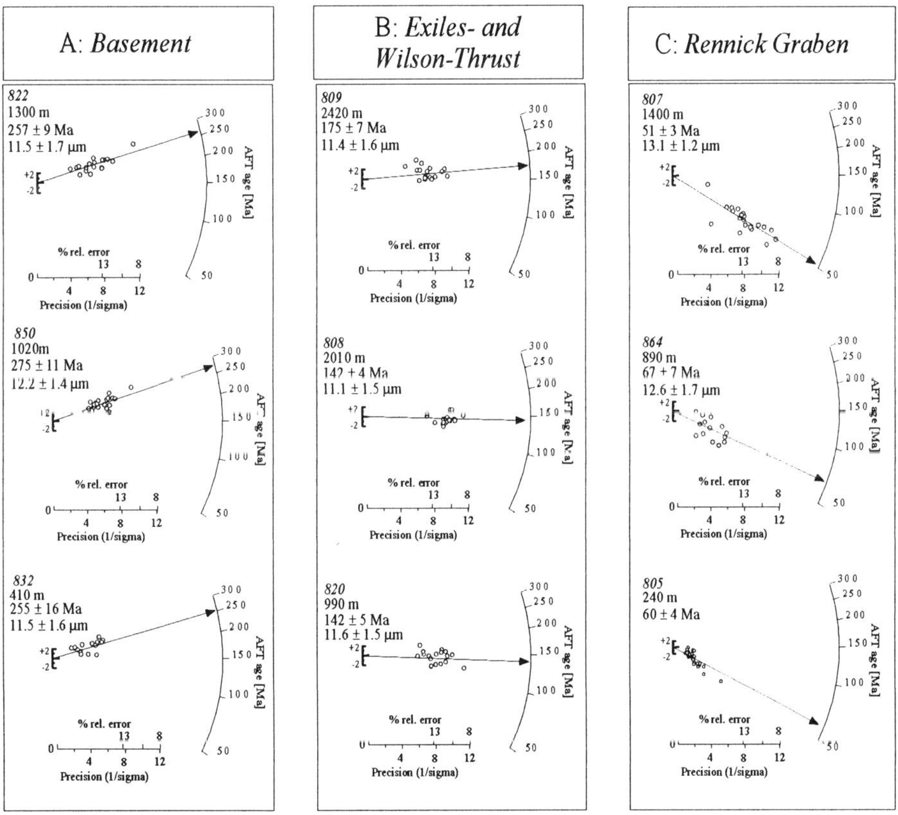

Figure 4. Representative data from Oates Land. A: The radial plots of samples with a greater distance to the thrust system have single grain ages older than the Jurassic Ferrar event. B: Along the Exiles and Wilson thrust the apparent apatite fission track ages correlate with the Jurassic thermal event or post-date these event by ~40 Ma. C: The young AFT ages along the Rennick Graben are correlate with the Cenozoic denudation/uplift phase of Oates Land.

Another explanation accounts for the apparent AFT ages between 100 and ~50 Ma along the western flank of the Rennick Graben. In the age-sample elevation profile (Fig. 3) is no "break in slope" at around 50 Ma as it is determined by various authors for NVL (eg. Fitzgerald & Gleadow, 1988; Balestrieri et al., 1994), but the ages are similar to those for the regional ages of these authors. We assume that the observed AFT ages ~50 Ma represent samples which layed below a fossil apatite PAZ zone with zero AFT ages prior to the Cenozoic uplift. We suggest that older samples which layed at varying levels in a pre-uplift apatite PAZ were completely eroded since the Cenozoic

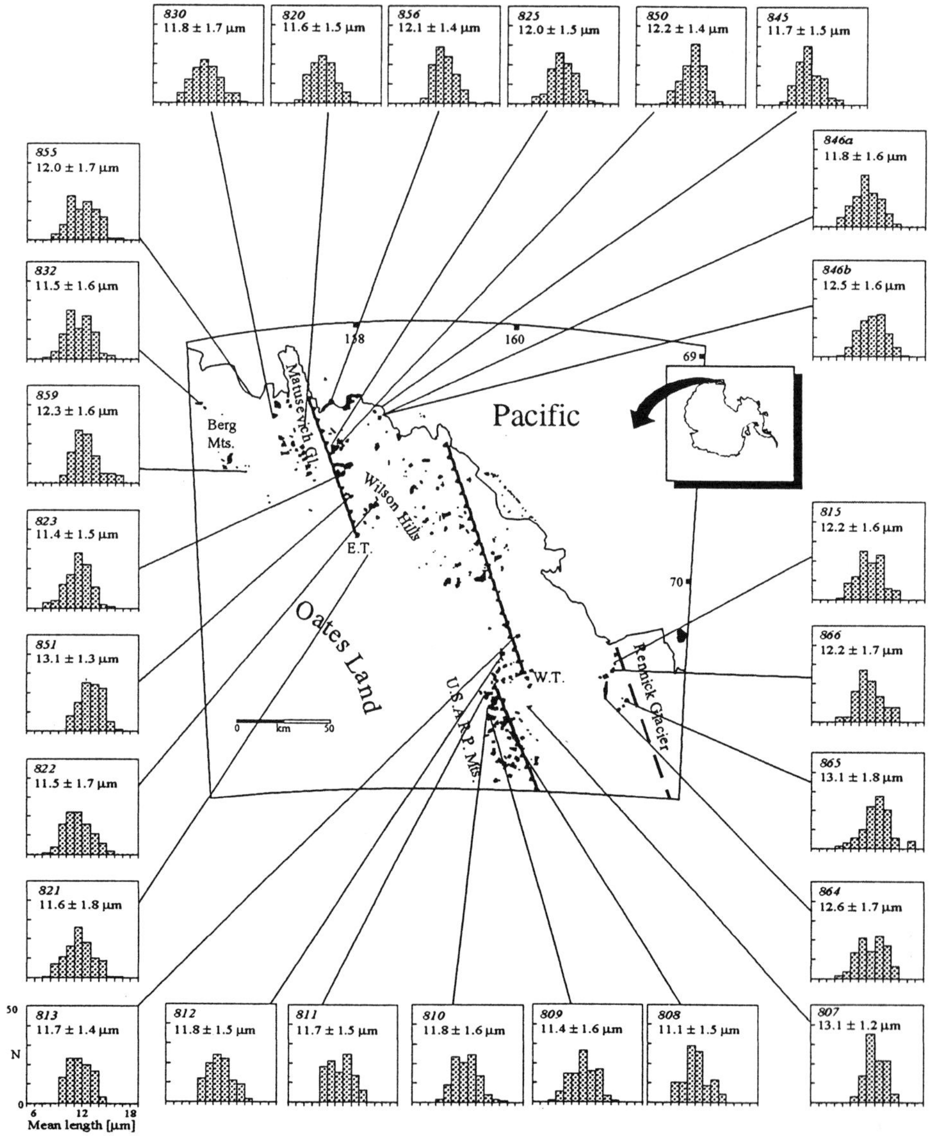

Figure 5. AFT confined length distributions plotted according to the sample location. Distributions normalised to 100.

denudation/uplift. Therefore, it was not possible to reconstruct the recent elevation of these uplifted PAZ in Oates Land. If we assume that uplift began at the same time in Oates Land as in the rest of NVL (~50 Ma) and the calculated palaeo-gradient of ~25 °C for southern Victoria Land (Fitzgerald & Gleadow, 1988) is equal in Oates Land, it is possible to estimate crudely the apparent amount of denudation/uplift. The same AFT ages at around 50 Ma could be estimated on samples with elevations between ~200 and 1400 m. Assuming an annual surface temperature of 0 °C and a palaeo-surface elevation of between 0 and 500 m, than the apparent denudation/uplift here has been in the order of 3.7 up to 5.4 km since the Cenozoic. It must be pointed out that the calculated apparent amount of denudation/uplift in the Rennick area of Oates Land is only a vague estimation, caused by the limited data set.

5. Summary and Conclusions of the thermal history

The AFT results from Oates Land are similar to the results of previous AFT analysis in NVL and southern Victoria Land (eg. Fitzgerald & Gleadow, 1988; Wagner et al., 1989; Fitzgerald & Stump, 1991; Stump & Fitzgerald, 1992; Balestrieri et al., 1994a, 1994b; Lisker, 1996). West of the Rennick Graben the oldest AFT ages in Oates Land and NVL were determined. These ages pre-date the Jurassic Ferrar Event and we suggested that these samples were affected by the Jurassic thermal event by varying degrees resulting in apparent AFT ages younger then the emplacement ages of the granitoid samples. We assume that the tholeiitic plutonism and volcanism associated with the emplacement of the Ferrar Dolerite and extrusion of Kirkpatrick Basalt at 179 ± 2 Ma (Kyle et al., 1981) caused an overprint of the basement of Oates Land. This event was certainly almost extensional, and Schmidt & Rowley (1986) suggest the formation of a continental rift during the Jurassic. It was shown by Dalziel et al. (1987) and Dalziel & Pankhurst (1987) that the emplacement of granite and gabbro of similar age (180 Ma) in the Ellsworth-Whitemore Mountain block was the result of an early thermal event during the Jurassic that may have triggered the initial breakup of Gondwana. Woolfe & Barrett (1995) suggested that the same may be true for the Ferrar Group in southern Victoria Land. This event may be influenced by thermal uplift caused by a late Triassic to early Jurassic mantle plume and subsequent development of a continental failed rift during the mid Jurassic in the Ross Sea sector (cf. Fig. 6, in Woolfe & Barrett, 1995).

Along the Wilson and Exiles thrust the effect of a thermal overprint was much more intensive. The apparent AFT ages between 190 and ~140 Ma date the Ferrar Event or post-date these event by ~40 Ma. It was assumed that a second thermal event caused the apparent AFT ages slightly younger than ~180 Ma (Ferrar Event).

The variety of the thermal evolution present in Oates Land can be traced in a plot of apatite ages against the mean track lengths (Fig. 6). The extreme left of the diagram starts with the original slow cooling ages of the granitoids. Then the ages decrease with increasing amounts of annealing due to Jurassic magmatism. The mean track lengths initially decrease and then increase representing short partially annealed pre-Jurassic tracks and relatively longer post-Jurassic tracks. Post-Jurassic the samples slowly cooled again and the apparent AFT ages decrease with increasing amount of annealing due to a Cretaceous (~100 Ma) event. The mean track lengths initially

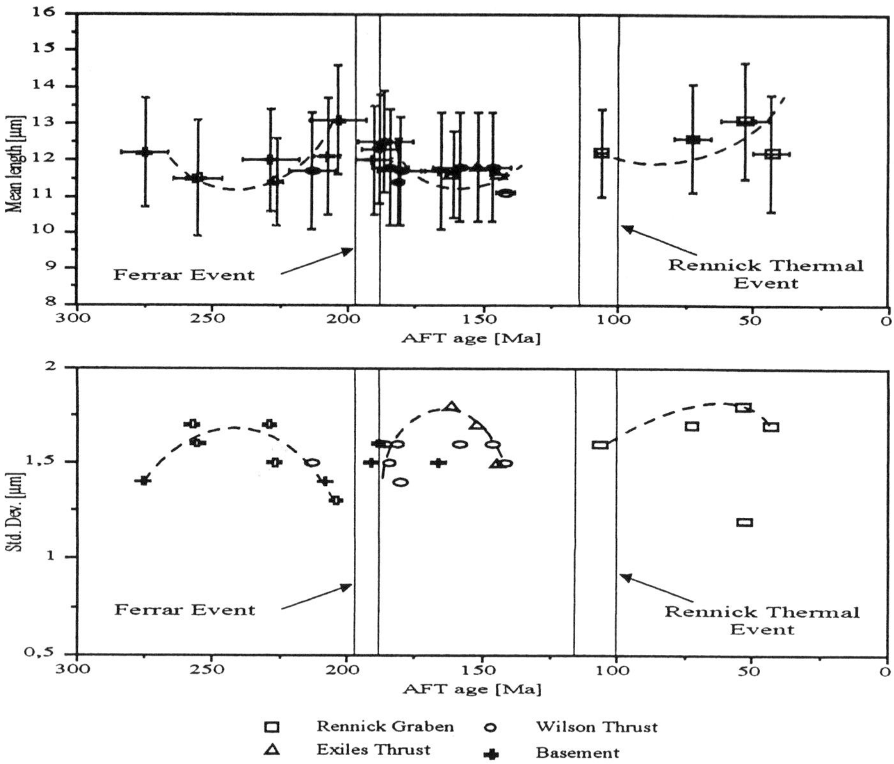

Figure 6. Apatite fission track data from Oates Land. Relationships are shown between mean track length (± 1σ) and fission track age (±1σ) and track length standard deviation and fission track age (±1σ). This plot summarises the thermal histories existing in Oates Land. See text for explanation.

decrease and then increase also representing short partially annealed pre-Cretaceous tracks and relatively longer post-Cretaceous tracks. The Cretaceous event obviously affected the samples along the Wilson and Exiles thrust much more than the samples with with a greater distance to these faults.

Tessensohn (1994) pointed out a three stage post-Cretaceous evolution for NVL. A first phase of continental shearing (±100 Ma) supports the formation of the Rennick basins onshore and the Victoria Land basin offshore. This initial event was not directly dated but there are several indications in the Jurassic volcanic rocks of the Rennick Graben and its flanks (120-90 Ma K/Ar ages, Kreutzer et al., 1981) and anomalous age data from the Mesa Ranges (Elliot & Foland, 1986; Mensing et al., 1988; Fleming et al., 1993; Foland et al., 1993). The second phase (± 60 Ma) marks the final separation of Antarctica from Australia/Tasmania. Hinz et al. (1990) deduced from data of the Tasmanian side an age of about 53 Ma for these phase of separation. During the Cenozoic the Ross sea rifting accelerated and marks the third phase. This late stage is characterised by the development of N-S orientated extensional features and NE-SW directed faults (Tessensohn, 1994). The tectonic scenario was associated by rift-related alkali volcanism and plutonism. Previous AFT data indicated that at this time (~50 Ma) the major uplift phase of the TAM took place (e.g. Fitzgerald et al., 1986; Fitzgerald & Gleadow, 1988; Fitzgerald & Gleadow, 1990; Fitzgerald, 1992; Balestrieri et al., 1994a). The Cenozoic denudation/uplift phase of Oates Land is visualised by the apparent AFT ages on the extreme right of the diagram (Fig. 6).

These samples with AFT ages at around ~50 Ma tend to longer mean track lengths and small standard deviations. This indicates a period of rapid uplift.
For different areas of NVL different amounts of uplift have been proposed (Fitzgerald & Gleadow, 1988; Balestrieri et al., 1994b; Lisker, 1996). The amount of denudation/uplift decrease from eastern to western NVL from ~10 km (Mt. Murchison) to ~2 km (northern area of Oates Land). It is evident from the AFT results that Oates Land western of the Rennick Graben follows the denudation/uplift history which was suggested by Fitzgerald & Gleadow (1988) for the areas western of the Rennick Graben and for the rest of NVL. The trend of older apparent AFT ages from the southern area of the Wilson terrane (Fitzgerald & Gleadow, 1988) could also be observed continuously throughout Oates Land by our AFT results. Furthermore, we can show that a Cretaceous thermal event influenced the basement of Oates Land which was first suggested by Lisker (1996) for the northern part of Oates Land and was not described for other regions of NVL before.

Acknowledgments

This work has been carried out with the financial support of the University of Bremen and grants of the Deutsche Forschungsgemeinschaft. Helpful discussions with J. Jacobs and F. Lisker are gratefully acknowledged. The field work was done by M. Olesch during the GANOVEX V Antarctic expedition. We are thankful for the support of the expeditionleader N.W. Roland, Bundesanstalt für Geowissenschaften, Hannover. We thank P.G. Fitzgerald and A.J. Hurford for their careful review and for constuctive comments.

References

Adams, C.J., 1986. Age and ancestry of metamorphic rocks in the Daniels Range, USARP Mountains, Antarctica.- In: Stump, E. (ed.), *Geological investigations in northern Victoria Land.* Washington, D.C., American Geophysical Union, 25-38.

Armienti, P., Ghezzo, C., Innocenti, F., Mannetti, P., Rocchi, S. and Tonarini, S., 1989. Genesis and evolution of Granite Harbour Intrusives of the Wilson terrane: Geochemical and isotopic evidence.- *Earth Science in Antarctica*, 2nd Meeting, Siena 1988, Proceedings, 133-137.

Babcok, R.S., Plummer, C.C., Sheraton, J.W. and Adams, C.J., 1986. Geology of the Daniels Range, northern Victoria Land, Antarctica.- In: Stump, E. (ed.), *Geological investigations in northern Victoria Land.* Washington, D.C., American Geophysical Union, 1-24.

Balestrieri, M.L., Bigazzi, G., Ghezzo, C. and Lombardo, B., 1994a. Fission track dating of apatites from the Granite Harbour Intrusives Suite and uplift-denudation history of the Transantarctic Mountains in the area between the Mariner and David Glaciers (northern Victoria Land, Antarctica).- *Terra Antartica*, **1**, 82-87.

Balestrieri, M.L., Bigazzi, G., Ghezzo, C. and Lombardo, B., 1994b. A review of apatite Fission Track Data from Northern Victoria Land and a First Indication of a Late Cretaceous Uplift Phase.- *Terra Antartica*, **1**, 539-540.

Borg, S.G., Stump, E., Chappell, B.W., Mc Culloch, M.T., Wyborn, D., Armstrong, R.L. and Holloway, J.R., 1987. Granitoids of Northern Victoria Land, Antarctica: Implications of chemical and isotopic variations to regional crustal structure and tectonics.- *American Journal of Science*, **287**, 127-169.

Burret, C.F, Findlay, R.H., 1984. Cambrian and Ordovician conodonts from the Robertson Bay Group, Antarctica and their tectonic significance.- *Nature*, **307**, 723-726.

Carmigiani, L., Ghezzo, L., Gosso, G., Lombardo, B., Meccheri, M., Montrasio, A., Pertusati, P.C. and Salvini, F., 1989. geology of the Wilson terrane area between David and Mariner glaciers, Victoria Land (Antarctica).- *Società Geologica Italiana*, Memorie, **33**, 77-98.

Dalziel, I.W.D. and Pankhurst, R.J., 1987. Tectonic development of Wset Antarctica and its relation to East Antarctica: Joint US/UK program 1986-87.- *Antarctic Journal of the United States* (1987 review), **22**, 50-51.

Dalziel, I.W.D., Garrett, S.W., Grunow, A.M., Pankhurst, R.J., Storey, P.C. and Vennum, W.R., 1987. The Ellworth-Whitemore Mountains crustal block: Its role in the tectonic evolution of West antarctica.- In: Mc Kenzie (ed.), *Gondwana Six: structure, Tectonics and Geophysics*.- American Geophysical Union, geophys. Monogr., 40, 173-182.

Elliot, D.H., 1992. Jurassic magmatism and tectonism associated with Gondwanaland breakup: an Antarctic perspective.- In: Storey, B.C., Alabaster, T., Pankhurst, R.J. (eds.). *Magmatism and the Causes of Continental Break-up*.- Geological Society Special Publication, No. 68, 165-184.

Elliot, D.H. and Foland, K.A., 1986. K-Ar age determinations of the Kirkpatrick Basalt, Mesa Range.- In: Stump, E. (ed.), *Geological investigations in northern Victoria Land*. Washington, D.C., American Geophysical Union, 279-288.

Findlay, R.H., 1987. A review of the problems important for interpretation of the Cambro-Ordovician palaeogeography of northern Victorian Land (Antarctica), Tasmania, and New Zealand. In: McKenzie, G.D. (ed.). *Gondwana six: structure, Tectonics, and Geophysics, Geophysical Monograph*, 40, 49-66. Washington, American Geophysical Union.

Fitzgerald, P.G., 1992. The Transantarctic Mountains of southern Victoria Land: the application of apatite fission track analysis to a rift shoulder uplift.- *Tectonics*, **11**, 634-662.

Fitzgerald, P.G., 1994. Thermochronological constraints on post-paleozoic tectonic evolution of the Central Transantarctic Mountains, Antarctica.- *Tectonics*, **13**, 818-836.

Fitzgerad, P.G. and Gleadow, A.J.W., 1988. Fission-track geochronology, tectonics and structure of the Transantarctic Mountains in Northern Victoria Land, Antarctica.- *Chemical Geology (I.G.S.)*, **73**, 169-198.

Fitzgerald, P.G. and Gleadow, A.J.W, 1990. New approaches in fission track geochronology as a tectonic tool: Examples from the Transantarctic Mountains.- *Nuclear Tracks*,**17**, 351-357.

Fitzgerald, P.G. and Stump, E., 1991. Uplift history of the Transantarctic Mountains from Victoria Land (70 °) to scott Glacier (86°): evidence from fission-track analysis.- *Sixth Int. Symp. on Antarctic Earth Sciences, 9.-13. Sept. 1991, Saitama, Japan*, Abstracts, 1p.

Fitzgerald, P.G., Sandiford, M., Barrett, P.J. and Gleadow, A.J.W., 1986. Asymmetric extension in the Transantarctic Mountains and Ross Sea Embayment.- *Earth Planet. Sci. Lett.*, **81**, 67-81.

Fleming, T.H., Elliot, D.H., Foland, G.A., Jones, L.M. and Bowman, J.R., 1993. Disturbance of Rb-Sr and K-Ar isotopic systems in the Kirkpatrick Basalt, north Victoria Land, Antarctica: implications for middle Cretaceous tectonism.- In: Findlay, R.H., Unrug, R., Banks, M.R. and Veevers, J.J. (eds.). *Gondwana Eight*, Balkema, Rotterdam, 411-423.

Flöttmann, T. and Kleinschmidt, G., 1991a. Opposite thrust systems in northern Victoria Land, Antarctica: imprints of Gondwana's Paleozoic accretion.- *Geology*, **19**, 45-47.

Flöttmann, T. and Kleinschmidt, G., 1991b. Kinematics of major structures of North Victoria and Oates Land, Antarctica.- *Mem. Soc. Geol. It.*, **46**, 273-282.

Flöttmann, T. and Kleinschmidt, G., 1993. The structure of Oates land and Implications for the structural style of northern Victoria Land, Antarctica, *Geologisches Jahrbuch*, **E47**, 419-436.

Foland , K.A., Fleming, T.H., Heimann, A. and Elliot, D.H., 1993. Potassium-Argon dating of fine-grained basalts with massive Ar loss: Application of the $^{40}Ar/^{39}Ar$ technique to plagioclase and glass from the Kirkpatrick Basalt, Antarctica.- *Chemical Geology (I.G.S.)*, **107**, 173-190.

Galbraith, R.F., 1981. On statistical Models for Fission track counts.- *Mathematical Geology*, **13**, 471-478.

Galbraith, R.F. and Laslett, G.M., 1993. Statistical models for mixed fission track ages.- *Nuclear Tracks*, **21**, 459-470.

Gallagher, K., Sambridge, M. and Drijkoningen, G., 1991. Genetic Algorithms: An Evolution from Monte Carlo Methods for Strongly Non-Linear Geophysical optimization Problems.- *Geophysical Research Letters*, Vol. 18, No. **12**, 2177-2180.

Gallagher, K., 1995. Evolving temperature histories from apatite fission-track data.- *Earth Planet. Sci. Lett.*, **136**, 421-435.

Gleadow, A.J.W., 1981. Fission track dating methods: What are the real alternatives?.- *Nuclear Tracks*, **5**, 3-14.

Gleadow, A.J.W. and Lovering, J.F., 1978. Thermal history of granitic rocks from western Victoria: a fission-track dating study.- *Geolgical Society of Australia Journal*, **25**, 323-340.

Gleadow, A.J.W. and Fitzgerald, P.G., 1987. Uplift history and structure of the Transantarctic Mountains: New evidence from fission track dating of basement apatites in the dry Valleyes area, southern Victoria Land.- *Earth Planet. Sci. Lett.*, **82**, 1-14.

Green, P.F.,1981. A new look at statistics in the fission track dating.- *Nuclear Tracks*, **5**, 77-86.

Green, P.F., 1986. On the thermo-tectonic evolution of Northern England: Evidence from fission track analysis.- *Geolgical Magazine*, **123**, 493-506.

Hinz, K., Hemmerich, M., Salge, U. and Eiken, O., 1990. Structures in rift-basin sediments on the conjugate margins of western Tasmania, south tasman Rise, and Ross Sea, Antarctica.- In: Bleil, U. And Thiede, J. (eds.), *Geological History of the Polar Oceans*.- Arctic versus Antarctic, 119-130.

Hurford, A.J. and Green, P.F., 1983. The zeta age calibration of fission track dating.- *Isotope Geosci.*, **1**, 285-317.

Kreuzer, H., Höhndorf, A., Lenz, H., Vetter, U., Tessensohn, F., Müller, P., Jordan, H., Harre, W. and Besang, C., 1981. K/Ar and Rb/Sr dating of Igneous Rocks from North Victoria Land, Antarctica.- *Geolgisches Jahrbuch*, **B41**, 267-273.

Kyle, P.R., Elliot, D.H. and Sutter, J.F., 1981. Jurassic Ferrar Supergroup Tholeiites from the Transantarctic Mountains, Antarctica and their relationship to initial fragmentation of Gondwana.- In: Cresswell, M.M. and Vella, P. (eds.), *Gondwana Five*, Balkema, Rotterdam, 283-287.

Laslett, G.M., Gleadow, A.J.W. and Duddy, I.R., 1984. The Relationships between Fission Track Length and Track Density in Apatite.- *Nuclear Tracks*, **9**, 29-38.

Laslett, G.M., Green, P.F, Duddy, I.R. and Gleadow, A.J.W., 1987. Thermal modelling of fission tracks in apatite: A quantitative analysis.- *Chemical Geology*, **65**, 1-13.

Lisker, F., 1996. Geodynamics of the West Antarctic Rift System based on Apatite Fission Tracks.- *Berichte z. Polarforschung*, **198**, 108 pp.

Mensing, T.M., Faure, G., Jones, l.M. and Hoefs, J., 1988. Evidence for Cretaceous volcanism in northern Victoria Land, Antarctica, and implications for a mantle enrichment event during the Ross Orogeny based on chemical and isotopic data from the Kirkpatrick Basalt in the Mesa Range.- *7th Gondwana Symposium*, Sao Paulo, Brazil (abstrct).

Neaser, C.W., 1979. Fission track dating and geologic annealing of fission tracks. In: Jäger, E. and Hunziker, J.C. (eds.), *Lectures in isotope Geology*.- Springer, New York, 154-169.

Olesch, M., Roland, N.W., Fenn, G., Krauss, U., 1996. Petrogenesis of Granitoid Rocks of Oates Land, Antarctica.- *Geologisches Jahrbuch*, **B89**, 195-245.

Oliver, R.L., 1972. Some aspects of Antarctic-Australian geological relationships. In Adie, R.J. (ed.), *Antarctic Geology and Geophysics*, 858-864. Oslo: Universistetsforlaget.

Roland, N.W. and Tessensohn, F., 1987. Rennick faulting- an early phase of Ross sea Rifting.- *Geologisches Jahrbuch*, **B66**, 203-229.

Schmidt, D.L. and Rowley, P.D., 1986. Continental rifting and transform faulting along the Jurassic Transantarctic Rift.- *Tectonics*, **5**, 279-291.

Smith, A.G. and Drewry, D.J., 1984. Delyed phase change due to hot asthenosphere causes transantarctic uplift?.- *Nature*, **309**, 536-538.

Stern, T.A. and ten Brink,, U.S., 1989. Flexural uplift of the Transantarctic Mountains.- *Journal of Geophysical Research*, **94**, 10315-10330.

Stump, E. and Fitzgerald, P.g., 1992. Episodic uplift of the Transantarctic Mountains. *Geology*, **20**, 161-164.

Stump, E., White, A.J.R., Borg, S.G., 1986. Reconstruction of Australia and Antarctica: evidence from granites and recent mapping.- *Earth Planet. Sci. Lett.*, **79**, 348-360.

Tessensohn, F., 1994. The Ross Sea region, Antarctica: Structural Interpretation in Relation to the Evolution of the southern Ocean.- *Terra Antartica*, **1**, 553-558.

Vetter, U. and Tessensohn, f., 1987. S- and I-type granitoids of North Victoria Land, Antarctica, and their inferred geotectonic setting.- *Geologische Rundschau*, **76**, 233-243.

Wagner, G.A., Gleadow, A.J.W. and Fitzgerald, P.G., 1989. The significance of the partial annealing zone in apatite fission track analysis: projected track length meassurements and uplift chronology of the Transantarctic Mountains.- *Chemical Geology (I.G.S.)*, **79**, 295-305.

Weaver, S.D., Bradshaw, J.D. and Laird, M.G., 1984. Geochemistry of Cambrian Volcanics of the Bowers Supergroup and Implications for the Early Paleozoic Tectonic evolution of Northern Victoria Land, Antarctica.- *Earth Planet. Sci. Lett.*, **68**, 128-140.

Woolfe, K.J. and Barrett, P.J., 1995. Constraining the Devonian to triassic tectonic Evolution of the Ross Sea Sector- *Terra Antartica*, **2**, 7-21.

EFFECTS OF SURFACE COOLING ON APATITE FISSION-TRACK DATA: EVIDENCE FOR MIOCENE CLIMATIC CHANGE, NORTH SLOPE, ALASKA

PAUL B. O'SULLIVAN & RODERICK W. BROWN
VIEPS School of Earth Sciences
La Trobe University, Bundoora, Victoria, 3083, Australia

Abstract

Apatite fission-track (AFT) data from samples collected from the Exxon Alaska State C-1 well located along the northern coastline of Alaska, indicate cooling of the order of ~15-20°C occurred during the Miocene at some time between 20 and 10 Ma. There are three possible explanations for the recorded cooling: 1) a decrease in the geothermal gradient, 2) erosion, or 3) sub-surface cooling in response to a long-term decrease in the paleo-mean annual surface temperature. The first option is inconsistent with the fission-track and vitrinite reflectance data from the well. The second option is unlikely, since limited burial history analyses and palynology data from wells located in the region do not record any erosion during the Oligocene or Miocene. The third option seems most likely since a major climatic change occurred across the North Slope of Alaska during the late Cenozoic decreasing the mean annual surface temperature by between 10° and 20°C, resulting in subsurface cooling of a similar magnitude. These results indicate that: 1) subsurface AFT data can be used to determine the timing of cooling and/or heating which occurs in response to large magnitude (>~10°C) changes in long-term mean surface temperature; 2) the subsurface effects of long-term climatic changes must be taken into consideration when using AFT data to estimate denudation; and 3) the AFT data from the C-1 well place new constraints on the initial timing of the establishment of long-term arctic conditions in northern Alaska.

Keywords

Apatite fission-track thermochronology, climatic change, North Slope, Alaska.

1. Introduction

Interpretation of apatite fission-track (AFT) analyses of five samples from the Exxon Alaska State C-1 well located along the northern coastline of Alaska (Fig. 1), indicate cooling of the order of ~15-20°C occurred during the early to late Miocene at some time between ~20 and 10 Ma (O'Sullivan, 1996). However, detailed analysis of the subsurface stratigraphic record by Bird and Molenaar (1987) reveals no erosion occurred during that time. Furthermore, both AFT and vitrinite reflectance (R_0) data indicate that the well did not experience higher basal heat flow in the past (O'Sullivan, 1996). Therefore, an explanation for the recorded cooling, other than in response to erosion of the section or a decrease in the geothermal gradient, is required.

At some time during the middle to late Cenozoic, a major climatic change occurred throughout the North Slope of Alaska. This change resulted in a decrease in the estimated mean annual surface temperature from ~+5°C during the Late Cretaceous to

P. Van den haute and F. De Corte (eds.), Advances in Fission - Track Geochronology, 255-267.

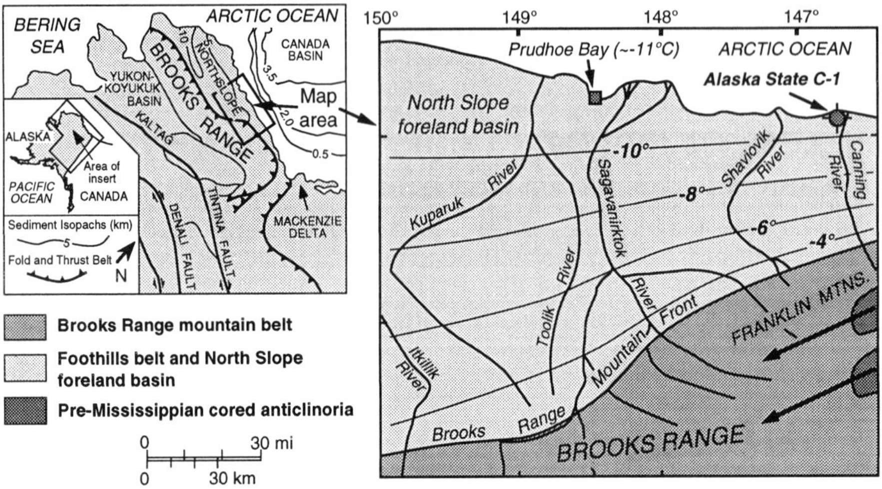

Figure 1. Index map showing the general locations of major physiographic provinces across northern Alaska, the location of the Alaska State C-1 well, and generalized contours of the estimated long-term mean annual surface temperatures (°C) determined by Lachenbruch (1988). Tectonic elements of northern Alaskan Cordillera include North Slope foreland basin and Brooks Range fold and thrust belt.

Eocene (e.g., Spicer, 1987; Clark, 1982), to present values of between –4°C along the northern flank of the Brooks Range, and –12°C along the present-day northern coastline (Fig. 1; Lachenbruch *et al.*, 1988). This change dropped temperatures from temperate to arctic conditions and introduced perennial ice cover over the Arctic Ocean (e.g., Clark, 1982, 1990; Carter *et al.*, 1986a,b; Dinter *et al.*, 1990; Marincovich *et al.*, 1990; Thiede *et al.*, 1990).

It is proposed here that the decrease in mean annual surface temperature of at least –17°C along the northern coast resulted in subsurface cooling of a similar magnitude within the upper few kilometers of the subsurface sequence, and that the cooling recorded by the AFT data from the C-1 well occurred in response to the initial onset of a major climatic change which affected the North Slope Alaska during the middle to late Cenozoic.

2. Tectonic Framework

The northern part of the Alaskan Cordillera has long been interpreted as a component of a major Jurassic-Cretaceous convergent margin compressional orogen (e.g., Mayfield *et al.*, 1988; Moore *et al.*, 1994). During uplift of the Brooks Range to the south, material was shed from the northward-verging deformed rocks to be deposited into the North Slope foreland basin (Fig. 1, Mayfield *et al.*, 1988; Bird and Molenaar, 1992, and references therein). Subsequently, the geologic relationships across much of the North Slope have been complicated by continued advancement of the Brooks Range fold and thrust belt (e.g., Mayfield *et al.*, 1988; O'Sullivan, 1996). Sediments derived from the Brooks Range and deposited along the then southern flank of the foreland basin were subsequently uplifted and deformed during two major episodes of Tertiary deformation, at ~60±4 Ma and ~25±3 Ma (e.g., O'Sullivan, 1996). Along the northern flank of the foreland basin the geologic relationships are relatively uncomplicated because the region has not experienced extensive post-depositional deformation (Mayfield *et al.*, 1988; Bird and Molenaar, 1992; O'Sullivan, 1996).

3. Middle to Late Cenozoic Change in Climatic Conditions

Besides the effects of Tertiary episodes of tectonism and associated erosion, the recent thermal history of the North Slope foreland basin is further complicated because the region experienced a significant change in climatic conditions at some time since the late Eocene (e.g., Clark, 1990). During the late Maastrichtian through to the late Eocene, the North Slope foreland basin experienced temperate climatic conditions with mean annual surface temperatures of ~+5°C (e.g., Spicer, 1987; Marincovich *et al.*, 1990; Thiede *et al.*, 1990). Immediately prior to the late Eocene the region was still experiencing temperate conditions, a fact supported by reports of Paleogene vertebrates, terrestrial and aquatic invertebrates, and plants in the Arctic (West *et al.*, 1977; Marincovich and Zinsmeister, 1985; Wolfe, 1985). However, the present-day estimate for the mean annual surface temperature along the northern coastline is approximately −12°C (Lachenbruch *et al.*, 1988). Therefore, associated with this long-term climatic change, the northern coastline of Alaska has experienced a net decrease of ~17°C in the mean annual surface temperature.

The timing of the initial change in climatic conditions is poorly constrained (e.g., Clark, 1990; Dinter *et al.*, 1990). Clark (1982, 1990) and Thiede *et al.* (1990) proposed that the transition from ice-free to arctic conditions within the Arctic Ocean to the north of Alaska occurred during the middle or late Cenozoic, based on the arrangement and size of the continents during that time as well as the offshore stratigraphic record. During the middle or late Cenozoic, the continents were tectonically rearranged, opening up gateways to paleocirculation within the oceans resulting in less heat being transported towards the poles (e.g., Berggren, 1982). Furthermore, evidence from limited stratigraphic cores recovered from the Arctic Ocean indicate that perennial sea ice was not present prior to the late Eocene, but was present by the late Miocene. Unfortunately, post-Eocene to late Miocene sediments have not yet been recovered from the ocean drilling program so the late Miocene age for establishment of perennial sea ice on the Arctic Ocean may be regarded as a minimum age (Thiede *et al.*, 1990).

In apparent conflict to the ideas of Clark (1982, 1990) and Thiede *et al.* (1990), Carter *et al.* (1986a,b) and Dinter *et al.* (1990) propose that the present-day arctic climate was not established in the region of the Arctic Ocean until the early Pliestocene (~1.5 Ma). In support of this they cite marine fossils from northern Alaska which suggest the marine environment between ~1.5-2.5 Ma was warm enough to significantly restrict the extent of sea ice which might have been present. However, Dinter *et al.* (1990) also point out that as far back as ~3.5 Ma (limit of their study), the onshore Alaskan coastal plain was experiencing a much colder climate than suggested from the marine fossils: one similar to present-day eastern Siberia. They also report that by 3.5 Ma the onshore coastal plain contained significant layers of permafrost, though they were probably not as continuous as those today. Therefore, it is likely that the initial decrease in mean annual surface temperature, which would accompany the transition from temperate to arctic conditions, had occurred prior to 3.5 Ma. Once again, this age should be considered a minimum estimate for the initiation of colder temperatures.

Thus, as proposed by Clark (1982, 1990), Clark and Morris (1985) and Thiede *et al.* (1990), it is possible that cooling within the region of the Arctic Ocean began as early as the early to middle Miocene, and conditions essentially the same as exist today were first established by the late Miocene. Subsequent short-term variations since the initial establishment of arctic conditions in northern Alaska could easily explain the presence of marine fossils indicative of slighly warmer climates in the marine environment.

4. Thermal History of the Alaska State C-1 Well

AFT results from five samples were recently obtained from the Alaska State C-1 well, located along Alaska's northern coastline (Figs. 1, 2). Complete sample information (depth, temperature, lithology, etc.), as well as analytical results, descriptions of both the methods used to model the AFT data and detailed interpretation of the AFT data were previously presented in O'Sullivan (1996). Therefore, only a summary of the results from the Alaska State C-1 well is presented below.

Based on interpretation of the AFT data from the Alaska State C-1 well, O'Sullivan (1996) proposed that the section is currently at near maximum temperatures (less than ~95°C), as would be expected if the region had been experiencing continual deposition (e.g., Bird and Molenaar, 1987). However, two samples from Upper Cretaceous-aged rocks at ~2774 m (present-day corrected downhole temperature of ~60°C) and 3491 m (present-day corrected downhole temperature of ~80°C), each contain a distinct population of young single-grain ages of ~20 Ma (Fig. 3), that are not easily explained by a simple constant-burial/heating model. These young ages could either represent grains which have been totally reset and subsequently cooled, thus recording an episode of cooling at some time between ~10 and 20 Ma, or are reduced ages which have not been totally reset by higher paleotemperatures and therefore represent a maximum age of cooling from higher temperatures. Though inconclusive by themselves, the few confined tracks present in these young grains (n=5) are all long (>14 μm) which is not inconsistant with the grains having cooled rapidly from paleotemperatures high enough to have totally reset the individual grain ages.

To test whether there was a relationship between apatite chemistry and grain-age, all 40 grains dated from the Upper Cretaceous-aged samples were microprobed. The results show that the young grains were Fl-rich (<0.2 weight percent Cl) and would therefore be expected to be reset at lower paleotemperatures than the more Cl-rich grains in the same samples (Fig. 3). This is consistent with the observation that fission-tracks within Fl-rich apatites are more readily annealed than tracks within Cl-rich apatites at similar temperatures (e.g., Green *et al.*, 1985). Therefore, we propose that the apatite ages for the Fl-rich grains were either totally reset, or significantly reduced prior to an episode of cooling at some time since ~20 Ma. The second option is more likely since there is no evidence to indicate that these samples were previously exposed to paleotemperatures necessary to totally reset the apatite age.

However, these rocks must have previously been exposed to higher paleotemperatures since the present-day temperatures can not account for the amount of annealing necessary to significantly reduce the grain ages. Based on modelling of the AFT data from the entire well, O'Sullivan (1996) proposed the section cooled at least ~15-20°C from elevated paleotemperatures at some time during the latest Miocene (Fig.

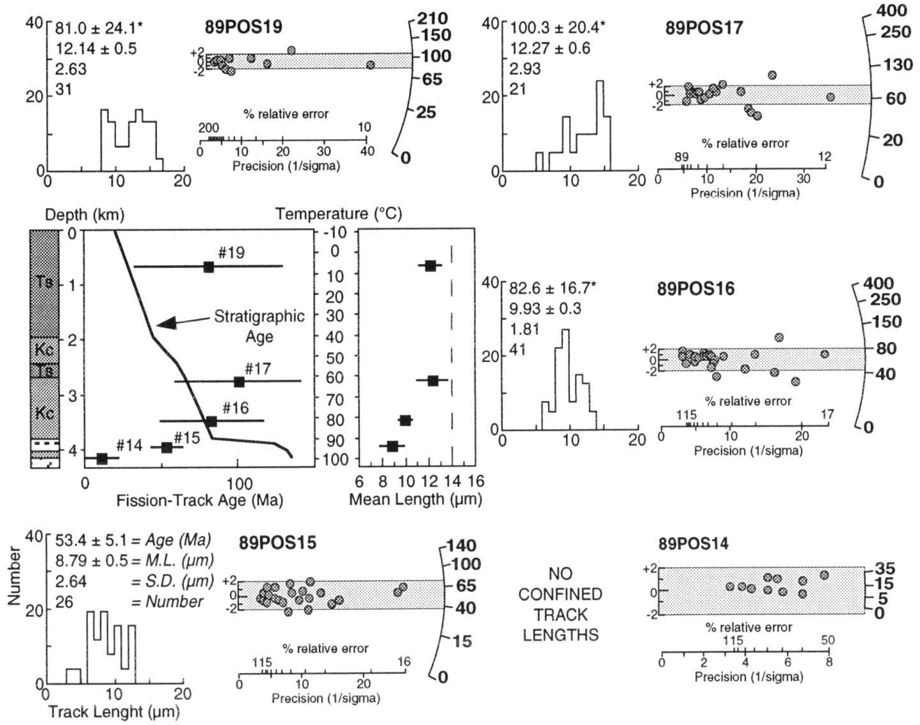

Figure 2. Composite figure showing fission-track parameters plotted against sample depth, present-day temperature for samples, and AFT data (error bars are ±2s) from the Exxon Alaska State C-1 well. Sample 14 did not contain any confined fission-tracks. AFT data includes histograms of fission-track length distributions and radial plots (e.g., Galbraith, 1988), showing single grain ages. Information shown with each AFT length distribution includes fission-track age (±1s), mean confined-track length (±1s), standard deviation, and number of confined tracks measured. An asterisk (*) denotes a sample that contains multiple-grain-age populations. Track length distributions are normalized to 100 tracks. Symbols: Ts, Tertiary Sagavanirktok Formation; Kc, Cretaceous Canning Formation.

4), and that cooling from higher temperatures was necessary in order to preserve the young ages and long confined track lengths within these grains. There are three possible explanations for the recorded cooling; 1) as a result of erosion, 2) a decrease in the geothermal gradient, or 3) sub-surface cooling in response to a long-term decrease in the mean annual surface temperature during the middle or late Cenozoic. The first option can be ruled out on the basis of detailed burial history analyses and palynology data from wells located along the northern coast of Alaska that do not record any erosion along the north flank of the basin in the region of the C-1 well during Oligocene and Miocene time (Bird and Molenaar, 1987, and their plate #1). The second option is inconsistent with the AFT and R_O data from the well. These indicate that the maximum paleogeothermal gradient experienced by the section was similar to the present-day value of ~27°C/km^{-1} with no evidence that the section experienced higher basal heat flow in the past (O'Sullivan, 1996). This leaves cooling in response to long-term climatic

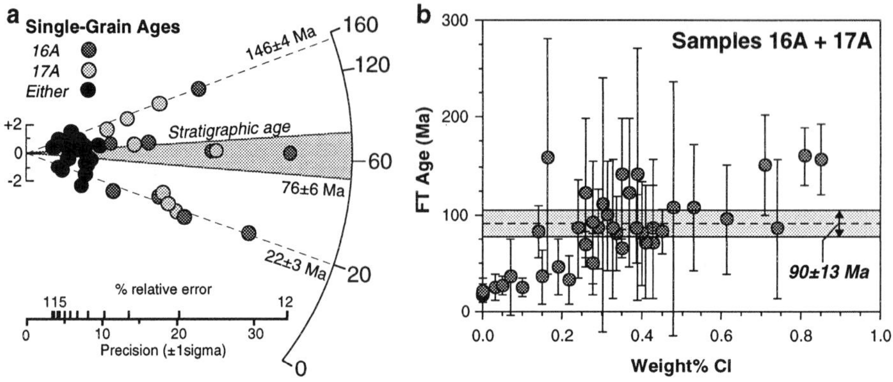

Figure 3. (a) Radial plot of the combined single-grain age data from the two Upper Cretaceous samples (16A and 17A) discussed in the text. Each sample contains a wide range in single grain ages with the youngest two components (~22 and ~76 Ma) being significantly younger than the stratigraphic age of the sample (90±13 Ma). The age of the youngest population of grains was estimated using the technique described by Sambridge and Compston (1994) and assuming the data can be treated as a sample from a three component mixture. The estimated ages of the three component mixture are 22±3 Ma, 76±6 Ma, and 146±4 Ma (±1s); compared with the samples stratigraphic age of 90±13 Ma. (b) Weight percent of chlorine plotted against the apparent apatite fission-track age for each of 40 grains dated from samples 89POS16A and 89POS17A. Error plotted at ±2s for each sample.

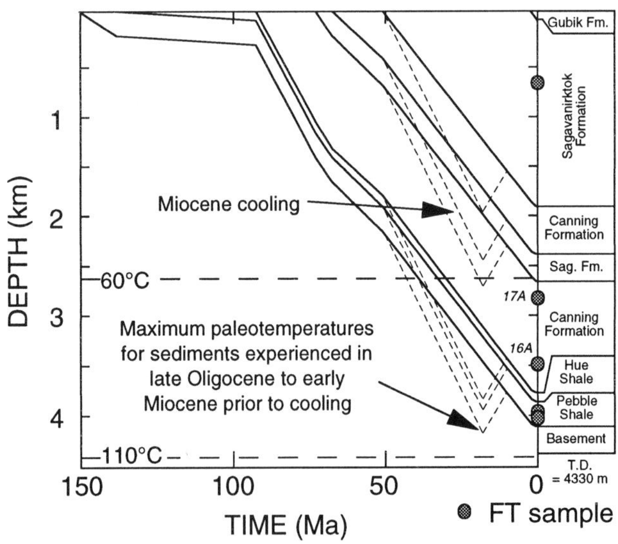

Figure 4. Proposed depositional history for Exxon Alaska State C-1 well based on AFT, R_O, and stratigraphic parameters (modified after O'Sullivan, 1996). Solid lines represent true burial history, whereas dashed lines follow the thermal history in relation to prevailing down-hole temperatures based on the modelling of the AFT parameters.

change (decrease in mean surface temperature) as the only remaining possibility.

5. Effect of Climatic Change on Sub-surface Temperatures

It is well recognized that short term climatic changes such as glacial retreat from higher latitudes will affect the geothermal gradient at shallow depths (e.g., Beck, 1977; Gretener, 1981). The depth to which events like the retreat of glaciers (scale of ~10^3 years) will affect the geothermal gradient is of the order of ~500 m (Beck, 1977). However, it is significant that Beck (1977) also concluded that longer-term events would affect temperatures at greater depths. Therefore, the long-term climatic change experienced by the North Slope during the middle to late Cenozoic (scale of ~10^6 years) is likely to have influenced the distribution of temperatures at depths much greater than 500 m. To test this we used simple analytical and numerical calculations to demonstrate that the decrease in the mean surface temperature of –17°C, documented for the northern coast of Alaska, was sufficient to cause similar amounts of cooling at depths of several kilometers within the crust.

5.1. QUANTITATIVE ASSESSMENT OF HEAT TRANSFER

In order to make any quantitative calculations of heat transfer within the crust it is necessary to formulate an appropriate thermal model which can be described using the mathematical equations which govern the transfer of heat (e.g., Carslaw and Jaeger, 1959). A reasonable first order thermal model suitable for describing temperatures within the shallow crust under changing surface conditions is a simple solid slab with a uniform thermal conductivity and with the lower surface maintained at a constant temperature. A constant temperature boundary at the base of the lithosphere is a reasonable assumption since convection within the underlying asthenosphere should maintain a near constant temperature at the interface with the mechanical boundary of the lithosphere (e.g., Parsons and Sclater, 1977). In any case, the conditions imposed at the lower boundary (120 km) do not significantly affect the near surface temperatures (~5 km).

The time needed to conduct heat (or a change in temperature) through a thickness of lithosphere and a measure of the change in temperature that occurs at various depths can be estimated using a one dimensional analytical solution to the heat conduction equation (Carslaw and Jaeger, 1959; p. 9, Equation 4);

$$\nabla^2 T - \frac{1}{\kappa}\left(\frac{\partial T}{\partial t}\right) = 0 \tag{1}$$

where;

$$\kappa = \frac{k}{\rho c} \tag{2}$$

where T is temperature, t is time, κ is the thermal diffusivity, k is the thermal conductivity, ρ is density, and c is the heat capacity. The change in temperature throughout the thickness of an infinite slab, resulting from an instantaneous temperature change at one surface of the slab, can be calculated using the appropriate analytical solution to equation 1 (Carslaw and Jaeger, 1959, section 12.6, equation 10);

$$T(t,z) = \frac{T_o z}{L} + \frac{2T_o}{\pi} \sum_{n=1}^{\infty} \frac{(-1)^n}{n} \exp\left(\frac{\kappa n^2 \pi^2 t}{L^2}\right) \sin\left(\frac{n\pi z}{L}\right) \tag{3}$$

where T_o is the temperature imposed at the surface of the slab of thickness L, z is depth and t is time. The insulating effect of the slab causes a delay between the onset of the thermal anomaly at one surface and the time at which the anomaly reaches its maximum value at the other. The results of this calculation clearly demonstrate that an instantaneous decrease in the surface temperature of ~20°C causes a rapid decrease in sub-surface temperatures at depths 5 km (Fig. 5).

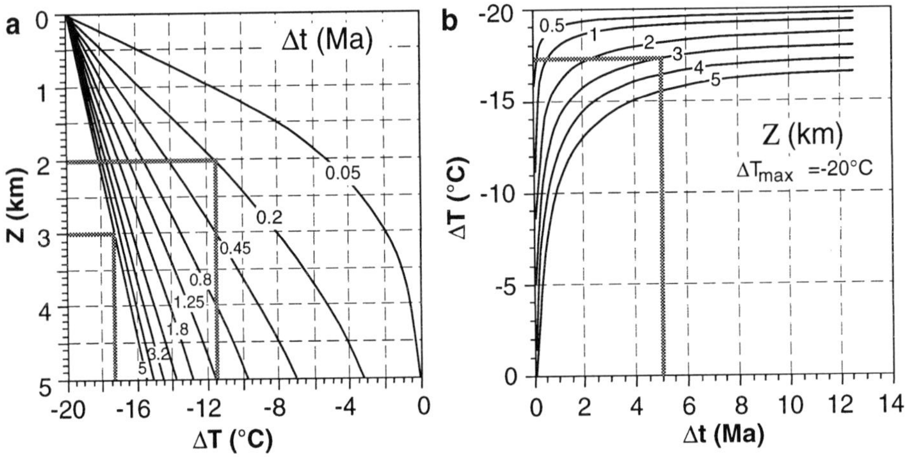

Figure 5. (a) Plot of the temperature change versus depth (shown for a range of times between 0.05 Ma to 5 Ma) caused by an instantaneous temperature change of –20°C at the surface. The curves were determined using equation 3 for a slab thickness (L) of 30 km and a thermal diffusivity (k) of 0.8x10-6 m2/s^{-1}. The plot shows that after 0.2 Ma the temperature at ~2 km depth will have decreased by ~11° and after 5 Ma the temperature at ~3 km depth will have decreased by ~17°. (b) Plot of the temperature change versus time (for a range of depths between 0.5 and 5 km). The plot shows that a decrease in surface temperature of 20°C will cause a 17°C decrease in temperature at a depth of 3 km after ~5 Ma.

A one-dimensional numerical solution to equation 1 (Brown *et al.*, 1994), was used to investigate the effect of a non-instantaneous temperature change at the surface. The surface temperature was varied from +5°C to –12°C over periods of 1, 5, 10 and 20 Ma using a cosine function centered at 20 Ma. The results of these calculations are shown in Figure 6a. The numerical model illustrates that the magnitude of the steady state temperature change at depth is independent of the time over which the surface temperature is changed. The model was deliberately scaled to simulate the conditions relevant to interpreting the thermal history recorded by fission-track data from the Alaska State C-1 well.

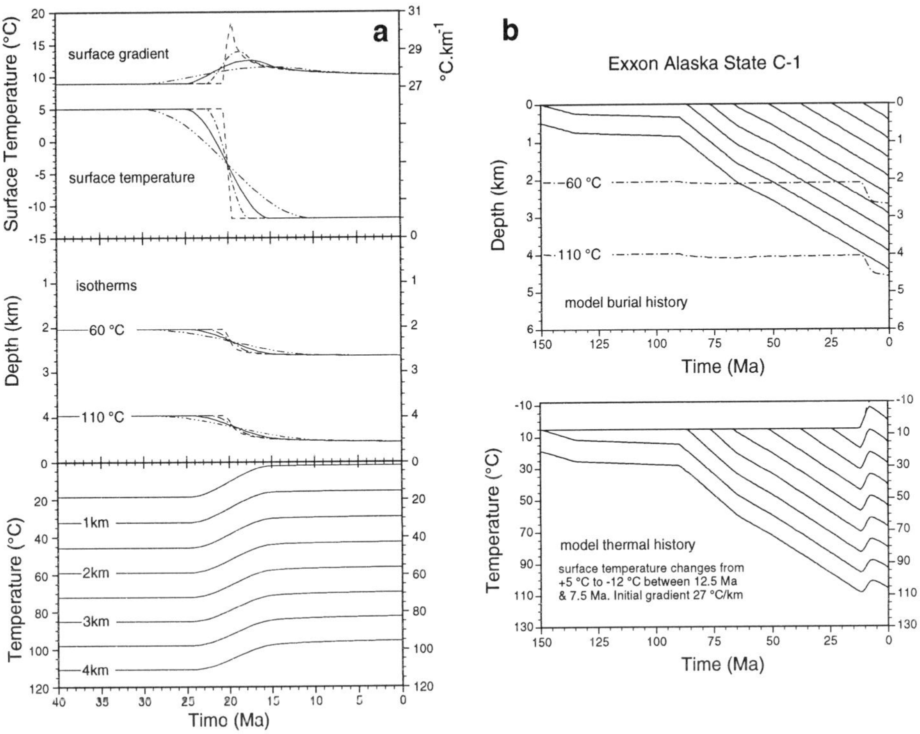

Figure 6. (a) Results of the numerical model which demonstrate that the maximum (steady state) temperature change at depth is independent of the duration of the temperature change effected at the surface. The effects of changing the surface temperature from +5°C to −12°C over times of 1, 5, 10 and 20 Ma (using a cosine function centred at 20 Ma) are shown. (b) The change in temperature through time, predicted by the 10 Ma model, are shown for a range of depths in the lower panel. This style of thermal history would predict erosion of 630 m of section over 10 Ma (for a thermal gradient of 27°C/km^{-1}) if the surface temperature was assumed to have remained constant.

The numerical model was also used to predict the thermal history for a series of hypothetical samples, with a regular spacing of 500 m, using the burial history for the Alaska State C-1 well (Fig. 6b). In the models shown in Figure 6b the surface temperature change of −17°C occurs between 12.5 Ma and 7.5 Ma. If the surface temperature change occurs earlier than ~20 Ma then the model predicts maximum temperatures occurring at present. So for the maximum paleotemperatures to have been greater than present temperatures within the well, the change in mean surface temperature must have occurred at or later than ~20 Ma.

5.2. ESTIMATING RESPONSES TO CHANGES IN SURFACE TEMPERATURES

After maximum paleotemperatures have determined for fission-track and vitrinite reflectance samples from a vertical sequence, these values can be plotted relative to depth to provide a maximum paleotemperature profile (e.g., Bray *et al.*, 1992). From this profile, a paleogeothermal gradient at the time of maximum paleotemperatures can be estimated as well as the amount of section that has been removed since maximum paleotemperatures were reached. In a vertical section that has been hotter in the past, the paleogeothermal gradient can be compared to the present-day geothermal gradient, allowing interpretation of the cause of the high paleotemperatures, and the cause of the subsequent cooling to the present temperatures. Figure 7a shows a simple example where heating and subsequent cooling were caused solely by deep burial followed by denudation with no change in mean annual surface temperature or basal heat flow. In this case, the present gradient would remain the same as the maximum paleogeothermal gradient, but the paleotemperature profile trend would be translated towards higher temperatures. Therefore, the amount of removed section can be estimated by dividing the amount of cooling by the geothermal gradient.

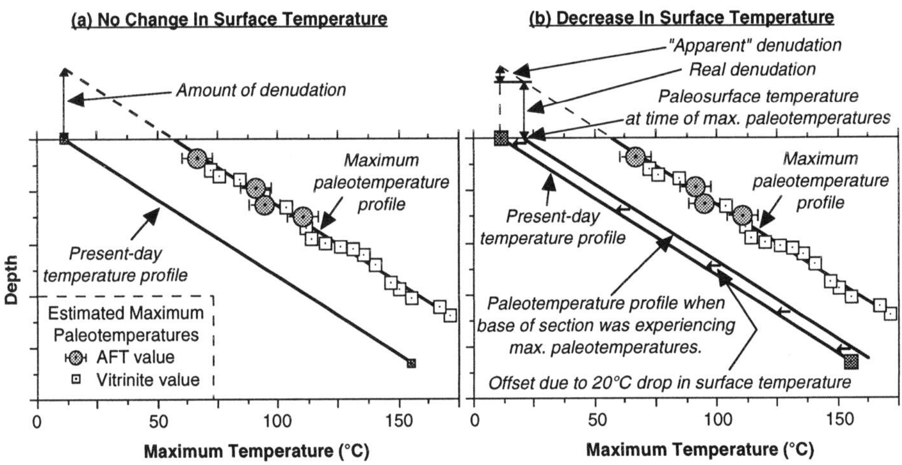

Figure 7. Hypothetical case to illustrate estimation of paleogeothermal gradients and the cause of heating and subsequent cooling from paleotemperature profiles. In each case a paleotemperature profile is shown which is similar to the present-day profile, indicating that heat flow has not changed since maximum paleotemperatures were reached. However, the maximum profile is offset to higher temperatures indicating that the section has been hotter in the past. Therefore, heating in the past was caused by deep burial prior to cooling by denudation, and the amount of removed section can be estimated by dividing the amount of cooling by the geothermal gradient. (a) shows the simple case where mean annual surface temperatures have not changed since the time at which the section reached maximum paleotemperatures and all cooling has resulted from denudation. (b) shows the case where the mean annual surface temperature has decreased significantly since the time at which the section reached maximum paleotemperatures. In this case a component of "apparent" denudation is introduced.

Figure 7b shows a more complicated example in which the mean annual surface temperature has decreased significantly since maximum paleotemperatures were reached (i.e., the North Slope of Alaska). If the surface temperature at the time the section began to cool from maximum temperatures is not known, and thus the present surface temperature is used, the amount of cooling which has occurred since maximum paleotemperatures were reached will be overestimated by a value of "apparent" denudation. The value of "apparent" denudation can be estimated by dividing the amount of cooling (difference between paleo- and present-day mean annual surface temperatures) by the geothermal gradient. The paleo-surface temperature is that which was present at the time when the subsurface section started to cool from maximum temperatures. The case shown in Figure 7b reiterates the point that when calculating the amount of denudation in regions which have experienced significant changes in climate, it is important to know the time at which the climate changed, the time when cooling from maximum paleotemperatures began, and the surface temperature at that time.

6. Conclusions

The results of simple analytical and numerical calculations demonstrate that a long-term change (10^7 years) in surface temperature causes changes of a similar magnitude at depths of several kilometres within the crust. The AFT and R_0 data from the Alaska State C-1 well require ~15-20°C of cooling during the mid to late Miocene and, since there is no evidence for either significant erosion or a change in basal heat flow, we believe that the cooling was caused by the reported decrease in the mean annual surface temperature associated with the change from temperate to arctic conditions. Apatite fission-track data from subsurface samples can thus be used to determine the timing of cooling and/or heating which occurs in response to changes in long-term climatic conditions. These particular data also provide important new evidence to constrain the timing of the initial change from temperate to arctic climatic conditions within northern Alaska to the middle to late Miocene. A final implication from this study concerns estimating the amount of erosion from fission-track data. If a decrease in surface temperature resulted in cooling of the stratigraphic section at depth, then the component of cooling related to this decrease in surface temperature must be considered when estimating depths of erosion from thermochronologic data. This is particularly important when working in arctic and antarctic regions were Cenozoic climatic changes may have caused significant changes in the mean annual surface temperatures (e.g., Harwood, 1985; McKelvey *et al.*, 1990).

Acknowledgments

This work was made possible with partial funding by the Minerals Management Service Continental Margins Project (Cooperative Agreement No. 14-12-0001-30432) and the industry sponsors of the Tectonics and Sedimentation Research Group at the University of Alaska-Fairbanks. P.B. O'Sullivan acknowledges the support of a LaTrobe University Post-Graduate Scholarship and R.W. Brown an Australian Endeavour Research Fellowship. Fission-track research at LaTrobe University is supported by the Australian Research Council and Australian Institute for Nuclear Science and Engineering. We thank David Foster and John Murphy for helpful discussions during early preparation of this manuscript, and Tony Hurford, Andrew Carter, and Stuart Thomson for their constructive reviews.

References Cited

Beck, A.E. (1977): Climatically perturbed temperature gradients and their effect on regional and continental heat-flow means. *Tectonophysics* **41**, 17-39.

Berggren, W.A. (1982): Role of ocean gateways in climatic change. *Climate in Earth history; Studies in Geophysics:*, National Academy Press: pp. 118-125.

Bird, K.J. and Molenaar, C.M. (1987): Stratigraphy. In: *Petroleum Geology of the northern part of the Arctic National Wildlife Refuge, Northeastern Alaska* (eds Bird, K.J. and Magoon, L.B.), U.S. Geological Survey Bulletin 1778, pp. 37-59.

Bird, K.J. and Molenaar, C.M. (1992): The North Slope foreland basin, Alaska. In: *Foreland Basins and Foldbelts* (eds Macqueen, R.W. and Leckie, D.A.), American Association of Petroleum Geologists Memoir 55, 363-393.

Bray, R.J., Green, P.F. and Duddy, I.R. (1992): Thermal history reconstruction using apatite fission track analysis and vitrinite reflectance: A case study from the East Midlands of England and the southern North Sea. In: *Exploration Britain: Geological Insights For the next Decade* (ed Hardman R.S.P.). Geological Society Of London Special Publication 67: pp. 3-25.

Brown, R.W., Gallagher, K. and Duane, M.J., 1994): A quantitative assesment of the effects of magmatism on the thermal history of the Karoo sedimentary sequence., *African Journal of Earth Science*, **18**, 227-243.

Carslaw, H.S. and Jaeger, J.C. (1959): *Conduction of heat in solids*. Oxford University Press, New York. 386 pp.

Carter, L.D., Brigham-Grette, J. and Hopkins, D.M. (1986a): Late Cenozoic marine transgressions of the Alaskan Arctic Coastal Plain. In: *Correlation of Quaternary deposits and events around the margin of the Beaufort Sea* (eds Heginbottom, J.A. and Vincent, J.S.), Contributions from a joint Canadian-American workshop, April, 1984: Geological Survey of Canada Open-File Report 1237, pp. 21-26.

Carter, L.D., Brigham-Grette, J., Marincovich, L. Jr., Pease, V.L., and Hillhouse, J.W. (1986b): Late Cenozoic Arctic Ocean sea ice and terrestrial paleoclimate. *Geology*, **14**, 675-678.

Clark, D.L. (1982): Origin, nature, and world climate effect of Arctic Ocean ice-cover. *Nature* **300**, 321-325.

Clark, D.L. (1990): Arctic Ocean ice cover; geologic history and climatic significance. In: *The Arctic Ocean Region* (eds Grantz, A., Johnson, L. and Sweeney, J.F.), Geological Society of America, The Geology of North America, v. L: 53-62.

Clark, D.L. and Morris, T.H. (1985): Arctic Ocean sediment texture and the Pleistocene climate cycle. *Geological Society of America Abstracts with Programs* **17**, pp. 547.

Dinter, D.A., Carter, L.D. and Brigham-Grette, J. (1990): Late Cenozoic geologic evolution of the Alaskan North Slope and adjacent continental shelves. In: *The Arctic Ocean Region.* (eds Grantz, A., Johnson, L. and Sweeney, J.F.), Geological Society of America, The Geology of North America, v. L: 459-490.

Galbraith, R. (1988): Graphical display of estimates having differing standard errors. *Technometrics* **30**, 271-281.

Green, P.F., Duddy, I.R., Gleadow, A.J.W., Tingate, P.T. and Laslett, G.M. (1985): Fission-track annealing in apatite--track length measurements and the form of the Arrhenius plot. *Nuclear Tracks* **10**, 323-328.

Gretener, P.E. (1981): *Geothermics: Using temperature in hydrocarbon exploration*. American Association of Petroleum Geologists Education Course Note Series #17. 156 pp.

Harwood, D.M. (1985): Late Neogene climatic fluctuations in the southern high latitudes: implications of a warm Pliocene and a deglaciated Antarctic continent. *South African Journal of Science* **81**, 239-241.

Lachenbruch, A.H., Sass, J.H., Lawver, L.A., Brewer, M.C., Marshall, B.V., Munroe, R.J., Kennelly, J.P., Galanis, S.P. and Moses, Jr., T.H. (1988): Temperature and depth of permafrost on the Arctic slope of Alaska. In: *Geology and exploration of the National Petroleum Reserve in Alaska, 1974 to 1982* (ed Gryc, G.), U. S. Geological Survey Professional Paper 1399: pp. 645-656.

Marincovich, L. Jr. and Zinsmeister, W.J. (1985): Early Tertiary climates of the Arctic Ocean. *Geological Society of America Abstracts with Programs* **17**, pp. 653.

Marincovich, L. Jr., Brouwers, E.M., Hopkins, D.M. and McKenna, M.C. (1990): Late Mesozoic and Cenozoic paleogeographic and paleoclimatic history of the Arctic Ocean Basin, based on shallow-water marine faunas and terrestrial vertebrates. In:*The Arctic Ocean Region* (eds Grantz, A., Johnson, L. and Sweeney, J.F.), Geological Society of America, The Geology of North America, v. L, pp. 403-426.

Mayfield, C.F., Tailleur, I.L. and Ellersiech, I. (1988): Stratigraphy, structure, and palinspastic synthesis of the western Brooks Range, northwestern Alaska. In: *Geology and exploration of the National Petroleum Reserve in Alaska, 1974 to 1982* (ed Gryc, G.), U. S. Geological Survey Professional Paper 1399, pp. 143-186.

McKelvey, B.C., Webb, P.N., Harwood, D.M. and Mabin, M.C.G. (1990): The Dominion Range Sirius Group: a record of the late Pliocene–early Pleistocene Beardmore Glacier. In: *Geological Evolution of Antarctica* (eds Thomson, M.R.A., Crame, J.A. and Thomson, J.W.), Cambridge University Press, pp. 675-683.

Moore, T.E., Wallace, W. K., Bird, K. J., Karl, S. M., Mull, C. G., and Dillon, J. T. (1994): Geology of northern Alaska, in *The Geology of North America*, vol. G1, *The Geology of Alaska*, edited by G. Plafker, and H. Berg, pp. 49-140, Geol. Soc. of Am., Boulder, Colo.

O'Sullivan, P.B. (1996): Late Mesozoic and Cenozoic thermal evolution of the North Slope foreland basin, Alaska. In: *Thermal evolution of sedimentary basins in Alaska* (eds Johnsson, M.J., and Howell, D.G.), U.S. Geological Survey Bulletin 2142, pp. 45-79.

Parsons, B. and Sclater, J.G. (1977): An analysis of the variation of ocean floor bathymetry and heat flow with age. *Journal of Geophysical Research* **82**, 803-827.

Sambridge, M.S. and Compston, W. (1994): Mixture modelling of zircon ages. *Earth and Planetary Science Letters* **128**, 373-390.

Spicer, R.A. (1987): Late Cretaceous floras and terrestrial environment of northern Alaska. In: *Alaskan North Slope Geology* (eds Tailleur, I.L. and Weimer, P.), Society of Economic Paleontologists and Mineralogists Pacific Section Publication 50: pp. 497-512.

Thiede, J., Clark, D.L. and Herman, Y. (1990): Late Mesozoic and Cenozoic paleoceanography of the northern polar oceans. In: *The Arctic Ocean Region* (eds Grantz, A., Johnson, L. and Sweeney, J.F.), Geological Society of America, The Geology of North America, v. L, pp. 427-458.

West, R.M., Dawson, M.R. and Hutchinson, J.H. (1977): Fossils from the Paleogene Eureda Sound Formation, N.W.T., Canada; Occurrence, climatic, and paleogeographic implications. In: *Paleontology and plate tectonics* (ed West, R.M.), Milwaukee Public Museum Special Publications in Biology and Geology **2**, 77-93.

Wolfe, J.A. (1985): Probabilities of high-latitude glaciers during the Tertiary. *Geological Society of America Abstracts with Programs* **17**, pp. 753.

GEOMORPHIC FACTORS IN THE INTERPRETATION OF FISSION-TRACK DATA

MICHAEL A. SUMMERFIELD
Department of Geography, University of Edinburgh, Edinburgh, EH8 9XP, United Kingdom
RODERICK W. BROWN
School of Earth Sciences, Victorian Institute of Earth and Planetary Sciences, La Trobe University, Bundoora, Victoria 3083, Australia

1. Abstract

Fission-track analysis, along with other thermochronologic techniques, is now routinely used to infer tectonic 'uplift' events, yet the basis for linking crustal thermal histories with such events is rarely made explicit. We evaluate some problems involved in the application of fission-track data to tectonic studies, focussing on a range of geomorphic factors that need to be considered before inferring a tectonic episode from a crustal cooling event recorded by such data. Tectonic uplift involves work against gravity and must therefore be defined in terms of vertical movement with respect to the geoid. Unfortunately, such absolute displacements of the surface (surface uplift) or rocks within the crust (crustal uplift) cannot be directly inferred from fission-track data since these record the cooling history of rocks in the relative frame of reference of movement towards the landsurface; that is, they record denudation. Making inferences about a tectonically-driven uplift event from fission-track data requires a link to be established with the associated denudational (crustal cooling) response. It is generally assumed that pulses of denudation result from tectonically-driven surface uplift events, but such an assumption is problematic for two reasons. First, an increase in surface elevation does not necessarily lead to an increase in denudation rate since erosional energy is controlled by local slope rather than absolute elevation above base level. Secondly, several other factors, such as climate, lithology and patterns of landscape and drainage development can affect spatial and temporal patterns of denudation. Any attempt to relate cooling histories derived from fission-track data to tectonic uplift events requires an adequate understanding and assessment of these geomorphic factors.

2. Introduction

Over the past decade, the primary intended application of fission-track analysis has been the dating of tectonic events. It is now widely, although not universally, ackowledged that fission-track data cannot provide direct information on crustal deformation in the absolute frame of reference of the geoid; rather such data reflect cooling of the crust with respect to the relative frame of reference of the landsurface and therefore record denudation (uplift of rocks relative to the surface) rather than surface or crustal uplift with respect to the geoid (Clark and Jäger, 1969; Summerfield, 1989, 1991a; England and Molnar,

P. Van den haute and F. De Corte (eds.), Advances in Fission - Track Geochronology, 269-284.

1990; Brown, 1991). However, fission-track analysis continues to be used to date tectonic events because of the virtually universal assumption that episodes of accelerated denudation (and hence crustal cooling) arise as a direct, and essentially geologically contemporaneous, response to tectonic uplift events. This assumption is reflected in the frequent use of the phrase 'uplift and erosion' in fission-track studies.

There is no doubt that in some cases denudational episodes recorded by fission-track data have indeed resulted from essentially contemporaneous tectonic uplift events. Our key point here, however, is that this is not necessarily always the case, and that a more sophisticated approach to understanding the factors that control spatial and temporal variations in denudation in different morphotectonic settings is required if fission-track data is to be correctly interpreted; in other words we argue that those using fission-track data in tectonic studies should at least consider the geomorphic factors that can influence the denudational response to tectonic events and qualify their interpretations accordingly. Whilst we focus on fission-track data, our points are also relevant to the interpretation of thermochronologic data in general. Our discussion is confined to the role of sub-aerial denudation since other aspects of the interpretation of fission-track data have been considered elsewhere (Brown *et al.*, 1994a; Stüwe *et al.*, 1994 ; Brown and Summerfield, 1997). We begin by briefly clarifying the definition of those concepts integral to the analysis of the relative and absolute vertical movements of the earth's surface and crust.

3. Definitions, Frames of Reference and Assumptions

Since uplift involves vertical movement parallel to the gravity vector, the geoid provides the fundamental datum against which vertical motions can be measured (Fig. 1). The geoid defines the vertical position of the continent-ocean boundary, and its projection within the continents provides the most consistent definition of ultimate base level for denudation. Upward movement of the earth's surface with respect to a datum of known elevation is here termed *surface uplift*, and in the context of the regional scale of fission-track studies this refers to changes in *mean* surface elevation over an extensive area of the order of $>10^3$ km^2. As well as changes in the elevation of the earth's surface with respect to the geoid, it is also necessary to consider changes in the elevation of points within the crust with respect to this absolute datum. We term such upward movement *crustal uplift* (Summerfield, 1991a).

Over geological time scales it is rarely possible to relate changes in the elevation of the earth's surface, or points within the crust, to the absolute datum represented by the geoid. In such circumstances all we can do is record changes in the movement of rocks relative to the earth's surface at a particular point without knowing how that point may have moved relative to the geoid. This upward movement of rocks relative to the landsurface, is referred to here as *denudation*. If unqualified, the term denudation is invariably taken to refer to the effects of surface processes only. However, large-scale crustal displacements can also give rise to crustal stripping, a process frequently referred to as tectonic denudation. This is clearly a potentially significant mechanism in the interpretation of fission-track data, but in the following discussion reference to 'denudation' should be taken as excluding this mechanism. The term denudation is preferred here to erosion as it unambiguously encompasses the aggregate effects of all surface processes leading to the downslope movement of material. It is also used in preference to 'exhumation', as used by England and Molnar (1990), since this term has long been used in a much more specific sense to refer to the exposure of a pre-existing

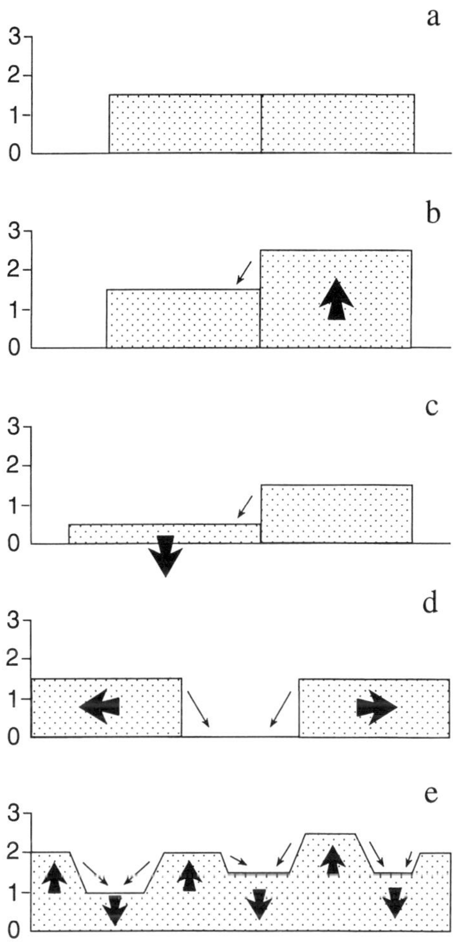

Figure 1. Schematic representation of the implications of absolute and relative frames of reference for the interpretation of crustal cooling in response to denudation (units are arbitrary). (a) shows an initial condition of a landsurface at a certain absolute elevation. In (b) one crustal block is elevated both in absolute terms and with respect to an adjacent block; the fall in base level due to the relative movement between the two blocks would be expected to promote a denudational (crustal cooling) event in the elevated block. (c) is the same as (b) except that one block has subsided both in the absolute frame of reference, and with respect to the adjacent block. Although tectonically distinct in terms of vertical movement with respect to the geoid, scenarios (b) and (c) would produce an identical denudational (crustal cooling) response in the block which stands *relatively* higher. In (d) continental break-up in a region of significant residual elevation would lead to a fall in base level and an episode of denudation and crustal cooling along the newly formed continental margin even in the absence of a 'tectonic uplift' event. In (e) relative movements between internally-drained adjacent blocks would produce a pattern of local denudation of topographic highs and sedimentation in intervening basins with no change in mean regional elevation.

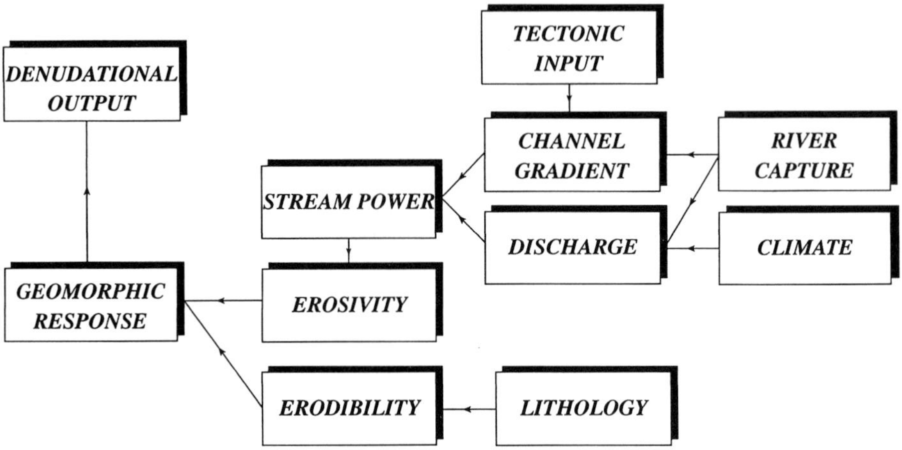

Figure 2. Schematic representation of the way the denudational response to tectonics is mediated by the geomorphic system. Note that tectonics is only one of a number of factors that influence denudation rate.

local relief actually decreases at higher elevations due to the presence of an extensive inter-montane plateau. In high-elevation passive margin settings characterized by a high plateau separated from the coast by a major escarpment there is little, or no, relationship between local relief and elevation (Summerfield, 1991b). The clear implication of this for the interpretation of fission-track data is that regional surface uplift can occur without leaving any thermochronologic record if the morphological setting is such that there is no significant increase in local relief and therefore no significant denudational response.

A further point related to the interpretation of elevation and local relief is the use, by some researchers, of the existence of erosion surfaces and associated weathering deposits in the interpretation of surface elevation change in fission-track studies. For instance, saprolites and lateritic soils have been cited by Bohannon *et al.* (1989) as evidence for low elevation in the Red Sea region until the mid-Miocene . Low relief surfaces can, however, form irrespective of elevation since rates of fluvial incision can be predominantly controlled by regional or local base levels. For example, high level erosion surfaces are continuing to develop in southern Africa at elevations in excess of 1000 m. Since regolith accumulation rate depends on both the rate of weathering and rate of removal, thick weathering mantles can develop on such high elevation surfaces since local relief, and therefore rates of denudation, are low.

This effect is dramatically illustrated by the Tibetan Plateau, approximately a half of which is internally drained and therefore completely disconnected from external base-level changes (Fig. 3). Within the internally-drained area, spatial and temporal patterns of denudation will reflect localized faulting and folding and be unrelated to regional surface elevation changes. It is clear that particular care must also be exercised in interpreting fission-track data for locations from the Tibetan Plateau that may have been internally drained in the past. In fact it is possible that the complex patterns of drainage evolution evident in mountain systems, such as the Himalayas, can themselves generate major

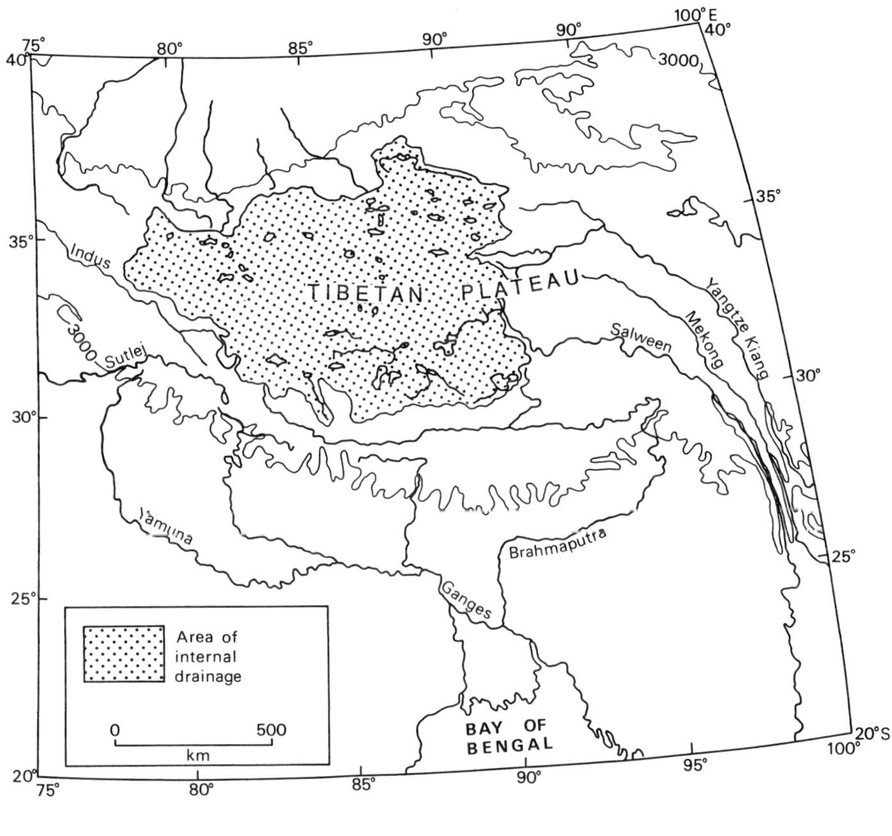

Figure 3. Location of internally-drained area within the Tibetan Plateau. This extensive region, which is well above the 3000 m contour (indicated on map), is presently disconnected from base-level changes related to the absolute datum of sea level. The distribution and areal extent of internal drainage has probably changed substantially over the past several million years, with consequential implications for the interpretation of thermochronologic data from this region.

surface, landscape or feature that has been previously buried (Bates and Jackson, 1987).

It is apparent from the distinction drawn above between absolute and relative frames of reference that no direct inference about surface elevation change can be made from denudational histories constructed from fission-track data (Fig. 1). However, it has been almost universal in fission-track studies to assume that all denudational (cooling) episodes must have occurred in direct response to 'uplift' events and are geologically contemporaneous with them. To take just one example, amongst many, in discussing a peak of erosion (denudation) which affected part of the Tibetan Plateau about 18 Ma BP, Copeland *et al.* (1987) have referred to a "pulse of uplift [being] responsible for this erosion". The assumption that it is crustal uplift that creates the regional surface uplift that promotes the denudational episodes recorded by fission-track data rests on two further assumptions. One is that an episode of crustal uplift that results in regional surface uplift will inevitably promote a rapid increase in denudation rate. The other is that a regional increase in surface elevation is the only way in which an increase in denudation rate can be achieved, and that, therefore, since fission-track data record denudational (cooling) histories they also record tectonic uplift events. We now examine each of these assumptions, which are central to conventional interpretations of fission-track data, and demonstrate that they can only be evaluated through an understanding of the geomorphic factors controlling denudation rates and modes of landscape evolution.

4. Factors Controlling Denudation Rates

The assumed link between regional surface uplift and denudation rests on the following argument. A cooling episode arises from the upward movement of rocks relative to the surface as a result of denudation. This pulse of denudation is assumed to be caused by an increase in regional elevation resulting from the crustal uplift rate exceeding the denudation rate. Since regional (as opposed to local) increases in surface elevation can only be generated by either an increase in crustal thickness, or by mantle processes, it is assumed that the cooling event must have a primary tectonic cause. Consideration of the way in which the denudational response to tectonic uplift is mediated by the geomorphic system (Fig. 2) reveals that there are two difficulties with this line of reasoning .

The first of these is that elevation itself cannot influence denudation rates, although a direct relationship between elevation and denudation has frequently been assumed (Lambeck and Stephenson, 1986; Pitman and Golovchenko, 1991; Lorenzo and Vera, 1992). Erosivity in fluvial systems is a function of stream power which, in turn, is a function of channel gradient and discharge. Mean gradient is closely related to local relief (defined as the difference between maximum and minimum elevation over a specified area) (Ahnert, 1970; Summerfield, 1991b), and the latter variable, which is much more readily quantified over large areas, has been shown to be fairly strongly correlated with denudation rate (Summerfield and Hulton, 1994). Elevation can therefore only legitimately be used as a predictor of denudation rate if it can be shown to be strongly correlated with local relief. The availability of digital topographic data now makes it possible to explore the relationship between modal elevation and local relief on a global basis. Such an analysis reveals highly variable relationships depending on the terrain being considered (Summerfield and Hulton, 1994). In some catchments draining orogenic belts local relief does indeed increase fairly consistently with elevation, but in others, such as the Brahmaputra basin draining the Himalayan-Tibetan Plateau orogen, denudational episodes capable of being recorded as cooling events in fisson-track data. This could occur where headward erosion breaches a plateau rim and and the newly-captured drainage begins rapidly to incise into the previously internally-drained, elevated surface. In the case of the Tibetan Plateau the available relief is of the order of 4 km and capture of internal drainage here would create extremely high gradient river systems. On the basis of rates of denudation recorded elsewhere for basins of comparably high relief (Summerfield, 1991a; Summerfield and Hulton, 1994) rates of regional denudation well in excess of 1000 m Ma^{-1} would be likely to arise from such a drainage capture episode. If isostatic rebound in response to denudational unloading is included this implies that depths of crustal stripping of well in excess of 4 km would occur in just a few million years and this would generate a crustal cooling event comparable to (and in terms of fission-track data indistinguishable from) rapid tectonic uplift and resulting denudation.

It might be argued that this possibility can be ruled out for the major transverse rivers of the Himalayas since these are antecedent, and thus the denudational events precipitated by the formation of their associated valleys do not post-date the tectonic uplift forming the orogen. Harrison *et al.* (1992), for instance, acknowledge that the development of the Himalayan-Tibetan Plateau drainage is an important factor in their interpretation that the rapid cooling they identify from thermochronologic data resulted from denudation occurring soon after uplift. They appear to view an antecedent origin for the majority of transverse Himalayan rivers as firmly established. However, neither they, nor some others using thermochronologic data to interpret the tectonic history of orogens, seem to appreciate the difficulties involved in establishing unequivocally the origin of transverse drainage, especially in mountain ranges where several kilometres of denudation has occurred during drainage development. The possibly complex origin of transverse rivers was demonstrated by Oberlander (1965) in his seminal analysis of the drainage of the Zagros Mountains, and his subsequent interpretation of the Himalayan drainage casts severe doubt on the antecedence model (Oberlander, 1985). More recently, Fielding (in press) has concluded that the question of drainage antecedence in the Himalayas is still open, while Hovius (1995) has favoured a headward erosion, rather than antecedence, model for many Himalayan transverse rivers on the basis of a detailed analysis of drainage and channel characteristics.

Examination of the drainage of the Andes shows that a potentially complex denudational response is not confined to the Himalayas and Tibetan Plateau. Internal drainage is extensive in the central Andes, occupying virtually the whole of the range above 3000 m in some locations (Kennan, in press) (Fig. 4). Although this internally-drained area, which was probably established by about ~15 Ma BP in the southern Altiplano (Vandervoort *et al.*, 1995), is currently isolated from external base-level changes and therefore denudationally unresponsive to regional surface uplift, there is clearly the possibility in the future of progressive headward erosion by a number of major river systems draining to the east which could generate pulses of greatly increased denudation along the plateau edge (Fig. 5). Indeed, there is some evidence that the eastern flank of the Altiplano has retreated towards the plateau interior during the Neogene (Masek *et al.*, 1994), and the incision of gorges since 10 Ma BP in the Coastal Cordillera of northern Chile appears to be a result of a change from internal to external drainage rather than rapid uplift of the coastal block (Kennan,in press).

We conclude from the examples of the Tibetan Plateau and the Altiplano that the processes of headward erosion and drainage capture may provide an alternative explanation for 'uplift' (denudational) events recorded in thermochronologic data in mountain belts traditionally ascribed to tectonic events alone. This is most likely to be

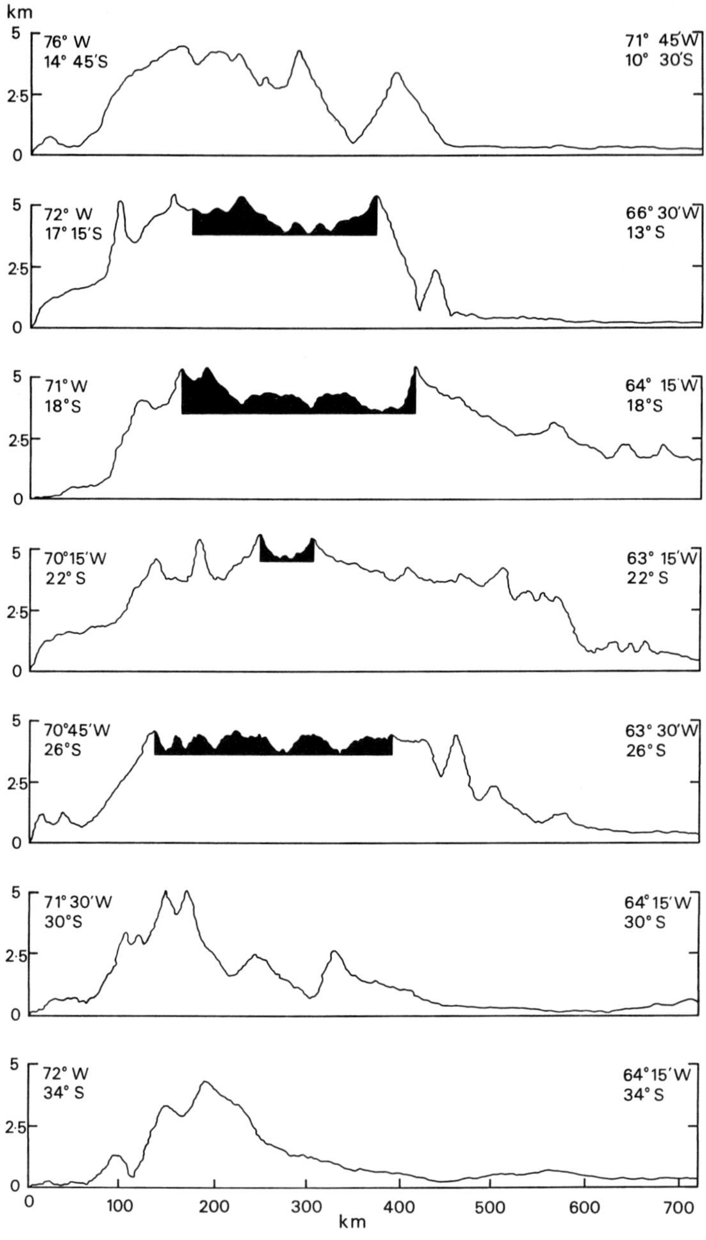

Figure 4. Distribution of internal drainage within the Andes illustrated by topographic cross-sections from 12° to 34° S. Note that internal drainage is confined to the highest elevations (see Fig. 5).

the case in regions peripheral to areas of internal drainage since it is here that there is the highest probability of previous drainage capture events. Again, we reiterate that we are not arguing that this explanation will apply in all cases. Our point is that a phase of accelerated denudation promoted by drainage capture should be considered when interpreting fission-track data in relevant morphotectonic settings, and that tectonic uplift should not automatically be assumed to be the immediate cause of denudation and crustal cooling. If drainage capture cannot be excluded as a possible mechanism for denudation and crustal cooling then great care should be taken in 'dating' tectonic uplift events using fission-track data since, on the basis of the timing of Tibetan Plateau and Altiplano uplift, denudation (and consequential crustal cooling) resulting from drainage capture can post-date tectonic uplift by >10 Ma.

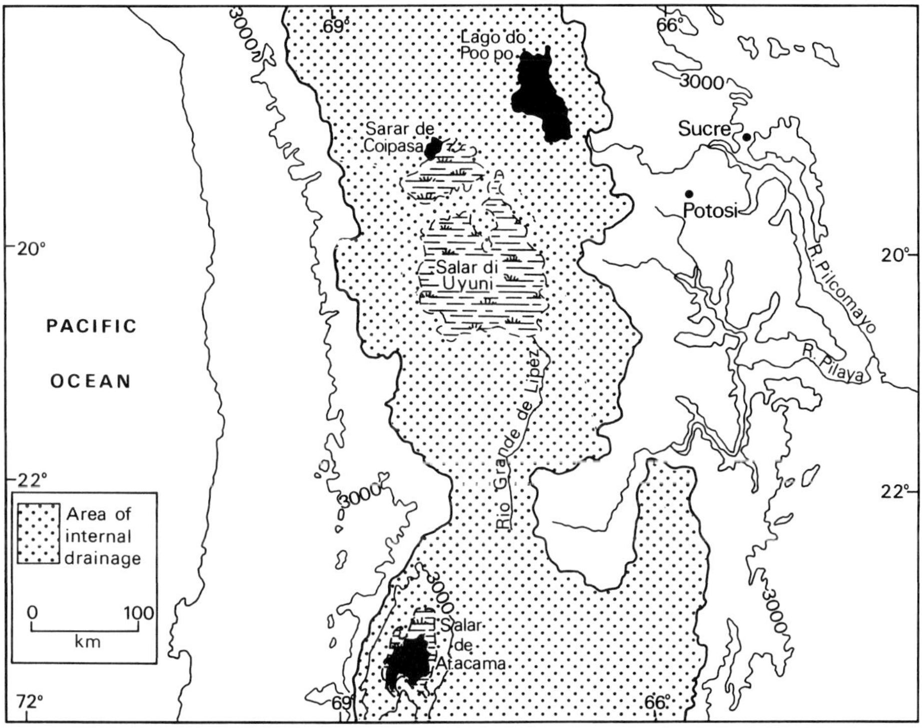

Figure 5. Map of internal drainage within the central Andes showing its distribution with respect to the 3000 m contour. Note how headward erosion and drainage capture by steep gradient, eastward-flowing rivers would result in a denudational pulse (and crustal cooling) along the eastern flank of the internally-drained area without any directly associated tectonic uplift event. A possible example of this process is represented by the embayment formed by the headwater tributaries of the Rio Pilaya which appears to have developed at the expense of the area of internal drainage.

The second problem with the assumed link between regional surface uplift and denudation is that factors other than relief can affect rates of denudation (Summerfield, 1991a) (Fig. 2). Numerous surveys have pointed to the role of climate in influencing denudation rates (Wilson, 1973; Ohmori, 1983; Jansson, 1988), and although the emphasis in more recent studies has been on the role of topographic factors (Pinet and Souriau, 1988; Milliman and Syvitski, 1992; Summerfield and Hulton, 1994) the possibility that a sustained and major change in climate could significantly affect denudation rates should at least be considered in the interpretation of fission-track data. This is particularly relevant given the major modifications of regional climates that undoubtedly accompanied the break-up of Pangaea, and the more recent significant shifts in global climate during the late Cenozoic.

In addition to erosivity, rates of denudation are moderated by erodibility, and over geological time spans this is largely controlled by lithology. Variations in rock type have generally been regarded as causing only local variations in denudation rates which are not usually significant at the regional scale. Nonetheless, since fission-track data monitors denudation at the local scale, lithological factors should be taken into account in their interpretation. This is especially the case where a rock unit of significant thickness overlies another unit of markedly different erodibility. A possible example of this situation may have occured in passive margin settings where thick sequences of flood basalts originally overlay much more readily erodible sedimentary strata. River incision would be expected to have been initially relatively slow in the resistant basalts, but would probably have accelerated significantly once the more easily erodible underlying sedimentary units were reached. The fission-track data in this case would record an episode of accelerated cooling once the basalt sequence had been penetrated, but no regional surface uplift or associated tectonic event would have precipitated the cooling event. This kind of situation may well apply to the denudational history of parts of southern Africa where the prominence of the Karoo flood basalts as an escarpment forming lithology indicates their resistance relative to the underlying Karoo sedimentary strata. Lithological factors may also play a role in promoting variations in rates of fluvial incision in orogenic belts, particularly when flysch-like sediments or schists are exposed (Oberlander, 1965, 1985).

In highlighting factors other than tectonically-driven surface uplift as a possible cause of denudational episodes recorded as cooling events by fission-track data we are not suggesting that these factors are always important; rather, we are arguing that such factors should be considered, and that a tectonic interpretation should not be automatically accepted.

5. Morphological Constraints on Uplift-Denudation Relationships

In addition to considering factors other than regional surface uplift that can affect denudation rates, any assessment of fission-track data must also consider the mode of landscape response associated with the denudational episode recorded as a cooling event. Central to this assessment is the notion of landscape sensitivity (Brunsden and Thornes, 1979) since this plays a crucial role in determining how rapidly a tectonically-driven surface uplift event will produce the denudational response that is recorded by fission-track data. Tectonic uplift can leave a denudational imprint because the temporary excess of crustal uplift over denudation promotes surface uplift and a fall in regional or ultimate base level which increases channel gradient and local relief. Such a base-level change will propagate through a drainage basin along river channels, but the rate at which this occurs will depend crucially on the strength of coupling between basin system components, and particularly on the linkage of valley-side slopes and river channels. Two end members of landscape sensitivity can be identified: 1) mobile, fast-responding systems which react quickly to externally generated impulses, such as base-level changes, and readily adjust to a new equilibrium; and 2) insensitive, slowly-responding systems which react only very gradually to external changes.

5.1 OROGENIC SETTINGS

Many orogens are morphologically sensitive in that a fall in base level generated by an increase in regional elevation usually leads to a rapid response by the denudational system. This can be achieved because the geomorphic system has a high degree of connectivity with changes being transmitted rapidly through drainage basins due to the maintenance of threshold valley-side slope gradients by channel incision. The precise nature of this response will depend on the spatial pattern of surface elevation change, and it may not be uniformally instantaneous, even on a geological time scale. Clearly mountain belts can only attain a high mean elevation if there is a lag between an increase in the rate of crustal uplift and the resulting increase in the rate of denudation. According to Ahnert (1984) this lag will be in part a function of the width of a mountain range. Assuming that the rate of crustal uplift is relatively constant across a mountain belt the denudational response to the resulting fall in base level will initially be focussed at the margins, and will then propagate towards the main drainage divide by knickpoint retreat. Consequently, it will take longer for this headward erosion to reach the axis of a broad mountain belt than a narrow one and so the divide of a broad range can attain a greater elevation before the rate of denudation matches the rate of crustal uplift and a new steady-state topography is established. Empirical support for this conceptual model comes from the strong correlation reported by Ahnert (1984) between mountain range elevation and width, a relationship which also supports the idea that many mountain belts are in an approximate morphological steady state.

The relevance of notions of landscape sensitivity to the interpretation of fission-track data is that it provides a basis for assessing the temporal coincidence of denudational episodes and the tectonic events which may have given rise to them. In general the high rates of crustal uplift (as calculated from rates of crustal shortening) attained in some mountain belts, such as the Southern Alps (Tippett and Kamp, 1995; Beaumont *et al.*, 1996), would very rapidly generate unrealistic topography unless the denudational system responded promptly. In broader mountain belts, however, there could be a significant lag between an increase in the elevation of the axis of the range and a denudational response capable of being recorded by fission-track data.

The denudational and morphological response to regional surface uplift becomes potentially much more complicated, however, in the case of complex mountain systems incorporating extensive plateaus, especially if they are internally drained at some stage of their development. In Figure 1(e), for instance, the area is internally drained so the denudational (crustal cooling) response depends only on local conditions and would be independent of absolute elevation; if the horizontal scale of the topography were sufficient, crustal cooling could be recorded by fission-track data from topographic highs but would not represent a *regional* cooling (denudational) event. Such an area will, however, experience regional denudation and crustal cooling when the internal drainage is eventually captured as external base-level changes would then be transmitted across it.

5.2 PASSIVE MARGIN SETTINGS

In contrast to most orogenic terrains, passive margin and cratonic settings are generally characterized by low morphological sensitivity (Summerfield, 1991a). This is well illustrated by the geomorphic response to continental break-up. If rifting occurs at a location where the surface elevation is significantly above sea level then the newly created continental margins will possess high local relief in the zone where the elevated landsurface abuts the newly-formed coast (see Figure 1(d)). A high surface elevation may be related to tectonic mechanisms directly associated with rifting, but even in the absence of a specific regional surface uplift mechanism there may be a significant residual elevation in regions originally located deep in the interior of a supercontinent. In fact, it is highly likely that many regions in the interior of supercontinents which subsequently become sites of continental rifting and break-up would have had elevations significantly above sea level prior to the rifting and subsidence associated with the break-up event. This is because a number of factors would have militated against denudation and the reduction of elevation in such regions. They would, for instance, have been remote from continental coastlines and therefore the gradients of rivers draining them would have been low and their erosive power minimal. Extreme continentality of climate leading to predominantly arid climates and low runoff would also have contributed to slow rates of denudation and elevation reduction. Low annual precipitation would also have encouraged the development of internal drainage, as still exists over extensive areas of northern and southern Africa, and this would have further attenuated regional denudation and the reduction of elevation. Some support for this scenario comes from the stratigraphic record of some of the Gondwana continents prior to break-up. Across southern Africa, for example, terrestrial sedimentation prevailed from the mid-Permian until break-up in the Jurassic, and the occurrence of lacustrine deposition (Yemane, 1993) demonstrates the existence of internal drainage for a part of this time.

The high-elevation rifted margins resulting from continental break-up will be characterized by an interior plateau of predominantly low local relief flanked by a steep escarpment, or series of escarpments (Summerfield, 1985; Gilchrist and Summerfield, 1990). Irrespective of the mechanisms responsible for the high initial elevation, the establishment of new coastlines and base levels for continental drainage will result in an episode of accelerated denudation along the new continental margin, an outcome amply demonstrated by the patterns of crustal cooling recorded by fission-track data from virtually all passive margins (Brown *et al.*, 1990, 1994b; Gleadow and Brown, in press). As a result of isostatic uplift a considerable depth of denudation can result from the reduction to base level of the edge of even a relatively low plateau. For instance, in the case of a passive margin with an initial elevation of 1 km, crustal stripping to a depth of around 5 km would be required to reduce elevation to sea level. Such a depth of denudation would be recorded as a significant cooling event by fission-track data, although it may not *neccesarily* be connected to any tectonically-driven surface uplift event. Obviously tectonic mechanisms have promoted the lithospheric extension responsible for continental break-up, and hence the creation of new, lower base levels leading to accelerated denudation; however, such a denudational episode *could* occur without any active tectonic uplift.

The possible patterns of landscape evolution for a specific passive margin could result in a wide range of denudational histories depending, in particular, on the style of drainage evolution (Summerfield, 1991a, 1996). Where a major drainage system establishes a connection to a new continental coastline soon after break-up, then a pattern of accelerated denudation and associated crustal cooling would be expected to penentrate a significant distance inland. If, however, there were a long delay before interior drainage was captured by aggressive headward eroding coastal rivers, then accelerated denudation and marked crustal cooling immediately after break-up would be confined to a relatively narrow coastal zone. The interior regions would be expected to record a later phase of accelerated denudation and crustal cooling.

An example of a delayed denudational response is provided by the Lesotho Highlands located ~200 km inland of the south-east Drakensberg margin of southern Africa. Extrapolation of likley depths of denudation across the Lesotho Highlands on the basis of apatite fission-track data from deep boreholes both seaward and inland of the basalt outcrop forming the present highlands indicate less than 1 km of denudation since the basalt was emplaced in the Jurassic (Brown, 1992). Low rates of denudation over the past ~100 Ma are confirmed by the relatively shallow depth of exposure of kimberlite pipes in the Lesotho Highlands. Despite, therefore, the close proximity to the Indian Ocean of an area of high elevation (>2000 m) for apparently the last ~130 Ma, there has been minimal denudation during this period. The present relatively high local relief of the Lesotho Highlands, however, suggests that current rates of denudation are probably >100 m Ma^{-1} (Summerfield, 1991b). Clearly such high rates have only prevailed recently given the modest aggregate depth of denudation over the past ~130 Ma. A possible explanation for this mode of landscape development is that recent high rates of denudation only became established once the headwaters of the Orange River had cut through the basalt into the more readily erodible sediments of the underlying upper Karoo sequence. Nevertheless, irrespective of the mechanisms involved, the phase of accelerated denudation now affecting the Lesotho Highlands has occurred some 130 Ma *after* the tectonic event that gave rise to the formation of the south-east African margin, and there is no convincing evidence that it has resulted from recent tectonically-driven surface uplift.

The Lesotho Highlands example should provide a salutary warning for the interpretation of fission-track data on other passive margins since, although an initial phase of rapid denudation essentially coincident with rifting and continental break up is widely observed in these morphotectonic settings, more complex denudational responses should also be anticipated and discrete cooling episodes may not have a direct tectonic cause. For instance, the apparent episodic 'uplift' (denudation) suggested for the Transantarctic Mountains (Stump and Fitzgerald, 1992), as well as other rifted margins, *may* be due at least in part to morphologically-controlled changes in spatial patterns of denudation through time. Clearly other lines of evidence are required in deciding the relative role of active tectonics as opposed to the response of the sub-aerial denudational system in such cases.

6. Conclusions

Several points emerge from our analysis of the geomorphic factors that need to be considered in the interpretation of fission-track data. First, it is critical that terminology relating to relative and absolute vertical movements of the landsurface and rocks within the crust is defined precisely and applied consistently. Secondly, since factors other than tectonic uplift can engender a phase of accelerated denudation and associated cooling of the shallow crust, it cannot be assumed that all cooling episodes recorded by fission-track data are directly related to tectonic events. Thirdly, the lack of any simple relationship between elevation and denudation rate means that an increase in the former as a result of tectonic uplift may not necessarily lead to a denudational (crustal cooling) response; landscape sensitivity and differences in modes of landscape evolution both have to be taken into account. Fourthly, major drainage capture events, especially on the flanks of intermontane plateaus are as yet an unexplored means of generating major denudational and crustal cooling episodes. Fifthly, continental rifting and break-up may precipitate regional-scale cooling along passive margins even in the absence of accompanying tectonically-driven, surface uplift if there is significant pre–break-up residual elevation and therefore a significant drop in elevation from the continental interior to the newly created coastline. Finally, an emphasis on the role of denudation rather than 'uplift' in determining fission-track thermal histories points to the currently untapped potential of such data in determining rates of denudation and sediment transport in regional and global mass balance studies.

7. Acknowledgements

Some of the ideas contained in this paper were developed as part of research projects supported by Natural Environment Research Council grants GR3/6693 and GR3/9128 (MAS) and an Endeavour Research Fellowship (RWB). Initial research collaboration was assisted by a Royal Society Anglo-Australian Visiting Fellowship (MAS) and a British Council Visiting Bursary (RWB).

8. References

Ahnert, F. (1970): Functional relationships between denudation, relief, and uplift in large mid-latitude drainage basins. *Amer. J. Sci.* **268**, 243-263.

Ahnert, F. (1984): Local relief and the height limits of mountain ranges. *Amer. J. Sci.* **284**, 1035-1055.

Bates, R.L. and Jackson, J.A. eds (1987): *Glossary of Geology*, 3rd edn, American Geological Institute, Alexandria.

Beaumont, C., Kamp, P.J.J., Hamilton, J. and Fullsack, P. (1996): The continental collision zone, South Island, New Zealand: Comparison of geodynamical models and observations. *J. Geophys. Res.* **101**, 3333-3359.

Bohannon, R.G., Naeser, C.W., Schmidt, D.L. and Zimmermann, R.A. (1989): The timing of uplift, volcanism, and rifting peripheral to the Red Sea: A case for passive rifting? *J. Geophys. Res.* **94**, 1683-1701.

Brown, R.W. (1991): Backstacking apatite fission-track 'stratigraphy': A method for resolving the erosional and isostatic rebound components of tectonic uplift histories. *Geology* **19**, 74-77.

Brown, R.W. (1992): *A Fission Track Thermochronology Study of the Tectonic and Geomorphic Development of the Sub-aerial Continental Margins of Southern Africa*, PhD thesis, La Trobe University.

Brown, R.W., Gallagher, K. and Duane, M.J. (1994a): A quantitative assessment of the effects of magmatism on the thermal history of the Karoo sedimentary sequence. *Afr. J. Earth Sci.* **18**, 227-243.

Brown, R.W., Rust, D.J., Summerfield, M.A., Gleadow, A.J.W. and De Wit, M.C.J. (1990): An early Cretaceous phase of accelerated erosion on the south-western margin of Africa: Evidence from apatite fission track analysis and the offshore sedimentary record. *Nucl.Tracks and Radiat. Meas.*, **17**, 339-350.

Brown, R.W., Summerfield, M.A. and Gleadow, A.J.W. (1994b): Apatite fission track analysis: Its potential for the estimation of denudation rates and implications for models of long-term landscape development. In: *Process Models and Theoretical Geomorphology* (ed. Kirkby, M.J.), Wiley, Chichester, 23-53.

Brunsden, D. and Thornes, J.B. (1979): Landscape sensitivity and change. *Inst. Brit. Geogr. Trans.* **NS4**, 463-484.

Clark, S.P. and Jäger, E. (1969): Denudation rate in the Alps from geochronologic and heat flow data. *Amer. J. Sci.* **267**, 1143-1160.

Copeland, P., Harrison, T.M., Kidd, W.S.F., Ronghua, X. and Yuquan, Z. (1987): Rapid early Miocene acceleration of uplift in the Gangdese Belt, Xizang (southern Tibet), and its bearing on accommodation mechanisms of the India-Asia collision. *Earth Planet. Sci. Lett.* **86**, 240-252.

England, P. and Molnar, P. (1990): Surface uplift, uplift of rocks, and exhumation of rocks. *Geology* **18**, 1173-1177.

Fielding, E.J. (in press): Morphotectonic evolution of the Himalayas and Tibetan Plateau. In: *Geomorphology and Global Tectonics* (ed. Summerfield, M.A.), Wiley, Chichester.

Gilchrist, A.R. and Summerfield, M.A. (1990): Differential denudation and flexural isostasy in formation of rifted-margin upwarps. *Nature* **346**, 739-742.

Gleadow, A.J.W. and Brown, R.W. (in press): Fission track thermochronology and the long-term denudational response to tectonics. In: *Geomorphology and Global Tectonics* (ed. Summerfield, M.A.), Wiley, Chichester.

Harrison, T.M., Copeland, P. Kidd, W.S.F. and Yin, A. (1992): Raising Tibet. *Science* **255**, 1663-1670.

Hovius, N. (1995): *Macro Scale Process Systems of Mountain Belt Erosion and Sediment Delivery to Basins*, D.Phil. thesis, University of Oxford.

Jansson, M.B. (1988): A global survey of sediment yield. *Geogr. Annal.* **70A**, 81-98.

Kennan, L. (in press): Large-scale geomorphology of the Andes: interrelationships of tectonics, magmatism and climate. In: *Geomorphology and Global Tectonics* (ed. Summerfield, M.A.), Wiley, Chichester.

Lambeck, K. and Stephenson, R. (1986): The post-Palaeozoic uplift of history of south-eastern Australia. *Austr. J. Earth Sci.* **33**, 253-270.

Lorenzo, J.M. and Vera, E.E. (1992): Thermal uplift and erosion across the continent-ocean transform boundary of the southern Exmouth Plateau. *Earth Planet. Sci. Lett.* **108**, 79-92.

Masek, J.G., Isacks, B.L., Gubbels, T.L. and Fielding, E.J. (1994): Erosion and tectonics at the margins of continental plateaus. *J. Geophys. Res.* **99**, 13941-13956.

Milliman, J.D. and Syvitski, J.P.M. (1992): Geomorphic/tectonic control of sediment discharge to the ocean: The importance of small mountainous rivers. *J. Geol.* **100**, 525-544.

Oberlander, T. (1965): *The Zagros Streams*, Syracuse Geographical Series No. 1., Department of Geography, Syracuse University.

Oberlander, T.M. (1985): Origin of drainage transverse to structures in orogens. In: *Tectonic Geomorphology* (eds Morisawa, M. and Hack, J.T.), Allen and Unwin, Boston, pp. 155-182.

Ohmori, H. (1983): Erosion rates and their reltaion to vegetation from the viewpoint of world-wide distribution. *Bull. Dept. Geog. Univ. Tokyo* **15**, 77-91.

Pinet, P. and Souriau, M. (1988): Continental erosion and large-scale relief. *Tectonics*, **7**, 563-582.

Pitman, W.C. III and Golovchenko, X. (1991): The effect of sea level changes on the morphology of mountain belts. *J. Geophys. Res.* **96**, 6879-6891.

Stump, E. and Fitzgerald, P.G. (1992): Episodic uplift of the Transantarctic Mountains. *Geology* **20**, 161-164.

Stüwe, K., White, L. and Brown, R. (1994): The influence of eroding topography on steady-state isotherms. Application to fission track analysis. *Earth Planet. Sci. Letts*, **124**, 63-74.

Summerfield, M.A. (1985): Plate tectonics and landscape development on the African continent. In: *Tectonic Geomorphology* (eds Morisawa, M. and Hack, J.T.), Allen and Unwin, Boston, pp. 27-51.

Summerfield, M.A. (1989): Tectonic geomorphology: convergent plate boundaries, passive continental margins and supercontinent cycles, Prog. Phy. Geog. 13, 431-441, 1989.

Summerfield, M.A. (1991a): *Global Geomorphology* Longman, London and Wiley, New York.

Summerfield, M.A. (1991b): Sub-aerial denudation of passive margins: regional elevation versus local relief models, Earth Planet. Sci. Lett. 102, 460-469.

Summerfield, M.A. (1996): Tectonics, geology, and long-term landscape development. In: *The Physical Geography of Africa* (eds Adams, W.M., Goudie, A.S. and Orme, A.R.), Oxford University Press, Oxford, pp. 1-17.

Summerfield, M.A. and Hulton, N.J. (1994): Natural controls of fluvial denudation rates in major world drainage basins. *J. Geophys. Res.* **99**, 13871-13883.

Tippett, J.M. and Kamp, P.J.J. (1995): Quantitative relationships between uplift and relief parameters for the Southern Alps, New Zealand, as determined by fission track analysis. *Earth Surf. Processes Ldfms* **20**, 153-175.

Vandervoort, D.S., Jordan, T.E., Zeitler, P.K. and Alonso, R.N. (1995): Chronology of internal drainage development and uplift, southern Puna plateau, Argentine central Andes. *Geology* **23**, 145-148.

Wilson, L. (1973): Variations in mean annual sediment yield as a function of mean annual precipitation. *Amer. J. Sci.* **273**, 335- 349.

Yemane, K. (1993): Contribution of Late Permian palaeogeography in maintaining a temperate climate in Gondwana. *Nature* **361**, 51-54.

Section IV

Glass Studies

POTENTIAL GLASS AGE STANDARDS FOR FISSION-TRACK DATING: AN OVERVIEW

M.L. BALESTRIERI[a,b], G. BIGAZZI[b], V. BOUŠKA[c], E. LABRIN[d], J. C. HADLER N.[e], N. KITADA[f], A.M. OSORIO A.[g], G. POUPEAU[d], K. WADATSUMI[h] AND A. ZÚÑIGA[e]

[a] *Department of Earth Sciences, University of Pisa, Pisa, Italy*
[b] *Institute of Geochronology and Isotope Geochemistry, CNR, Pisa, Italy*
[c] *Institute of Geochemistry, Charles University, Prague, Czech Republic*
[d] *Institut Dolomieu, Joseph Fourier University, Grenoble, France*
[e] *Institute of Physics, University of Campinas, Campinas SP, Brazil*
[f] *Geo-Research Institute, Osaka Soil Test Laboratory, Japan*
[g] *Faculty of Sciences and Technology, S. Paulo State University, Presidente Prudente, Brazil*
[h] *Okayama University of Science, Faculty of Science, Department of Applied Science, Japan*

Abstract. In this work four glasses proposed as reference materials for fission-track dating were investigated. Samples from various localities from southern Bohemia and Moravia have been studied in order to identify the Moldavite-bearing deposit(s) with negligible track annealing. The remaining three glasses, JAS-G1 (Japan), Roccastrada (Italy) and Macusanite (Peru) yield thermally lowered fission-track ages, and need application of correction techniques. Therefore, following the requirement that a standard should fulfil: 'no corrections should be necessary in obtaining the fission-track age', they should be excluded. However, we considered that their potentiality as reference materials was worthy to be investigated: accuracy may be greatly improved when various reference samples with characteristics similar to those of real samples are available. All these glasses yielded reproducible plateau ages in agreement with available independent ages. The ζ factors computed using track densities corresponding to plateau age determinations are reciprocally well consistent and substantially agree with those of the FCT and Durango apatites, although the data suggest a slight systematical deviation. The Moldavites least affected by track annealing are those from the Jankov deposit (Middle Miocene): this glass appears the most convincing potential age standard among those studied here. In principle, the studied glasses form a very promising set of reference materials for standardisation of techniques, interlaboratory comparison and age calibration itself.

1. Introduction

In 1990 the Fission Track Working Group of the I.U.G.S. Subcommission on Geochronology decided that the only reliable way for calibrating all fission-track dating systems is the use of age standards (Hurford, 1990). When the ζ calibration is adopted, the ζ factor is directly determined through age standard analyses (Hurford and Green, 1983). For an absolute approach, the chosen calibration system has to be checked against age standards, and it can be considered reliable only when it produces age results consistent with their independently determined reference age.

Experience has shown that different mineral species may yield different ζ factors (Green, 1985). Therefore, when a given phase is analysed, standards of the same phase have to be used as reference, either when the ζ calibration or the absolute approach are

P. Van den haute and F. De Corte (eds.), Advances in Fission - Track Geochronology, 287-304.

adopted. Age standards are currently available for apatite, zircon and sphene. The only one glass age standard recommended by the I.U.G.S. Subcommission on Geochronology is Moldavite but *only splits from a single Moldavite* (Hurford, 1990). However, a supplier for this standard does not exist yet. Other glasses, namely: the obsidian JAS-G1 (Japan), the rhyolitic glass Roccastrada (Italy) and the obsidian Macusanite (Peru), have also been suggested as putative age standards or as useful reference samples for analytical procedure standardisation.

In this work the reliability of all these glasses as reference materials is investigated.

2. The four potential glass age standards

JAS-G1. The youngest of these glasses, the obsidian JAS-G1 from central Japan, was proposed by Wadatsumi *et al.* (1994) and Kitada and Wadatsumi (1995a, 1995b), who give detailed information on its location and characteristics. Its availability is unlimited, and splits of JAS-G1 are supplied upon request by K. Wadatsumi and N. Kitada. Its reference age, 0.947 ± 0.005 Ma (Kitada and Wadatsumi, 1995a), based on Ar-Ar determinations on glass, appears well documented.

Roccastrada. The glass Roccastrada (central Italy) was proposed as a useful reference material for analytical procedure standardisation by Bigazzi *et al.* (1993). Although its reference age, 2.444 ± 0.015 Ma, based on Ar-Ar determinations on three high purity sanidine samples and supported by K-Ar on glass and whole rock-biotite Rb-Sr ages (Bigazzi *et al.*, 1993), appears also well documented, confirmation by further determinations by other laboratories is still desirable. Availability of this glass is also unlimited. The Roccastrada glass consists of a population of shards of various sizes, therefore, its analysis presents more difficulties compared to obsidians. However, following the I.U.G.S. Subcommission on Geochronology, analysis of reference samples not so easy to analyse but which are similar to many real samples is recommended strongly: this glass meets this requirement, as a population of glass shards is the only datable phase of many tephra.

Macusanite. Macusanite glass (south-eastern Peru) was proposed by McCorkell and Naeser (1988) and more recently it was studied by Poupeau *et al.* (1992, 1993). In contrast to the first two glasses, Macusanite exists as pebbles of various sizes not found *in situ*, but only in river banks and fluvio-glacial deposits around and in the Macusani volcanic field. Only small glass fragments were found *in situ* as inclusions in two ignimbrites (Cheilletz *et al.*, 1992; Sabil, 1995). Experimental data indicate that this glass was produced by different eruptions during a relatively wide time span (Poupeau *et al.*, 1993). Therefore, the reference glass should consist in splits from Macusanites from the same eruption. Such a sample already exists: the Macusanite proposed by McCorkell and Naeser (1988), a 220g cobble large enough to meet the requirements of the fission-track community. The reference age of this glass is not yet satisfactorily well constrained. McCorkell and Naeser (1988) report an average age of 4.13 ± 0.20 Ma, whereas four K-Ar ages on glass determined on four different pebbles, including McCorkell and Naeser's Macusanite, have a mean of 5.59 ± 0.06 Ma (Poupeau *et al.*, 1992). In this work only these four independently dated pebbles are considered, as their ages are mutually consistent, and the above mean is adopted as a preliminary reference age, assuming that these pebbles are due to the same eruption.

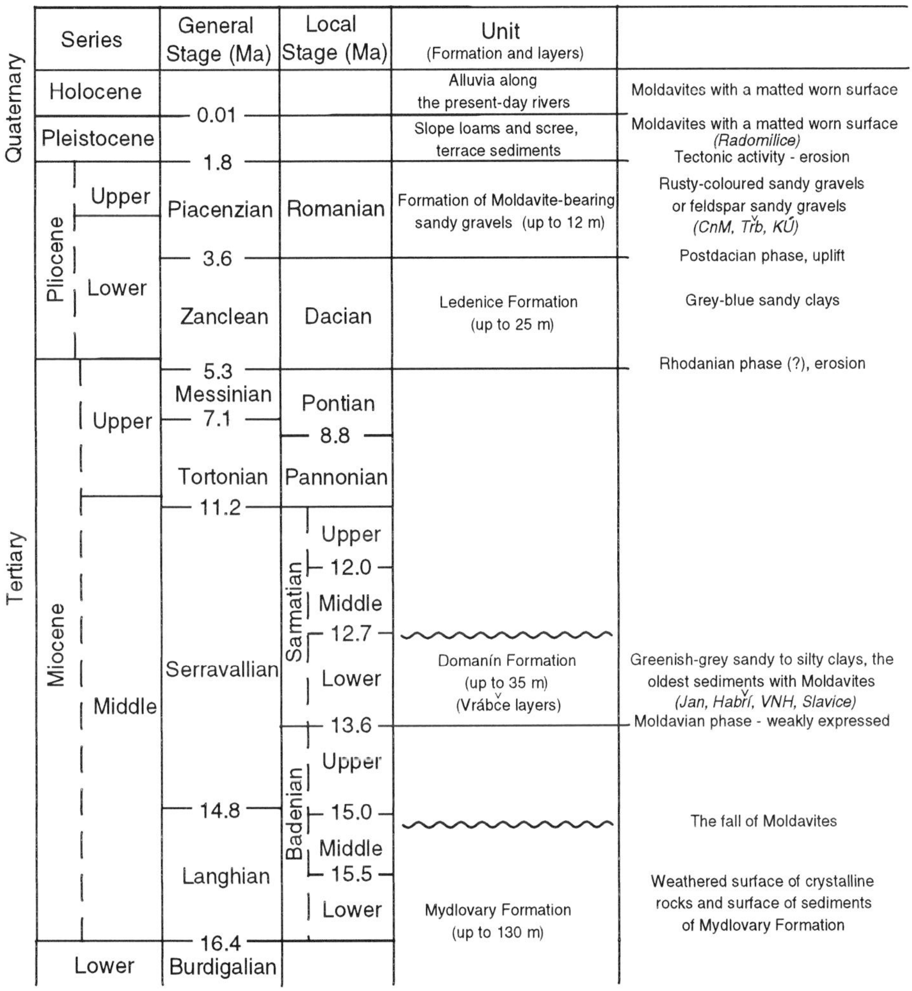

CnM: Chlum nad Malší; *Třb*: Třebanice; *Ú*: Kamenný Újezd; *Jan*: Jankov; *VNH*: Vrábče Nová Hospoda

Figure 1. Chrono-stratigraphic position of the Moldavite-bearing sediments in southern Bohemia, including those considered in this work. The Moravian Moldavites are found in a similar stratigraphic position (figure redrawn after Bouška, 1994).

Moldavite. Moldavite glass was studied by Wagner (1966) as soon as the fission-track dating method was developed. Its significance as a potential age standard was confirmed by McCorkell and Naeser (1988), during the Besançon fission-track dating workshop. Moldavite is a tektite found in sedimentary deposits covering a long time span from the Middle Miocene up to the Holocene (Fig. 1). Moldavite-bearing sediments were recognized in a wide area of central Europe, in southern Bohemia (Fig. 2) and Moravia (Czech Republic), in Lusatia (Germany) and in the Horn region (Austria). A detailed review on history, location and characteristics of Moldavites is given by Bouška (1994). Several K-Ar and Ar-Ar data are available for this tektite. Wagner and Van den haute (1992) report in their list of age standards two age determinations: 15.21 ± 0.15 Ma (Ar-Ar on glass, Staudacher *et al.*, 1982) and 15.1 ± 0.7 Ma (K-Ar on glass, Gentner *et al*, 1967), respectively. The first of these two determinations is adopted here as a preliminary reference age for its higher precision. However, this age may slightly overestimate the true formation age of Moldavites. Storzer *et al.* (1995) quote for Moldavite glass an age of 14.82 ± 0.32 Ma, mean of 17 K-Ar and Ar-Ar determinations reported by different authors between 1963 and 1993.

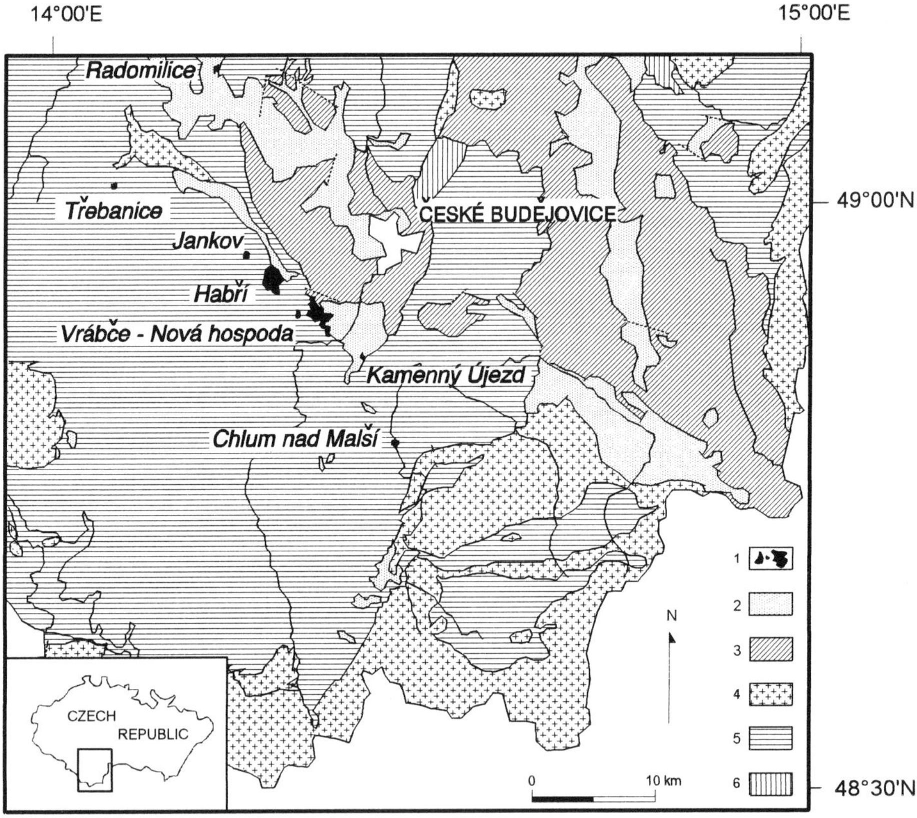

Figure 2. Simplified geological map of southern Bohemia showing location of the Moldavite-bearing deposits studied in this work. The only Moravian deposit considered in this work, Slavice, is located around 5 km south of the town of Třebíč (for the geological map of Moravia see Bouška, 1994). 1: Moldavite localities; 2: Tertiary sediments; 3: Upper Cretaceous sediments; 4: Granitoids of the Central Pluton; 5: Metamorphous rocks of the Moldanubicum; 6: Permian sediments (figure redrawn after Bouška, 1994).

Among the four glasses discussed in this paper, only Moldavite appears to have chance of fulfilling one of the basic requirements propounded by G.A. Wagner in 1980, during the fission-track dating workshop held at Pisa (Hurford and Green, 1981): *no corrections should be necessary in obtaining the fission-track age*. Whereas some Moldavite samples yielded spontaneous to induced track-size ratios (Ds/Di) close to 1, which indicates, following Storzer and Wagner (1969), a negligible amount of track annealing, the remaining three glasses *certainly* need age corrections, as they have Ds/Di significantly lower than 1 (Fig. 3). Therefore, it could be argued that the glasses JAS-G1, Roccastrada and Macusanite should be excluded. However, long experience in fission-track dating of glass suggested us that the potentials of these glasses as reference samples was worthy to be investigated, for several reasons:

(1) tektite is a special glass: most of the real samples important for dating are volcanic glasses, and sometimes fission-track dating is the only technique that can be reliably applied to them. Therefore, it is desirable to have standards similar to normal volcanic glasses.

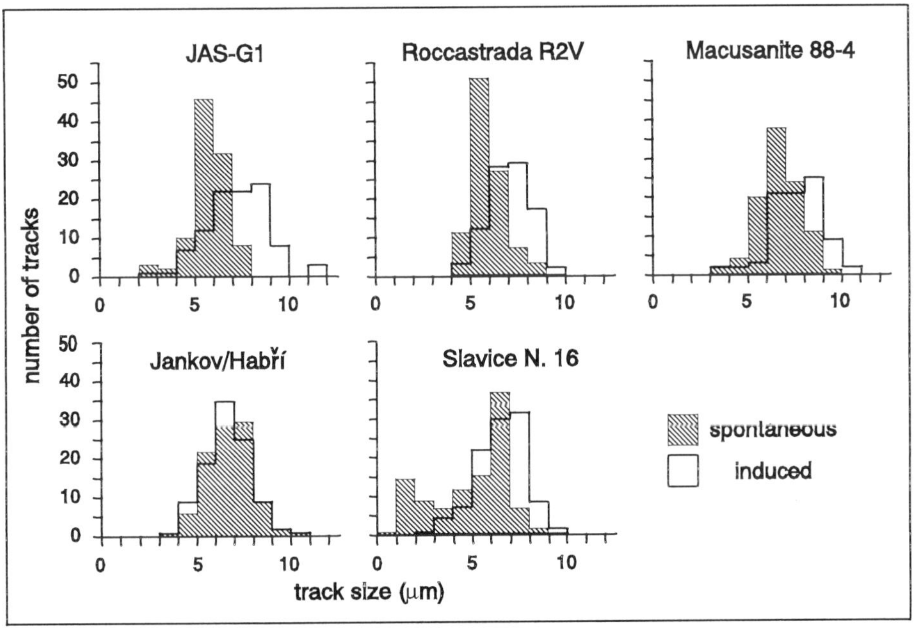

Figure 3. Spontaneous and induced track size distributions of four samples of the glasses studied in this work. JAS-G1, Roccastrada and Macusanite show a significant reduction of the spontaneous track sizes compared to the induced tracks. In contrast, spontaneous and induced track sizes are very similar in the Moldavite Jankov/Habří. Whereas in the first three glasses the spontaneous track were subjected to significant annealing, they remained fairly undisturbed over geological times in the Jankov/Habří Moldavite sample. The bimodal size distribution of spontaneous tracks of Moldavite Slavice N.16 indicates that this sample experienced a peculiar thermal history (see text).

(2) although the exact meaning of corrected fission-track ages is still considered somewhat controversial, experimental evidence indicates that the size-correction method (Storzer and Wagner, 1969) as well as the plateau method (Storzer and Poupeau, 1973) yield reliable formation ages of glasses (Komarov and Raykhlin, 1977;

Arias *et al.*, 1981; Naeser *et al.*, 1981; Obradovich *et al.*, 1982; Storzer and Wagner, 1982; Westgate, 1989), at least in case of moderate spontaneous-track annealing and simple thermal histories.

(3) it is desirable to have more than one reference sample. It is also desirable that these samples cover a wide range of ages and have characteristics similar to those of the various natural glasses.

(4) does a volcanic glass *totally* unaffected by track annealing exist ? Ds/Di values close to 1 were determined only in very few samples. All of them were very young and showed very low track densities. They cannot be advocated as age standards, as it is in practice impossible to determine their age with sufficient precision.

3. Our sample-set

Several samples of glasses JAS-G1, Roccastrada and Macusanite were available to the fission-track groups involved in this research. Only two splits of Moldavite had been previously analysed by the Campinas and Pisa fission-track groups. One of them, supplied by C.W. Naeser (McCorkell and Naeser, 1988), had revealed a slight annealing of spontaneous tracks. Its provenance was unknown to us. The second sample, a few grams of Moldavite labelled 'Jankov/Habří', had been supplied by V. De Michele (Civic Museum of Natural Sciences, Milan, Italy). Its spontaneous tracks appeared to be undisturbed. An apparent age in close agreement with the reference age had been measured (Bigazzi and De Michele, 1996). Another sample of unknown provenience, showing a certain amount of track annealing, had been analysed by the Grenoble group. One of us (G. Bigazzi) decided to visit V. Bouška in order to find an answer to two main questions:

(1) is it possible to find a Moldavite large enough to be distributed as a standard to the fission-track community ?

(2) fission-track analyses of Moldavites revealed a variable degree of track annealing. Moreover, Lange *et al.* (1995) showed that size-corrected ages of two Moldavites from terrace sediments of the Elbe River (Lusatia, Germany) were significantly lower than the ages measured using the Ar-Ar method. Which are the good Moldavite samples that contain undisturbed tracks ?

The meeting at Prague was very clarifying and revealed that the requirement *only splits from a single Moldavite* was hard to be fulfilled. Moldavites large enough are very rare, guarded in museums and private collections.

The different amount of annealing exhibited by different samples very likely depends on two factors: (1) the peculiar history experienced by the single Moldavite and (2) a differential track retentivity due to variable chemical-physical properties.

The oldest Moldavite-bearing deposits (Domanín formation, lower Sarmatian, Fig. 1) are only 1-2 Ma younger than the probable age of the impact which produced this tektite. Moldavites from these deposits are sharp-edged irregular fragments, with a sharp, deep sculpture, as they were corroded over a long period of time.

Other Moldavite-bearing sediments were deposited during a long period of time, in the Pliocene and Pleistocene. Samples from these younger sediments have shapes that are characteristic of the peculiar deposits, indicating different histories of reworking and transportation.

Several samples were selected from sediments with different characteristics, in order to identify those that are the least affected by track annealing. In principle, the best deposits regarding this question should be the oldest ones (Fig. 1). A number of different pieces from such a deposit(s) might be a reasonable substitute of *only one Moldavite*. Although found in much younger deposits, the Třebanice Moldavites could also be good, as they do not appear to have experienced transportation. On the contrary, those from Chlum nad Malší and Radomilice are well rounded. The Komenný Újezd deposit contains abundant fragments of charcoal, indicating that the Moldavites may have experienced high temperatures during Pliocene times.

The exact provenance of the previously analysed sample Jankov/Habří is somewhat controversial: Jankov and Habří are two different deposits located at a distance of a few kilometres from each other (Fig. 2). According to Dr. De Michele (personal communication) Jankov is the most likely true provenience of that Moldavite.

4. Methods

The four glasses were analysed with the standard techniques described in Poupeau *et al.* (1992, 1993), Bigazzi *et al.* (1993) and Kitada and Wadatsumi (1995a, 1995b).

Ages were determined using the calibration system known as the 'absolute approach'. The plateau method, preferred to the size-correction method for its higher precision, was routinely applied for correcting thermally lowered ages. The ζ values referred to the sandwich glass NIST SRM 612 - muscovite were also computed.

For the age calculation, the fission-track groups involved in this research used the following parameters: $\lambda_F = 7.03\text{x}10^{-17}$ a^{-1}; $\lambda = 1.55125\text{x}10^{-10}$ a^{-1}; $\sigma = 5.802\text{x}10^{-22}$ cm^2; $^{238}U/^{235}U = 137.88$. The NIST glass standard SRM 962a (Carpenter, 1984) was used for neutron fluence (Φ) determinations, considering as reference the great mean of the calibration fluence values corresponding to the RT3 and RT4 positions and the Au and Cu monitors (Campinas, Grenoble and Pisa) or the Au monitor alone (Osaka). Age in close agreement with the reference ages were obtained by the Pisa group (Bigazzi and Oddone, 1990; Bigazzi *et al.*, 1993) and by the Grenoble group (Lelarge, 1993; Sabil, 1995) on Fish Canyon Tuff (FCT) and Durango age standard apatites using this calibration system.

Van den haute *et al.* (1988) have shown that NIST glass may yield unreliable Φ values. These authors used metal monitors whose nuclear parameters are very accurately known (Au and Co foils), and they obtained mutually consistent Φ determinations. In addition, they showed that their Φ values yielded reliable ages on age standards when a ^{238}U fission decay constant $\lambda_F = 8.46\text{x}10^{-17}$ a^{-1} was adopted. Although this calibration system is physically more convincing than the traditional one adopted in this work, fluence determinations referred to Au and Co foils require an easy access to the reactor facilities and an intimate co-operation with the reactor scientists for measurements of neutron fluences for each irradiation or for calibrating by multiple irradiations a U-doped glass 'transfer' standard. This is difficult to accomplish for many fission-track geochronologists. The NIST standards are the only glasses that are supplied with pre-irradiated wafers: Φ values are easily measured by the fission-track geochronologists. For this reason, although the absolute approach adopted here is not absolute anymore as it is based on a combination of Φ and λ_F operational values which yield right age values

TABLE 1. Fission-track data of glasses proposed as reference materials. ζ factors referred to standard apatites are also shown

Glass	ρ_s (x10^{-3})	Apparent Age ± 1 s.d. (Ma)	Ds/Di ± 1 s.d.	Plateau Age ± 1 s.d. (Ma)	P(χ^2)	ζ apparent ± 1 s.d.	ζ plateau ± 1 s.d.	ζ apatites
JAS-G1 n.ir. 4	1,80 - 3,00	0.68 ± 0.05 (6.9%) n. 4	0.78 ± 0.02 (2.3%)	0.96 ± 0.02 (1.8%) n. 4	75% >10% m.v. 3%	474 ± 23 (4.9%)	321 ± 16 (5.1%)	
(Osaka group) n.ir. 10	1,40	0.75 ± 0.04	0.77 ± 0.01	1.00 ± 0.03 (3.3%) n. 11			355 ± 12 (3.3%)	FCT Pisa: 328 ± 10
Roccastrada n.ir. 4; n.s. 4	18,7 - 22,8	1.87 ± 0.09 (4.7%) n. 10	0.84 ± 0.02 (2.7%)	2.51 ± 0.07 (2.6%) n. 13	95%>10% m.v. 8%	416 ± 21 (5.0%)	310 ± 10 (3.2%)	Grenoble: 315 ± 8
Macusanite* n.ir. 9; n.s. 4	16,4 - 30,5	4.12 ± 0.54 (13.1%) n. 17	0.79 ± 0.05 (6.6%)	5.77 ± 0.36 (6.3%) n. 22	92%>10% m.v. 4%	420 ± 18 (6.7%)	316 ± 12 (3.9%)	DURANGO Pisa:
Moldavite n. ir. 4; n.s. 15	11,4 - 16,7	14.1 ± 0.7 (5.3%) n. 16	0.92 ± 0.04 (3.9%)	15.2 ± 0.4 (2.5%) n. 17	79%>10% 2 val. < 1%	347 ± 18 (5.2%)	317 ± 9 (2.8%)	336 ± 16 Grenoble:
Jankov n.s. 4		14.9 ± 0.3 (1.8%) n. 5	0.96 ± 0.03 (3.5%)	15.2 ± 0.2 (1.5%) n. 6				311 ± 8
(Grenoble group) n.ir. 4; n.s. 4	13.5 - 16.2	14.7 ± 1.2 (7.9%) n. 5	0.86 ± 0.05 (6.0%)	15.6 ± .8 (5.2%) n. 3		330 ± 27 (8.1%)	306 ± 16 (5.2%)	

ρ_s: natural spontaneous track density (cm^{-2}); s.d.: standard deviation; P(χ^2) : probability of obtaining Chi-square value for the induced track count distribution tested against a Poisson distribution (75% > 10%: 75% of the P(χ^2) values are > 10%; m.v.: minimum P(χ^2) determined value); n.ir.: number of irradiations; n.: number of determinations; n.s.: number of samples; *: only four samples from Chilcuno Chico (3) and Caluyo Mayo (1) dated using the K-Ar method have been considered. For glasses JAS-G1 and Moldavite data obtained by the Osaka and Grenoble groups, respectively, are separately reported, as they were independently determined (apparent age and Ds/Di ratio obtained at Osaka are reported ± the standard error of the unique determination). Data regarding JAS-G1 (first row), Roccastrada and Macusanite were produced in close cooperation by the Campinas, Grenoble and Pisa groups, with cross-cheking of samples. Finally, data regarding the Jankov deposit (including the Moldavite labelled Jankov/Habřří) produced at Pisa are pointed out. Data regarding single determinations are reported in Kitada and Wadatsumi (1995a) for JAS-G1 (Osaka group). For Roccastrada and Macusanite, detailed tables were published by Bigazzi et al. (1993) and Poupeau *et al.* (1992, 1993), respectively.

on age standards, it is still used by most fission-track groups working with glass. According to De Corte *et al.* (this volume), the new IRMM U-doped glass for fission-track analysis is the final solution to all problems of an absolute approach based on the real physical values of Φ and λ_F.

Samples analysed by the Pisa and Campinas groups were irradiated in the Lazy Susan (LS, Cd ratio 6.5 for Au and 48 for Co) and Thermal Column (TC, Cd ratio 85.3 for Au and 643 for Co) facilities of the Triga Mark II reactor of LENA, University of Pavia (Italy). The Grenoble group used the position Canal P1 (Cd ratio 3,000 for Au and 12,000 for Co) of the Orphée reactor (Saclay, France). The Osaka group used positions Pn-1 (Cd ratio for Au 6.8) and Tc-Pn (Cd ratio for Au 145.1) of the Kyoto University reactor. Neutron flux spatial gradients are negligible in irradiation positions LS, TC (Pavia, Bigazzi *et al.*, 1988; Balestrieri *et al.*, 1996), Pn-1 and Tc-Pn (Kyoto, unpublished results). Multiple irradiations using several NIST SRM 612 glasses have shown that axial gradients are negligible in Canal P1 (Saclay) facility, whereas a radial gradient of around 8% cm^{-1} was detected. To minimise the effect of this gradient, samples and dosimeters were piled in the axial part only of the irradiation tube.

Glasses were etched with 20% HF at 40°C at Campinas and Pisa, with 20% HF at 20°C or 20% HF at 40°C at Grenoble and with 20% HF at room temperature at Osaka. Etching times were adjusted in order to produce induced track-sizes of 6-8 μm. Induced tracks in muscovite were etched for 8 min with 40% HF at 40°C at Campinas and Pisa, with 40% HF at 20°C for 70 min at Grenoble and with 48% HF at 25°C for 20 min at Osaka. Track-sizes were measured using a projection tube and a digitizing tablet at Grenoble, a Leitz 16x moving-line eyepiece or a Leica Microvid equipment at Campinas and Pisa and a computerised image processing system for fission-track dating at Osaka. Between 100 and 400 tracks were measured for each mean size determination. Errors of Ds/Di ratios are between 1.5% and 2.5%.

5. results and discussion

A synthetic overview of the results is shown in Table 1. Detailed analytical data on fission-track dating of Moldavites are shown in Table 2. Analytical data on the ζ factor measurements are reported in Table 3. Ages (± 2σ) normalised to the reference ages are shown in Fig. 4. Finally, the range of ζ values is illustrated in Fig. 5.

The single and mean apparent ages (Table 1, Fig. 4) of glasses JAS-G1, Roccastrada and Macusanite are significantly lower than the reference ages, whereas Bohemian Moldavites (specially those from Jankov), show apparent ages close to the reference age (Table 2). This is consistent with the measured Ds/Di values, that are significantly lower than 1 for the first three glasses ad close to 1 for Bohemian Moldavites (at least for those measured at Pisa. The Grenoble group obtained lower Ds/Di values).

Age determinations were also less reproducible for the Macusanites than for the remaining glasses. The reason of this lower reproducibility is not clear. All volcanic glasses showed induced track count distributions consistent with Poisson distributions. On the contrary, in some Moldavites tracks did not appear to be homogeneously distributed (Table 1, Table 2).

The mean plateau ages of all glasses are well consistent with the reference ages.

For all glasses the standard deviation of the age values also greatly reduces when the plateau method is applied. This is consistent with a differential annealing of the spontaneous tracks within a same glass.

TABLE 2. (A) Fission-track dating of Moldavites; determinations made at Pisa

Sample heating	ϕ ($x10^{15}$) (cm^{-2})	ρ_S (cm^{-2}) ($x10^3$)	N_S	ρ_I (cm^{-2}) ($x10^5$)	N_I	$p(\chi^2)$ (%)	Ds/Di	Age ± 1 σ (Ma)
MA P26	3.01	12.5	1111	1.640	2553	49	0.90	13.8 ± 0.5
4 h 250°C		12.5	1112	1.405	2363	88	1.02	16.1 ± 0.6
Jankov/Habří								
JH P37-1	1.09	15.8	609	0.679	1293	48	0.99	15.2 ± 0.7
4 h 220°C		12.2	1129	0.520	1240	81	1.00	15.3 ± 0.6
4 h 250°C		13.1	607	0.559	1332	36	1.01	15.3 ± 0.8
JH P39	1.77	14.2	1152	1.020	1122	77	0.98	14.7 ± 0.6
4 h 220°C		11.8	1079	0.831	1443	93	0.99	15.0 ± 0.6
Radomilice								
N.02 P41-1	2.72	16.7	1287	1.889	1587	29	0.90	14.4 ± 0.5
4 h 220°C		15.9	1103	1.693	1430	< 1	1.03	15.2 ± 0.7
N.04 P41-1	2.72	12.9	1100	1.478	1241	23	0.92	14.3 ± 0.6
4 h 220°C		11.0	1061	1.204	1268	15	0.99	14.9 ± 0.6
N.06 P41-1	2.72	13.0	1155	1.493	1253	45	0.92	14.2 ± 0.6
4 h 220°C		12.3	1092	1.325	1118	8	0.98	15.1 ± 0.6
Chlum nad Malší								
N.28 P41-1	2.72	13.4	1089	1.642	1372	64	0.90	13.3 ± 0.5
4 h 220°C		12.9	646	1.368	1156	12	1.01	15.2 ± 0.7
Třebanice								
N.20 P41-1	2.72	12.9	1146	1.652	1375	48	0.89	12.7 ± 0.5
4 h 220°C		13.5	1039	1.454	1227	50	1.01	15.0 ± 0.6
Kamenný Újezd								
N.14 P41-1	2.72	13.3	1132	1.528	1604	89	0.89	14.2 ± 0.6
4 h 220°C		11.6	1033	1.291	1358	54	1.01	14.7 ± 0.6
Vrábče Nová Hospoda								
N.08 P41-1	2.72	14.7	1134	1.660	1395	69	0.96	14.4 ± 0.6
4 h 220°C		14.0	1083	1.485	1255	54	1.01	15.4 ± 0.6
N.09 P41-1	2.72	14.4	1112	1.832	1526	72	0.87	12.8 ± 0.5
4 h 220°C		14.0	1082	1.530	1290	53	1.01	14.9 ± 0.6
N.11 P41-1	2.72	13.2	665	1.594	1333	81	0.90	13.5 ± 0.6
4 h 220°C		14.0	1244	1.433	1518	92	1.01	15.9 ± 0.6
N.18 P41-1	2.72	13.2	1027	1.587	1328	95	0.89	13.6 ± 0.6
4 h 220°C		12.4	1055	1.368	1152	97	1.00	14.8 ± 0.6
Jankov								
N.22 P41-1	2.72	13.3	1027	1.488	1252	4	0.92	14.5 ± 0.7
4 h 220°C		12.0	1063	1.303	1923	3	0.99	14.9 ± 0.6
N.24 P41-1	2.72	11.4	1098	1.243	1310	< 1	0.97	14.9 ± 0.6
4 h 220°C		10.1	1051	1.186	1374	8	0.99	15.1 ± 0.6
N.25 P41-1	2.72	16.5	1270	1.782	1403	88	0.92	15.0 ± 0.6
4 h 220°C		15.2	1172	1.584	1340	14	1.01	15.6 ± 0.6
Slavice								
N.16 P41-1	2.72	5.23	505	1.673	1332	3	0.76*	5.1 ± 0.3
(&)		3.59						3.5 ± 0.2

TABLE 2.. (B) Fission-track dating of Moldavites; determinations made at Grenoble

Sample	heating	ϕ (x10^{15}) (cm^{-2})	ρ_s (cm^{-2}) (x10^3)	N_S	ρ_I (cm^{-2}) (x10^5)	N_I	Ds/Di	Age $\pm$ 1 σ (Ma)
Mol 1993		1.56	14.4	1757	0.949	1736	0.86	14.2 ± 0.6
	2 h 225°C		17.1	1433	1.000	2282	1.01	15.3 ± 0.8
Chlum nad Malší								
N.28		1.19	15.0	1740	0.818	2818	0.82	13.1 ± 0.5
	2 h 250°C		13.9	1743	0.598	2630	1.01	16.5 ± 0.7
Třebanice								
N.20		1.19	16.2	1758	0.726	2584	0.94	15.9 ± 0.6
Jankov								
N.22		1.19	14.8	1480	0.722	2415	0.81	14.6 ± 0.8
N.24		1.19	13.6	895	0.612	1978	0.86	15.7 ± 0.7
	2 h 250°C		9.78	1262	0.467	2688	1.01	14.9 ± 0.6
Slavice								
N.16		1.19	5.50	624	0.717	1978	0.94	5.4 ± 0.3

Heating: thermal treatment imposed for the plateau age determination. ϕ: neutron fluence. ρ_I: induced track density. N_S (N_I): spontaneous (induced) track counted. *: spontaneous track sizes distribute in two groups with Ds/Di ratios of 0. 35 and 0.95, respectively. (&): data regarding the 'large tracks' population (see text). Errors of ages are propagation of counting errors of spontaneous and induced tracks. Errors of neutron fluences are between 1.7% and 2%.

The ζ factors of the first three glasses (Table 1, Fig. 5), when computed using the track densities of the unheated samples, are significantly higher than those commonly found for the sandwich glass NIST SRM 612 - muscovite, using the apatite age standards and the population method (commonly between 310 and 350; those determined by the Pisa and Grenoble groups are shown in Table 1 and Fig. 5). The ζ factors for Moldavites are more close to those of apatites. After the thermal treatment imposed for the plateau age determinations, the four glasses produce ζ factors that are mutually quite consistent and that are comparable to those obtained for apatite.

A few comments can be made with respect to the Moldavites. Only the samples from Jankov (including the sample labelled Jankov/Habří) showed a very small or negligible degree of track annealing (Table 1, Table 2). The results are somewhat different from expected. The Moravian Moldavite from Slavice (not considered in Table 1, Fig. 4 and Fig. 5), which was expected to show undisturbed tracks, yielded a very low apparent age (5.1 ± 0.3 Ma). In addition, it showed a bimodal spontaneous track-size distribution which indicates a complex thermal history (Fig. 3). After their fall, Moldavites were redeposited in Middle Miocene shallow sediments. Some of these deposits were eroded or reworked in Pliocene-Pleistocene (this is the case of the Slavice deposit), and the Moldavites were exposed to natural events. Very likely during reworking the Slavice sample experienced a short heating process which produced a significant annealing of its pre-existing tracks. Considering the track density of the undisturbed tracks (these are the larger tracks that form the peak located on the right in Fig. 3), this thermal event can be dated around 3.5 Ma. The analysis made by the

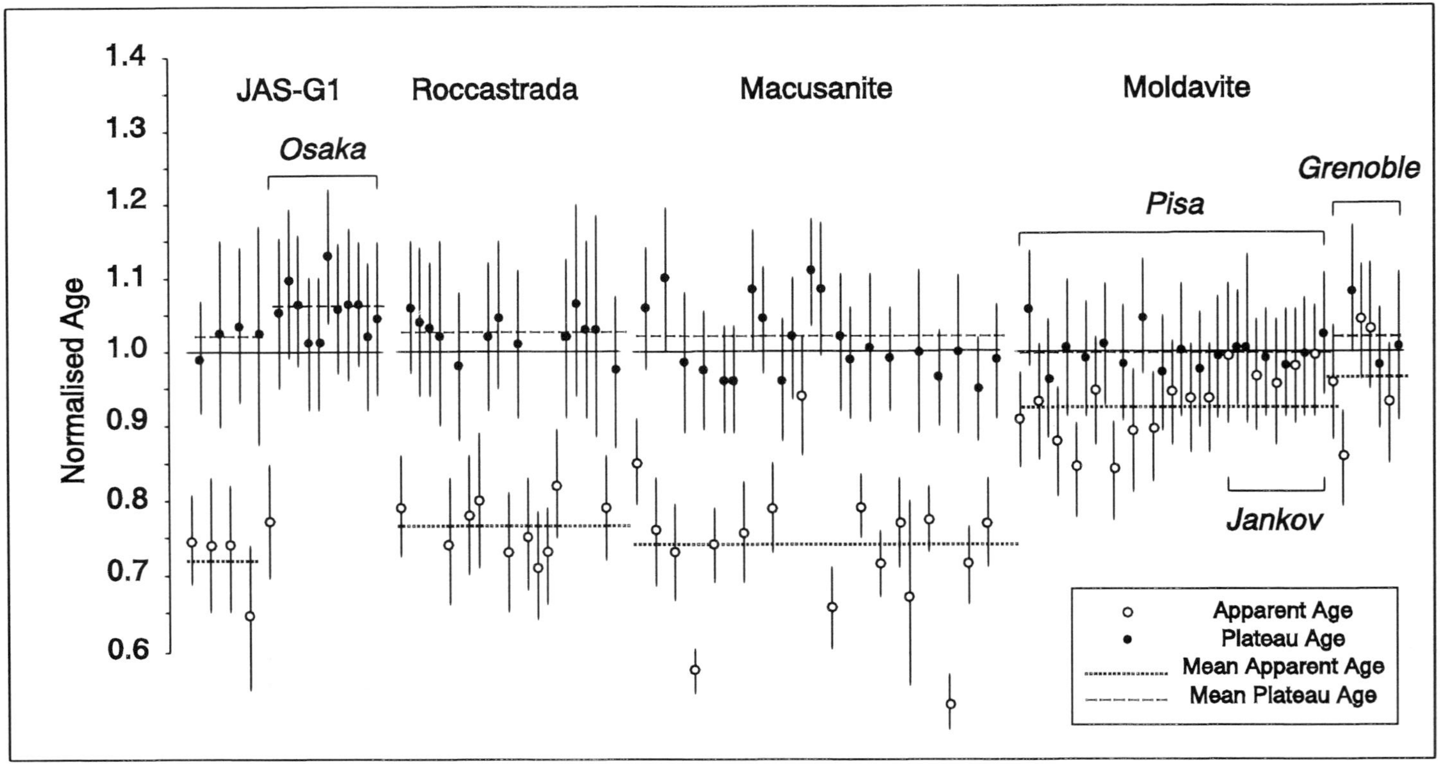

Figure 4. Normalised fission-track apparent and plateau ages (± 2 σ) of the four glasses JAS-G1, Roccastrada, Macusanite and Moldavite. The independent reference Ar-Ar or K-Ar age is assumed as 1. Whereas apparent ages of glasses JAS-G1, Roccastrada and Macusanite are significantly lower than 1, those of Moldavites are close to 1 (specially for Jankov deposit, including the Moldavite labelled Jankov/Habří). All plateau ages are consistent with the reference age. Where not explicitly indicated, data are an average result of the Campinas, Grenoble and Pisa fission-track groups. The Slavice Moldavite is not considered here (see text).

Grenoble group on a different split confirmed the age measured at Pisa, but did not detect the bimodal shape of the track-size distribution. Cross-checking of the two splits analysed by the two groups and further investigations are in progress. A better knowledge of the thermal history experienced by this Moldavite may be geologically important (the Slavice deposit had been considered reworked during the Quaternary).

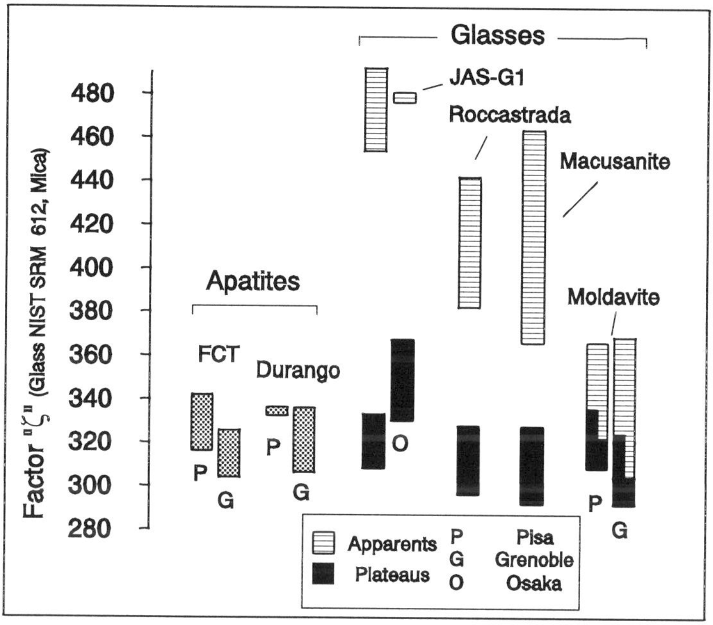

Figure 5. Range of the ζ factors determined for the sandwich glass NIST SRM 612 - muscovite using as reference the four glasses studied here. Track densities corresponding to apparent age as well as to plateau age determinations were considered. ζ factors referred to apatites determined at Pisa (P) and Grenoble (G) are also shown for comparison. Only 'plateau' ζ factors are consistent with those of apatites, with the exception of Moldavite, for which the ranges of the 'apparent' and 'plateau' ζ factors partially overlap. Where not explicitly indicated, data are an average result of the Campinas, Grenoble and Pisa fission-track groups.

With regard to the subject of this work, sample Slavice N.16 demonstrates (as well as the two samples from Lusatia studied by Lange *et al.*, 1995) that some Moldavites may have experienced peculiar thermal histories, and hence they have to be excluded from a set of potential age standards. This, in turn, confirms that an analysis of Moldavites from different deposits with different characteristics has to be performed in order to select the best age standard candidates.

The Chlum nad Malší, Radomilice and Kamenný Újezd deposits also yielded surprising results. Moldavites from the first two deposits are well rounded and indicate long-distance transportation, whereas at Kamenný Újezd the sediments contain abundant charcoal fragments. These factors had led to believe that Moldavites might be affected by a larger amount of spontaneous track annealing. Actually, on samples from these

TABLE 3. Determination of the factor ζ (determinations made at Pisa/Campinas and Grenoble)

Sample	Heating	ρ_S (x10^3)	N_S	ρ_I (x10^5)	N_I	ρ_D (x10^5)	N_D	$\zeta \pm 1\sigma$
GLASS								
JAS - G1								
P 36-1	ambient	2.38	367	1.877	1101	1.634	3227	458 ± 29
	5h 220°C	1.94	300	1.112	1310			332 ±22
P 41-1	ambient	1.79	220	4.761	1107	5.135	4758	491 ± 37
	4h 220°C	1.86	237	3.117	1453			309 ± 21
Roccastrada								
R2V P 26	ambient	20.7	640	19.39	1477	5.72.9	3539	399 ± 20
	1h 220°C	19.4	792	13.46	2573			296 ± 13
	2h 220°C	17.7	579	12.55	2655			303 ± 15
	4h 220°C	16.4	839	11.70	2015			305 ± 14
	4h 220°C	16.5	354	11.90	1491			307 ± 19
R2V P 29	ambient	20.4	530	0.432	800	0.117	2127	420 ± 25
	4h 200°C	24.0	592	0.397	1180			329 ± 18
R2V P 30	ambient	18.7	416	15.90	2485	5.110	3952	406 ± 22
	4h 180°C	19.6	605	12.66	1985			309 ± 15
R2V P41-1	ambient	22.8	423	19.22	1502	5.135	4758	401 ± 23
	4h 200°C	18.0	556	12.30	1156			325 ± 17
Roc 3 P 26	ambient	20.1	545	19.66	1534	5.729	3539	418 ± 22
Roc 10 P 26 (1)	ambient	19.5	543	20.29	1582			443 ± 23
Roc 10 P 26 (2)	ambient	20.0	596	20.33	3170			433 ± 21
Roc 11 P 26	ambient	21.2	543	19.01	2969			382 ± 19
	1h 220°C	18.4	434	13.32	2096			309 ± 17
	2h 220°C	19.1	305	13.67	2622			306 ± 19
	4h 220°C	15.8	202	11.26	2086			307 ± 23
Macusanite								
M 1 P29	ambient	27.9	1292	0.249	1156	0.117	2127	407 ± 19
M 1 P 30	ambient	26.1	2117	10.56	2340	5.118	3952	442 ± 15
	3h 250°C	16.5	1276	4.828	2410			319 ± 12
M 2 P 26	ambient	27.5	1914	11.44	2895	5.729	3539	405 ± 14
	3h 250°C	16.3	1261	5.452	2168			326 ±13
Ma 11 P 26	ambient	26.2	1214	11.84	1869			441 ± 18
	3h 250°C	13.2	1019	4.473	2667			331 ± 13
88-4 P 26	ambient	24.4	1414	10.60	2012			424 ± 16
	4h 220°C (1)	17.7	1366	5.952	1479			328 ± 13
	4h 200°C (2)	18.3	1062	6.149	2455			327 ± 13
88-4 P 29	ambient	22.0	1188	0.231	1158	0.117	2127	465 ± 22
	4h 220°C	22.1	1025	0.154	775			309 ± 16
88-4 P 30	ambient	24.9	770	9.532	1605	5.118	3952	418 ± 19
	3h 250°C (1)	15.8	1095	4.183	2102			291 ± 12
	3h 250°C (2)	15.6	1207	4.328	3463			302 ± 11
88-4 P 36-1	ambient	29.8	1152	3.143	1063	1.634	3227	366 ± 17
	5h 220°C	24.7	1810	2.197	1698			309 ± 12
	5h 250°C	23.8	1104	2.171	1163			316 ± 14
88-4 P41-1	ambient	25.0	1160	9.426	1345	5.135	4758	410 ± 17
	4h 220°C	20.9	1215	6.175	1331			321 ± 14
Moldavite								
MA P 26	ambient (1)	11.8	1040	1.613	2400	5.729	3539	365 ± 15
	ambient (2)	12.5	1111	1.638	1554			348 ± 15
	4h 250°C	13.2	1021	1.470	1547			298 ± 13

TABLE 3. Determination of the factor ζ (continued).

Sample	Heating	ρ_S (x10[3])	N_S	ρ_I (x10[5])	N_I	ρ_D (x10[5])	N_D	$\zeta \pm 1\sigma$
Moldavite								
Mol 1993 (G)	ambient	14.4	1757	0.949	1736	2.950	1881	339 ± 14
	2h 250°C	17.1	1433	1.000	2282			302 ± 12
Rad 02 P 41-1	ambient	16.7	1287	1.889	1587	5.135	4758	336 ± 14
	4h 220°C	15.9	1103	1.693	1430			316 ± 15
Rad 04 P 41-1	ambient	12.9	1100	1.478	1241			338 ± 15
	4h 220°C	11.0	1061	1.204	1268			325 ± 14
Rad 06 P 41-1	ambient	13.0	1155	1.493	1253			341 ± 15
	4h 220°C	12.3	1092	1.325	1118			320 ± 14
CnM 28 P 41-1	ambient	13.4	1089	1.642	1372			363 ± 16
	4h 220°C	12.9	646	1.368	1156			315 ± 16
CnM 28 (G)	ambient	15.0	1740	0.818	2818	2.249	10709	369 ± 12
	2h 250°C	13.9	1743	0.598	2630			292 ± 9
Třb 20 P 41-1	ambient	12.9	1146	1.652	1375	5.135	4758	380 ± 16
	4h 220°C	13.5	1039	1.454	1227			321 ± 14
Třb 20 (G)	ambient	16.2	1758	0.726	2584	2.249	10709	303 ± 10
	2h 250°C	13.2	1046	0.497	3577			255 ± 9
KÚ N.14 P 41-1	ambient	13.3	1132	1.528	1604	5.135	4758	340 ± 14
	4h 220°C	11.6	1033	1.291	1358			329 ± 14
VNH 08 P 41-1	ambient	14.7	1134	1.660	1395	5.135	4758	335 ± 14
	4h 220°C	14.0	1083	1.485	1255			314 ± 14
VNH 09 P 41-1	ambient	14.4	1112	1.832	1526			377 ± 16
	4h 220°C	14.0	1082	1.530	1290			324 ± 14
VNH 11 P 41-1	ambient	13.2	665	1.594	1333			357 ± 18
	4h 220°C	14.0	1244	1.433	1518			303 ± 12
VNH 18 P 41-1	ambient	13.2	1027	1.587	1328			354 ± 16
	4h 220°C	12.4	1055	1.368	1152			327 ± 15
Jan 22 P 41-1	ambient	13.3	1027	1.488	1252			332 ± 17
	4h 220°C	12.0	1063	1.303	1923			323 ± 13
Jan 22 (G)	ambient	14.8	1803	0.722	2415	2.249	10709	330 ± 11
Jan 24 P 41-1	ambient	11.4	1098	1.243	1310	5.135	4758	324 ± 16
	4h 220°C	10.1	1051	1.186	1374			320 ± 14
Jan 24 (G)	ambient	13.5	895	0.612	1978	2.249	10709	306 ± 13
	2h 250°C	9.78	1262	0.467	2688			323 ± 11
Jan 25 P 41-1	ambient	16.5	1270	1.782	1403	5.135	4758	321 ± 13
	4h 220°C	15.2	1172	1.584	1340			309 ± 13
APATITE *								
FCT P 26 (1)	—	175.4	1375	11.59	5818	5.729	3539	333 ± 14
P 26 (2)	—	169.8	1351	11.62	6665			322 ± 15
P 30 (1)	—	170.5	829	10.31	2424	5.118	3952	329 ± 17
P 30 (2)	—	173.0	1103	10.03	2403			316 ± 16
P 36 - 1	—	179.5	760	3.599	1411	1.634	3227	342 ± 20
Durango P 31	—	160.9	1577	7.887	1546	4.586	3187	336 ± 15

ρ_D, N_D: track density (cm^{-2}) and number of tracks counted on the muscovite detector attached to NIST SRM 612 glass. *Rad*: Radomilice, *CnM*: Chlum nad Malší, *Třb*: Třebanice, *KÚ*: Kamenný Újezd, *VNH*: Vrábče Nová Hospoda, *Jan*: Jankov. (G): determinations made at Grenoble. *: ζ determinations made at Grenoble are reported in Lelarge (1993) and Sabil (1995). Errors are propagation of counting errors.

deposits were determined mean apparent ages of 13.3 ± 0.5 Ma, 14.3 ± 0.3 Ma and 14.2 ± 0.6 Ma, and Ds/Di mean values of 0.80, 0.91 and 0.89, respectively. The Moldavites from Třebanice (Pliocene, no long-distance transportation) and, specially, from Vrábče Nová Hospoda (Middle Miocene), that had been considered to contain possibly undisturbed tracks, yielded mean apparent ages significantly lower than expected (12.7 ± 0.5 Ma, with Ds/Di = 0.89, and 13.6 ± 0.3 Ma, with Ds/Di = 0.91).

6. Conclusions

The results of this study indicate that the glasses considered here are very promising reference materials for interlaboratory comparisons and analytical procedure standardisation. They cover a range of ages and have characteristics which include those of most real samples to be studied. JAS-G1 and Roccastrada are more difficult to analyse because the first has a low track density, whereas the second consists of a population of glass shards. However, they may be very useful, as many samples important for dating have such characteristics. A precaution to be made for JAS-G1, Macusanite and Moldavite is to use *fragments from a single split for spontaneous and induced track density determinations.*

Can the glasses studied here be considered as age standards for fission-track dating ? The main reason why an ideal standard should contain undisturbed tracks is that a general consensus on the reliability of age correction techniques does not yet exist. The best glasses in relation to the requirement *no annealing* are certainly Moldavites. Among those studied here, the Jankov deposit seems to be the best promising potential age standard. Anyway, as all Moldavites analysed in this work (excepted Slavice) yielded well consistent plateau ages, they should be considered somewhat equivalent, provided that the plateau technique is applied. In this respect it has to be taken into account that a slight amount of track annealing can not be definitely excluded for Jankov Moldavites.

Glasses similar to most real samples (generally, volcanic glasses) and totally unaffected by partial annealing of the spontaneous tracks and with an adequate track density appear to be extremely rare. Therefore, the glasses studied here which present a significant degree of annealing, although, at the present stage, they cannot be considered real age standards, may turn very useful also for testing age correction procedures. Our data proved the validity of the plateau method in a number of cases, although this technique undoubtedly requires further investigation.

The independent reference ages of all glasses should be better constrained by further analyses. As a matter of fact, Fig. 4 suggests a slight (by around 2%) systematical deviation of the plateau ages of glasses JAS-G1, Roccastrada and Macusanite from the reference ages adopted in this study. Moldavites analysed at Pisa did not show the same behaviour. However, if the true age of Moldavite is slightly lower than that one adopted in this work, the normalised plateau age of this glass also will become larger than 1 in Fig. 4. Moreover, the ζ factors referred to the glasses, although consistent with those referred to the FCT and Durango apatites, suggest a slight systematical deviation (Fig. 5). Such systematical variations of ζ factors could arise from different causes. First of all taking into account the quite different characteristics of glass and apatite with respect to track and bulk etching, a complete agreement of the ζ factors for both materials should not be expected. Further on, the fact that the spontaneous tracks of FCT and Durango apatites exhibit some annealing may also play role. To further investigate this problem, repeated fission-track analyses performed on glass samples with more precisely known reference ages are required.

At the present stage, the Jankov Moldavite is the only glass which appears to meet the requirements for an age standard, and it represents a good starting-point, at least when the uncertainty on its reference age will be overcome. For this reason a batch of 200 g of Moldavites from the Jankov Middle Miocene deposit have been selected for distribution upon request to colleagues who intend to test it (Balestrieri *et al.*, 1997). These Moldavites may be a reasonable substitute for the *splits from a single piece of Moldavite*.

Acknowledgements. We are grateful to Charles W. Naeser and Vincenzo De Michele which supplied part of the samples investigated in this study. We are also grateful to Dr. Peter Van den haute and to an anonymous referee for careful reviewing of the manuscript and for their helpful suggestions.

References

Arias C., Bigazzi G. and Bonadonna FP. (1981): Size correction and plateau age in glass shards. *Nucl. Tracks Radiat. Meas.* **5**, 129-136.

Balestrieri M.L., Bigazzi G. and Oddone M. (1996): The thermal column: a new irradiation position for fission-track dating in the University of Pavia Triga Mark II nuclear reactor. *J. Radioanal. Nucl. Chem., Letters* **213**, 99-108.

Balestrieri M.L., Bigazzi G. and Bouška V. (1997): Jankov Moldavite: a potential glass standard for fission-track dating. *On Track* **7**, 13-16.

Bigazzi G. and De Michele V. (1996): New fission-track age determinations on impact glasses. *Meteoritics & Planet. Sci.* **31**, 234-236.

Bigazzi G. and Oddone M. (1990): Datazione dello standard internazionale apatite del Fish Canyon Tuff. *Rend. Soc. Geol. It.* **13**, 133-134.

Bigazzi G., Dompier E., Hadler Neto J.C., Poupeau G. and Vulliez E. (1988): A reactor intercalibration for fission track-dating and uranium micromapping. *Nucl. Tracks Radiat. Meas.*, **15**, 755-757.

Bigazzi G., Bonadonna FP., Laurenzi M.A. and Tonarini S. (1993): A test sample for FT dating of glass shards. *Nucl. Tracks Radiat. Meas.* **21**, 489-497.

Bouška V. (1994): *Moldavites. The Czech tektites*. Stylizace, Prague.

Carpenter B.S. (1984): Calibrated glass standards for fission track use. Nat. Bur. Stand. Spec. Publ. 260-92.

Cheilletz A., Clark A.H., Farrar E., Arroyo P.G. and Pichavat M. (1992): Volcano-stratigraphy and 39Ar/40Ar geochronology of the Macusani ignimbrite field: monitor of the Miocene geodynamic evolution of the Andes of SE Peru. *Tectonophysics* **205**, 307-327.

De Corte F., Bellemans F., Van den haute P., Ingelbrecht C. and Nicholl C. (this volume): A new U doped glass certified by the European Commission for the calibration of fission-track dating (this issue).

Gentner W., Kleinmann B. and Wagner G.A. (1967): New K-Ar and fission track ages of impact glasses and tektites. *Earth planet. Sci. Lett.* **2**, 83-86.

Green P.F. (1985): Comparison of zeta calibration baselines for fission-track dating of apatite, zircon and sphene. *Chem. Geol. (Isot. Geosc. Sect.)* **58**, 1-22.

Hurford A.J. (1990): Standardization of fission track dating calibration: Recommendation by the Fission Track Working Group of the I.U.G.S. Subcommission on Geochronology. *Chem. Geol. (Isot. Geosci. Sec.)* **80**, 171-178.

Hurford A.J. and Green P.F. (1981): Standards dosimetry and the uranium-238 λ_f decay constant: a discussion. *Nucl. Tracks Radiat. Meas.* **5**, 73-75.

Hurford A.J. and Green P.F. (1983): The zeta age calibration of fission-track dating. *Chem. Geol. (Isot. Geosci. Sec.)* **1**, 285-317.

Kitada N. and Wadatsumi K. (1995a): Proposal for a glass age standard "JAS-G1". *Bulletin of Liaison and Informations, IUGS Subcomission on Geochronology* **13**, 23-27.

Kitada N. and Wadatsumi K. (1995b): Annealing properties of natural glass fission tracks at the controlled temperature and its implication on the isothermal plateau dating technique. *J. of Geosciences Osaka City University* **38**, 1-12.

Komarov A.N. and Raykhlin A.I. (1977): Comparison of fission-track and potassium-argon dating of impactites. *Doklady of the Russian Akademy of Sciences, Earth Sci. Sections* **228**, 35-38.
Lange J.-M, Bollinger K., Horn P., Jessberger E.K. and Storzer D. (1995): Moldavites from Lusatia (Germany). III: Sr-isotope, 40Ar/39Ar and fission-track studies. *Lunar Planet. Sci.* **26**, 823-824.
Lelarge V. M.L. (1993): Thermochronologie par la méthode des traces de fission d'une marge passive (dome de Ponta Grossa, SE Brésil) et au sein d'une chaîne de collision (zone externe de l'arc alpin, France). Ph. D. Thesis, Univ. Grenoble I.
McCorkell R. and Naeser C.W. (1988): Natural glass inter-laboratory reference materials for fission-track dating. 6th Int. Workshop on Fission-Track Dating, 5-9 Sept. 1988, Besançon, France. Abstract, 1p. Reproduced in *Nucl. Tracks Radiat. Meas.* **17**, 422.
Naeser C.W., Briggs N.D., Obradovich J.D. and Izett G.A. (1981): Dating of tephra. Geochronology of Quaternary tephra deposits. In: *Tephra Studies* (eds Self S. and Sparks R.S.J.), D. Reidel Pub. Co., Dordrecht, pp. 13-47.
Obradovich J.D., Naeser C.W., Izett G.A., Pasini G. and Bigazzi G. (1982): Age constraints on the proposed Plio-Pleistocene boundary stratotype at Vrica, Italy. *Nature* **298**, 55-59.
Poupeau G., Sabil N., Villa I.M., Bigazzi G., Vatin-Perignon N., Flores P., Pereyra P., Salas P. and Arroyo G. (1992): Fission-track and K-Ar ages of "Macusanite" obsidian glasses, (SE Peru): geodynamic implications. *Tectonophysics* **205**, 295-305.
Poupeau G., Labrin E., Sabil N., Bigazzi G., Arroyo P.G. and Vatin-Perignon N. (1993): Fission track dating of 15 Macusanite glass pebbles from the Macusani volcanic field (SE Peru). *Nucl. Tracks Radiat. Meas.* **21**, 499-506.
Sabil N. (1995): La datation par traces de fission: aspects methodologiques et applications thermochronologiques en contexte alpin et de marge continentale. Ph. D. Thesis, Univ. Grenoble I.
Staudacher Th., Jessberger E.K., Dominik B., Kirsten T. and Schaeffer O.A. (1982): 40Ar-39Ar ages of rocks and glasses from the Nördlinger Ries crater and the temperature history of impact breccias. *J. Geophys.* **51**, 1-11.
Storzer D. and Poupeau G. (1973): Géochronologie. Ages plateaux de minéraux et verres par la méthode des traces de fission. *Compt. Rend. Acad. Sci., Paris* **D276**, 137-139.
Storzer D. and Wagner G.A. (1969): Correction of thermally lowered fission track ages of tektites. *Earth Planet. Sci. Lett.* **5**, 463-468.
Storzer D. and Wagner G.A. (1982): The application of fission track dating in stratigraphy: a critical review. In: *Numerical Dating in Stratigraphy* (ed. Odin G.B.), John Wiley & Sons Ltd., London, pp. 199-222.
Storzer D., Jessberger E.K., Kunz J. and Lange J.-M. (1995): Synopsis von Spaltspuren- und Kalium-Argon-Datierungen an Ries-Impaktgläsern und Moldaviten. *Exkurs. f. u. Veröfftl. GGW* **195**, 79-80.
Van den haute P., Jonckheere R. and De Corte F. (1988): Thermal neutron fluence determination for fission-track dating with metal activation monitors: a re-investigation. *Chem. Geol. (Isot. Geosci. Sec.)* **73**, 233-244.
Wadatsumi K., Kitada N., Bigazzi G., Deino A., Laurenzi M., Kowallis B., Nagao K. and Masuda H. (1994): Obsidian from Japan as a 'putative' glass age standard "JAS-G1". ICOG 8, Berkeley, California, USA, June 5-11 1994. *Abstracts*, 345.
Wagner G.A. (1966): Altersbestimmungen an Tektiten und anderen natürlichen Gläsern mittels Spuren der spontanen Kernspaltung des Uran^{238} ("fission-track"-Methode). *Z. Naturforschung* **21a**, 733-745.
Wagner G.A. and Van den haute P. (1992): *Fission track dating*. Kluwer Academic Publishers, Dordrecht.
Westgate J.A. (1989): Isothermal plateau fission track ages of hydrated glass shards from silicic tephra beds. *Earth Planet. Sci. Lett.* **95**, 226-234.

AGE DETERMINATION OF OBSIDIAN SOURCE SAMPLES FROM NORTH QUEENSLAND AND NEW SOUTH WALES, AUSTRALIA

R. BONETTI[(1)], A. GUGLIELMETTI[(1)], F. MALERBA[(1)], E. MIGLIORINI[(1)], M. ODDONE[(2)], R. BIRD [(3)]

(1) *Istituto di Fisica Generale Applicata dell' Universita' di Milano, Via Celoria, 16 20133 Milano, Italy*

(2) *Dipartimento di Chimica dell'Universita' di Pavia, Via Taramelli, 12 27100 Pavia, Italy*

(3) *Australian Nuclear Science and Technology Organization, Menai, NSW, Australia*

ABSTRACT

By means of the fission track method very old obsidian samples from various Australian sites have been dated. The track fading of these samples has been corrected with the plateau age method and with the track size correction method. Ages from 85.5 Ma to 92.3 Ma have been obtained.

KEYWORDS

Fission Track Dating, Obsidian, Australia, Fading Correction.

I INTRODUCTION

Fission track dating of obsidians has been widely used in the last two decades both for provenance studies of prehistoric artifacts and for establishing the chronology of the related volcanic activity. Intense investigation has been carried out e.g. in the Mediterranean area (Bigazzi et al., 1982) and more recently extended to nearby regions (Bigazzi et al., 1990; Bigazzi et al., 1993) and in other parts of the world such as western Asia, the Americas and the Pacific. However, this was not the case for Australia where only recently obsidian artifacts and raw material sources have been discovered and studied by means of trace element analysis (Bird et al., 1987; David et al., 1992).

It is purpose of the present paper to provide the first age determination of Australian obsidians, thus contributing to a more complete characterization of sources for the study of both the provenance of prehistoric artifacts and of their geological history.

II OBSIDIAN SOURCES IN AUSTRALIA

It is believed that as a consequence of continental drift, which moved Australia northwards from the original Gondwana, volcanic activity arose in the eastern part of the continent, with decreasing age from north to south. Three different obsidian locations have been found (or hypothesized) up to now. One is in Far North Queensland (FNQ), where obsidian flows have been discovered at Nolan's Creek and Elisabeth

P. Van den haute and F. De Corte (eds.), Advances in Fission - Track Geochronology, 305-312.

Creek, two sites a few km apart, about 200 km west from the coast. Artifacts have also been found a few km apart from the sources, at a site called Echidna's Rest; see Fig. 1. PIXE/PIGME analysis performed at the Australia Nuclear Science and Technology Organization (David et al., 1992) has shown a general similarity with the Nolan's Creek obsidian and a definitely different composition from that of the widely studied Melanesian sources.

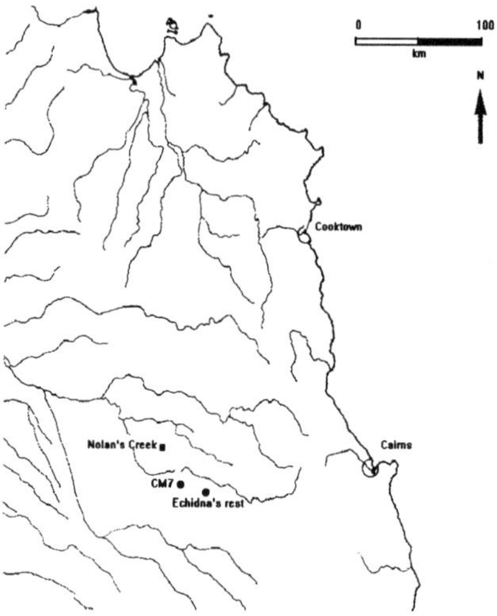

Figure 1. Map of north Queensland showing locations of the Nolan's Creek obsidian source and archaeological site mentioned in the text.

The second location is in the Mount Warning area, in northern New South Wales (NSW); see Fig. 2.

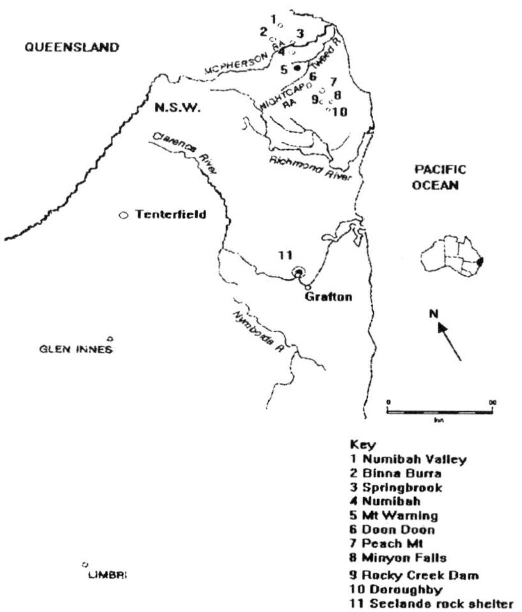

Figure 2. Map of Mt Warning area in northern New South Wales

PIXE/PIGME analysis (Bird et al., 1987) has shown that this obsidian forms a well defined compositional group, different from any of the other known sources in the nearby southern Pacific. No confirmed artifacts have been found, probably because of the badly weathered nature of most of this obsidian, often almost completely devitrified. The third location is supposed to be in the south-east part of the continent, where several artifacts have been found. However, no source has been located yet (Duerden et al., 1987).
In the present work we analyze by means of the Fission Track (FT) method four samples (AU602 ,AU603 ,AU612 ,AU662) from the FNQ area and two samples (AU101, AU102) from NSW. Location and sampling of the analyzed material is described in more detail in Bird et al. (1987) and David et al. (1992).

III EXPERIMENTAL RESULTS

The six obsidian samples were divided in two fractions, one of which was irradiated at the Lazy Susan facility (Cd ratios 6.5 for gold and 48 for cobalt) of the Triga Mark II Reactor of the LENA Laboratory of the University of Pavia, Italy. Metallic monitors and CN1 and CN2 standard glasses, together with calibrated mica and Makrofol foils acting as external detectors, were also irradiated for thermal neutron fluence determination; results are shown in Table 1.

Table 1. Measurement of thermal neutron fluence.

Monitor	external detector	B (10^9 n t^{-1})	track density on ext. detector (10^5 t cm^{-2})	thermal neutron fluence (10^{15} n cm^{-2})
Au wire	-	-	-	2.399± 0.034
Co wire	-	-	-	2.392± 0.025
CN1 glass	mica	1.873± 0.025	13.43± 0.26	2.515± 0.036
	makrofol	1.863± 0.023	13.24± 0.26	2.468± 0.058
CN2 glass	mica	1.985± 0.007	12.34± 0.25	2.449± 0.051
	makrofol	2.016± 0.041	12.31± 0.25	2.482± 0.072

Calibration of glass dosimeters (B value measurement) was performed by using cobalt and gold monitors (R. Bonetti et al., 1994).
Thermal neutron fluence value used in age measurements has been obtained as weighted mean of the values listed in the last column.

After irradiation, the irradiated and non-irradiated fractions were prepared for etching and microscope observation. However, the standard technique consisting of mounting samples in epoxy resin, followed by grinding and polishing did not work for NSW samples due to their almost completely crystallized condition.

In this case, an appropriate technique had to be developed which consisted in cutting samples in thin slices with a diamond wheel, gluing them onto microscope slides and finally carefully polishing the cut surface. Etching was subsequently performed with 20% HF at T = 40°C for 70''-120'' and samples were observed with total 500x magnification for spontaneous and induced fission track identification. Results are shown in Table 2.

It is important to point out that the spontaneous to induced track size ratio, Ds/Di, comes out to be mostly smaller than unity. This indicates considerable fading of spontaneous tracks, producing lowered fission track ages. Standard thermal based correction techniques have therefore been applied. However, the peculiar conditions of the samples (crystallization and devitrification for NSW samples and evidence for hydration for FNQ ones) imposed restrictions and concern. Experimentation at different temperatures and times has shown indeed that only a low-temperature isothermal treatment was tolerable for the FNQ samples, while NSW samples could not allow any temperature increase.

Table 2. Results of FT dating of Australian obsidian samples.

Sample id.	n_s (t)	Spontaneous track density ($x10^4$ t cm^{-2})	n_{i+s} (t)	Induced track density ($x10^5$ t cm^{-2})	Neutron fluence ($x10^{15}$ n cm^{-2})	Ds/Di	Age ± σ (Ma)	Heating hours at 140 °C	Age,corr.± σ (Ma)
FNQ AU602	2533	8.17± 0.16	3736	3.51± 0.07	2.431± 0.016	0.63± 0.06	33.9± 1.1	60	86.9 3.8
FNQ AU603	1390	7.25± 0.19	3666	3.23± 0.07	2.431± 0.016	0.54± 0.05	32.7± 1.3	-	-
FNQ AU612	278	10.86± 0.65	1002	4.13± 0.15	2.431± 0.016	0.68± 0.07	38.2± 3.2	(*)	92.3± 10.6
FNQ AU662	122	4.24± 0.38	608	1.69± 0.09	1.644± 0.014	0.53± 0.10	24.7±3.1	70	85.5 5.9
NSW AU101	273	2.32± 0.14	898	2.57± 0.09	2.431± 0.016	0.79± 0.06	13.1± 1.0	-	-
NSW AU102	95	1.48± 0.15	760	2.23± 0.09	2.431± 0.016	-	9.7± 1.1	-	-

n_s, n_{i+s} represent the number of spontaneous and induced+spontaneous tracks being counted, respectively; the induced track density has been evaluated by means of the population subtraction method.

Ages are calculated by using the following constants: $\lambda_\alpha = 1.551x10^{-10} a^{-1}$; $^{238}U/^{235}U = 137.88$; $\sigma_f = 580.2$ b; $\lambda_f = 7.03x10^{-17} a^{-1}$.

Thermal neutron fluence is determined from both calibrated mica and makrofol detectors in contact with glass dosimeters (Corning CN1; CN2) and two metallic monitors (Co; Au).

All corrected ages are weighted means of values in the plateau region.

(*) In this case the corrected age is determined by using the correction curve of sample AU602.

Both the "plateau" technique and the "correction curve" technique (Storzer and Wagner, 1969; Storzer and Poupeau, 1973; Arias et al., 1981) have been applied to FNQ samples. The results of the isothermal treatment at T = 140°C from 10 to 60 h is shown in Fig. 3.

As clearly seen, the plateau age of 85 Ma is reached at 50 h, this time consistently corresponding to reaching the condition Ds/Di =1 (lower part of Fig. 3). Figs. 4a and 4b how the track size distributions before and after the thermal treatment, respectively. Corrected ages are shown in the last column of Table 2.

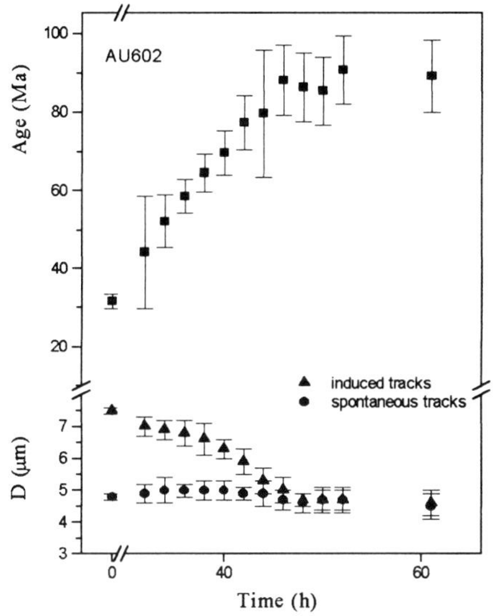

Figure 3. Plateau age for Australian obsidian AU602 obtained by using isothermal treatment at T = 140 °C (upper). Induced and spontaneous fission-track diameters at each step of thermal treatment (lower).

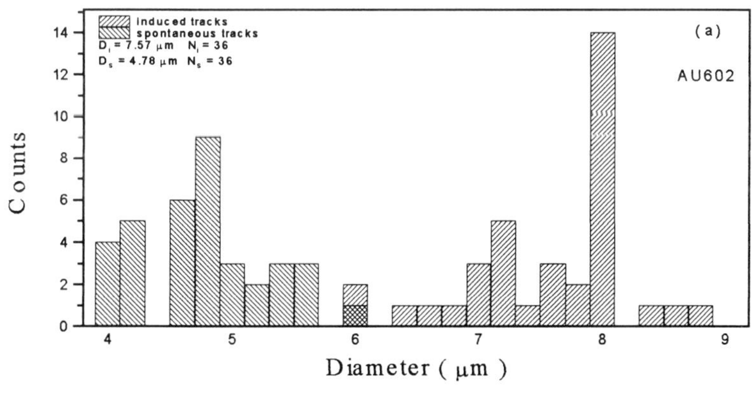

Figure 4. (a) Size distribution histogram of fission track diameters in AU602: sample not subjected to any thermal treatment; Ds and Di are the average spontaneous and induced track diameters, respectively; N is the number of tracks being counted.

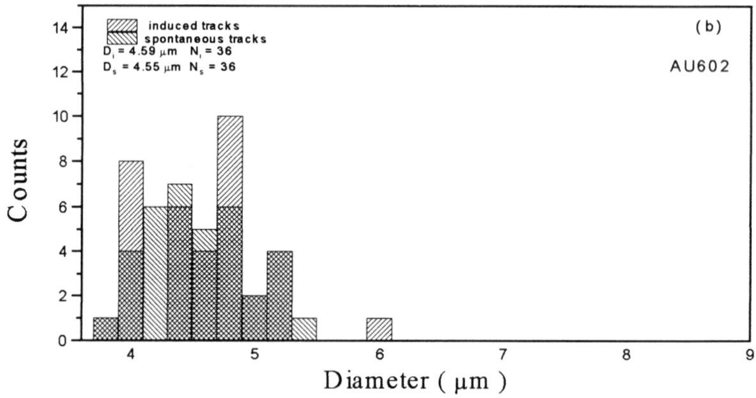

Figure 4. (b) Size distribution histogram of fission track diameters in AU602: sample heated at T = 140 °C for 63 hours; symbols have the same meaning as in Figure 4 (a).

From the thermal treatment data the activation energy of the FNQ obsidian was extracted. Since it is defined as the minimum energy required to start the annealing process, the activation energy provides a measure of track stability for a given detector. Following the formalism of Singh and Virk (1989), the activation energy can be determined from a plot of ln(1-r), with r being the track diameter reduction, versus the inverse of temperature for a constant heating time. The result, obtained by linear interpolation of data shown in Fig. 5, comes out to be 0.27 eV.

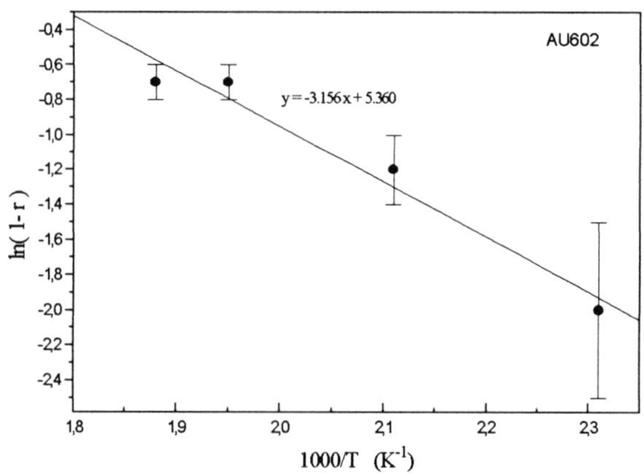

Figure 5. Plot of ln(1-r) vs inverse of temperature (in K) for AU602 sample; $r=D/D_0$ is the track diameter reduction (D_0 is the diameter of unannealed tracks).

IV DISCUSSION AND CONCLUSIONS

As Table 2 shows, not only has the fading correction considerably increased the sample ages, but also has decreased their apparent spread. The three age-corrected FNQ samples exhibit now a common age, of the order of 85-90 Ma, despite the fact that they have been obtained from two physically different sources (AU612 from Elisabeth Creek, AU602 and 662 from Nolan's Creek) as demonstrated by their rather different Uranium content (see Table 2).

Such common age seems to suggest a common volcanic activity in the FNQ region as responsible for the various obsidian flows.

Comparison of ages obtained for the two sets of NSW and FNQ samples supports the hypothesis, outlined in Section I, that the volcanic activity is older in the northern part of the continent. Moreover, such ages are relatively consistent with independent geological information: 100 Ma for the volcanics of FNQ (David et al., 1992) and 21 Ma for the rhyolites of NSW (Bird et al., 1987).

An interesting remark from the point of view of the methodology of FT is that the ages obtained for the FNQ obsidians are among the oldest ever dated with this method in volcanic glasses. The present result shows indeed that good quality obsidians as old as 100 Ma are able to retain the memory of the information accumulated, despite the occurrence of physical/chemical processes such as thermal fading, hydration, leaching and so on. This conclusion is supported by the activation energy value we determined for FNQ obsidians. The fact that the reported value, 0.27 eV, lies within the range (0.15-0.3 eV) typically measured for much younger volcanic glasses (Sandhu et al., 1993), might be regarded as a proof of the persisting good quality of this extremely old natural track detector.

However, this remark cannot be generalized to any kind of obsidian, as clearly shown by the problems we had in analyzing the NSW samples, for which a thermal-based age correction has not even been applied. This is most probably due to the peculiar composition of the raw material, which exhibits an unusually high Si and Al content (Bird et al., 1987).

In conclusion, we have measured for the first time the age of very old Australian obsidians with the FT method. Apart from the geological and methodological consequences outlined above, this work yields complementary information in respect to trace element analysis and can be useful to follow better the diffusion of obsidian within prehistoric cultures.

ACKNOWLEDGEMENTS

We are indebted to G. Bigazzi for useful discussions and continuous encouragement.

REFERENCES

Arias C., Bigazzi G. and Bonadonna F.P. (1981). Size Correction and Plateau Age in Glass Shards. Nucl. Tracks, 5: 129-136.

Bigazzi G., Bonadonna F.P. and Radi G. (1982). Fission Track Dating of Obsidian and Prehistory. Workshop Track Dating, Abstracts. Fifth International Conference on Geocronology, Cosmochronology and Isotope Geology. Nikko National Park Japan, 30 June 1982, pp. 1-4

Bigazzi G., Márton P., Norelli P. and Rozloznik L. (1990). Fission Track Dating of Carpathian Obsidians and Provenance Identification. Nucl. Tracks and Radiat. Meas., 17 No.3: 391-396.

Bigazzi G., Yegingil Z., Ercan Z., Oddone M. and Özdogan M. (1993). Fission Track Obsidian in Central and Northern Anatolia. Bull. Volc., 55: 588-595.

Bird J. R., Duerden P., Wall T., Clayton E., McBryde I. (1987). Characterization of Seelands artefacts. In: W.R. Ambrose, J.M.J. Mummery (Editors), Archeometry: Further Australasian Studies. The Australian National University Press, Canberra, pp.250-260.

Bonetti R., Bongiorno L., Guglielmetti A., Malerba F. and Oddone M. (1994). Datazione di Ossidiane con il Metodo delle Tracce di Fissione. Report INFN/TC-94/72.

David B., Bird J. R., Fullagar R. and Little L. (1992). Glassy Obsidian Artefacts from North Queensland. The Artefact,15: 25-30.

Duerden P., Clayton E. and Bird J. R. (1987). Obsidian Composition Catalogue. In: W.R. Ambrose, J.M.J. Mummery (Editors), Archeometry: Further Australasian Studies. The Australian National University Press, Canberra, p.232.

Sandhu A. S., Westgate J. A. and Alloway B. V. (1993). Optimizing the Isothermal Plateau Fission Track Dating Method for Volcanic Glass Shards. Nucl. Tracks and Radiat. Meas., 21 No.4: 479-488.

Singh G. and Virk H. S. (1989). Heavy Ion Radiation Damage in Glass Detectors. Nucl. Instr. and Meth., B44: 103-106.

Storzer D. and Wagner G. A. (1969). Correction of Thermally Lowered Fission Track Ages of Tektites. Earth Planet. Sci. Lett., 5: 463-468.

Storzer D. and Poupeau G. (1973). Ages-Plateaux de Mineraux et Verres par la Methode des Traces de Fission. Compt. Rend. Acad. Sci., Paris D276: 137-139.

FISSION TRACK DATING AND PROVENIENCE OF ARCHAEOLOGICAL OBSIDIAN ARTEFACTS IN COLOMBIA AND ECUADOR

O. DORIGHEL[a,b], G. POUPEAU[a], L. BELLOT-GURLET[a], E. LABRIN[a]

[a] *Groupe de Géophysique Nucléaire, UPRES-A n°5025 of CNRS and Université Joseph Fourier, Institut Dolomieu, 15 rue Maurice Gignoux, 38031 Grenoble, France.*

E-mail : poupeau@ujf-grenoble.fr

[b] *Université Paris I-Sorbonne, Paris, France.*

Abstract.- We dated by the fission track method eighteen samples of obsidian glass. Of these, seventeen were artefacts collected in prehispanic archaeological sites from Ecuador and Colombia and one comes from a secondary obsidian source located in southern Colombia. When our data are compared to (i) the fission track ages of obsidian from volcanic sources and (ii) the PIXE chemical composition of the same artefacts and of samples from obsidian sources, eight discrete age/composition groups appear.

Fifteen artefacts might come from a single unknown source; three others from sources in the Sierra de Guamani (East of Quito): one from the Mullumica rhyolitic flow and two from the Quiscatola/Yanaurcu sources. A fourth artefact might come from the South Colombia source of Rio Hondo. At least five other sources are necessary to account for the variety of ages and compositions found in the artefact collection studied.

Key words : fission tracks; obsidian; provenience; Colombia; Ecuador.

1. Introduction

Very shortly after the fission track dating method was proposed, it used to be applied to volcanic glasses for archaeological purposes, such as to determine the ages of man-made glasses, fired stones, palaeolithic marker horizons, etc (see early reviews by Fleischer *et al.*, 1975; Wagner, 1978 and more recently by Wagner and Van den Haute, 1992). Its application to obsidians is mostly useful in the search of provenience of the raw material used for archaeological tool making. This has been successfully realized in Japan (Suzuki, 1973) and the western (Bigazzi *et al.*, 1993a) and eastern (Bigazzi *et al.*, 1993b, 1994) Mediterranean basin.

In the Andean area, the first glasses to be dated by fission tracks were obsidian pebbles once supposed to be tektites : "americanites" from Colombia (Storzer and Poupeau, 1973) and Peruvian macusanites (Fleischer and Price, 1964; Naeser *et al.*, 1980; Miller and Wagner, 1981; Miller *et al.*, 1990; Poupeau *et al.*, 1992, 1993; Balestrieri *et al.*, 1997). Relatively few other obsidians from geological contexts from these two countries and from Ecuador were also dated by this method (Miller and Wagner, 1981; Arias *et al.*, 1986;

P. Van den haute and F. De Corte (eds.), Advances in Fission - Track Geochronology, 313-324.

Figure 1. Location of the archaeological sites and obsidian sources dated by fission tracks and/or analysed by PIXE.

Bigazzi *et al.*, 1992). Interest in comparing the fission track ages of obsidian archaeological artefacts with that of potential volcanic sources for provenience studies started with the dating of samples from Colombia, Ecuador and Peru (Miller and Wagner, 1981; Dorighel *et al.*, 1994).

In the frame of a project on the sourcing of archaeological obsidian artefacts from prehispanic sites in the Andes, we are characterizing selected samples both by their fission tracks record and their chemical composition, the latter determined by PIXE (particle induced X-ray emission), a technique allowing eventually non-destructive analysis (Dorighel *et al.*, 1996a). The data obtained, for Colombia and Ecuador, show that the obsidian artefacts studied thus far belong to discrete age-composition groups presumably related to specific sources (Dorighel *et al.*, 1996b; Bellot-Gurlet *et al.*, 1996; Poupeau *et al.*, 1996). It is the purpose of this article to present new fission track ages of artefacts from Colombia and Ecuador and one source sample from Colombia, and to show how the coupling of fission track dating with PIXE analysis may provide a better resolution in provenience studies.

2. Sampling

2.1. ARTEFACTS

The artefacts we worked on come mostly from Ecuadorian archaeological sites. Four of these, La Cantera, Maria Cecilia, La Propicia and La Tolita-Walberto, are located in the northeastern coastal province of Esmeraldas; three are from the intra-cordilleras altiplano, Huaca Rumi at Quito and La Florida in its vicinity, while Tababuela is in a more northeasterly position, some 50 km from Quito (figure 1). Finally one sample, from the Maria Olivia site, comes from southern Ecuador, in the Guayas basin, about 150 km north of Guayaquil.

Samples from Colombia collected in coastal sites are those of La Miranda and La Remigia, near to the border with Ecuador, and La Cocotera, at some 200 km north. The sierra site of La Elvira is located in the Cauca valley, near to the city of Popayan, dominated in its southern part by the Purace volcano. El Pital is a site located at an altitude of more than 1 000 m on the Pacific flank of the western Cordillera, near to the small town of Darien, in the Calima river canyon. All samples were provided by the archaeologists in charge of these Colombian and Ecuadorian sites.

The sites selected belong to the so-called Regional Developments to Integration (following the Ecuadorian terminology) Periods, ranging from 500 BC to 1500 AD. Only one site, La Elvira, might pertain (C. Gnecco, pers. comm.) to the earlier Pre-ceramic Period, on the basis of a typological attribution based on projectile points. The archaeological pieces are mostly unwanted materials remaining after flaking operations. Others are small flakes or blades. Their largest dimension rarely exceeds 4 to 5 cm and their thickness 5 to 10 mm.

2.2. VOLCANIC SOURCE SAMPLE

One obsidian source suspected to have been used by early men in the high Cauca valley of Colombia is that of Rio Hondo, located about 26 km SE of Popayan (figure 1). It is a secondary source, located in a gully and composed of obsidian pebbles originating most probably from the Purace volcano and embedded in bank sediments. We analyzed one sample from this source.

3. Fission Track Dating

3.1. ANALYTICAL PROCEDURES

Some obsidian samples are not convenient for fission track dating, according to their content in various kinds of impurities (bubbles, microcrystals), lack of transparency, presence of track-like defects "revealed" or enlarged by the track etchant, or devitrification effects (Poupeau *et al.*, 1997). Therefore, there is a biasing in the sample selection for fission track dating, the most glassy and transparent samples being in general preferentially used.

Slices ~1 mm thick of each sample were cut using either a wire or circular diamond saw. One slice was systematically kept for PIXE analysis (after diamond polishing), others were chosen, taking into account the above-mentioned criteria for fission track dating.

Our dating procedures were described in Dorighel *et al.* (1994) and Poupeau *et al.* (1997). Slices of obsidian glasses with their unetched content of spontaneous tracks were irradiated in the P1 position (Cd ratio for Au of ~3300) of the Orphée nuclear reactor of the *Centre d'Etudes Nucléaires* of Saclay (France) together with three NIST glass standards of the 962a/612 series. After neutron exposure, one irradiated and one non-irradiated slices were mounted together in an epoxy disc and polished with diamond powders. Fission tracks were etched with a 20% concentration HF solution at 40°C for times ranging from 90 to 195 seconds. The ages where therefore determined with the subtraction method (Gleadow, 1981).

Fission track diameter measurements showed that spontaneous tracks were always shorter, on the average, than induced tracks, as the D_s/D_i spontaneous/induced mean track

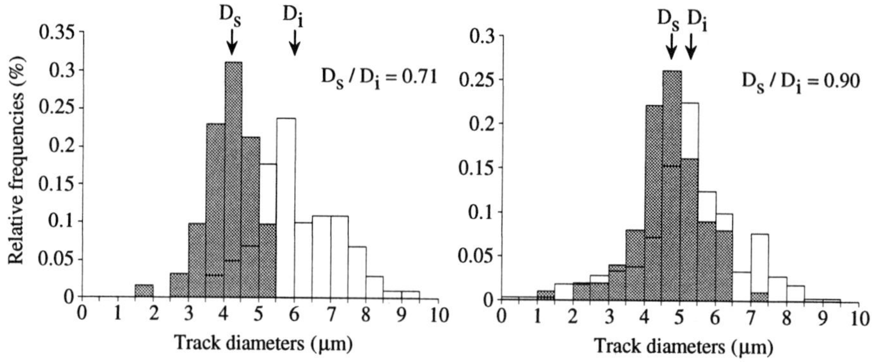

Figure 2. Histograms of spontaneous and induced track diameters distributions in two obsidian artefacts: at left, sample RMGb, from the archaeological site of La Remigia (Colombia); at right, sample S42, from the Maria Olivia site (Ecuador). The spontaneous track diameters histograms are stippled. D_S and D_i, respectively mean spontaneous and induced track diameters.

diameters ratios were always lower than unity. The samples were therefore treated with the plateau dating method of Storzer and Poupeau (1973). For eight samples, we first determined, as in Dorighel *et al.* (1994) an apparent age, before any thermal treatment, then heated together their irradiated and unirradiated slices for 1 h at 205°C before a new dating/track diameter measurements step. This thermal treatment was sufficient to bring the D_s/D_i ratio to unity, i. e. to the plateau condition. Two samples were heated in different conditions, respectively 2h at 205°C for the Rio Hondo source sample HON2c and 2h at 250°C for the La Miranda artefact MIRb.

Ages and diameter measurements were made with a 100x dry objective and 10x oculars in transmitted light. For track diameters determinations, we used a Zeiss-Jena Jenapol microscope with a design tube projecting track images on an electronic tablet wired on a microcomputer. Ages were determined both with this microscope and a Leitz Orthoplan.

3.2. RESULTS

Seventeen artefacts and one source obsidian were dated. As shown in tables 1 and 2, all samples exhibited various degrees of natural track fading, with D_s/D_i ratios varying from 0.64 to 0.95 (figure 2).

Three observers participated to the measurements. As the precision on these ages depends critically on the number of fossil tracks counted, generally lower than 700 (and down to less than 100), due to the low track densities, mostly <1000/cm^2, the plateau age of eight samples was determined on the same mounts by two observers. In all cases, the results were in agreement to within the statistical uncertainties (2σ). Sample MIRb from La Miranda, dated twice by one observer from a single mount, showed good reproducibility. Sample WLB1 presents the youngest apparent age, with 0.019±0.005 Ma. As the number of fossil tracks presently measured is very low, it is not significant to report its D_s/D_i ratio value. However, from the relative sizes of spontaneous and induced tracks, there is no reason to suspect a D_s/D_i ratio significantly lower than those observed for other Ecuadorian artefacts.

TABLE 1. Fission track ages of obsidian artefacts

Archaeological site	Sample	Laboratory annealing	ρ_s t/cm² (N_s)	ρ_i 10^5 t/cm² (N_i)	Φ 10^{15} n/cm² (N_m)	Age Ma	Age 1σ	Obs.	D_s/D_i	Obs.
			Coastal sites							
Colombia										
La Cocotera	COCO	none	688 *(93)*	2.27 *(3010)*	1.0728 *(16391)*	0.19 ±	0.02	1	0.85	1
		1h/205°C	616 *(119)*	1.46 *(2523)*	1.0728 *(16391)*	**0.27 ±**	**0.03**	2	1.06	1
			662 *(70)*	1.46 *(2523)*	1.0728 *(16391)*	**0.29 ±**	**0,03**	2		
La Miranda	MIRa	1h/205°C	445 *(88)*	1.11 *(1477)*	1.188 *(10709)*	**0.28 ±**	**0.03**	1	1.05	1
	MIRb	2h/250°C	308 *(100)*	1.11 *(1477)*	1.188 *(10709)*	**0.20 ±**	0.02	1	1.07	1
			300 *(100)*	1.15 *(2435)*	1.188 *(10709)*	**0.18 ±**	0.02	2		
La Remigia	RMGa	1h/205°C	651 *(99)*	1.32 *(1758)*	1.188 *(10709)*	**0.35 ±**	**0.04**	1	0.99	1
			560 *(98)*	1.32 *(2272)*	1.188 *(10709)*	**0.30 ±**	**0.03**	2		
Ecuador										
La Tolita/Walberto	WLB1	none	68 *(14)*	2.53 *(2006)*	1.188 *(10709)*	0.019 ±	0.005	2	n. m.	
	WLB5	none	537 *(27)*	1.71 *(2256)*	1.188 *(10709)*	0.22 ±	0.04	3	0.81	3
		1h/205°C	550 *(92)*	1.32 *(2988)*	1.188 *(10709)*	**0.30 ±**	**0.03**	2	0.97	3
			600 *(60)*	1.56 *(1561)*	1.188 *(10709)*	**0.27 ±**	**0.04**	3		
	WLB6	none	545 *(62)*	1.92 *(2534)*	1.188 *(10709)*	0.20 ±	0.03	3	0.84	3
		1h/205°C	350 *(56)*	1.02 *(2153)*	1.188 *(10709)*	**0.25 ±**	**0.03**	2	1.00	3
			411 *(43)*	1.29 *(1292)*	1.188 *(10709)*	**0.23 ±**	**0.04**	3		
Maria Cecilia	MCLa	none	775 *(123)*	2.50 *(2983)*	1.0728 *(16391)*	0.20 ±	0.02	2	0.81	2
La Cantera	EO2	1h/205°C	595 *(68)*	1.36 *(1805)*	1.0728 *(16391)*	**0.28 ±**	**0.03**	1	1.02	1
La Propicia	EO12	none	764 *(97)*	2.60 *(3453)*	1.0728 *(16391)*	0.19 ±	0.02	1	0.85	1
		1h/205°C	622 *(87)*	1.37 *(1819)*	1.0728 *(16391)*	**0.29 ±**	**0.03**	1	1.05	1
			540 *(92)*	1,53 *(3273)*	1.0728 *(16391)*	**0.23 ±**	**0.03**	2		
			Sites from the Sierra							
Colombia										
El Pital	ELPI2	1h/205°C	10020 *(477)*	1.48 *(2089)*	1.188 *(10709)*	**4.81 ±**	**0.27**	1	1.001	2
			10140 *(1247)*	1.52 *(3218)*	1.188 *(10709)*	**4.74 ±**	**0.17**	2		
	ELPI1		8740 *(786)*	1.49 *(2087)*	1.188 *(10709)*	**4.17 ±**	**0.19**	1	1.14	1
			8280 *(1139)*	1.47 *(2881)*	1.188 *(10709)*	**4.00 ±**	**0.15**	2		
La Elvira	ELV5	none	10690 *(311)*	3.60 *(490)*	1.0728 *(16391)*	1.91 ±	0.14	1	0.67	1
		1h/205°C	9850 *(521)*	1.74 *(2345)*	1.0728 *(16391)*	**3.63 ±**	**0.19**	2	1.03	1
Ecuador										
La Florida	FLO2	1h/205°C	4540 *(180)*	1.93 *(2615)*	1.188 *(10709)*	**1.67 ±**	**0.13**	1	1.02	1
			4470 *(626)*	2.07 *(3075)*	1.188 *(10709)*	**1.54 ±**	**0.07**	2	1.03	2

TABLE 1 (continued)

Archaeological site	Sample	Laboratory annealing	ρ_s t/cm² (N_s)	ρ_i 10^5 t/cm² (N_i)	Φ 10^{15} n/cm² (N_m)	Age Ma	1σ	Obs.	D_s/D_i	Obs.
Huaca Rumi	P156d	none	640 *(127)*	2.34 *(2789)*	1.0728 *(16391)*	0.18 ±	0.02	2	0.96	2
		1h/205°C	n. m.	n. m.		—			0.96	2
Tababuela	TBB4	1h/205°C	669 *(108)*	1.59 *(2096)*	1.152 *(10384)*	**0.29 ±**	**0.03**	1	1.05	1
			500 *(102)*	1.306 *(1418)*	1.152 *(10384)*	**0.26 ±**	**0.03**	2		2
Guayas Basin, Ecuador										
Maria Olivia	S42	1h/205°C	621 *(102)*	1.36 *(1804)*	1.152 *(10384)*	**0.31 ±**	**0.03**	1	1.12	1
			640 *(120)*	1.26 *(2505)*	1.152 *(10384)*	**0.35 ±**	**0.03**	2	1.10	2

ρ_s, and ρ_i, respectively sample spontaneous and induced track densities ; Φ, neutron fluence (neutron/cm²), calculated from the induced track density relative to the neutron glass monitors according to Bigazzi *et al.* (1988a) and Balestrieri *et al.* (1997); N_s, N_i and N_m are the respective number of tracks counted ; D_s and D_i, mean spontaneous and induced track diameters. Ages were calculated with $\sigma=580.2\times10^{-24}$ cm^{-2}, $\lambda_f=7.03\ 10^{-17}$ a^{-1} and $^{235}U/^{238}U=7.252\times10^{-3}$. Plateau ages are in **bold caracters**. As fission track counts are Poissonally distributed, relative uncertainties on densities are given by : $\sqrt{\frac{1}{N}}$ and on ages by : $\frac{\sigma t}{t} = \sqrt{\frac{1}{N_s} + \frac{1}{N_i} + \frac{1}{N_m}}\ \frac{\rho_s + \rho_i}{\rho_i}$

The uncertainties on D_s/D_i are around 2-3%.

Observers: 1, Olivier Dorighel ; 2, Erika Labrin ; 3, Elena Bartolomé.

4. Discussion

All artefacts have apparent and plateau ages older than 10^4 a and will thus be considered in the following as obsidian formation ages. The apparent and plateau ages we measured are in the same range as found by other authors on obsidians from Colombia and Ecuador (Miller and Wagner, 1981; Arias *et al.*, 1986; Bigazzi *et al.*, 1992, 1997). However, before comparing data obtained in these laboratories, one must ascertain whether comparison of individual ages are significant.

TABLE 2. Fission track age of the Rio Hondo source sample HON2c

Laboratory annealing	ρ_s t/cm² (N_s)	ρ_i 10^5 t/cm² (N_i)	Φ 10^{15} n/cm² (N_m)	Age Ma	1σ	Obs.	D_s/D_i	Obs.
none	7788 *(206)*	2.98 *(3239)*	1.0728 *(16391)*	1.67 ±	0.12	1	0.64	1
2h/205°C	9208 *(995)*	1.63 *(2275)*	1.0728 *(16391)*	**3.63 ±**	**0.12**	1	0.98	1
	9472 *(927)*	1.71 *(2503)*	1.0728 *(16391)*	**3.56 ±**	**0.15**	2		
	9200 *(1552)*	1.701 *(2620)*	1.0728 *(16391)*	**3.47 ±**	**0.10**	2		
			Weighed age :	**3.54 ±**	**0.07**			

Footnotes : see table 1

4.1. COMPARABILITY OF DATA OBTAINED IN DIFFERENT LABORATORIES

The laboratories which dated Colombian and Ecuadorian obsidians, in Heidelberg (Miller and Wagner, 1981), Pisa and Campinas (Arias *et al.*, 1986; Bigazzi *et al.*, 1992, 1997) and Grenoble use different nuclear reactor facilities, different calibrations for absolute neutron dosimetries and different sets of standard samples. As a consequence, they also use different operational values for the spontaneous fission decay constant : 8.46×10^{-17} a^{-1} in Heidelberg, or 7.03×10^{-17} a^{-1} in our, Campinas and Pisa laboratories.

The Grenoble and Pisa laboratories started fission track ages crossed determinations on glasses long ago (Bigazzi *et al.*, 1988a, b). In general, the plateau ages determined in these two laboratories agree within the statistical uncertainties (Poupeau *et al.*, 1992, 1993; Balestrieri *et al.*, 1997). No such intercomparison program was led with the Heidelberg laboratory. However, moldavite fission-track ages determined in Heidelberg (Wagner, 1966; Storzer and Wagner, 1977) are in good agreement with ours (Balestrieri *et al.*, 1997).

The only Andean material on which an age comparison can be made is the lava flow of Mullumica. Two uncorrected ages of 0.19 Ma and 0.20 Ma (no uncertainty reported) ages obtained for this flow by Miller and Wagner (1981) are slightly higher than eight uncorrected ages measured on four other samples of the same lava unit by Bigazzi *et al.* (1992), all in the 0.15-0.17 Ma with typical 1σ uncertainties of 0.01 to 0.02 Ma. Part of this difference might be attributable to a lower degree of track annealing in the Miller and Wagner samples, as the plateau ages obtained by Bigazzi *et al.* on these samples are in the 0.17 to 0.20 Ma range (with uncertainties of ±0.02 to 0.03 Ma).

As the Mullumica datings by Bigazzi *et al.* (1992) were the results of a Campinas-Pisa collaboration, we will consider below that direct comparisons of obsidian glass plateau ages determined independently in Campinas, Heidelberg, Pisa and Grenoble are justified.

4.2. SOURCE SAMPLES

The Rio Hondo age of 3.54±0.07 Ma (table 2) is the first one determined for a sample from this secondary obsidian source of Colombia. The only other Colombian obsidian dated by fission tracks to our knowledge, the Americanite treated by Storzer and Poupeau (1973), of unknown origin, presents an older age, of 6.52±0.20 Ma and thus must come from another primary source.

We did not date any source sample from Ecuador, but Bigazzi *et al.* (1992) did it on six obsidian sources from the Sierra de Guamani. They found concordant plateau ages at 0.18±0.01 Ma for the Callejones source, one obsidian from the Potrerillos volcano and the Mullumica flow. Other ages were older : 0.39-0.47 Ma for one sample from Yurac Paccha, 0.84-0.85 Ma for the El Tablon flow, 1.30-1.68 Ma for the Quiscatola-Yanaurcu source. Slightly older plateau ages, of 1.78 Ma and 1.81 Ma, were obtained by Miller and Wagner (1981) for two geologic obsidians from the region of Quito. In the absence of information about their precise origin or chemistry, it is not possible to infer whether or not they might come from the Quiscatola-Yanaurcu volcanic remnants.

TABLE 3. Artefact plateau-ages / composition groups and potential origins

Sample	Age Ma	Age ±2σ	Site	Location	Chrono-cultural phase	Geochemical group§	Age/composition group	Source ?	Ref.
MIRb	0.19*	0.03	La Miranda	coastal (Co)	Regional Developments	Mullumica	Mullumica	**Mullumica**	1
LT1	0.30	0.06	La Tolita/Cancha H9	coastal (Ec)	Regional Developments	Mullumica	0.25-0.30 Ma/Mullumica	unknown	2
LT2	0.29*	0.04	La Tolita/Cancha H9	coastal (Ec)	Regional Developments	Mullumica	0.25-0.30 Ma/Mullumica	unknown	2
LT3	0.28*	0.04	La Tolita/Cancha H9	coastal (Ec)	Regional Developments	Mullumica	0.25-0.30 Ma/Mullumica	unknown	2
LT4	0.25*	0.04	La Tolita/Cancha H9	coastal (Ec)	Regional Developments	Mullumica	0.25-0.30 Ma/Mullumica	unknown	2
LT7	0.28*	0.04	La Tolita/Cancha H9	coastal (Ec)	Regional Developments	Mullumica	0.25-0.30 Ma/Mullumica	unknown	2
LT8	0.295*	0.04	La Tolita/Cancha H9	coastal (Ec)	Regional Developments	Mullumica	0.25-0.30 Ma/Mullumica	unknown	2
I1	0.25	0.06	Inguapi	coastal (Co)	Regional Developments	Mullumica	0.25-0.30 Ma/Mullumica	unknown	1
COCO	0.28*	0.04	La Cocotera	coastal (Co)	Regional Developments	Mullumica	0.25-0.30 Ma/Mullumica	unknown	1
MIRa	0.28	0.06	La Miranda	coastal (Co)	Regional Developments	Mullumica	0.25-0.30 Ma/Mullumica	unknown	1
WLB5	0.29*	0.05	La Tolita/Walberto	coastal (Ec)	Regional Developments	Mullumica	0.25-0.30 Ma/Mullumica	unknown	1
WLB6	0.24*	0.06	La Tolita/Walberto	coastal (Ec)	Regional Developments	Mullumica	0.25-0.30 Ma/Mullumica	unknown	1
EO2	0.28	0.06	La Cantera	coastal (Ec)	Regional Developments	Mullumica	0.25-0.30 Ma/Mullumica	unknown	1
EO12	0.25*	0.04	La Propicia	coastal (Ec)	Regional Developments	Mullumica	0.25-0.30 Ma/Mullumica	unknown	1
TBB4	0.27*	0.04	Tababuela	sierra (Ec)	Regional Developments	Mullumica	0.25-0.30 Ma/Mullumica	unknown	1
S42	0.33*	0.05	Maria Olivia	Guayas basin (Ec)	Regional Developments	Mullumica	0.25-0.30 Ma/Mullumica	unknown	1
RMGa	0.32*	0.03	La Remigia	coastal (Co)	Regional Developments	Tumaco	0.25-0.30 Ma/Tumaco	unknown	1
LT6	0.42	0.08	La Tolita	coastal (Ec)	Regional Developments	Callejones	0.42 Ma/Callejones	unknown	2
LT5	1.65	0.32	La Tolita	coastal (Ec)	Regional Developments	Quiscatola-Yanaurcu	Quiscatola-Yanaurcu	**Quiscatola-Yanaurcu**	2
FLO2	1.57*	0.12	La Florida	sierra (Ec)	Integration	Quiscatola-Yanaurcu	Quiscatola-Yanaurcu	**Quiscatola-Yanaurcu**	1
I2	2.58	0.24	Inguapi	coastal (Co)	Regional Developments	Quiscatola-Yanaurcu	2.60 Ma/Quiscatola-Yanaurcu	unknown	2
I3	2.60*	0.16	Inguapi	coastal (Co)	Regional Developments	Quiscatola-Yanaurcu	2.60 Ma/Quiscatola-Yanaurcu	unknown	2
ELV5	3.63	0.38	La Elvira	sierra (Co)	Preceramic/Formative	Rio Hondo	Rio Hondo	**Rio Hondo**	1
ELPI1	4.07	0.25	El Pital	sierra (Co)	Integration	Rio Hondo	Rio Hondo ?	**Rio Hondo ?**	1
ELPI2	4.76	0.30	El Pital	sierra (Co)	Integration	Rio Hondo	4.8 Ma/Rio Hondo	unknown	1

*, Weighed ages; §, As defined by Bellot-Gurlet *et al.* (1997); 1, This work; 2, Dorighel *et al.* (1994)

4.3. ARTEFACTS

All the samples we dated by fission-tracks were also chemically characterized by PIXE, where their contents in 13 major, minor and trace elements were determined (Dorighel *et al.*, 1996a; Bellot-Gurlet *et al.*, 1996, 1997). We also analyzed by PIXE aliquots of the source samples dated by Bigazzi *et al.* (1992) and the Rio Hondo sample of table 2. From such data, it appeared previously from a more limited data set that Colombian and Ecuadorian artefacts could be attributed to 5 compositional groups, 3 of which corresponding to the composition of obsidians from the Sierra de Guamani and one to the South Colombia Rio Hondo source (Bellot-Gurlet *et al.*, 1997). Fission track data and geochemical analysis allow to define independently age and compositional groups. To date, we obtained plateau ages on 25 artefacts. These ages are reported with other relevant data : site location, chrono-cultural period, the geochemical (Bellot-Gurlet *et al.*, 1997) and age/composition groups to which they belong, and proposals about potential sources. The age/PIXE results are discussed below following the increasing age classification of column 1:
0.19 Ma.- This group is composed of only one artefact from the La Miranda site, in the Nariño coastal province of SW Colombia. Its age and chemical composition are compatible with an origin from the Mullumica flow of the Sierra de Guamani, near to Quito.
0.24-0.33 Ma.- Fifteen samples belong to this age group. This is by far the most populated age group. The samples composing this group were collected in seven archaeological sites along the Colombian-Ecuadorian coastal regions, in one site (Tababuela) from the northern Ecuadorian altiplano and at Maria Olivia (Guayas Basin). Their chemical composition would be compatible with a Mullumica source, but their ages exclude this source. Thirteen samples are in the age range 0.25-0.30 Ma. Given the uncertainties in fission track dating, it is thus very likely that all obsidians from this group come from a single, yet unknown, source.
0.32 Ma.- This is the age of sample RMGa from La Remigia, one artefact belonging to a group of three archaeological pieces from three sites located in the same coastal area of SW Colombia. These pieces form a very distinct compositional group (Bellot-Gurlet *et al.*, 1997). Although having an age concordant with the above group, RMGa must therefore come from another unknown source. Dating of the two others artefacts from this Tumaco chemical group will show whether or not they form an homogeneous age/composition group.
0.42 Ma.- Only one sample, from the La Tolita coastal site defines this group. Compositionally, it is similar to the Callejones source of the Sierra de Guamani, but its plateau age of 0.42±0.08 Ma is more than two times higher than that of this source, of age 0.17-0.19 Ma (Bigazzi *et al.*, 1992). Another unknown source must be invoked for the origin of this material.
1.6 Ma.- Two artefacts present an age around 1.6 Ma. One is again coming from the La Tolita coastal site, while the other was sampled in the Quito area. The plateau ages we found are in the range of that found by Bigazzi *et al.* (1992) for two samples of the Quiscatola-Yanaurcu source of the Sierra de Guamani, with ages 1.52±0.16 and 1.49±0.08 Ma for one sample from Cerro Yanaurcu and 1.68±0.18 Ma, 1.64±0.14 Ma and 1.30±0.22 Ma for another from Loma Quiscatola. These samples were taken about 4 km apart from

each other. Bigazzi *et al.* dated also a third sample from this area, coming from the secondary deposit of Rodeo Corrales, on which plateau ages of 1.38±0.19 Ma and 1.37±0.19 Ma were obtained. We analyzed by PIXE splits of these three source samples and observed that they presented same composition as the two artefacts of this age group. Based on age and compositional similarities, we attribute to our artefacts a "Quiscatola-Yanaurcu" origin.

2.6 Ma.- Two artefacts from the Colombian coastal site of Inguapi are composing this group, compositionally belonging to the Quiscatola-Yanaurcu group. However, they are older by about 1 Ma than this source, and one more unknown source is needed to account for these samples.

3.6-4.1 Ma.- Two Colombian artefacts, from respectively La Elvira and El Pital, have plateau ages in this range and belong to the Rio Hondo compositional group. However, whereas the La Elvira artefact age is concordant with that of the Rio Hondo source, the El Pital one is only marginally compatible with it. Thus, if we consider the La Elvira obsidian may come from the Rio Hondo source, the question of the origin of the El Pital sample remains open, awaiting in particular further age determinations on various pebbles from this secondary source.

4.8 Ma.- This age was found for one artefact from El Pital. Although its PIXE composition is the same as that found for the Rio Hondo source, it must be, due to its older age, related to an unknown origin.

Thus, the analysis of our data shows that a provenience study based on either age or PIXE chemical composition could not lead to unequivocal source attribution (Bellot-Gurlet *et al.*, 1997). For instance, samples with plateau ages in the range 0.25-0.30 Ma may belong to two discrete compositional groups; similarly, samples presenting a chemical composition compatible with a Mullumica flow origin may have plateau ages concordant or significantly older than that of this flow, thus implying a different origin.

5. Conclusion

Obsidian is a raw material of the lithic industry present in many archaeological sites of Andean countries. Knowing the provenience of this material, for sites located far from volcanic areas, would be an important contribution towards the understanding of goods procurement and ancient trade traditions in prehispanic times. Bigazzi *et al.* (1992), working on the obsidian sources from the Sierra de Guamani, near to Quito (Ecuador) showed that volcanic sources of obsidians might be distinguished by their age and composition. Following this approach, we determined plateau ages on 25 obsidian artefacts from 13 archaeological sites of Colombia and Ecuador (Dorighel *et al.*, 1994 and this work) and one obsidian source from Colombia (this work) and characterized by PIXE the chemical composition of these samples and aliquots of those dated by Bigazzi *et al.* (1992), (Dorighel *et al.*, 1996a; Bellot-Gurlet *et al.*, 1997).

We found that the artefacts we analyzed can be grouped into only eight age/composition groups. The artefacts of the most numerically important of these groups, with as much as 15 samples, are nearly all found in the archaeological sites of the La Tolita-Tumaco culture, which flourished for about 1000 years, from about 500 BC to 500 AD along the north Ecuador-south Colombia coasts. The origin of this material is however unknown.

From table 3, it appears that an origin may be proposed for only five artefacts: two from respectively the La Tolita/Cancha H9 and La Miranda coastal sites and one from La Florida (near to Quito), might come from the Sierra de Guamani; two others, from La Elvira (Popayan), and El Pital Colombian sites could be attributed to the Rio Hondo source. Three artefacts are only characterized by their apparent ages (table 2), one around 0.019 Ma (La Tolita/Walberto), the two others around 0.2 Ma (Huaca Rumi and Maria Cecilia). Whereas the latter uncorrected ages are rather common for artefacts in the area considered, the first one is by far the youngest we measured in an artefact with a moderate spontaneous track fading. Interestingly enough, this apparent age is concordant with that recently obtained (Bigazzi *et al.*, 1997, sample CTX 73) for an obsidian from the Cotopaxi volcano (about 50 km SE of Quito).

Considering that very little is known about potential obsidian sources in Colombia and Ecuador, we find these results very encouraging. Presently, additional PIXE chemical composition have been obtained in our laboratory on more than 150 artefacts and their fission track dating is in progress. These as yet unpublished data confirm that a limited number of sources might account for the origin of these obsidians.

Acknowledgements.- We are deeply indebted to Giulio Bigazzi (Pisa, Italy) for sending us splits of his Sierra de Guamani source samples and for many exchanges and discussions; to C. Gnecco for the Rio Hondo source sample; to our colleagues archaeologists Monica Bolaños, Instituto Nacional del Patrimonio Cultural, Quito, Ecuador; Jean-François Bouchard, UPR 312 du CNRS, Paris, France; José Echeverria, Pontifical Universidad Catolica del Ecuador, Ibarra, Ecuador; Nicolas Guillaume-Gentil, Séminaire de Préhistoire, Université de Neuchatel, Switzerland; Mercedes Guinea, Universidad Complutense, Madrid, Spain; Cristobal Gnecco, Universidad del Cauca, Popayan, Colombia; Diogenes Patiño, Temple University, Philadelphia, USA; Hector Salgado, INCIVA, Museo Arqueologico CALIMA, Darien, Colombia; Alexandra Yepez, Museo del Banco Central, Quito, Ecuador, who provided us samples and allowed us to publish our data. We thank Elena Bartolomé (Zaragoza, Spain) for her contribution to this project. One of us (O. Dorighel) acknowledges the Institut Français d'Etudes Andines and Victor Latorre (Lima), Director of Multiciencias, for their financial support during a field trip in the Summer of 1995. G. Poupeau thanks the Zaragoza and Joseph Fourier Universities for supporting Ms. E. Bartolomé during a four months stay in his laboratory through an Erasmus European Union fellowship. Finally, G. Poupeau wishes especially to thank Marc Tardy, Director of UPRES-A n°5025 of CNRS, for his constant support and encouragements. This work was partially financed by the GDR n°1033 of CNRS "Méthodes Nucléaires en Archéologie".

References

Arias C., Bernardes C., Bigazzi G., Bonadonna F. P., Cesar M. F., Hadler N. J. C., Lattes C. M. G., Oliveira J. V., Osorio A. A. M. and Radi G. (1986): Identificaçao da proveniencia de manufaturados de obsidianas atraves da dataçao com o metodo do traço de fissao. *Ciencia e Cultura*, **38**, 285-308.

Balestrieri M. L., Bigazzi G., Bouska V., Labrin E., Hadler N. J. C., Kitada N., Osorio A. A. M., Poupeau G., Kiyoshi W. and Zuniga A. (1997): Potential glass age standards for fission track dating : an overview. This volume.

Bellot-Gurlet L., Calligaro T., Dorighel O., Dran J.-C., Poupeau G. and Salomon J. (1996): On the coupling of PIXE and fission track dating for obsidian sourcing. *Intern. Assoc. for Obsidian Studies Bull.*, **16**, 4-7.

Bellot-Gurlet L., Poupeau G., Dorighel O., Calligaro T., Dran J.-C. and Salomon T. (1997): A PIXE/fission track dating approach to sourcing studies of obsidian artefacts in Colombia and Ecuador. *J. Archaeological Sci.*, submitted.

Bigazzi G., Dompnier E., Hadler N. J. C., Poupeau G. and Vuillez E. (1988a): A reactor intercalibration for fission track dating and Uranium micromapping. *Nucl. Tracks and Radiat. Meas.*, **15**, 755-757.

Bigazzi G., Hadler N. J., Norelli P., Osorio A. A. M., Paulino R., Poupeau G. and Stella de Navia L (1988b): Dating of glass : the importance of correctly identifying fission tracks. *Nucl. Tracks and Radiat. Meas.*, **15**, 711-714.

Bigazzi G., Coltelli M., Hadler N. J. C., Osorio A. A. M., Oddone M. and Salazar E. (1992): Obsidian-bearing lava flows and pre-Colombian artefacts from the Ecuadorian Andes : First new multidisciplinary data. *J. South Amer. Earth Sci.*, **6**, 31-32.

Bigazzi G., Meloni S., Oddone M. and Radi G. (1993a): Nuovi dati sulla diffusione dell'ossidiana negli insediamenti preistorici italiani. In: *Papers of the 4th International Conference of Italian Archaeology*, (eds Herring E., Whitehouse R. and Wilkins J.), Accordia Research Centre. 3- New Developments in Italian Archaeology, 9-13.

Bigazzi G., Yegingil Z., Ercan T., Oddone M. and Özdogan M. (1993b): Fission track dating obsidians in Central and Northern Anatolia. *Bull. Volcanol.*, **55**, 588-595.

Bigazzi G., Yegingil Z., Ercan T., Oddone M. and Özdogan (1994): Provenance studies of prehistoric obsidian artefacts in Eastern Anatolia : first results from an interdisciplinary research. *Miner. Petrog. Acta*, **37**, 17-36.

Bigazzi G., Coltelli M., Hadler N. J. C. and Osorio A. A. M. (1997): Provenance studies of obsidian artefacts using the fission-track analysis in South America: an overview. In: *Proc. of the 49th International Congress of Americanists*, July 7-11, 1997, Quito, Ecuador.

Dorighel O., Poupeau G., Bouchard J.-F. and Labrin E. (1994): Datation par traces de fission et étude de provenance d'artefacts en obsidienne des sites archéologiques de La Tolita (Equateur) et Inguapi (Colombie). *Bull. Soc. Préhist. Fr.*, **91**, 133-144.

Dorighel O., Bellot-Gurlet L., Poupeau G., Bouchard J.-F. and Widemann F. (1996a): Premiers résultats sur la provenance des artefacts en obsidienne de quelques sites archéologiques côtiers de Colombie et d'Equateur. *Revue d'Archéométrie*, Supplément, 59-64.

Dorighel O., Bellot-Gurlet L., Poupeau G., Bouchard J.-F., Calligaro T., Dran J.-C. and Salomon J. (1996b): L'obsidienne des sites archéologiques côtiers préhispaniques de la région nord-Equateur/sud-Colombie : caractérisation par PIXE, traces de fission et provenance. In: *Proc. XIIIth Congrès International des Sciences Préhistoriques et Protohistoriques*, Forli, September 10-15, 1996, in press.

Fleischer R. L. and Price P. B. (1964): Fission track evidence for the simultaneous origin of tektites and other natural glasses. *Geochim. Cosmochim. Acta*, **28**, 755-760.

Fleischer R. L., Price P. B. and Walker R. M. (1975): *Nuclear Tracks in Solids; Principles and Applications.* University of California Press, Berkeley, pp. 604.

Gleadow A. J. W. (1981): Fission track dating methods : what are the real alternatives? *Nucl. Tracks*, **5**, 3-14.

Miller D. S. and Wagner G. (1981): Fission-track ages applied to obsidian artifacts from South America using the plateau-annealing and the track-size age-correction techniques. *Nucl. Tracks*, **5**, 147-155.

Miller D. S., Eby N., McCorkell R., Rosenberg P. E. and Suzuki M. (1990): Results of an interlaboratory comparison of fission track ages for the 1988 Fission Track Workshop. *Nucl. Tracks Radiat. Meas.*, **17**, 237-245.

Naeser C. W., Izett G. A. and Obradovitch J. D. (1980): Fission-track and K-Ar ages of natural glasses. *U. S. Geol. Survey Bull. 1489*, 1-31.

Poupeau G., Sabil N., Villa M., Bigazzi G., Vatin-Pérignon N., Flores P., Pereyra P., Salas G. and Arroyo G. (1992): Fission-track and K-Ar ages of "Macusanite" obsidian glasses, SE Peru. Geodynamic implications. *Tectonophysics*, **205**, 295-305.

Poupeau G., Labrin E., Sabil N., Bigazzi G., Arroyo G. and Vatin-Pérignon N. (1993): Fission-track dating of 15 macusanite glass pebbles from the Macusani volcanic field (SE Peru). *Nucl. Tracks Radiat. Meas.*, **21**, 499-506.

Poupeau G., Bellot-Gurlet L., Dorighel O., Calligaro T., Dran J.-C. and Salomon J. (1996): PIXE et traces de fission : une approche des réseaux d'échanges de l'obsidienne dans l'aire andine préhispanique (Colombie, Equateur). *Compt. Rend. Acad. Sci. Paris*, **322**, série IIa, 443-450.

Poupeau G., Bigazzi G., Bellot-Gurlet L. and Dorighel O. (1997): Obsidian fission-track dating in archaeology. In: *Synthèse sur l'Obsidienne au Proche-Orient : du Volcan à l'Outil (Géologie, Géochimie, Datation, Archéologie)* (eds Cauvin M.-C., Gourgaud A., Gratuze B., Poidevin J.-L. and Poupeau G.), BAR Intern. Series, Tempus Reparatum, in press.

Storzer D. and Poupeau G. (1973): Ages-plateaux de verres et minéraux par la méthode des traces de fission. *Compt. Rend. Acad. Sci. Paris*, **276D**, 137-139.

Storzer D. and Wagner G. (1977): Fission track dating of meteorite impacts. *Meteoritics*, **12**, 368-369.

Suzuki M. (1973): Chronology of prehistoric human activity in Kanto, Japan. Part I : framework for reconstructing prehistoric human activity in obsidian. *J. Fac. Sci. Tokyo Univ.*, **IV**, 241-318.

Wagner G. (1966): Age measurements on tektites and other natural glasses by tracks from spontaneous fission of ^{238}U. *Z. Naturf.*, **21**, 733-745.

Wagner G. (1978): Archaeological applications of fission track dating. *Nucl. Track Detection*, **2**, 51-63.

Wagner G. and Van den Haute P. (1992): *Fission-track dating.* Kluwer Academic Publishers, Dordrecht, pp. 285.

Technical note:

A BETTER WAY TO SEPARATE APATITE FROM ZIRCON USING CONSTRICTION TUBES

T A. DUMITRU AND D F. STOCKLI
Department of Geological and Environmental Sciences
Stanford University, Stanford, California 94305-2115 USA

Abstract Most fission track dating laboratories currently separate apatite (density ≈3.1-3.2 g/cm^3) from zircon (≈4.5 g/cm^3) in separatory funnels using the heavy liquid methylene iodide (MEI, density 3.32 g/cm^3). We have found it better to undertake these separations using MEI in glass constriction tubes nested inside 15 ml glass test tubes. This procedure is considerably more efficient than the separatory funnel method provided sample sizes are smaller than about 250 mg.

1. Introduction

Fission track dating requires the separation of concentrates of the trace minerals apatite, zircon, and sphene from rock samples (e.g., Wagner and Van den haute, 1992). Original rock samples are generally 0.5 to 10 kg in mass, while optimal final mineral separates are about 10 mg. Recovery of the final separates generally requires an involved sequence of rock grinding, water panning, magnetic, heavy liquid, and sieving procedures (e.g., Hauff and Airey, 1980; Gleadow, 1984; Grist and Ravenhurst, 1992).

One of the later of these steps is the separation of apatite (density ≈3.1-3.2 g/cm^3) from zircon (≈4.5 g/cm^3) using the heavy liquid methylene iodide (MEI, density 3.32 g/cm^3). In most fission track laboratories, this step is currently undertaken in small (e.g., 60 or 125 ml) separatory funnels. However, alternate methods using constriction tubes have been described for similar applications (e.g., Parfenoff *et al.*, 1970; Gautier and von Pechmann, 1984; Mange and Maurer, 1992) and would appear to offer several advantages. We have recently worked towards optimizing these methods for fission track use and have found them superior to separatory funnel methods.

2. Starting material

This description assumes that earlier steps have yielded a separate of apatite plus zircon plus minor impurities. Generally, this requires that previous steps have included (1) removal of fine grinding dust by washing or panning, (2) removal of low-density minerals such as quartz and feldspar with a heavy liquid of density 2.8 to 3.05 g/cm^3 (bromoform, acetylene tetrabromide, sodium polytungstate, or lithium metatungstate), and (3) removal of ferromagnetic minerals such as biotite, amphiboles, pyroxenes, and sphene with a magnetic separator. At this point, the separate is generally about 5 to 250 mg in mass with a grain size of about 50 to 250 μm.

P. Van den haute and F. De Corte (eds.), Advances in Fission - Track Geochronology, 325-330.

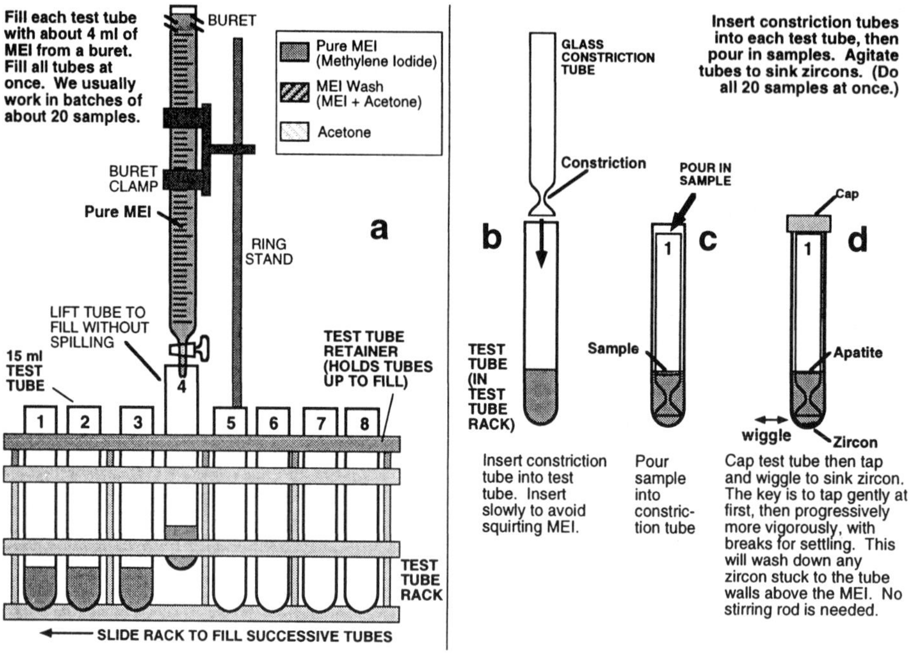

Figure 1: Summary of constriction tube separation procedure.

3. Overall Procedure

Figure 1 illustrates the constriction tube procedure used to separate apatite from zircon and the Appendix lists the required equipment.

4. Important Details

The procedure in Figure 1 is straightforward, but attention to numerous details is necessary to undertake the process efficiently and with a reasonable degree of safety:

(1) MEI is toxic and acetone is toxic and highly flammable. Use of a proper fume hood, gloves, eye protection, etc. is necessary. It is generally regarded that use of MEI requires training under the direct supervision of an experienced worker. Hauff and Airey (1980) is one source of practical safety information. Some of the key guidelines are that all work be performed with proper gloves in a fume hood, that care be used to avoid getting even the tiniest droplets of MEI on gloves or skin, and that all potentially contaminated apparatus and filter papers be handled with forceps. Filter papers retain trace MEI even after careful washing, so samples must be transferred to clean weighing paper using forceps before removal from the hood, and filter papers properly disposed of as hazardous waste. We recommend testing the procedures described here with water in your laboratory before attempting with MEI.
(2) Constriction tubes must be constructed to fit loosely within test tubes, to avoid squirting MEI when they are inserted and removed (Fig. 1b and 1f). Insert constriction tubes into test tubes slowly to keep the upper parts of the inner walls of the

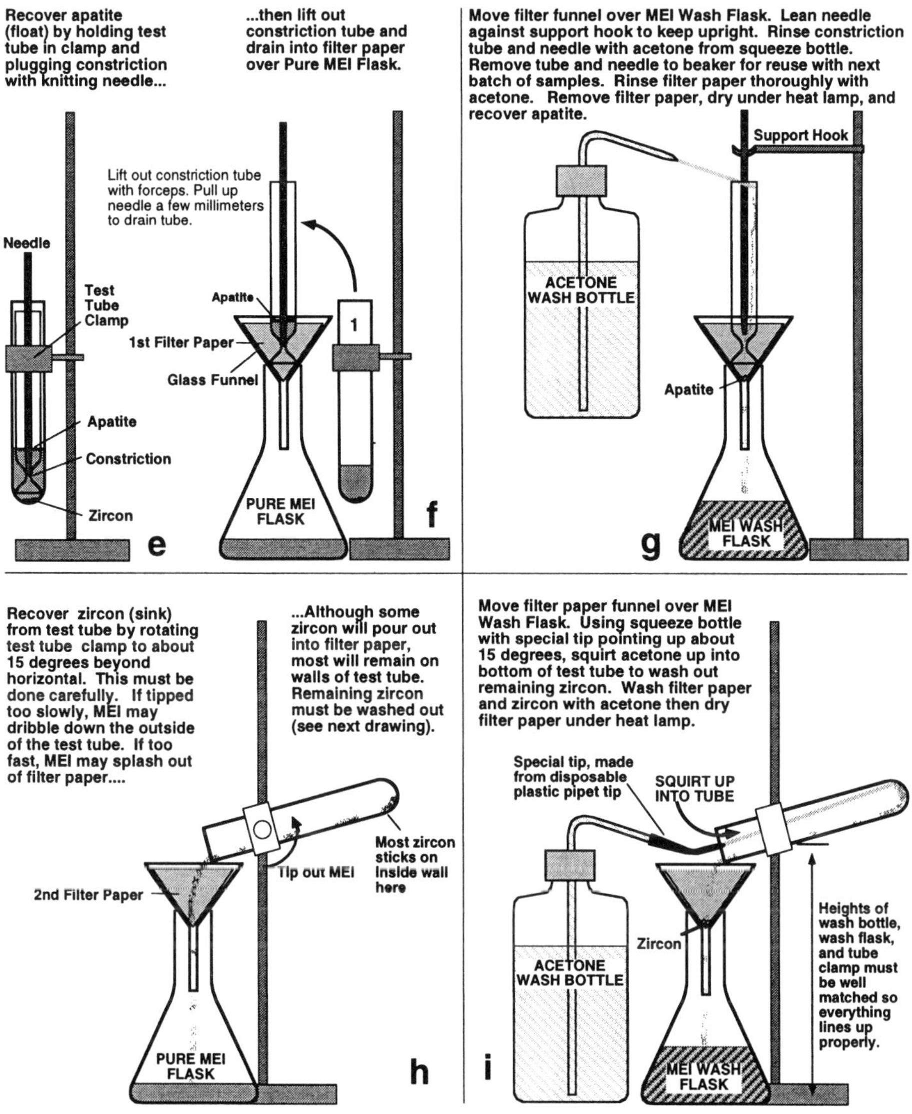

Figure 1 (continued).

constriction tubes dry so samples will not stick when you first pour in samples.

(3) Be sure all apparatus are resistant to MEI and acetone. For example, polypropylene centrifuge tubes might be used rather than glass test tubes, but polypropylene absorbs traces of MEI.

(4) Do not overfill test tubes with MEI. The MEI should only extend a centimeter or so above the constriction (Fig. 1b). Done this way, agitating the tubes does not wet the upper parts of the constriction tubes with MEI (Fig. 1d). The tubes may then be handled with forceps without getting MEI on the forceps.

(5) Figure 1d shows how to agitate the test tubes to allow the zircons to sink. The key

is to agitate the tube very gently at first, then progressively more vigorously, with breaks for settling. This will wash down any zircon stuck to the tube walls above the MEI, with each agitation washing the tube walls a little higher. Unlike separatory funnels, there is no need to use a stirring rod or to wash down the inner walls with a jet of MEI from a wash bottle.

(6) We use standard Teflon-coated knitting needles to block the constrictions. These usually work well, but sometimes the seal will leak slowly, especially if a few grains are trapped in the constriction. Try twisting the needle. Use care not to drip MEI (Fig. 1e and 1f).

(7) Use needle-tipped forceps to remove the constriction tube from the centrifuge tube (Fig. 1f). Lift part way out and allow time for MEI to drain off the outside of the tube. If you were to lose your grip, MEI would spill, so test procedures carefully with water. Some wide tipped forceps tend to chip the glass, so we prefer needle tips.

(8) When you recover the zircon by tipping the test tube, be sure to use a clamp to hold the test tube (Fig. 1h). Do not tip by hand. It is too easy to dribble MEI down the outside of the test tube. It is important to tip the tube smoothly at the right speed so all the MEI will pour cleanly into the filter paper without splashing. Practice with water.

(9) To clean out the test tube with acetone, you need a wash bottle with a special tip to squirt up into the bottom of the tube (Fig. 1i). This must be designed carefully to hit the bottom of the tube cleanly. The height of the tip must match the height of the test tube mouth and the height of the filter funnel in the MEI wash flask. We make tips by bending disposable plastic pipet tips on a hot plate. These tips then fit tightly onto the ends of the normal wash bottle tips (see Appendix).

(10) When using the wash bottle to squirt up into the bottom of the test tube, MEI and acetone may splatter out (Fig. 1i). Again, practice to learn the proper squeeze pressure to reliably avoid splattering.

(11) This method works well only with small samples, smaller than about 0.1 cm^3 ($\approx$250 mg). Larger samples do not separate cleanly. Larger samples may be run in larger tubes, or split into multiple tubes. Commonly larger samples are composed primarily of large contaminating composite or diagenetic grains and quickly sieving with 125 to 250 μm sieve cloth may greatly reduce the sample size.

(12) All apparatus that contact sample material must be properly cleaned before use with the next sample to prevent cross contamination. All apparatus that contact MEI must be thoroughly cleaned with acetone before removal from the fume hood. We generally leave apparatus in the fume hood overnight after cleaning to allow any trace residual MEI to evaporate. Filter papers in particular must be rinsed several times with acetone, thoroughly dried under a heat lamp, then disposed of properly as hazardous waste.

5. Minimization of MEI Loss and Hazardous Waste Generation

With this procedure, we no longer attempt to reclaim MEI from MEI wash for reuse. Recovery has generally been done by mixing MEI wash with large volumes of water to remove the acetone (e.g., Hauff and Airey, 1980). Current environmental regulations in California require that we dispose of such water as hazardous waste because it contains significant MEI and acetone in solution. Instead, we attempt to minimize the volume of MEI wash we create and minimize the amount of MEI lost into wash. The much smaller volume of MEI wash is then disposed of properly as hazardous waste.

In routine use, MEI loss into wash averages about 1.2 cm^3 per sample in our laboratory. Most of this appears to be MEI absorbed into the filter paper fibers. We previously attempted to reduce absorption by using a vacuum to draw the MEI through the filter papers quickly, but this proved not to be very effective.

6. Comparison to other methods

We consider this method significantly better than the separatory funnel method currently in use in most fission track laboratories. The main advantages are the efficiencies of working in batches of about twenty samples at once, the smaller volumes of MEI that need to be handled, the avoidance of stirring rods and wash bottles of MEI to rinse down grains from separatory funnel walls, the lower MEI loss rates, the smaller volumes of MEI wash generated, and the ease of cleaning as compared to separatory funnels with stopcocks.

We sometimes use this same procedure with MEI diluted with acetone to a density of about 3.05 g/cm^3 to separate apatite from trace low density impurities such as quartz or carbonates that were not totally removed earlier by the main 2.8-3.05 g/cm^3 heavy liquid step. This is straightforward because we have the equipment already set up. However, for most separations at densities of less than about 3.15 g/cm^3, other procedures using aqueous solutions of low-toxicity lithium metatungstate (LMT) or sodium polytungstate (SPT) are probably better (e.g., Morrow and Webster, 1989; Mange and Maurer, 1992). We have investigated using very concentrated LMT and SPT solutions, as well as suspensions of micron-size tungsten carbide powder in SPT (suggested by the SPT manufacturer) to separate apatite from zircon, but unfortunately these liquids would not reach high enough density (or severely clogged filter papers in the case of tungsten carbide).

7. Disclaimer and Safety Warning

As mentioned, MEI and acetone are hazardous and should be used only by properly trained workers. Consult other sources for safety information. Use these methods at your own risk. This information is provided without warrantee of any kind. No person or organization takes any responsibility for the safety or efficacy of these procedures. Practice carefully with water before attempting with MEI. This description is intended for skilled, trained worker who have previous experience using MEI and acetone with separatory funnels to separate apatite from zircon. Others must obtain appropriate training from experienced workers. Use this information as a guide, but confirm carefully that it is reliable and safe for your particular laboratory set up.

8. Appendix: Parts List

This is a list of most of the equipment we use. We buy most of it from VWR Scientific Products Corporation, a large U.S. supplier which serves most countries (telephone 1-908-757-4045 or 1-800-932-5000; web site http://www.vwrsp.com).

•Test tubes ("culture tubes"), glass, VWR #60825-618, 16 mm outside diameter by 100 mm long, $60 per 1000. We discard tubes and tube caps after one use.

•Test tube caps, plastic, "Uniflex", VWR #60828-760, for 16 mm tubes, $45 per 1000
•Test tube rack, polypropylene, "Nalgene Unwire", VWR# 66023-542, for 16 mm tubes
•Test tube rack retainer, "Nalgene Unwire", VWR#60914-766, for 16 mm tubes
•Buret, glass, approximately 75-100 ml capacity, about $65
•Buret clamp, about $25
•Ring stand for buret clamp
•Marking pen to label test tubes
•Methylene iodide, about 4 ml per sample
•Acetone
•Two squeeze wash bottles for acetone, 500 ml size (e.g., VWR# 16651-965)
•Pipet tips, polypropylene, to bend on hotplate to make special tip for upward squirting wash bottle. About $25 per pack of 1000, although you only need a few to successfully make one properly bent one. Standard size "1-200 µl. Universal Fit" pipet tips (e.g., VWR#53508-808) will fit the above wash bottles well.
•Constriction tubes, custom made by laboratory glassblower from glass tubing (12 mm outside diameter, 1 mm wall thickness); constriction inner diameter should match needle tip diameter (about 3 mm), constrictions slopes should be ≈45°, about $6 each (buy 20).
•Knitting needles to block constriction, about 150-200 mm long and 4 mm in diameter, from local store, about $1 each (buy 20). Most knitting needles have nicely tapered ends and are Teflon coated, which make them ideal for this purpose.
•Rotating test tube clamp or swivel utility clamp
•Ringstand for test tube clamp
•One or two Erlenmeyer flasks for MEI wash, about 250 ml (height needs to match acetone wash bottle, see Fig. 1i).
•One or two Erlenmeyer flasks for pure MEI, identical to above
•Four conical funnels, glass
•Filter papers for funnels, two per sample
•Ringstand with rings to hold funnels between uses. Various beakers to hold used tubes and needles.
•Improvised support hook on ringstand (Fig. 1g)
•Fume hood, chemical safety gloves, eye protection, lab coat and apron, chemical spill control kits, hazardous waste basket, and other chemical safety and fire safety items
•Forceps, heat lamp, weighing paper, binocular microscope

References

Gautier A.M. and von Pechmann E. (1984): Mineral separation by centrifugation with heavy liquids: Improvement of a method. *Schweiz. Miner. Petrogr. Mitt.*, **64**, 458-464.
Gleadow A.J.W. (1984): *Fission Track Dating Methods--II: A Manual of Principles and Techniques.* Workshop on Fission Track Analysis: Principles and Applications, James Cook University, Townsville, Australia, 50 pp.
Grist A.M. and Ravenhurst C.E. (1992): Mineral separation techniques used at Dalhousie University. In *Low Temperature Thermochronology* (eds Zentilli M. and Reynolds P.H.), Mineralogical Association of Canada Short Course Handbook **20**, 203-209.
Hauff P.L. and Airey J. (1980): The handling, hazards, and maintenance of heavy liquids in the geological laboratory. *U.S. Geological Survey Circular* **827**, 24 pp.
Mange M.A. and Maurer H.F.W. (1992): *Heavy Minerals in Colour*, Chapman & Hall, London, 147 pp.
Morrow J.R. and Webster G.D. (1989): A cryogenic density separation technique for conodont and heavy mineral separations. *J. Paleon.* **63**, 953-955.
Parfenoff A., Pomerol C., and Tourenq J. (1970): *Les Minéraux en Grains: Méthods d'Étude et Détermination*, Masson et Cie, Paris, 578 pp.
Wagner G. and Van den haute P. (1992): *Fission-Track Dating*, Kluwer Academic, Dordrecht, 285 pp.

AUTHOR INDEX

Solid Earth Sciences Library

1. E.I. Galperin: *The Polarization Method of Seismic Exploration*. 1983 ISBN 90-277-1555-6
2. S. Hastenrath: *The Glaciers of Equatorial East Africa*. 1984 ISBN 90-277-1572-6
3. R.K. Verma: *Gravity Field, Seismicity and Tectonics of the Indian Peninsula and the Himalayas*. 1985 ISBN 90-277-1864-4
4. R. Haenel, L. Rybach and L. Stegena (eds.): *Handbook of Terrestrial Heat-Flow Density Determination*. With Guidelines and Recommendations of the International Heat-Flow Commission. 1988 ISBN 90-277-2589-6
5. E. Roth and B. Poty (eds.): *Nuclear Methods of Dating*. 1989 ISBN 0-7923-0188-9
6. G. Wagner and P. Van den haute: *Fission-Track Dating*. 1992 ISBN 0-7923-1624-X
7. J. Lima-de-Faria: *Structural Mineralogy*. An Introduction. 1994 ISBN 0-7923-2821-3
8. I.M. Varentsov: *Manganese Ores of Supergene Zone: Geochemistry of Formation*. 1996 ISBN 0-7923-3906-1
9. A.A. Meyerhoff, I. Taner, A.E.L. Morris, W.B. Agocs, M. Kamen-Kaye, M.I. Bhat, N.C. Smoot and Dong R. Choi; D. Meyerhoff Hull (ed.): *Surge Tectonics: A New Hypothesis of Global Geodynamics*. 1996 ISBN 0-7923-4156-2
10. P. Van den haute and F. De Corte (eds.): *Advances in Fission-Track Geochronology*. A selection of papers presented at the International Workshop on Fission-Track Dating, Ghent, Belgium (1996). 1998 ISBN 0-7923-4904-0

Kluwer Academic Publishers – Dordrecht / Boston / London